T0358607

Urban Drainage

This new edition of a well-established textbook covers the health, environmental, and engineering aspects of the management of rainwater and wastewater in areas of human development. *Urban Drainage* deals comprehensively not only with the design of new systems but also with the analysis and upgrading of existing infrastructure. Keeping its balance of *principles, practice, and research*, the fifth edition has had the most comprehensive update of any edition so far. It includes a new chapter on urban drainage planning, some significant restructuring of others and the introduction of new topics, including emerging contaminants, wastewater surveillance, AI, digital twins, and cyber-physical security. It also addresses current concerns about climate change impacts and intermittent wastewater pollution, and new ideas about sustainable and resilient systems. In all cases, the aim is to provide comprehensive, authoritative, and evidence-based content prioritising innovation, improved methods, and solutions.

This is an essential text for undergraduates and graduate students, lecturers, and researchers in water engineering, environmental engineering, public health engineering, engineering hydrology, and related non-engineering disciplines. It also serves as a dependable and up-to-date reference for drainage engineers in water service providers, local authorities, and consulting engineers. Throughout the text, extensive examples are used to support and demonstrate the key issues.

David Butler is Professor of Water Engineering and Co-director of the Centre for Water Systems at the University of Exeter, UK.

Christopher Digman is an Executive Technical Director at Stantec and a Visiting Professor at the University of Sheffield, UK.

Christos Makropoulos is a Professor in the School of Civil Engineering of the National Technical University of Athens (NTUA), Greece, and the Director of NTUA's Laboratory of Applied Hydraulics.

John W. Davies is a Lecturer in Civil Engineering at the University of Plymouth, UK.

Urban Drainage

Fifth Edition

David Butler, Christopher Digman,
Christos Makropoulos, and John W. Davies

CRC Press
Taylor & Francis Group
Boca Raton London New York

CRC Press is an imprint of the
Taylor & Francis Group, an **informa** business

MATLAB® and Simulink® are trademarks of The MathWorks, Inc. and are used with permission. The MathWorks does not warrant the accuracy of the text or exercises in this book. This book's use or discussion of MATLAB® or Simulink® software or related products does not constitute endorsement or sponsorship by The MathWorks of a particular pedagogical approach or particular use of the MATLAB® and Simulink® software.

Fifth edition published 2025
by CRC Press
2385 NW Executive Center Drive, Suite 320, Boca Raton FL 33431

and by CRC Press
4 Park Square, Milton Park, Abingdon, Oxon, OX14 4RN

CRC Press is an imprint of Taylor & Francis Group, LLC

© 2025 David Butler, Christopher Digman, Christos Makropoulos and John W. Davies

First edition published by E & F N Spon 2000
Fourth edition published by CRC Press 2018

Reasonable efforts have been made to publish reliable data and information, but the author and publisher cannot assume responsibility for the validity of all materials or the consequences of their use. The authors and publishers have attempted to trace the copyright holders of all material reproduced in this publication and apologize to copyright holders if permission to publish in this form has not been obtained. If any copyright material has not been acknowledged please write and let us know so we may rectify in any future reprint.

Except as permitted under U.S. Copyright Law, no part of this book may be reprinted, reproduced, transmitted, or utilized in any form by any electronic, mechanical, or other means, now known or hereafter invented, including photocopying, microfilming, and recording, or in any information storage or retrieval system, without written permission from the publishers.

For permission to photocopy or use material electronically from this work, access www.copyright.com or contact the Copyright Clearance Center, Inc. (CCC), 222 Rosewood Drive, Danvers, MA 01923, 978-750-8400. For works that are not available on CCC please contact mpkbookspermissions@tandf.co.uk

Trademark notice: Product or corporate names may be trademarks or registered trademarks and are used only for identification and explanation without intent to infringe.

ISBN: 978-1-032-52835-9 (hbk)
ISBN: 978-1-032-51331-7 (pbk)
ISBN: 978-1-003-40863-5 (ebk)

DOI: 10.1201/9781003408635

Typeset in Sabon
by SPi Technologies India Pvt Ltd (Straive)

Contents

8 Hydraulic features   163

Preface

As you read this fifth edition of *Urban Drainage*, we felt it was a good moment to reflect on where we have come from and where we are going to. The book had its genesis on a train journey in September 1995 when we (David Butler and John Davies) discussed the need for a contemporary textbook on Urban Drainage. There was really nothing else on the market at the time. We debated what should be included, and sketched-out the main chapter headings. We had in mind a text that would be suitable reading for final year undergraduate and postgraduate university students. The book itself has gone along a journey to the present time, although referring back to our early notes, most chapters still survive in this fifth edition albeit in significantly expanded and updated form. The first edition of the book was actually written in the late 1990s and was published in 2000. It received encouraging reviews and a positive response from readers. As a consequence, it was decided to produce a second edition which was published in 2004 and then a third published in 2011. It had become clear by this time that although the book was initially aimed at students, our readership included many practising engineers and was also increasingly being cited by researchers. To reflect this and to strengthen our team, we invited Christopher Digman and Christos Makropoulos to join us in producing the fourth edition of the book published in 2018. This fifth edition completes our journey for now and it remains to be seen whether this is the terminus or just an intermediate station leading to a further destination.

It is interesting to review how the contents and themes within the book have changed throughout the five editions. A salutary example of this is "flooding" which was hardly mentioned in the first edition. It wasn't until the third edition that sewer flooding had its own chapter. Climate change was not referred to at all in the first edition. Other topics such as design, operation, and modelling have been more consistently present throughout. This fifth edition has had the most comprehensive update of any edition so far. It includes a new chapter on urban drainage planning, some significant restructuring of others, and the introduction of new topics, including emerging contaminants, wastewater surveillance, AI, digital twins, and cyber-physical security. It also addresses current concerns about climate change impacts and intermittent wastewater pollution, and new ideas about sustainable and resilient systems. In all cases, the aim is to provide authoritative, evidence-based, and quantitative content prioritising improved methods and solutions.

Readership

As with previous editions, we cover engineering and environmental aspects of the drainage of rainwater and wastewater from areas of human development. We present basic principles and engineering best practices. The principles are essentially universal but, in this book, are mainly illustrated by UK and European practice with examples of international best practice. We have also included introductions to current developments and recent research.

The book is still primarily intended as a text for students pursuing undergraduate and postgraduate courses in Civil or Environmental Engineering. However, we believe that the content is of sufficient rigour and thoroughness to also be of value to practising engineers and researchers in the field. The basic principles of drainage include wider environmental and societal issues, and these are of significance not only to engineers but also to all with a serious interest in the urban environment, such as students, researchers, and practitioners in environmental science, environmental management, technology, policy and planning, geography, and health studies. These wider issues are covered in particular parts of the book, deliberately written for a wide readership (indicated in the table below). The material makes up a significant portion of the book, and if these sections are read together, they should provide a coherent and substantial insight into an important and timely environmental topic (see table below).

The book is divided into 24 chapters, with numerical examples throughout, and problems at the end of each chapter. Comprehensive reference lists that point the way to further, more detailed information, support the text. Our aim has been to produce a book that is both comprehensive and accessible, and to share our conviction with all our readers that urban drainage is a subject of extraordinary variety, challenge, and interest.

Recommended reading for a wider readership

Chapter	Section(s)
1	All
2	2.5, 2.6
11	11.1, 11.2, 11.4, 11.5, 11.6
12	12.1
13	13.1, 13.2
16	All
17	17.1, 17.2
18	18.1, 18.2
19	All
20	20.1, 20.2
23	All
24	All

Acknowledgements

We remain hugely grateful to those who gave us support in writing the previous editions of this book and to all our colleagues in the urban drainage field around the world.

For help with the fifth edition, we particularly thank Andrew Daugherty, Environment Agency; Peter Fay, Esh Group; Andrew Arnison, Ken Gedman, Elliot Gill, Dr. Bruce Horton, Tony Hughes, Prof. Adrian Johnson, Joanna Kelsey, Gwen Rhodes, Chris Roxborough, John Titterington, Stantec; Prof. Guangtao Fu, University of Exeter; Nick Orman, WRc.

Notation

a	constant
a_{50}	effective surface area for infiltration
A	catchment area, cross-sectional area, plan area
A_b	area of base
A_D	impermeable area from which runoff received
A_{gr}	sediment mobility parameter
A_i	impervious area
A_o	area of orifice
A_p	gully pot cross-sectional area
$API5$	FSR 5-day antecedent precipitation index
ARF	FSR rainfall areal reduction factor
b	width of weir, sediment removal constant, constant
Base Capex	capex that maintains long-term asset capability
b_p	width of Preissman slot
b_r	sediment removal constant (runoff)
b_s	sediment removal constant (sweeping)
B	flow width
B_c	outside diameter of pipe
B_d	downstream chamber width (high side weir), width of trench at top of pipe
Botex	base expenditure
B_u	upstream chamber width (high side weir)
c	concentration, channel criterion, design number of appliances, wave speed
c_0	dissolved oxygen concentration
c_{0s}	saturation dissolved oxygen concentration
Capex	capital expenditure
c_v	volumetric sediment concentration
c_w	weir height
C	runoff coefficient consequence of an occurrence
C_d	coefficient of discharge
C_v	volumetric runoff coefficient
C_R	dimensionless routing coefficient
$C1\text{-}C4$	empirical coefficients
d	depth of flow, depression storage
d'	sediment particle size
d_c	critical depth
d_m	hydraulic mean depth
d_n	normal depth

d_u	upstream depth
d_1	depth upstream of hydraulic jump
d_2	depth downstream of hydraulic jump
d_{50}	sediment particle size larger than 50% of all particles
D	internal pipe diameter, rainfall duration, wave diffusion coefficient, longitudinal dispersion coefficient
D_o	orifice diameter
D_{gr}	sediment dimensionless grain size
D_P	gully pot diameter
DWF	dry weather flow
e	voids ratio, sediment accumulation rate in gully
E	specific energy, gully hydraulic capture efficiency, industrial effluent flow rate
$EBOD$	Effective BOD_5
f	soil infiltration, rate potency factor
f_c	soil infiltration capacity
f_o	soil initial infiltration rate
f_s	number of sweeps per week
f_t	soil infiltration rate at time t
F_m	bedding factor
F_r	Froude number
Fs	specific force
F_{se}	factor of safety
g	acceleration due to gravity
G	water consumption per person
G'	wastewater generated per person
h	head
h_a	acceleration head
h_f	head loss due to friction
h_L	total head loss
h_{local}	local head loss
h_{max}	depth of water gully pot trap depth
H	total head, difference in water level, height of water surface above weir crest, depth of cover to crown of pipe
H_c	height of culvert (internal vertical dimension)
H_{min}	minimum difference in water level for non-drowned orifice
i	rainfall intensity
i_e	effective rainfall intensity
i_n	net rainfall intensity
I	inflow rate, pipe infiltration rate, rainfall depth
IF	effective impervious area factor
$IMKP$	maximum rainfall density over 5 minutes
j	time
J	housing density, criterion of satisfactory service, empirical coefficient
k	constant
k_b	effective roughness value of sediment dunes
k_{DU}	dimensionless frequency factor
k_L	local head loss constant
k_s	pipe roughness
k_T	constant at T °C

k_1	depression storage constant
k_2	Horton's decay constant
k_3	unit hydrograph exponential decay constant
k_4	pollutant washoff constant
k_5	amended pollutant washoff constant
k_{20}	constant at 20°C
K	routing constant, constant in CSO design (Figure 13.14), Rankine's coefficient, empirical coefficient
K_{LA}	volumetric reaeration coefficient
K_{PBOD}	BOD$_5$ potency factor
K_{Pn}	pollutant n's potency factor
L	length, load rate, gully spacing
L_E	equivalent pipe length for local losses
L_I	initial gully spacing
L_w	weir length
m	Weibull's event rank number, reservoir outflow exponent
M	mass, empirical coefficient
M_s	mass of pollutant on surface
MT-D	FSR rainfall depth of duration D with a return period T
n	number, Manning's roughness coefficient, porosity
n_{DU}	number of discharge units
N	total number
$NAPI$	net antecedent precipitation index
O	outflow rate
$Opex$	operational expenditure
p	pressure, probability of appliance discharge, BOD test sample dilution, projection ratio
P	wetted perimeter, perimeter of infiltration device, population, power, probability, height of weir crest above channel
P_d	downstream weir height (high side weir)
P_F	peak factor
PF	porosity factor
PI	precipitation index
P_s	surcharge pressure
P_u	upstream weir height (high side weir)
$PIMP$	FSR percentage imperviousness
PR	WP percentage runoff
q	flow per unit width, appliance flow rate
Q	flow rate
Q_1	upstream flow rate
Q_2	downstream flow rate
$Q80$	nonparametric 80% exceeded flow rate
Q_{av}	average flow rate
Q_b	gully bypass flow rate, brownfield flow rate
$QBAR_{rural}$	IoH24 mean annual peak flow rate
Q_c	gully capacity
Q_d	continuation flow rate (high side weir)
Q_f	pipe-full flow rate
Q_g	greenfield flow rate
Q_{max}	maximum flow

Q_{min}	minimum flow
Q_o	wastewater baseflow
Q_p	peak flow rate
Q_r	runoff flow rate
Q_r	wastewater spill flow rate
Q_u	inflow (high side weir)
$\bar{Q}$	gully approach flow
$\bar{Q}_L$	limiting gully approach flow
r	probability that an event will equal or exceed the design storm at least once in N years, number of appliances discharging simultaneously, FSR ratio of 60 min to 2 day 5-year return period rainfall
r_b	oxygen consumption rate in the biofilm
r_s	oxygen consumption rate in the sediment
r_{sd}	settlement deflection ratio
r_w	oxygen consumption rate in the bulk water
R	hydraulic radius, ratio of drained area to infiltration area, total risk
R_e	Reynolds number
Res	urban drainage resilience index
$RMED$	FEH median of annual rainfall maxima
s	ground slope
S	storage volume, soil moisture storage depth
S_c	critical slope
SCF	seasonal correction factor
S_d	sediment dry density
S_f	hydraulic gradient or friction slope
S_G	specific gravity
S_o	pipe, or channel bed, slope
$SAAR$	FSR standard average annual rainfall
SMD	FSR soil moisture deficit
$SOIL$	FSR soil index
SPR	standard percentage runoff
t	time, pipe wall thickness
t'	duration of appliance discharge
t_c	time of concentration
t_e	time of entry
t_f	time of flow
t_F	mean duration of flooding
$Totex$	total expenditure
T_p	time for hydrograph to reach peak value
t_p	time to peak
T	rainfall event return period, wastewater temperature, pump cycle (time between starts)
T'	mean interval between appliance use
T_a	approach time
T_c	time between gully pot cleans
u	unit hydrograph ordinate
U^*	shear velocity
$UCWI$	FSR urban catchment wetness index
v	mean velocity
v_c	critical velocity

v_f	pipe-full flow velocity
V_F	total flood volume
v_{GS}	gross solid velocity
v_L	limiting velocity without deposition
v_{max}	maximum flow velocity
v_{min}	minimum flow velocity
v_t	threshold velocity required to initiate movement
V	volume
V_f	volume of first flush
V_I	inflow volume
V_o	outflow volume, baseflow volume in approach time
V_t	basic treatment volume
w	channel bottom width, pollutant-specific exponent
W	width of drainage area, pollutant washoff rate
W_b	sediment bed width
W_c	soil load per unit length of pipe
W_{csu}	concentrated surcharge load per unit length of pipe
W_e	effective sediment bed width external load per unit length of pipe
W_s	settling velocity
W_t	crushing strength per unit length of pipe
W_w	liquid load per unit length of pipe
x	DWF pollutant concentration, longitudinal distance, return factor
X	chemical compound, pollutant concentration
X_1	upstream pollutant concentration
X_2	downstream pollutant concentration
y	depth
Y	chemical element
Y_d	downstream water depth (high side weir)
Y_u	upstream water depth (high side weir)
z	potential head, side slope
Z	reflectivity
Z	index of hydrogen sulphide generation
Z'	modified Z formula index
Z_1, Z_2	pollutant specific constant, FSR growth factor
$\bar{Z}$	depth to centroid of flow cross section
α	channel side slope angle to horizontal, number of reservoirs, turbulence correction factor, empirical coefficient
	wastewater DWF multiple
β	empirical coefficient
	overflow DWF multiple
γ	empirical coefficient
	watercourse DWF multiple
ε	empirical coefficient, gully pot sediment retention efficiency
ε'	gully pot cleaning efficiency
ζ	sediment washoff rate
η	sediment transport parameter pump efficiency
θ	transition coefficient for particle Reynolds number, angle subtended by water surface at centre of pipe, Arrhenius temperature correction factor

Φ	vertical slope angle
κ	sediment supply rate
λ	friction factor
λ_b	friction factor corresponding to the sediment bed
λ_c	friction factor corresponding to the pipe and sediment bed
λ_g	friction factor corresponding to the grain shear factor
μ	coefficient of friction
μ'	coefficient of sliding friction
ν	kinematic viscosity
ρ	density
τ_b	critical bed shear stress
τ_o	boundary shear stress
γ	unit weight, temperature correction factor
χ	surface sediment load
χ_u	ultimate (equilibrium) surface sediment load
ω	counter
ψ	shape correction factor for part-full pipe

Units are not specifically included in this notation list but they have been included in the text. Some symbols have multiple definitions and these are shown separated by commas.

Abbreviations

AA	Annual allowance
AI	Artificial intelligence
AEP	Annual exceedance probability
AMP	Asset management planning
ANN	Artificial neural network
AOD	Above ordnance datum
ARF	Areal reduction factor
ASCE	American Society of Civil Engineers
ASR	Aquifer storage recovery
ATU	Allylthiourea
AWWA	American Waterworks Association
BAU	Business as usual
BeST	Benefits of SuDS Tool
BGI	Blue–green infrastructure
BHRA	British Hydrodynamics Research Association
BMP	Best management practice
BOD	Biochemical oxygen demand
BRE	Building Research Establishment
BS	British standard
BSM	Benchmark simulation model
CA	Cellular automata
CAD	Computer-aided drawing/design
CAPEX	Capital expenditure
CARP	Comparative acceptable river pollution procedure
CBOD	Carbonaceous biochemical oxygen demand
CCA	Climate change allowance
CCD	Charge coupling device
CCTV	Closed-circuit television
CCW	Consumer Council for Water
CDM	Construction (design and management) regulations
CEC	Council of European Communities
CEN	European Committee for Standardisation
CESMM4	Civil Engineering Standard Method of Measurement, fourth edition
CFD	Computational fluid dynamics
CFU	Colony forming unit
CIWEM	Chartered Institution of Water and Environmental Management
CIH	Critical input hyetograph

CIPP	Cured in place pipe (lining)
CIRIA	Construction Industry Research and Information Association
CMIP	Coupled model intercomparison projects
CNN	Convolutional neural network
COD	Chemical oxygen demand
COVID-19	Coronavirus disease 2019
CPA	Cyber-physical attack
CPS	Cyber-physical system
CSO	Combined sewer overflow
CSV	Comma-separated value
DBP	Dibutyl phthalate
DCLG	Department for Communities and Local Government
DD	Directional drilling
DDF	Depth-duration-frequency
DDT	Dichlorodiphenyltrichloroethane
DEFRA	Department for Environment, Food and Rural Affairs
DEHP	Di-(2-ethylhexyl) phthalate
DEM	Digital elevation model
DG5	Ofwat performance indicator (properties at risk of flooding)
DL	Deep learning
DO	Dissolved oxygen
DoE	Department of the Environment
DoS	Denial of service
DOT	Department of Transport
DN	Nominal diameter
DSD	Drop size distribution
DTM	Digital terrain model
DU	Discharge unit
EA	Environment Agency
EC	*Escherichia coli*
ECOsan	Ecological sanitation
EDM	Event duration monitor (CSO)
EEA	European Environment Agency
EEC	European Economic Community
EFRA	Exceedance flood risk assessment
EGL	Energy grade line
EMC	Event mean concentration
EM-DAT	Emergency event database
EN	European standard
EO	Earth observations
EPA	Environmental Protection Agency (US)
EPS	Ensemble prediction systems
EQO	Environmental quality objectives
EQS	Environmental quality standards
ESM	Earth system model
EWPCA	European Water Pollution Control Association
EWS	Early warning systems
EU	European Union
FC	Faecal coliform
FEH	Flood Estimation Handbook

FFC	Flood Forecasting Centre
FFG	Flash flood guidance
FGS	Flood guidance statement
FIO	Faecal indicator organism
FIS	Fundamental intermittent standards
FOG	Fats, oils, and grease
FORGEX	FEH focused rainfall growth curve extension method
FS	Faecal streptococci
FSR	Flood Studies Report
FWR	Foundation for Water Research
GC	Gene copy
GCM	General circulation model
GHG	Greenhouse gas
GI	Gastrointestinal illness
GIS	Geographical information system
GL	Ground level
GLUE	Generalised likelihood uncertainty estimation
GMT	Greenwich mean time
GPRS	General packet radio service
GPS	Global positioning system
GRA	Global resilience analysis
GRP	Glass-reinforced plastic
GSI	Green stormwater infrastructure
GSM	Global system for mobile communications
HF	High frequency
HDPE	High-density polyethylene
HGL	Hydraulic grade line
HMSO	Her Majesty's Stationery Office
HR	Hydraulics Research
HRS	Hydraulics Research Station
HSE	Health and Safety Executive
IAHR	International Association of Hydraulic Engineering and Research
IAWPRC	International Association on Water Pollution Research and Control
IAWQ	International Association on Water Quality
ICBM	Integrated component-based model
ICE	Institution of Civil Engineers
ICP	Inductively coupled plasma
ICT	Information and communications technology
IDF	Intensity–duration–frequency
IE	Intestinal *enterococci*
IL	Invert level
IPCC	Intergovernmental Panel on Climate Change
IoH	Institute of Hydrology
IoT	Internet of Things
IUD	Integrated urban drainage
IUDM	Integrated urban drainage model
IUWCM	Integrated urban water cycle model
IUWSM	Integrated urban water system model
IWA	International Water Association
IWEM	Institution of Water and Environmental Management

LC50	Lethal concentration to 50% of sample organisms
LID	Low-impact development
LiDAR	Light detection and ranging
LOD	Limit of deposition
LSTD	Limiting solids transport distance
LSTM	Long short-term memory
MAC	Maximum allowable concentration
MAFF	Ministry of Agriculture, Fisheries and Food
MC	Monte Carlo
MDG	Millennium development goal
MDPE	Medium density polyethylene
MH	Manhole
ML	Machine learning
MP	Microplastic
MPC	Model-based predictive control
MPN	Most probable number
MZC	Manhole zoom camera
NBOD	Nitrogenous biochemical oxygen demand
NERC	Natural Environment Research Council
NEU	Neighbourhood energy utility
NGO	Non-governmental organisation
NOAA	National Ocean and Atmospheric Administration
NOD	Nitrogenous oxygen demand
NRA	National Rivers Authority
NSTS	Non-Statutory Technical Standards (for SuDS)
NWC	National Water Council
NWP	Numerical weather predictions
NWS	National Weather Service (US)
O&M	Operation and management
Ofwat	Water Services Regulation Authority
OD	Outside diameter
OGC	Open geospatial consortium
Open MI	Open modelling interface and environment
OPEX	Operational expenditure
OPP	Orangi Pilot Project
OS	Ordnance Survey
PAH	Polyaromatic hydrocarbons
PCB	Polychlorinated biphenyl
PCC	Per capita consumption
PE	Polyethylene
PFA	Pulverised fuel ash
PFAS	Per- and polyfluoroalkyl substances
PFF	Pass forward flow
PFR	Property flood resilience
PID	Proportional-integral-derivative
PLC	Programmable logic controller
POST	Parliamentary Office of Science and Technology
PPE	Personal protective equipment
PVC-U	Unplasticised polyvinylchloride
PU	Polyurethane

QUALSOC	Quality Impacts of Storm Overflows: Consent Procedure
RBF	Radial basis functions
RCP	Representative concentration pathway
ReFH	Revitalised flood hydrograph model
RII	Rainfall-induced infiltration
RL	Reinforcement learning
RRL	Road Research Laboratory
RNA	Ribonucleic acid
RTC	Real-time control
RTU	Remote terminal units
RWH	Rainwater harvesting
SAR	Synthetic aperture radar
SAAR	Standard annual average rainfall
SARS-CoV-2	Severe acute respiratory syndrome coronavirus
SCADA	Supervisory control and data acquisition
SCM	Stormwater control measure
SDD	Scottish Development Department
SDI	Sustainable development indicators
SDG	Sustainable development goal
SEPA	Scottish Environmental Protection Agency
SMS	Short message service
SOD	Sediment oxygen demand
SG	Specific gravity
SM	Sewerage management plan
SPZ	Groundwater source protection zone
SQL	Structured query language
SRM	Sewerage Rehabilitation Manual
	Sewerage risk management
SS	Suspended solids
SSP	Shared socio-economic pathway
SRES	Special report on emissions scenarios
STC	Standing Technical Committee
STEPS	Short-term ensemble prediction system
SuDS	Sustainable drainage system
SWMP	Surface water management plan
SWMM	Stormwater Management Model
SWO	Stormwater outfall
TBC	Toxicity-based consents
TISCIT	Totally integrated sonar and CCTV inspection technique
TKN	Total Kjeldahl nitrogen
TNO	The Netherlands Organisation for Applied Scientific Research
TOC	Total organic carbon
TOTEX	Total expenditure
TRRL	Transport & Road Research Laboratory
TWL	Top water level
UAV	Unmanned aerial vehicle
UCWI	Urban catchment wetness index
UH	Unit hydrograph
UHF	Ultra-high frequency
UI	User interface

UK	United Kingdom
UX	User experience
UKCIP	UK Climate Impacts Programme
UKWIR	UK Water Industry Research
ULFT	Ultra-low-flush toilet
UPM	Urban pollution management
US	United States
UWOT	Urban water optioneering tool
UWWTD	Urban wastewater treatment directive
VCS	Virtual case study
VHF	Very high frequency
VIP	Ventilated improved pit latrine
VSAT	Very small aperture terminal
WAA	Water Authorities Association
WaSSP	Wallingford Storm Sewer Package
WaPUG	Wastewater Planning User Group
WBE	Wastewater-based epidemiology
WC	Water closet (toilet)
WEF	Water Environment Federation
WFD	Water Framework Directive
WMO	World Meteorological Organisation
WP	Wallingford Procedure
WPCF	Water Pollution Control Federation
WSA	Water Services Association
WSUD	Water-sensitive urban design
WTP	Wastewater treatment plant
WO	Welsh Office
WRAP	Winter rain acceptance potential
WRc	Water Research Centre

Chapter 1

Introduction

In this chapter, which forms an introduction and scene-setter for the whole book, we introduce exactly what urban drainage is (Section 1.1) and its relationship with urbanisation (Section 1.2) and climate change (Section 1.3). The next three sections (Sections 1.4 through 1.6) introduce in turn the particular need for this infrastructure system, the history of its implementation, and then some diverse geographical examples. The various types of urban drainage systems are introduced in Section 1.7 and placed in the wider setting of urban water systems in Section 1.8. The chapter concludes with a discussion of the changing context and drivers affecting the development of urban drainage (Section 1.9) and the future challenges it faces (Section 1.10).

The principles of hydrology, hydraulics, water science and so on are universal. However, they are applied differently typically depending on location. This book is set within a European context and based on urban drainage practice in the United Kingdom (UK) but includes examples and insights from around the world.

1.1 WHAT IS URBAN DRAINAGE?

Drainage systems are needed in developed urban areas because of the interaction between human activity and the natural water cycle. This interaction has two main forms: the abstraction of water from the natural cycle to provide a water supply for human life, and the covering of land with impermeable surfaces that divert rainwater away from the local natural system of drainage. These two types of interaction give rise to two types of water that require drainage.

The first type, *wastewater*, is water that has been supplied to support life, maintain a standard of living, and satisfy the needs of industry. After use, if not drained properly, it could cause pollution and create health risks. Wastewater contains dissolved material, fine solids, and larger solids, originating from toilets/water closets (WCs), from washing of various sorts, from industry, and from other water uses.

The second type of water requiring drainage, *stormwater*, is rainwater (or water resulting from any form of precipitation) that has fallen on a built-up area. If stormwater were not drained properly, it would cause inconvenience, damage, flooding, and further health risks. It contains some pollutants, originating from rain, the air, or the catchment surface.

Urban drainage systems handle these two types of water with the aim of managing the impact on human life and the environment. Thus, urban drainage has two *major* interfaces: with the public and with the environment (Figure 1.1).

In many urban areas, drainage is based on a completely artificial system of sewers: pipes and structures that collect and dispose of this water (the "grey system"). In contrast, isolated or low-income communities normally have no main drainage. Wastewater is treated locally

DOI: 10.1201/9781003408635-1

Figure 1.1 Interfaces with the public and the environment.

(or not at all), and stormwater is drained naturally into the ground. These sorts of arrangements have generally existed when the extent of urbanisation has been limited. However, as discussed in Section 1.7.4 and in Chapter 11 (Sustainable drainage systems), the current best practice is to use more natural and non-pipe-based drainage arrangements wherever possible (the "green system").

Urban drainage presents a classic set of modern environmental challenges: the need for cost-effective and socially acceptable technical improvements in existing systems, the need for assessment of the environmental and health impact of those systems, and the need to search for more sustainable and resilient solutions. As in all other areas of environmental concern, these challenges cannot be considered to be the responsibility of one profession alone. Policymakers, engineers, planners, builders, public health and environment specialists, together with all citizens, have a role. And these roles must be played in partnership. Engineers must understand the wider issues, while those who seek to influence policy must have some understanding of the technical problems. This book is intended as a source of information for all those with a serious interest in the urban environment.

1.2 EFFECTS OF URBANISATION

Let us consider further the effects of human development on the passage of rainwater. Urban drainage replaces one part of the natural water cycle and, as with any artificial system that takes the place of a natural one, it is important that the full effects are understood.

In nature, when rainwater falls on a natural surface, some water returns to the atmosphere through evaporation, or transpiration by plants; some infiltrates the surface and becomes groundwater; and some runs off the surface (Figure 1.2a). The relative proportions depend on the nature of the surface and vary with time during the storm. Surface runoff tends to increase as the ground becomes saturated. Both groundwater and surface runoff are likely to find their way to a river, but surface runoff arrives much faster. The groundwater will become a contribution to the river's general baseflow rather than being part of the increase in flow due to any particular rainfall.

Development of an urban area, involving covering the ground with artificial surfaces, has a significant effect on these processes. The artificial surfaces increase the amount of surface runoff in relation to infiltration, and therefore increase the total volume of water reaching the river during or soon after the rain (Figure 1.2b). Surface runoff travels quicker over hard surfaces and through sewers than it does over natural surfaces and along natural streams. This means that the flow will both arrive and die away faster; therefore, the peak flow will be greater (see Figure 1.3). (In addition, reduced infiltration means poorer recharge of groundwater reserves.)

This obviously increases the danger of sudden flooding of the river. It also has strong implications for water quality. The rapid runoff of stormwater is likely to cause pollutants

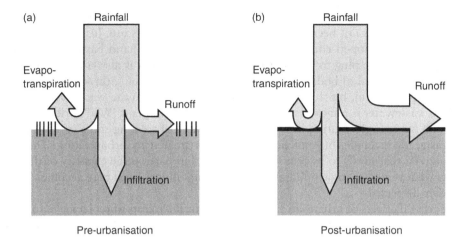

Figure 1.2 Effect of urbanisation on rate of rainfall.

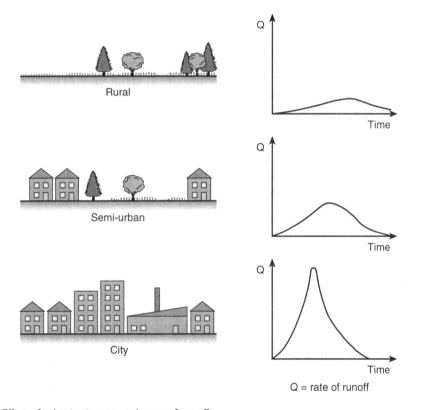

Figure 1.3 Effect of urbanisation on peak rate of runoff.

and sediments to be washed off the surface or scoured by the river. In an artificial environment, there are likely to be more pollutants on the catchment surface and in the air than there would be in a natural environment. Also, drainage systems in which there is mixing of wastewater and stormwater may allow pollutants and pathogens from the wastewater to enter the river.

The existence of wastewater in significant quantities is itself a consequence of urbanisation. Much of this water has not been made particularly "dirty" by its use. Just as it is a standard convenience in a developed country to turn on a tap to fill a hand basin, it is a standard convenience to pull the plug to let the water "disappear." Water is also used as the principal medium for the disposal of bodily waste, and varying amounts of bathroom litter, via WCs.

In a developed system, much of the material that is added to the water while it is being turned into wastewater is removed at a wastewater treatment plant (WTP) prior to its return to the urban water cycle. Nature itself would be capable of treating some types of material, bodily waste, for example, but not in the quantities created by urbanisation. The proportion of material that needs to be removed will depend in part on the capacity of the river to assimilate what remains. Some pollutants cannot be assimilated and may accumulate locally or be widely dispersed.

So, the general effects of urbanisation on drainage, or the effects of replacing natural drainage with (grey) urban drainage, are to produce higher and more sudden peaks in river flow, to introduce pollutants, and to create the need for end-of-pipe wastewater treatment.

1.3 EFFECTS OF CLIMATE CHANGE

In many ways, the effect of climate change on drainage systems and their performance is similar to that of urbanisation. Rainfall patterns are changing worldwide and in the UK, this has resulted in a rising historical trend of extreme rainfall events and the prediction that this will continue into the foreseeable future. In addition, more intense summer storms are also expected (see Chapter 4 for details). This will in turn generate greater runoff volumes and flow rates. The overall effect will be to put additional pressure on existing urban drainage systems and require new systems to have greater capacity just to provide similar levels of protection against flooding and pollution.

Drainage engineers must also play their part in minimising direct and indirect greenhouse gas (GHG) emissions which exacerbate climate change (see Chapter 15 for further discussion on energy use and net zero carbon emissions).

1.4 URBAN DRAINAGE PRIORITIES

1.4.1 Public health

In human terms, the most valuable benefit of an effective urban drainage system is the maintenance of public health. This particular objective is often *overlooked* in modern practice and yet is of extreme importance, particularly in protection against the spread of diseases.

Despite the fact that some vague association between disease and water had been known for centuries, it was in 1855 that a precise link was demonstrated. This came about as a result of the classic studies of John Snow in London concerning the cholera epidemic sweeping the city at the time. That diseases such as cholera are almost unknown in the industrialised world today is in major part due to the provision of centralised urban drainage (along with the provision of a microbiologically safe, potable supply of water).

Urban drainage has a number of major roles in maintaining public health and safety. Human excreta (particularly faeces) are the principal vector for the transmission of many communicable diseases. Urban drainage has a direct role in effectively removing excreta from the immediate vicinity of habitation. However, there are further potential problems in large river basins in which the downstream discharges of one settlement may become the upstream abstraction of another. In the UK, some 30% of water supplies are so affected. This indicates the vital importance of disinfection of water supplies as a public health measure. In recent years, the media spotlight has focussed on uncontrolled discharges into the environment from combined sewer overflows (CSOs) which discharge dilute but untreated wastewater into waterbodies that may be accessed or used by the public. The implications of this are covered in much more detail in Chapter 2 (Water quality) and Chapter 13 (Combined sewers and CSOs).

Also, of particular importance in tropical countries, standing water after rainfall can be largely avoided by effective drainage. This reduces the mosquito habitat and hence the spread of malaria and other diseases.

While many of these problems have apparently been solved, it is essential that in industrialised countries, as we look for ever more innovative sanitation techniques, we do not lose ground in controlling serious diseases. Sadly, while we may know much about waterborne and water-related diseases, some rank among the largest killers in societies where poverty and malnutrition are widespread. Millions of people around the world still lack any hygienic and acceptable method of excreta disposal. The issues associated with urban drainage in low-income communities are returned to in more detail in Chapter 23.

1.4.2 Minimising adverse impacts

It has already been stated that the basic function of urban drainage is to collect, convey and safely discharge wastewater and stormwater. In the UK and other developed countries, this has generally been taken to cover all wastewater and its contents. Most wastewater is treated at the downstream WTPs with the discharged effluent subject to legal permits (based on defined limits for various physical, chemical and biochemical parameters) discussed in more detail in Chapter 2. CSOs are also legally permitted to discharge but beyond screening and sometimes settlement, spill diluted wastewater depending on defined operating conditions. This can have an adverse impact on the receiving water (see Chapter 13 for more on this). Stormwater-only outfalls are typically not subject to permits for pollutant management and can freely discharge.

Storm drainage has traditionally been provided to remove water from surfaces (for storms up to a particular severity), especially roads, as *quickly* as possible thus causing the minimum of inconvenience. It is then disposed of, usually via a pipe system, to the nearest watercourse. Still, there is a risk of surface flooding from heavy rainfall and the effects of this need to be managed and mitigated as far as possible (see Chapter 12). So, drainage systems should protect people and property from stormwater damage and should also limit the impact of the drained flow on the receiving water. For this reason, and to achieve other environmental and social benefits, the application of more natural (green) methods of managing stormwater is being prioritised. These include infiltration, storage, and vegetated systems (discussed in full in Chapter 11), and the general intention is to attempt to reverse the trend illustrated in Figure 1.3: to decrease the peak flow of runoff and increase the time it takes to reach the watercourse. Water quality improvements can also be expected.

These tendencies towards reducing the dependence on grey engineering solutions to solve the problems created by urbanisation, and the philosophy that goes with them, are associated with the word *sustainability* and are further considered in Chapter 24.

As with other aspects of the environment, the nature of progress in relation to urban drainage, its consequences (especially flooding, pollution and health), desirability, limits and affordability, are being closely and frequently reassessed.

1.5 HISTORY

1.5.1 Ancient civilisations

There is plenty of archaeological evidence that ancient civilisations demonstrated no less skill in drainage provision than in other aspects of urban infrastructure. Here, we look briefly at some interesting examples of success in creating effective drainage systems, without modern scientific knowledge.

The Minoans on Crete (between about 3000 and 1000 BC) had some well-developed drainage systems, for stormwater and wastewater. At the Palace of Knossos, there is evidence of separate stone conduits for stormwater and wastewater, and even an early type of flushed toilet (De Feo et al., 2014).

At a similar time, in parts of what are now Afghanistan, Pakistan, and northwest India, the civilisation of the Indus Valley excelled in urban planning and infrastructure, including the provision of drainage. Typically, drainage channels were built into the centres of streets, and these carried both stormwater and wastewater. Wastewater did not flow directly into the channel. From each house, it flowed first via a pipe to a small pit where solids settled out; at a certain depth liquids overflowed into the channel. In some cases, the channels were covered with stones or bricks (Burian and Edwards, 2002).

Many towns and cities in ancient Greece (around 500–100 BC) had covered drainage channels built into the streets. An interesting detail of drainage in ancient Greece is that most Greek theatres, being large, impermeable, outdoor spaces, had well-thought-out arrangements for draining stormwater. Existing parts of the theatre's structure, including stairs and corridors, led the runoff towards the lower part of the theatre, the circular "orchestra," or main performing space. A duct, semi-circular in plan, sometimes covered, at the edge of the orchestra, collected the runoff and carried it away to an outfall (De Feo et al., 2014).

The Romans (particularly during the period from 500 BC to 300 AD) are well known for their public health engineering feats, particularly the impressive aqueducts bringing water into their cities; but they also excelled in systems to take water away. Besides the removal of water, they placed importance on systems for the reuse of rainwater. Urban drainage was achieved by sophisticated systems of open channels and buried sewers ("cloacae"). In Rome, these provided drainage for a population of up to 1 million, including 900 public baths. The largest and most well-known of the cloacae is the Cloaca Maxima, which drained the lowest-lying parts of Rome, including the Forum, to the River Tiber. Part of its function was to provide land drainage for marshy areas, and, in doing so, it reduced the presence of mosquitoes and therefore the risk of malaria. Construction of the Cloaca Maxima started in the sixth century BC, and it is still in use today.

1.5.2 Ancient to modern

After the Romans, urban drainage practices did not develop significantly for many centuries. In medieval Europe, drainage did not go far beyond the use of ditches that followed natural land drainage patterns. The English word *sewer* is derived from an Old French word, *esseveur*, meaning "to drain off," related to the Latin *ex-* (out) and *aqua* (water). The *Oxford English Dictionary* gives the earliest meaning as "an artificial water-course for draining marshy land and carrying off surface runoff into a river or sea."

It was not until the rapid growth of cities in the eighteenth and nineteenth centuries that things had to change. This was typically in response to serious problems with public health, leading to an urgent search for solutions and to developments in urban drainage practice. We consider the example of London.

1.5.3 London

In London, before the eighteenth century, sewers had the meaning given above and their alignment was loosely based on the natural network of streams and ditches that preceded them. In a quite unconnected arrangement, bodily waste was generally disposed of into cesspits (under the residence floor), which were periodically emptied. Flush toilets (discharging to cesspits) became quite common around 1770–1780, but it remained illegal until 1815 to connect the overflow from cesspits to the sewers. This was a time of rapid population growth and, by 1817, when the population of London exceeded one million, the only solution to the problem of under-capacity was to allow cesspit overflow to be connected to the sewers. Even then, the cesspits continued to be a serious health problem in poor areas, and, in 1847, 200,000 of them were eliminated completely by requiring houses to be connected directly to the sewers.

This moved the problem elsewhere – namely, the River Thames. By the 1850s, the river was filthy and stinking (Box 1.1) and directly implicated in the spread of deadly cholera.

There were cholera epidemics in 1848–1849, 1854, and 1867, killing tens of thousands of Londoners. The Victorian sanitary reformer Edwin Chadwick passionately argued for a dual system of drainage – one for human waste and one for rainwater: "the rain to the river and the sewage to the soil." He also argued for small-bore, inexpensive, self-cleansing sewer pipes in preference to the large brick-lined tunnels of the day. However, the complexity and cost of engineering two separate systems prevented his ideas from being put into practice. The solution was eventually found in a plan by Joseph Bazalgette to construct a number of "combined" interceptor sewers on the north and the south of the river to carry the contents of the sewers to the east of London. The scheme, an engineering marvel (Figure 1.4), was mostly constructed by 1875, and much of it is still in use today (Halliday, 2001).

Again, though, the problem had simply been moved elsewhere. This time, it was the Thames estuary, which received huge discharges of wastewater. Storage was provided to allow release on the ebb tide only, but there was no treatment. Downstream of the outfalls, the estuary and its banks were disgustingly polluted. By 1890, some separation of solids was carried out at works on the north and south banks, with the sludge dumped at sea. Biological treatment was introduced in the 1920s, and further improvements followed. However, it was not until the 1970s that the quality of the Thames was such that salmon were commonplace and porpoises could be seen under Blackfriars Bridge.

BOX 1.1   Michael Faraday's abridged letter to *The Times* of 7 July 1855

I traversed this day by steamboat the space between London and Hungerford Bridges [*on the River Thames*], between half-past one and two o'clock. The appearance and smell of water forced themselves on my attention. The whole of the river was an opaque pale brown fluid. The smell was very bad, and common to the whole of the water. The whole river was for the time a real sewer.

If there be sufficient authority to remove a putrescent pond from the neighbourhood of a few simple dwellings, surely the river which flows for so many miles through London ought not be allowed to become a fermenting sewer. If we neglect this subject, we cannot expect to do so with impunity; nor ought we to be surprised if, ere many years are over, a season give us sad proof of the folly of our carelessness.

Figure 1.4 Construction of Bazalgette's sewers in London. (From *The Illustrated* London News, 27 August 1859, reproduced with permission of The Illustrated London News Picture Library.)

The story of London's urban drainage continues to this day with the construction of the controversial £4.1 billion Thames Tideway Tunnel (or "super sewer" as it is known), which has the main purpose of reducing CSO spills into the river (Dolowitz et al., 2018; https://www.tideway.london/).

1.6 GEOGRAPHY

The main factors that determine the extent and nature of urban drainage provision in a particular region are wealth; the climate and other natural characteristics; the intensity of urbanisation; and history, legislation, and politics. The greatest differences are the result of differences in wealth. Most of this book concentrates on urban drainage practices in countries that can afford fully engineered systems. The differences in countries that cannot will be apparent from Chapter 23 where we focus on challenges and solutions for low-income communities.

Countries in which rainfall tends to be occasional and heavy have naturally adopted practices different from those in which it is frequent and generally relatively light. For example, it is common in Australia to provide "minor" (underground, piped) systems to cope with low quantities of stormwater, together with "major" (overground) systems for larger quantities. We return to this concept in Chapter 12. Other natural characteristics have a significant effect. Sewers in the Netherlands, for example, must often be laid in flat, low-lying areas and, therefore, must be designed to run frequently in a pressurised condition. The intensity of urbanisation and history of development have a strong influence on the percentage of the population connected to a main sewer system, which stands at 96% in the UK, for example.

Historical and political factors determine the age of the system (which is likely to have been constructed during periods of significant development and industrialisation), characteristics of operation such as whether or not the water/wastewater industry is publicly or privately financed, and strictness of statutory requirements for pollution control and the manner in which they are enforced. Countries in the European Union (EU) are subject to common requirements.

Boxes 1.2 through 1.4 present a selection of examples to give an idea of the wide range of different urban drainage issues throughout the world. International examples of good urban drainage practice are listed and described in PD ISO/TR 24539:2021 (2021).

BOX 1.2   Orangi Pilot Project, Pakistan

Orangi, a "katchi abadi," or squatter community, in Karachi, has a population of over one million. Until the 1980s, there were no sewers: people had to empty bucket latrines into the narrow alleys. In a special self-help programme, quite different from government-sponsored improvement schemes, the community built its own sewers, with no outside contractors. A small septic tank was placed between the toilet and the sewer to reduce the entry of solids into the pipe. The system had a simplified design and was built up alley-by-alley, as the people made a commitment to the improvements. This was a great success for community action and created major improvements in the immediate environment. In Orangi, about 100,000 households developed their own sanitation systems, and in 11 other towns in Pakistan, at least 40,000 households are known to have used the same approach (Hasan, 2006).

Growing from the work on sewers, the Orangi Pilot Project (OPP) has gone on to coordinate activity in a number of areas including housing, education, water supply, social development, and enterprise support (OPP-RTI, 2012).

> The OPP has supported one of the world's largest programmes for improved provision for sanitation in low income areas – in Orangi and in many other cities and small urban centres – as well as supporting improvements in other forms of infrastructure and in services
>
> (UN-Habitat, 2006).

"The growing influence of OPP can be seen not only in neighbourhood solutions but in developing city-wide policies and priorities" (UN-Habitat, 2003).

BOX 1.3   Belo Horizonte, Brazil

Belo Horizonte, the capital of the state of Minas Gerais, was the first modern "planned" city in Brazil. Streets are set out on a broadly spaced grid with ample green spaces. There is a high standard of drinking water provision with connections to virtually all residents. But although sewerage systems reach 92% of the population, there are serious issues with pollution and flooding of watercourses. This is partly because some of the main interceptor sewers needed to connect to the

WTPs have never been constructed. There has also been excessive development in flood-prone areas, and there is a serious problem with illicit interconnections between the wastewater and stormwater systems.

In the 1970s and 1980s, during rapid population growth, the emphasis was on heavily engineered river and drainage works. Since that time, water management processes have been democratised, with community participation in planning, in the development of infrastructure, and even in budgeting. Starting in 2006, Belo Horizonte was also a demonstration city for the SWITCH project, an EU-funded initiative for the practical application of research on sustainable urban water management (Knauer et al., 2011). The main areas of activity have been rainwater harvesting, providing infiltration and detention facilities, and developing wetlands. Sites have been developed to demonstrate the effectiveness of the approaches and to engage stakeholders (including the local community, local officials, and engineers). Many of the sites have deliberately been located at schools to maximise the learning impact. Belo Horizonte is judged to be achieving a genuine transition to integrated urban water management, which is bringing many benefits.

BOX 1.4   Sponge cities, China

An urban stormwater management national "Sponge City" Programme was initiated in China in 2014 and by 2020 a total of CNY 177 billion (about £20 billion) had been invested in 30 pilot cities (Figure 1.5) chosen by the central government (Li et al., 2017). The idea of sponge cities is to use sustainable drainage systems (SuDS) and other blue-green infrastructure such as green roofs, rain gardens, grass swales, and wetlands to absorb, retain and treat stormwater, only discharging it into the sewer and river as a last resort (Yin et al., 2021). Compared to the adoption of SuDS in other countries, however, the programme has been at an unprecedented speed and scale. The aim is to achieve the target that 80% of urban areas should have systems in place by 2030 to capture 70% of the stormwater experienced by the cities. This ambition is supported by a further 3-year pilot rollout phase from 2021 to 2023, with the subsequent group of 20 cities announced in June 2021 (Figure 1.5), each receiving funding of CNY 0.7–1.1 billion from the central government (Fu et al., 2022).

1.7 TYPES OF SYSTEM

Piped systems consist of drains carrying flow from individual properties, and sewers carrying flow from groups of properties or larger areas. The word *sewerage* refers to the whole centralised infrastructure system: pipes, manholes, structures, pumping stations, and so on. There are basically two types of conventional sewerage: a combined system in which wastewater and stormwater flow together in the same pipe, and a separate system in which wastewater and stormwater are kept in separate pipes.

Non-pipe systems (called SuDS or sustainable drainage systems in the UK) manage *stormwater* flows closer to the source at which they are generated (decentralised), using the infiltration and storage properties of semi-natural (green) features. Some schemes for reducing dependence on the main drainage also involve more localised collection and treatment of *wastewater*.

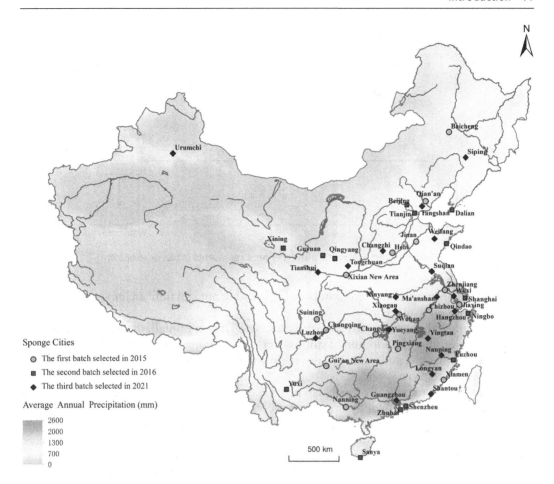

Figure 1.5 Fifty pilot cities chosen for the Chinese national Sponge City construction programme. (Reproduced from Fu, G., Zhang, C., Hall, J.W. and Butler, D. 2022. Are sponge cities the solution to China's growing urban flooding problems? WIREs Water, 10:e1613, under a CC BY 4.0 DEED. Attribution 4.0 International Licence.)

Together these form a very large asset base in many countries, requiring careful and sustained operation. In England & Wales, there is in excess of 570,000 km of public sewer (Discover Water, 2023) and a much longer length than this of connecting private drains associated with individual households and businesses. However, no national picture of the coverage of SuDS is currently available.

1.7.1 Combined systems

A combined sewer network is a complex (typically) branching system, and Figure 1.6 presents a simplification of a typical arrangement, showing a very small proportion of the branches. The figure is a plan of a town located beside a natural water system of some sort: a river or estuary, for example. Combined sewers convey both wastewater and stormwater together in the same pipe, and the ultimate destination is the WTP, located, in this case, a short distance out of the town. In the UK, continental Europe, and North America, most of the older sewerage is combined (some 30% of the total length of sewer in the UK). These types of systems

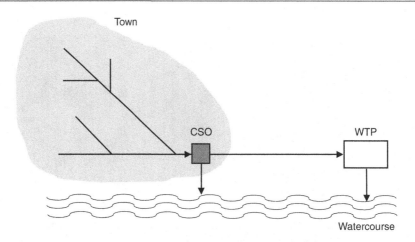

Figure 1.6 Combined sewer system schematic plan (grey area represents urban development).

are no longer constructed and have not been for decades, but still fulfil an important function and require careful management.

In dry weather, the system carries wastewater flow. During rainfall, the flow in the sewers increases as a result of the addition of stormwater. Even in quite light rainfall, the stormwater flows will predominate, and in heavy falls, the stormwater could be 50 or even 100 times the average wastewater flow.

It is simply not economically feasible to provide capacity for this flow along the full length of the sewers – which would, by implication, carry only a tiny proportion of the capacity most of the time. At the WTP, it would also be unfeasible to provide this capacity in the treatment processes. The solution was to provide structures in the sewer system that, during heavy rainfall, divert flows above a certain level out of the sewer system and into a natural watercourse. These structures are called combined sewer overflows (CSOs). A typically located CSO is included in Figure 1.6.

The key drawback of CSOs and by extension, combined sewers, is that large volumes of minimally treated wastewater can be discharged to the environment during heavy rainfall when they operate. To put it simply, CSOs cause pollution. Chapter 13 considers the design and operation of combined sewers and CSOs in more detail.

1.7.2 Separate systems

Most sewerage systems constructed in the UK since 1945 are separate (about 70%, by total length). Figure 1.7 is a sketch plan of the same town as shown in Figure 1.6, but this time, sewered using a separate system. Wastewater and stormwater are carried in separate pipes, usually laid side by side. Wastewater flows vary during the day, but the pipes are designed to carry the maximum flow all the way to the WTP. The stormwater is not mixed with wastewater and can be discharged to the watercourse at a convenient point. The first obvious advantage of the separate system is that CSOs, and the pollution associated with them, are avoided.

An obvious issue might be cost. It is true that the pipework in separate systems is more expensive to construct, but constructing two pipes instead of one does not cost twice as much. The pipes are usually constructed together in the same excavation. The stormwater pipe (the larger of the two) may be about the same size as the equivalent combined sewer, and the wastewater pipe will be smaller. So the additional costs are due to a slightly wider excavation and an additional, relatively small pipe.

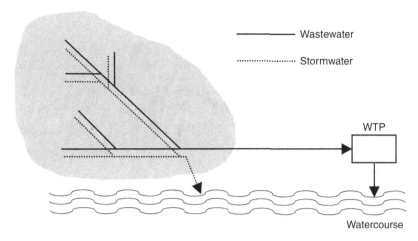

Figure 1.7 Separate sewer system schematic plan.

However, separate systems do have drawbacks of their own. These relate to the fact that perfect separation is effectively impossible to achieve. First, it is difficult to ensure that polluted flow is carried only in the wastewater pipe. Stormwater can be polluted for many reasons, including the washing-off of pollutants from the catchment surface. Second, it is very hard to ensure that no rainwater finds its way into the wastewater pipe. These points are considered in more detail in Chapter 5.

1.7.3 Mixed and partially separate systems

Some towns have systems that do not fall easily into the categorisation given above. An example is the "partially separate" systems, in which wastewater is mixed with some stormwater, while the remaining stormwater is conveyed by a separate pipe. Many other towns have mixed systems for more accidental reasons: for example, because a new town drained by a separate system includes a small old part drained by a combined system, or because misconnections resulting from ignorance, malpractice, or malfunctioning dual manholes have caused an unintended combination of the two types of flow.

1.7.4 Non-pipe systems

Non-pipe systems (SuDS) manage stormwater closer to its source of generation, typically on or near the ground surface, using a variety of devices (see Figure 1.8). This has clear advantages and disadvantages. Advantages include the ability to limit flows downstream when discharging into combined sewers and/or to contribute to local flood risk management, the ability to partially treat stormwater prior to discharge, and the opportunity to provide local reuse, amenity and biodiversity gains. The disadvantages involve the need for extra space, which is often at a premium in urban areas, and more practical matters concerning ownership and the maintenance of long-term performance. SuDS and stormwater management are considered in more detail in Chapter 11.

An initial comparison of the various systems is given in Table 1.1. We develop these ideas further throughout the remainder of this book.

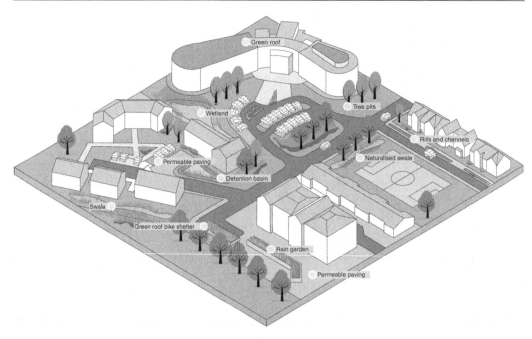

Figure 1.8 Non-pipe (SuDS) scheme. (Inspired by a graphic at https://www.susdrain.org/)

Table 1.1 Combined sewers, separate sewers, and SuDS compared

	Combined sewers	*Separate sewers*	*SuDS[a]*
CSOs	Required part of the system	Not required	Not required, and SuDS limit stormwater entering sewers, so reducing CSO operation
WTPs	Larger works inlets needed and stormwater storage required	Smaller works	Smaller works
Pumping	Higher if pumping of flow to treatment is necessary	Stormwater rarely pumped	Rarely required
Sediment	Deposition in pipes can lead to blockage and higher maintenance needs	More feasible to design for self-cleansing	Limits downstream discharge of sediment but requires on-site maintenance
Flooding (storm conditions)	If flooding occurs, foul conditions will be caused	Any flooding will be by stormwater only	Any flooding will be by stormwater only
Pipework	House drainage simple	More house drains, with risk of misconnections and are difficult to identify	Reduced pipework overall
Construction costs	Lower than separate system	Some extra cost	Similar to or lower than a separate system
Maintenance costs	Highest	Lower than combined	Different maintenance regimes required
Space requirements	Minimum underground space	Additional underground space due to two pipes	Higher above-ground space normally needed

(Continued)

Table 1.1 (Continued)

	Combined sewers	Separate sewers	SuDS[a]
Flow rate	Attenuation in pipes and tanks	Attenuation in pipes and tanks	Attenuation at source
Flow volume	No reduction	No reduction	Significant reduction
Treatment of stormwater	No treatment	Minimal treatment	Significant treatment
Long-term performance	Systems in operation for over a century	Systems in operation for many decades	Systems in operation for years
Amenity	No contribution	No contribution	Significant contribution if carefully designed
Biodiversity	No contribution	No contribution	Significant contribution if carefully designed

[a] A piped system for wastewater would also be required for this case.

1.8 URBAN WATER SYSTEM

In this section, we look at how urban drainage fits within the whole urban water system. Figure 1.9 is a diagrammatic representation of the urban water system with a general representation of the flow paths and the interrelationship of the main elements. Solid arrows represent intentional flows and dotted arrows unintentional ones. Heavy-bordered boxes indicate "sources," and dashed, heavy-bordered boxes show "sinks." This figure shows a separate sewer system, but a similar diagram can be produced for a combined system, and non-pipe systems can also be represented.

There are two main inflows. The first is rainfall that falls onto catchment surfaces such as "impervious" roofs and paved areas and "pervious" vegetation and soil. It is at this point that the quality of the flow is degraded as pollutants on the catchment surfaces are washed off. This is a highly variable input that can only be properly described in statistical terms (as considered in Chapter 4). The resulting runoff retains similar statistical properties to rainfall (Chapter 5). There is also the associated outflow of evaporation, whereby some water is removed from the system. Rainfall that does not run off will find its way into the ground and eventually the receiving water or infiltrate into the drainage system. The component that runs off is conveyed by the roof and highway drainage as stormwater directly into the storm sewer. Discharges from the storm sewer to the environment are intermittent and are statistically related to the rainfall inputs.

The second inflow is water supply. Water consumption is more regular than rainfall, although even here there is some variability (Chapter 3). The resulting wastewater is closely related in timing and magnitude to the water supply. The wastewater is conveyed by the building drainage directly to the foul sewer. An exception is when an industry treats its own waste separately and then discharges treated effluent directly to the receiving water. The quality of the water (originally potable) deteriorates during usage. The foul sewer conveys the wastewater to the WTP with patterns related to water consumption. Unintentional flow may leave the pipes via exfiltration to the ground (Chapter 3). At other locations, groundwater may act as a source and add water into the system via pipe infiltration (Chapter 3). Misconnections can cause unintentional mixing of the wastewater with stormwater in either pipe (Chapter 5). The treatment plant, in turn, discharges to the receiving water.

Non-pipe solutions are effectively included in the "SURFACES" components of Figure 1.9. Alterations to paved and roof surfaces such as porous pavements, rain gardens, and rainwater harvesting will reduce runoff rates into downstream combined sewers, local receiving waters,

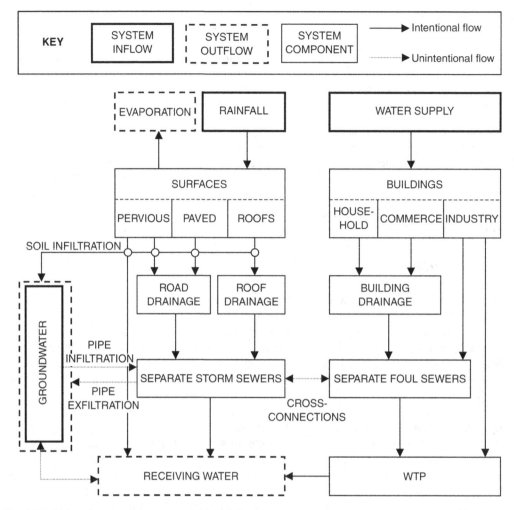

Figure 1.9 Urban drainage (separate sewers) and the urban water system.

or groundwater. This will mitigate flooding and reduce watercourse pollution. For separate systems, similar interventions can help eliminate the need for some storm sewers entirely.

1.9 CHANGING CONTEXT

Urban drainage has come a long way in a relatively short time. Figure 1.10 illustrates the changing context and drivers over the last 50 years or so. It is increasingly being realised that urban drainage sits within a wider context sometimes characterised as water cycle management, integrated urban water management, or sustainable water management (see Chapter 24 for more details). Furthermore, these areas also sit within wider urban and infrastructure planning (see Chapter 19). Such concepts recognise the "joined-up" nature of water systems and the fact that there is only one water cycle, however we might try to dissect it. In addition, it is increasingly being recognised that stormwater and wastewater are not nuisances or threats that should be immediately dealt with but are, in fact, resources that should be managed.

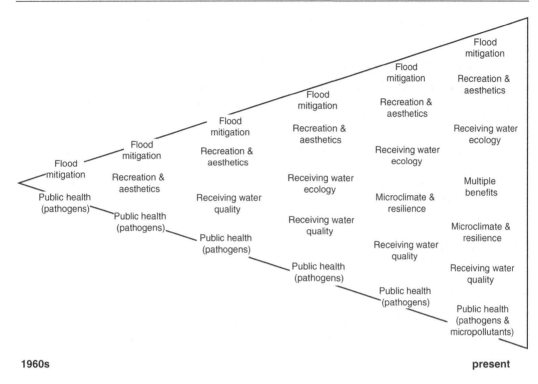

Figure 1.10 Changing context and drivers of urban drainage over time. (Adapted from Fletcher, T.D. et al. 2015. *Urban Water Journal*, 12(7), 525–542.)

1.10 FUTURE CHALLENGES

It is argued by some that water management, in general, and urban drainage, in particular, are in crisis with current approaches not fit for purpose. These approaches are effectively incurring increasing economic, social, and environmental costs, even in countries with a long tradition of successful practices (Larsen et al., 2016). Challenges are indeed many and include insufficient or deferred investment, an ageing and under-capacity built infrastructure, increasing population growth and urbanisation, the effects of climate change, achieving net zero carbon emissions, sewer flooding, pollution of marine and freshwater, emerging contaminants, public health concerns and increasing user expectations.

We hope that this book will help readers understand the strengths and weaknesses of our current systems, be inspired by how we can do things better, and look to the future with confidence.

PROBLEMS

1 Do you think urban drainage is taken for granted by most people in developed countries? Why? Is this a good or bad thing?
2 How does urbanisation affect the natural water cycle?
3 Some claim that urban drainage engineers, throughout history, have saved more lives than doctors and nurses. Can that be justified, nationally and internationally?

4 What have been the main influences on urban drainage engineers since the start of their profession?

5 "Mixing of wastewater and stormwater (in combined sewer systems) is fundamentally irrational. It is the consequence of historical accident, and remains a cause of significant damage to the water environment." Explain and discuss this statement.

6 Explain the characteristics of piped and non-piped systems. Discuss the advantages and disadvantages of each.

7 Describe how urban drainage systems interact with the overall urban *water* system. (Use diagrams.) How would you represent rainwater harvesting on the diagram?

8 In what way does urban drainage form a part of integrated urban water management?

9 What changes would you make to current urban drainage approaches to better meet future challenges?

KEY SOURCES

Guo, J.C.Y., Wang, W. and Li, J. 2023. *Urban Drainage and Storage Practices*. CRC Press.

Keller, M. 2023. Rethinking wastewater management. *ETH Magazine, GLOBE 2/23, Focus - Water: Life-giving. Contested. Destructive*, 26–28.

Tyson, J.M., Evans, T.D., and Orman, N. 2013. Urban Drainage and the Water Environment: A Sustainable Future? *A Review of Current Knowledge*. 2nd Edn., Foundation for Water Research, FR/R0011.

REFERENCES

Burian, S.J. and Edwards, F.G. 2002. Historical perspectives of urban drainage. *Global Solutions for Urban Drainage: Proceedings of the 9th International Conference on Urban Drainage, Portland, Oregon, September, on CD-ROM*.

De Feo, G., Antoniou, G., Fardin, H.F., El-Gohary, F., Zheng, X.Y., Reklaityte, I. 2014. The historical development of sewers worldwide. *Sustainability*, 6, 3936–3974.

Discover Water. 2023. *Collecting & treating sewage*. www.discoverwater.co.uk/treating-sewage

Dolowitz, D.P., Bell, S. and Keeley, M. 2018. Retrofitting urban drainage infrastructure: green or grey? *Urban Water Journal*, 15(1), 83–91.

Fletcher, T.D., Shuster, W., Hunt, W.F., Ashley, R., Butler, D., Arthur, S. 2015. SUDS, LID, BMPs, WSUD and more – The evolution and application of terminology surrounding urban drainage. *Urban Water Journal*, 12(7), 525–542.

Fu, G., Zhang, C., Hall, J.W. and Butler, D. 2022. Are sponge cities the solution to China's growing urban flooding problems? *WIREs Water*, 10, e1613.

Halliday, S. 2001. *The Great Stink of London: Sir Joseph Balzalgette and the Cleansing of the Victorian Metropolis*, Sutton Publishing.

Hasan, A. 2006. Orangi Pilot Project: The expansion of work beyond Orangi and the mapping of informal settlements and infrastructure. *Environment and Urbanization*, 18(2), 451–480.

Knauer, S., de Oliveira Nascimento, N., Butterworth, J., Smits, S. and Lobina, E. 2011. Towards Integrated Urban Water Management in Belo Horizonte, Brazil: A Review of the SWITCH Project. www.ircwash.org/resources/towards-integrated-urban-water-management-belo-horizonte-brazil-review-switch-project

Larsen, T.A., Hoffmann, S., Lüthi, C., Truffer, B. and Maurer, M. 2016. Emerging solutions to the water challenges of an urbanizing world. *Science*, 352, 6288.

Li, H., Ding, L., Ren, M., Li, C. and Wang, H. 2017. Sponge city construction in China: A survey of the challenges and opportunities. *Water*, 9, 1–17.

OPP-RTI. 2012. Orangi Pilot Project (OPP) – Institutions and Programs. http://www.opp.org.pk/opp-rti/

PD ISO/TR 24539: 2021. 2021. Service activities relating to drinking water supply, wastewater and stormwater systems — Examples of good practices for stormwater management.

UN-Habitat. 2003. *Water and Sanitation in the World's Cities: Local Action for Global Goals*, Earthscan.

UN-Habitat. 2006. *Meeting Development Goals in Small Urban Areas*, Earthscan.

Yin, D., Chen, Y., Jia, H., Wang, Q., Chen, Z., Xu, C., Li, Q., Wang, W., Yang, Y., Fu, G. and Chen, A. S. 2021. Sponge city practice in China: A review of construction, assessment, operational and maintenance. *Journal of Cleaner Production*, 280, 124963.

Chapter 2

Water quality

2.1 INTRODUCTION

The chemical and biological quality aspects of wastewater and stormwater conveyed in urban drainage systems are important for several reasons:

- Significant water quality changes can occur throughout the drainage system.
- Decisions made in the sewer system have significant effects on the wastewater treatment plant (WTP) performance.
- Direct discharges from drainage systems (e.g., combined sewer overflows (CSOs) and stormwater outfalls (SWOs) can have a serious pollution impact on receiving waters).

This chapter looks at the basic approaches to characterising wastewater and stormwater including outlines of the chemical basics (Section 2.2), the main water quality parameters and tests used in practice (Section 2.3), and common processes (Section 2.4). Typical test data are given in Chapters 5 and 7. It also considers the water quality impacts of discharges from urban drainage systems (Section 2.5) and discusses relevant legislation including water quality standards (Section 2.6).

2.2 BASICS

2.2.1 Strength

Water has been called the *universal solvent* because of its ability to dissolve numerous substances. The term *water quality* relates to all the constituents of water, including both dissolved substances and any other substances carried by the water.

The *strength* of polluted liquid containing a constituent of mass M in water of volume V is its *concentration* given by $c = M/V$, usually expressed in mg/L. This is numerically equivalent to parts per million (ppm) assuming the density of the mixture is equal to the density of water (1000 kg/m³). The plot of concentration c as a function of time t is known as a *pollutograph* (see Figure 20.4 for an example). Pollutant mass flow or flux is given by its *load rate* $L = M/t = cQ$, where Q is the liquid flow rate.

EXAMPLE 2.1

A laboratory test has determined the mass of a constituent in a 2 L wastewater sample to be 0.75 g. What is its concentration (c) in mg/L and ppm? If the wastewater discharges at a rate of 600 L/s, what is the pollutant load rate (L)?

DOI: 10.1201/9781003408635-2

Solution

$$c = \frac{M}{V} = \frac{750}{2} = 375\,\text{mg}\,/\,\text{L} = 375\,\text{ppm}$$
$$L = cQ = 0.375 = 600 = 225\,\text{g}\,/\,\text{s}$$

In order to calculate the average concentration, either of wastewater during the day or of stormwater during a rain event, the *event mean concentration* (EMC) can be calculated as a flow-weighted concentration c_{av}:

$$c_{av} = \frac{\sum Q_i c_i}{Q_{av}} \tag{2.1}$$

where c_i is the concentration of each sample i (mg/L), Q_i is the flow rate at the time the sample was taken (L/s), and Q_{av} is the average flow rate (L/s).

2.2.2 Equivalent concentrations

It is a common practice when dealing with a pollutant (X) that is a compound to express its concentration in relation to the parent element (Y). This can be done as follows:

$$\text{Concentration of compound } X \text{ as element } Y =$$
$$\text{concentration of compound } X \times \frac{\text{atomic weight of element } Y}{\text{molecular weight of compound } X} \tag{2.2}$$

The conversion of concentrations is based on the gram molecular weight of the compound and the gram atomic weight of the element. Atomic weights for common elements are given in standard texts (e.g., Droste, 2018).

Expressing substances in this way allows easier comparison between different compounds of the same element and more straightforward calculation of totals. Of course, it also means care needs to be taken in noting in which form compounds are reported (see Example 2.2).

EXAMPLE 2.2

A laboratory test has determined the mass of orthophosphate (PO_4^{3-}) in a 1 L stormwater sample to be 56 mg. Express this in terms of phosphorus (P).

Solution

Gram atomic weight of P is 31 g
Gram atomic weight of O is 16 g
Gram molecular weight of orthophosphate is $31 + (4 \times 16) = 95$ g
Hence, from Equation (2.2),

$$56\,\text{mg}\,PO_4^{3-}\,/\,L = 56 \times \frac{31\,\text{g}\,P}{95\,\text{g}\,PO_4^{3-}} = 18.3\,PO_4^{3-} - P\,/\,L$$

2.3 PARAMETERS

There is a wide range of quality parameters used to characterise wastewater, and these are described in the following section. Further details on these and many other water quality parameters and their methods of measurement can be found elsewhere (e.g., AWWA, 2023; SCA, 2024). Specific information on the range of concentrations and loads encountered in practice is given in Chapters 3 (wastewater) and 5 (stormwater).

2.3.1 Sampling and analysis

There are three main methods of sampling: grab, composite, and continuous. *Grab* samples are simply discrete samples of fixed volume taken to represent local conditions in the flow. They may be taken manually or extracted by an automatic sampler. A *composite* sample consists of a mixture of a number of grab samples taken over a period of time or at specific locations, taken to more fully represent the composition of the flow. *Continuous* sampling consists of diverting a small fraction of the flow over a period of time. This is useful for instruments that give almost instantaneous measurements, for example, pH and temperature. The Environment Agency provides further details (EA, 2021).

In sewers, where flow may be stratified, samples need to be taken throughout the depth of flow if a true representation is required. Mean concentrations can then be calculated by weighting with respect to the local velocity and area of flow.

In all of the tests available to characterise wastewater and stormwater, it is necessary to distinguish between precision and accuracy. In the context of laboratory measurements, *precision* is the term used to describe how well the analytical procedure produces the same result on the same sample when the test is repeated. *Accuracy* refers to how well the test reproduces the actual value. It is possible, for example, for a test to be precise, but inaccurate with all values closely grouped, but around the wrong value! Techniques that are both precise and accurate are required.

2.3.2 Solids

Solid types of concern in wastewater and stormwater can broadly be categorised into four classes: gross, grit, suspended, and dissolved (see Table 2.1). Gross and suspended solids may be further subdivided according to their origin as wastewater and stormwater.

2.3.2.1 Gross solids

There is no standard test for the gross solids found in wastewater and stormwater, but they are usually defined as solids (specific gravity [SG] = 0.9–1.2) captured by a 6 mm mesh screen (i.e., solids > 6 mm in two dimensions). Gross sanitary solids (also variously known as aesthetic, refractory, or intractable solids) include faecal stools, toilet paper, wet wipes,

Table 2.1 Basic classification of solids

Solid type	Size (μm)	SG (–)
Gross	> 6000	0.9–1.2
Grit	> 150	2.6
Suspended	≥ 0.45	1.4–2.0
Dissolved	< 0.45	–

and "sanitary refuse" such as women's sanitary protection, condoms, bathroom litter, and so on. Faecal solids and toilet paper break up readily and may not travel as far in the system as gross solids. Gross stormwater solids consist of debris such as bricks, wood, cans, paper, and so on.

There are two principal concerns with these solids. First, they are known to cause significant maintenance problems due to deposition and blockage (in pipes and pumping stations) and can cause blinding of screens at CSOs. Second, if discharged into the aqueous environment and onto riverbanks and beaches, their negative "aesthetic impact" is significant (further discussed in Section 2.5.3.5).

2.3.2.2 Grit

Again, there is no standard test for the determination of grit, but it may be defined as the inert, granular material (SG ≈ 2.6) retained on a 150-μm sieve. Grit forms the bulk of what is termed *sewer sediment*, and the nature and problems associated with this material are revisited in Chapter 16.

2.3.2.3 Suspended solids

The suspended solids (SSs) content is the solid matter (both organic and inorganic) maintained in suspension and retained when a sample is filtered (0.45 μm pore size). In the SS test, the residue is washed, dried, and weighed under standard conditions and expressed as a concentration. The accuracy of the SS test is approximately ±15%.

The finer fractions of suspended solids (< 63 μm) are extremely efficient carriers of pollutants, carrying greater than their proportionate share (see Sections 5.4.2 and 17.5.2). High concentrations may have a number of adverse effects on the receiving water, including increased turbidity, reduced light penetration, blanketing of the bed, and interference with many types of fish and aquatic invertebrates. Even after deposition, the pollutants attached to these enriched sediments still present a risk, since they can cause a "delayed" sediment oxygen demand (see Section 2.4.2), or may be resuspended at high flows.

By definition, solids not in suspension (i.e., with a diameter < 0.45 μm) are dissolved (see Example 2.3).

2.3.2.4 Volatile solids

The solids retained during the SS test can be ignited at 550°C in a muffle furnace. The residue is known as non-volatile or fixed material. The volatilised fraction (the volatile solids) gives an indication of the organic content of the SS.

2.3.3 Oxygen

A key to understanding the reactions occurring anywhere within the urban drainage system is the measurement and prediction of the oxygen levels in the aqueous phase. Dissolved oxygen (DO) levels depend on physical, chemical, and biochemical activities in the system.

2.3.3.1 Dissolved oxygen

Oxygen in water is only sparingly soluble. In equilibrium with air, the solubility of dissolved oxygen (DO) in water is referred to as its *saturation* value, and it decreases with the increase of both temperature and purity (salinity, solids content) and with the decrease in

Table 2.2 Dissolved oxygen concentration (under standard
conditions) in water as a function of temperature

Temperature (°C)	DO (mg/L)
0	14.62
5	12.80
10	11.33
15	10.15
20	9.17
25	8.38
30	7.63

EXAMPLE 2.3

In a standard laboratory test, a crucible and a filter pad are dried and their combined mass is measured at 64.592 g. A 250 mL wastewater sample is drawn through the filter under vacuum. The filter and residue are then placed on the crucible in an oven at 104°C for drying. The new combined mass is 64.673 g. The crucible and its contents are next placed in a muffle furnace at 550°C. After cooling, the combined mass is measured as 64.631 g. Determine (1) the suspended solids concentration of the sample and (2) the volatile fraction of the suspended solids.

Solution

Mass of suspended solids removed:

   Crucible + filter + solids = 64.673 g

   Crucible + filter = 64.592 g

   Mass of suspended solids = 0.081 g

   (1) Concentration of SS:

      81 (mg) / 0.250 (L) = 324 mg/L

Mass of volatile suspended solids removed:

   Initial crucible + filter + solids = 64.673 g

   Final crucible + filter + solids = 64.631 g

   Mass of volatile solids = 0.042 g

   (2) Volatile fraction of suspended solids:

      42 (mg) / 81 (mg) = 0.52

The concentration of DO is an excellent indicator of the "health" of a receiving water. All the higher forms of river life require oxygen. Coarse fish, for example, require in excess of 3 mg/L (see Table 2.3). In the absence of toxic impurities, there is a close correlation between DO and biodiversity.

Table 2.3 Oxygen requirements of fish species

Characteristic species	Minimum DO concentration (mg/L)	Minimum saturation (%)	Comment
Trout, bullhead	7–8	100	Fish require much oxygen
Perch, minnow	6–7	< 100	Need more oxygen for active life
Roach, pike, chub	3	60–80	Can live for long periods at this level
Carp, tench, bream	< 1	30–40	Can live for short periods at this level

Source: Adapted from Gray, N.F. 2010. *Water Technology. An Introduction for Environmental Scientists and Engineers.* Arnold.

atmospheric pressure (Table 2.2). Hence, warm water (even with no impurities) is, in effect, a pollutant source.

Dissolved oxygen (DO) can be measured analytically using the Winkler titration method. Titration is a laboratory technique where measured volumes of a reagent (the titrant) are incrementally added to a sample up to the equivalent amount of the constituent being analysed. Membrane electrodes are now more commonly and conveniently used both in the laboratory and for *in situ* measurements.

2.3.4 Organic compounds

Wastewater and stormwater contain significant quantities of organic matter in both particulate and soluble forms. Organic compounds in water are unstable and are readily oxidised either biologically or chemically to stable, relatively inert, end products such as carbon dioxide, nitrates, sulphates, and water. There are three main categories of biodegradable organics present:

- Carbohydrates such as sugars, starch, and cellulose
- Proteins, which are complex molecules built up of amino acids and urea
- Lipids and fats

The decomposition of organic matter by microorganisms consumes DO. In urban drainage systems, the main implication is oxygen depletion in

- Sewers, resulting in an anaerobic environment (considered in Chapter 18)
- Receiving waters (considered later in this chapter)

An indirect indication of the amount of organic material in a wastewater can be derived from one of two tests: the biochemical oxygen demand (BOD) or the chemical oxygen demand (COD). A third option is the total organic carbon (TOC) test, which gives a more direct measure of the carbon content of the sample under test.

2.3.4.1 Biochemical oxygen demand (BOD₅)

This test is a laboratory simulation of the microbial processes occurring in water contaminated with organic compounds. It measures the DO consumed in a sample diluted in a 300 mL bottle during a specified incubation period (usually 5 days at a temperature of 20°C in darkness). The DO is used by microorganisms as they break down organic material and certain inorganic compounds. Thus,

$$\text{BOD}_5 = \frac{(c_{\text{DOI}} - c_{\text{DOF}})}{p} \tag{2.3}$$

where P is the sample dilution (volume of sample/volume of the bottle), c_{DOI} is the initial dissolved oxygen concentration (mg/L), and c_{DOF} is the final dissolved oxygen concentration (mg/L).

Measured amounts of the wastewater sample are diluted with prepared water containing nutrients and DO. Seed microorganisms are added if insufficient amounts are available in the sample. Equation (2.3) assumes the dilution water has negligible BOD_5 (see Example 2.4). The test may also measure the oxygen used to oxidise reduced forms of nitrogen (nitrogenous demand – NBOD) unless an inhibitor (e.g., allylthiourea [ATU]) is used. The evolution of BOD with time is shown in Figure 2.1.

EXAMPLE 2.4

A laboratory test for BOD_5 (ATU) is carried out by mixing a 5 mL sample with distilled water into a 300 mL bottle. Prior to the test, the DO concentration of the mixture was 7.45 mg/L and after 5 days, it had reduced to 1.40 mg/L. What is the BOD_5 concentration of the sample?

Solution

Dilution, $p = 5 / 300 = 0.0167$

Equation 2.3:

$\text{BOD}_5 = (7.45 - 1.40) / 0.0167 = 363$ mg/L

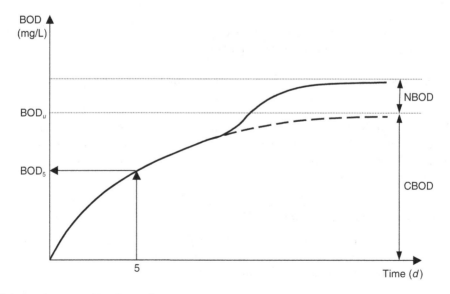

Figure 2.1 Development of biochemical oxygen demand (BOD) with time.

The test does not give a measure of the *total* oxidisable organic matter because of the presence of considerable quantities of carbonaceous matter resistant to biological oxidation. Over 5 days, only the readily biodegradable fraction of organic material present in the water will be broken down.

The test can be extended up to 10–20 days to reach the ultimate carbonaceous CBOD ≈ 1.5 BOD_5. BOD tests are subject to inhibition if the wastewater contains any toxic components (e.g., trace metals in runoff), and they should be seen as an indication rather than as an accurate determination.

BOD_5 is a common parameter used in the control of treated wastewater effluent quality through the setting and monitoring of discharge consent standards (e.g., 25 mg/L) (CEC, 1991a). Rivers are considered to be polluted if their BOD_5 exceeds 5 mg/L, and 3 mg/L is required for salmonid rivers (CEC, 1978).

2.3.4.2 Chemical oxygen demand (COD)

The COD test measures the oxygen equivalent of the organic matter that can be oxidised by a strong chemical oxidising agent (potassium dichromate) in an acidic medium. The standard test lasts for 3 hours using either a titrimetric or colorimetric method. Colorimetric test methods rely on measuring the intensity of light from colour changes in the reaction. Almost all organic compounds are oxidised. Some inorganics are also oxidised, but ammonium and ammonia are not. Thus, this measurement is a good estimation of the total content of organic matter.

One of the main limitations of the COD test is its inability to differentiate between biodegradable and biologically inert organic matter. However, if sufficient data are available, it is often possible to empirically relate the COD with BOD values, such as

$$c_{BOD} \approx a \times c_{COD} \tag{2.4}$$

where c_{BOD} is the BOD concentration (mg/L), c_{COD} is the COD (mg/L), and a is the 0.4–0.8 (–).

It should be stressed, however, that "a" will vary between wastewaters, and no universal relationship has been found between the two parameters. Nevertheless, it is a good indicator of wastewater treatability.

The COD of a sample can be further differentiated into several forms. The first major category is inert (suspended or soluble) material that is non-biodegradable within the timescale associated with an urban drainage system. The second is biodegradable matter, which in turn can be divided into readily and slowly degradable material. The former refers to material that can be immediately oxidised by microorganisms and the latter to matter, which is degraded more slowly. The approximate relationship between the COD fractions and BOD is summarised in Figure 2.2. The methods for characterising the various COD fractions and their effectiveness are summarised by Zhang et al. (2021).

2.3.4.3 Total organic carbon (TOC)

Unlike the BOD and COD tests, the TOC test directly measures the TOC content of a sample. The test is based on the fact that carbon dioxide (CO_2) is a product of combustion. It is carried out in an instrument containing a small furnace at 950°C in the presence of a catalyst (having first removed the *inorganic* carbon). After this, the CO_2 released is measured for the known volume of the sample. It can be performed very rapidly using a single instrument and is especially applicable in the analysis of small concentrations of organic matter.

COD is approximately related to TOC as follows (see Example 2.5):

$$COD \approx 2.5 \times TOC \qquad\qquad (2.5)$$

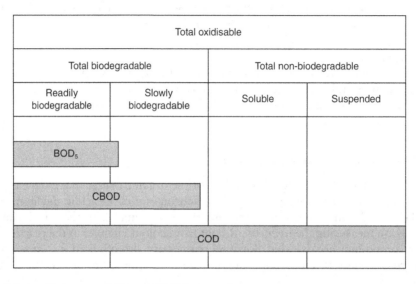

Figure 2.2 Relationship between BOD and COD for organic matter.

EXAMPLE 2.5

If organic material can be represented by the chemical formula $C_6H_{12}O_6$ (glucose), calculate the *theoretical* relationship between COD and TOC.

Solution

In the COD and TOC tests, organic material is oxidised to carbon dioxide and water:

$$C_6H_{12}O_6 + 6O_2 \rightarrow 6CO_2 + 6H_2O$$

From the above equation, each mole of organic material (molecular weight = 180) requires 6 moles of oxygen (molecular weight 32) for oxidation. As the weight in grams of a substance is the number of moles × molecular weight:

$$COD = \frac{6\,moles\,O_2}{1\,mole\,C_6H_{12}O_6} = \frac{6 \times 32}{1 \times 180} = 1.067\,gO_2\,/\,gC_6H_{12}O_6$$

Also, each mole of organic material contains 6 atoms of carbon, so

$$TOC \quad = \quad \frac{6\,atoms\,C}{1\,mole\,C_6H_{12}O_6} = \frac{6 \times 12}{1 \times 180} = 0.4\,gC\,/\,gC_6H_{12}O_6$$

$$\therefore COD \quad = \quad \frac{1.067}{0.4} \times TOC = 2.67\,TOC$$

2.3.5 Nitrogen

Nitrogen exists in four main forms: organic (in the protein that makes up much matter), ammonia (or ammonia salts), nitrite, and nitrate. Total nitrogen is the sum of all forms, although in wastewater and stormwater, organic and ammonia nitrogen make up most of the total. The concentration of nitrogen in domestic wastewater is usually related to the BOD_5.

Excessive levels of nitrogen discharged to receiving waters can promote the growth of undesirable aquatic plants such as algae and floating macrophytes. In severe cases, the receiving water can experience *eutrophic* symptoms such as water discolouration, odours, and depressed oxygen levels (considered in more detail later in this chapter).

2.3.5.1 Organic nitrogen (org.N)

Organic nitrogen includes such natural materials as proteins and peptides, nucleic acids and urea, and numerous synthetic organic materials, although not *all* organic nitrogen compounds.

Analytically, organic nitrogen and ammonia are determined together using the Kjeldahl method. In this test, the aqueous sample is boiled to remove any pre-existing ammonia and then digested, during which the organic nitrogen is converted to ammonia. The amount of ammonia produced is then determined as detailed in the next section.

2.3.5.2 Ammonia nitrogen (NH₃–N)

Ammonia nitrogen exists in solution in two forms – as the ammonium ion $\left(NH_4^+\right)$ and as ammonia gas (NH_3) – depending on the pH and temperature of the wastewater:

$$NH_3 + H^+ \rightleftharpoons NH_4^+ \tag{2.6}$$

At values of pH $\leq$ 7, virtually all the ammonia is present as ammonium. At pH 9, for example, 35% is present as NH_3.

Ammonia nitrogen is determined analytically by raising the pH and using a distillation process. The final measurement is then made by titration or by colorimetry. Sometimes organic nitrogen and ammonia nitrogen are determined together in the total Kjeldahl nitrogen (TKN) test, which is similar to the basic Kjeldahl method except any pre-existing ammonia is *not* removed.

Any unionised ammonia (NH_3) present in wastewater discharged to the environment is particularly toxic to fish, depending on the dissolved oxygen of the receiving water. Ammonia also exerts an oxygen demand during its conversion to nitrite and subsequently to nitrate.

2.3.5.3 Nitrite and nitrate nitrogen $\left(NO_2^- - N, NO_3^- - N\right)$

Nitrite is the intermediate oxidation state of nitrogen. It is relatively unstable and easily oxidised, and its presence shows that oxidation of nitrogenous matter is taking place. Nitrate forms the most highly oxidised state of nitrogen found in wastewater. Determination is usually by colorimetric methods.

2.3.6 Phosphorus

Phosphorus can be expressed as total, organic, or inorganic (ortho- and poly-) phosphorus. Organic phosphate is a minor constituent of wastewater and stormwater, with most phosphorus being in the inorganic form. Polyphosphates consist of combinations of phosphorus, oxygen, and hydrogen atoms. Orthophosphates (e.g., PO_4^{3-}, HPO_4^{2-}, $H_2PO_4^-$, H_3PO_4) are simpler compounds and may be in solution or attached to particles. Orthophosphates can be determined directly, but poly- and organic phosphates must first be converted to orthophosphates before determination.

Phosphorus-containing compounds are also implicated in receiving water eutrophication. Generally, phosphorus is the controlling nutrient in urban freshwater systems. Salmonid rivers have upper limits of 0.065 mg P/L (CEC, 1978).

2.3.7 Sulphur

Sulphurous compounds are found mainly in wastewater in the form of organic compounds and sulphates $\left(SO_4^{2-}\right)$. Under anaerobic conditions, these are reduced to sulphides (S^-), mercaptans, and certain other compounds. The principal product, hydrogen sulphide (H_2S), is formed mainly by biofilms on the walls of sewers and in sediment deposits.

Hydrogen sulphide is a flammable and very poisonous gas and, when escaping into the atmosphere, can cause serious odour nuisance. It is acutely toxic to aquatic organisms and could be a factor in fish kills near CSOs. Hydrogen sulphide in damp conditions can be oxidised biologically to sulphuric acid (H_2SO_4), which may cause serious damage to sewer materials, especially concrete (see Chapter 18).

2.3.8 Hydrocarbons

Hydrocarbons are organic compounds containing only carbon and hydrogen. They are classified into four groups based on molecular structure: aliphatic or straight chained, branch chained, aromatic (based on the benzene ring), and alicyclic. In this text, we are mainly concerned with the petroleum-derived group commonly found in stormwater, which includes petrol, lubricating and road oils. They are among the more stable organic compounds and are not easily biodegraded. Most have a strong affinity for suspended particulate matter. They are determined by extraction with carbon tetrachloride.

Hydrocarbons are lighter than water, and virtually insoluble, causing films and emulsions on the water surface and reducing atmospheric reaeration. Those in accumulated sediments can persist for long periods and exert a chronic impact on bottom-dwelling organisms, as well as can be remobilised by subsequent storm events.

2.3.9 FOG

FOG is a general term used to include the fats, oils, greases, and waxes of plant- or food-based origin present in wastewater. These lipids are complex organic molecules with building blocks of three, two, or one chain of fatty acids known as tri-, di-, and monoglycerides, respectively. These are very stable compounds and do not easily biodegrade naturally. They do not fully mix with water, and readily accumulate and float on water surfaces. However, in the presence of hot water and detergents, separation takes place at low rates. They are determined gravimetrically by extraction with trichlorotrifluroethane (Freon). Further details are given by Sultana et al. (2023).

Fats in sewer systems can cause blockages (see Chapter 18). When discharged into the environment, they cause films and sheens on the water surface, inhibiting oxygen transfer.

2.3.10 Heavy metals and synthetic compounds

A considerable number of heavy metals and synthetic organic and inorganic chemicals can be found in wastewater and stormwater. Among the many constituents of concern are metal species such as arsenic, cyanide, lead, cadmium, iron, copper, zinc, and mercury. Metals can exist in particulate, colloidal, and dissolved (labile) phases depending mainly on the prevailing redox and pH conditions. The concentration of metals is determined individually by atomic absorption spectrophotometry or jointly using multi-element equipment (e.g., ICP).

Metals in stormwater are predominantly in the particulate phase. This is important because the environmental mobility and bioavailability (and hence toxicity) of metals are highly related to their concentration in solution. Many (particularly the more soluble forms of zinc and copper) are known to have acute and chronic toxic effects on aquatic life and can inhibit biological processes at the WTP.

Herbicides and pesticides can be toxic to a variety of aquatic life at very low concentrations. Some of the more toxic varieties (e.g., chlorinated organics, DDT and PCBs) are no longer used, but their residues can still be found in the environment.

2.3.11 Microorganisms

Human excreta in wastewater and stormwater contains significant numbers of microorganisms (Garcia-Aljaro et al., 2018) some of which are pathogenic, and can cause a range of morbidities to people exposed through ingestion or contact (Sojobi and Zayed, 2022). They include the following:

- Bacteria such as salmonella, campylobacter, and some strains of *Escherichia coli* (*E.coli*) that can lead to gastrointestinal illnesses (GI) characterised by symptoms such as diarrhoea, vomiting, abdominal cramps, and fever.
- Viruses such as noroviruses and adenoviruses that are highly contagious and resilient in water environments and can cause acute gastroenteritis. Hepatitis A is also transmitted through contaminated water and can lead to acute liver infections and systemic illness.
- Protozoan parasites such as giardia and cryptosporidium that are resistant to traditional disinfection methods and can cause diarrhoeal diseases and GI upon ingestion or exposure.

Direct determination of the presence of these pathogenic microorganisms in wastewater and stormwater is not normally carried out unless there is a disease outbreak. Indeed, even microbiological indicator tests are not routinely undertaken. However, faecal indicator organisms (FIOs), such as total coliforms and faecal coliforms (*E. coli*), and faecal streptococci (FS) are known to occur in large numbers in both wastewater and stormwater. There is the added concern and increasing evidence that wastewater discharges contain bacteria that have acquired antimicrobial resistance (Joakim Larsson and Flach, 2022; Young et al., 2013).

During the global COVID-19 pandemic, many studies worldwide successfully monitored SARS-CoV-2 virus levels in wastewaters by measuring the virus's ribonucleic acid (RNA). The goal was to ascertain disease prevalence in a community without relying on clinical testing data (Polo et al., 2020). This is discussed further in Chapter 3 (Wastewater).

One of the major objectives of urban drainage systems (as discussed in Chapter 1) is the protection of public health, particularly reducing the risk associated with human contact with excreta. Wastewater, even when treated at WTPs, is not routinely disinfected (in Europe) before discharge to a receiving water unless it is classified as a bathing or shellfish water (see discussion following). Discharges from CSOs and SWOs are also not routinely disinfected, and bacterial standards for recreational activity can be violated during even modest storm events (Stott et al., 2018).

2.3.12 Priority substances

The Water Framework Directive (discussed in Section 2.6.1.3) defines a European priority list of 33 substances, extended to 45 substances in 2013, posing a threat to (or via) the aquatic environment as given in Table 2.4. These are defined as "substances or groups of substances that are toxic, persistent and liable to bioaccumulate, and other substances or groups of substances which give rise to an equivalent level of concern." The substances that are thought to pose the greatest threat are further identified as "priority" hazardous substances. Environmental quality standards (EQSs) for the concentrations of the priority substances in surface waterbodies were also set.

2.3.12.1 Phthalates

Phthalates are synthetic chemical plasticisers widely used in the manufacture of plastics and other consumer products. Di-(2-ethylhexyl)phthalate (DEHP) and dibutyl phthalate (DBP) are the most commonly used plasticisers in PVC products. DEHP is a priority hazardous substance. In water, phthalates biodegrade, adsorb onto sediments, and bioconcentrate in aquatic organisms. Various phthalates are classified as toxic to reproduction and have been

Table 2.4 EU hazardous and priority hazardous substances

Hazardous substance	Priority hazardous substance
251-835-4	Anthracene
Aclonifen	Brominated diphenylethers
Alachlor	Cadmium and its compounds
Atrazine	C_{10-13}-chloroalkanes
Benzene	DEHP – Di(2-ethylhexyl)phthalate
Bifeno	Dicofol
XChlorfenvinphos	Dioxins and dioxin-like compounds
Chlorpyrifos	Endosulfan
Cybutryne	Heptachlor and heptachlor epoxide
Cypermethrin	Hexabromocyclododecane
1,2-Dichloroethane	Hexachlorobenzene
Dichloromethane	Hexachlorobutadiene
Dichlorvos	Hexachlorocyclohexane
Diuron	Mercury and its compounds
Fluoranthene	Nonylphenols
Lead and its compounds	Pentachlorobenzene
Naphthalene	Perfluorooctane sulphonic acid
Nickel and its compounds	Polyaromatic hydrocarbons
Octylphenols	Quinoxyfen
Pentachlorophenol	Tributyltin compounds
Simazine	Trifluralin
Terbutryn	
Trichlorobenzenes	
Trichloromethane	

Source: Council of the European Communities (CEC). 2013. *Amending Directives 2000/60/EC and 2008/105/EC as regards Priority Substances in the Field of Water Policy*, Directive 2013/39/EU.

detected in wastewater and stormwater samples (Clara et al., 2010). Degraded phthalates are a source of micro- and nanoplastics.

2.3.12.2 PFAS

Per- and polyfluoroalkyl substances (PFAS) are a large class of contaminants of emerging concern, consisting of nearly 5000 human-synthesised chemicals which have been in use for decades in a variety of consumer and industrial applications. There is growing awareness that these so-called "forever chemicals" are broadly distributed in the environment, are linked to a range of adverse health outcomes and have negative impacts on ecosystems (Tavasoli et al., 2021). Release of PFAS into aquatic environments can occur through point and non-point sources including wastewater and stormwater (Ahrens and Bundschuh, 2014).

2.3.13 Micro- and nanoplastics

Microplastics (MPs) are commonly defined as synthetic polymer particles or fibres between 5 mm and 1 μm diameter or length, whereas nanoparticles range between 1 μm and 1 nm (Gillibert et al., 2019). MP particles derive from products to which they are added intentionally (e.g., cosmetics) or more commonly come from the breakdown of larger consumer or industrial products as well as painted surfaces, tyres, and synthetic fabrics. Nanoplastic particles are assumed to be formed mainly by the further erosion of MPs, and both are widely present in wastewater and stormwater. Particles are diverse in their size, shape, and colour (see Figure 2.3a and 2.3b respectively). Their common use has led to widespread contamination of the natural environment with evidence suggesting that nanoplastics have a higher

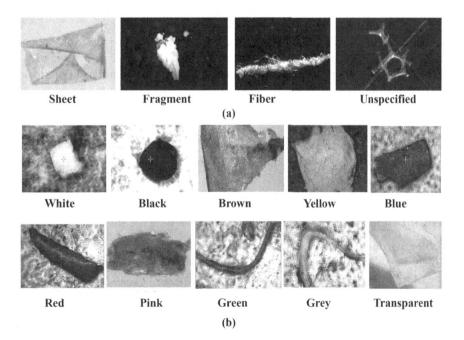

Figure 2.3 Typical characteristics of microplastics in wastewater. (Source: Reprinted from Kittipongvises, et al. 2022. Unravelling capability of municipal wastewater treatment plant in Thailand for microplastics: Effects of seasonality on detection, fate and transport. Journal of Environmental Management, 302, 113990, with permission from Elsevier.)

potential for ecotoxicity (Horton and Dixon, 2018; Sun et al., 2019). Evidence is also emerging (Zhou et al., 2023) that MPs discharged from CSOs can be a significant fraction of the total amount discharged via the urban wastewater system.

2.4 PROCESSES

2.4.1 Hydrolysis

Hydrolysis is an important precursor to both aerobic and anaerobic transformations. It consists of the natural reaction of large organic molecules with water in the presence of enzymes to produce smaller molecules that are potentially available for utilisation by bacteria. It is temperature-dependent.

Suspended organic particles are hydrolysed, bringing them into solution. For example, insoluble cellulose is slowly hydrolysed in two stages to form dextrose (Inman, 1979):

$$\left(C_6H_{10}O_5\right)n + nH_2O \rightarrow nC_6H_{12}O_6$$

Urea represents 80% of the nitrogen content of fresh wastewater. It is relatively rapidly hydrolysed to ammonia-nitrogen (by the enzyme *urease*) at a rate of 3 mg.N/L per hour at 12°C in stored samples. Polyphosphates will slowly hydrolyse to the orthophosphate form.

2.4.2 Aerobic degradation

Aerobic processes are those carried out by aerobic bacteria in the presence of free oxygen, whereby larger, but soluble, organic molecules are degraded to simple and stable end products. Such microorganisms may be freely suspended individually or in flocs in the flow, or be attached to pipe walls or sediment beds as biofilms.

The particular class of organism that carries out this reaction is aerobic *heterotrophic* bacteria. These take organic material as a "food" source to provide energy or to synthesise new bacterial cells. The end products of this reaction are CO_2, H_2O, and other oxidised forms such as nitrate, phosphate, and sulphate. This can be simply represented as

$$C,H,O,N,P,S + O_2 \rightarrow H_2O + CO_2 + NO_3^- + PO_4^{3-} + SO_4^{2-} + \text{new cells} + \text{energy}$$

Unless it is replaced, the oxygen in water can quickly be used up producing an environment hostile to aerobic microorganisms. Conditions in the bulk flow in gravity sewers are likely to be aerobic.

2.4.2.1 Nitrification

In an aerobic environment with low levels of organic material, heterotrophic microorganisms will no longer be able to thrive. However, another group of organisms known as *autotrophs* (nitrosomas and nitrobacter) can thrive, provided there is a sufficient source of oxygen. These utilise inorganic nutrients as an energy source (carbon dioxide and oxidised forms of nitrogen, phosphorus, and sulphur). Nitrification is the biological process by which ammonia (an inorganic nutrient) is converted first to nitrite and then to nitrate. This can be summarised as

$$NH_4^+ + 2O_2 \rightarrow NO_3^- + 2H^+ + H_2O + \text{newcells} + \text{energy}$$

The nitrification reaction consumes large quantities of oxygen and alkalinity.

It is unusual for nitrification to occur naturally in urban drainage networks, but it is possible towards the end of long, well-aerated outfall sewers, particularly in warmer climates.

2.4.3 Denitrification

In the denitrification process, nitrate is reduced to nitrogen gas by the same heterotrophic bacteria responsible for carbonaceous oxidation. Reduction occurs when dissolved oxygen levels are at or near zero. A carbon source must be available to the bacteria.

Denitrification can be stimulated in anaerobic sewer environments by adding a source of nitrate (see Chapter 18).

2.4.4 Anaerobic degradation

Anaerobic processes are those carried out by anaerobic bacteria in the absence of oxygen, whereby large organic molecules are degraded to simple organic gases as end products, resulting in a partial breakdown of the substrate.

The class of organisms known as anaerobic heterotrophic bacteria carries out this reaction. These take organic material as a food source to provide energy or to synthesise new bacterial cells. They must take their oxygen from dissolved inorganic salts; therefore, they produce reduced forms of end products together with methane and carbon dioxide (both GHGs). Although this is a three or more stage process, it can be simply represented as

$$C, H, O, N, P, S + H_2O \rightarrow CO_2 + CH_4 + NH_3 + H_2S + newcells + energy$$

The products of anaerobic processes are more objectionable and sometimes more dangerous than those of aerobic processes. For example, hydrogen sulphide is malodorous, and methane is combustible. Anaerobic conditions are likely to occur when sediment beds form in sewers and are common in pressurised rising mains.

2.5 RECEIVING WATER IMPACTS

All receiving waters can assimilate wastes to some extent, depending on their natural self-purification capacity. Problems arise when pollutant loads exceed this capacity, thus harming the aquatic ecology and restricting the potential use of the water (e.g., water supply, recreation, fisheries). Discharges in urban areas can be continuous or intermittent, but the latter is of relevance here. The aim of the urban drainage designer (in pollution terms) is, therefore, to ensure that intermittent CSO discharges comply with their regulatory permit (See Section 13.4.1).

2.5.1 Emissions

Urban drainage emissions can be categorised as follows:

- Intermittent discharges from CSOs consisting of a mixture of stormwater, domestic, commercial, and industrial wastewater with groundwater and sewer deposits
- Intermittent discharges from separate SWOs or direct stormwater discharges consisting mainly of runoff from urban surfaces including roads and highways

Other intermittent discharges can occur from overflows on storm tanks at WTPs.

Intermittent discharges are (or should be) dilute wastewater or stormwater only. However, apart from being screened and sometimes settled in storage, they are untreated, introducing contaminants and pathogens (discussed earlier in the chapter) directly into the water environment.

Intermittent discharges are particularly difficult to quantify and regulate because of their nature. Their acute (immediate) impact can only be measured during a spill event, and their chronic (long-term) effects are often difficult to isolate from background pollution. Therefore, design standards and performance criteria specifically tailored to intermittent discharges are needed. These are discussed in more detail in the following section.

2.5.2 Processes

The processes occurring in receiving waters subject to discharges from urban drainage systems include

- Physical: Transport, mixing, dilution, flocculation, erosion, sedimentation, thermal effects, and reaeration
- Biochemical: Aerobic and anaerobic oxidation, nitrification, and adsorption and desorption of metals and other toxic compounds
- Microbiological: Growth and die-off and toxicant accumulation

The extent and importance of individual processes will depend on the temporal and spatial scales as shown in Figure 2.4. For example, a shock load into flowing water in a river will transport downstream relatively quickly, interacting with the water column as it progresses. Thus, a significant length of water will be exposed to contamination for a short period. A load discharged to stagnant water in a lake will disperse more slowly and will generally persist for a longer period. We can identify three relevant timescales:

- Short-term (acute)
- Medium-term (delayed)
- Long-term (chronic, cumulative)

2.5.3 Specific impacts

Impacts can be divided into direct water quality effects (DO depletion, eutrophication, and toxics), public health issues, and aesthetic influences. These are summarised for various determinands and receiving water types in Table 2.5.

2.5.3.1 DO depletion

The most important DO-related phenomena caused by intermittent discharges (particularly CSOs) are the following:

- Mixing of low DO spills with the receiving water
- Degradation of discharged (dissolved and particulate) organic matter that exerts an *immediate* oxygen demand on the receiving water. In the case of a river, these occur in a plug moving downstream.
- *Delayed* sediment oxygen demand (SOD) caused by the deposited sediment and the scouring effect of discharges after the polluted plug has passed. Typical undisturbed

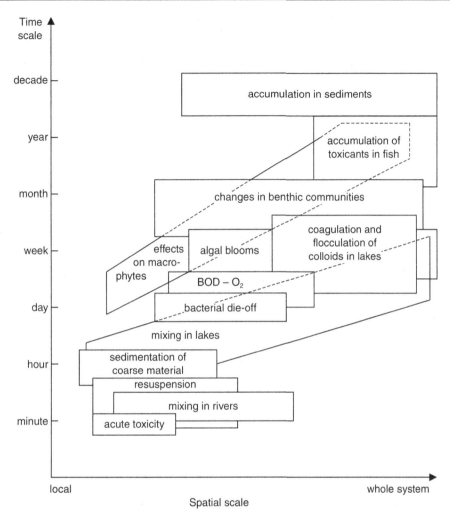

Figure 2.4 Time and spatial scales for receiving water impacts. (Based on Aalderink, R.H., and Lijklema, L. 1985. Water quality effects in surface waters receiving stormwater discharges, in *Water in Urban Areas, TNO Committee on Hydrological Research. Proceedings and Information*, 33, 143–159, with permission of the authors.)

SOD levels are 0.15–2.75 g/m^2.d, elevated to 240–1500 g/m^2.d during storm flow conditions (House et al., 1993). As the rate of decomposition is low, the area affected can be extensive. Sensitive benthic organisms are rapidly eliminated but are soon replaced by high-population densities of a few species tolerant of the low oxygen silty conditions.

The relative magnitude of these effects will depend on the specific circumstances of the discharge and the recipient. In larger rivers, immediate consumption dominates, whereas small rivers with flows <0.5 m^3/s often show depletion due to delayed consumption.

The most apparent consequence of reduced DO levels is fish kills. Additionally, odour problems may be experienced due to putrefaction.

Table 2.5 Qualitative assessment of receiving water impacts of urban discharges

Receiving water	Water quality			Public health		Aesthetics	
	Dissolved oxygen	Nutrients	Sediments	Toxics	Microbials	Clarity	Sanitary debris
Streams							
Steep	–	–	–	x	xx	–	xx
Slack	x	–	x	x	xx	–	xx
Rivers							
Small	xx	–	x	x	xx	–	xx
Large	x	–	x	x	xx	x	xx
Estuaries							
Small	x	x	x	x	xx	x	xx
Large	–	–	x	–	xx	x	xx
Lakes							
Shallow	x	xx	x	x	xx	x	xx
Deep	x	x	x	x	xx	x	xx

xx Probable; x Possible; – Unlikely.
Source: House, M.A. et al. 1993. *Water Science and Technology*, 27(12), 117–158.

2.5.3.2 Eutrophication

If large quantities of nutrients such as nitrogen or phosphorus are discharged into receiving waters, excessive growth of aquatic weeds and algae may occur. This can lead to

- Oxygen depletion
- Anaerobic conditions in bottom muds
- Fish kills
- Aesthetic problems

This is a long-term problem usually associated with shallow, stagnant waters such as lakes, estuaries, and the coastal zone, but rivers may also be affected. Intermittent discharges are usually a relatively small constituent of the total nutrient load.

2.5.3.3 Toxics

Intermittent discharges are a source of elevated levels of ammonia (toxic to fish), chlorides, metals, hydrocarbons, and trace organics, which cause toxic impacts. These may be either acute or chronic depending on the specific circumstances. The effect on the receiving water biota is to rapidly reduce species diversity and abundance leading to complete elimination at excessive concentrations. Downstream recovery is generally slower than the loss rate, with tolerant species returning, often at population densities greater than initial levels, due to lack of competition.

The difficulties in assessing toxic impacts based on single parameter values have led to the development of toxicity-based consents (TBCs). Typically, these are derived from laboratory and field ecotoxicological studies on fish and invertebrates. Results are expressed as LC50 values that indicate short-term lethal concentrations of a particular pollutant resulting in 50% mortality.

2.5.3.4 Public health

High concentrations of a variety of pathogens may be expected from both CSO and SWO discharges. Bacterial contamination is a relatively short-term problem as die-off usually occurs within several days, although this is a longer period than the discharge itself. In addition, bacteria tend to adhere to suspended solids. As many of these particles will settle, bacteria can become established in the receiving waterbed, considerably extending their survival times. Perkins et al. (2016) found that in freshwater microcosms, *enterococci* in sewage reduced in concentration from approximately 10^5 cfu/100 mL to less than 10^3 cfu/100 mL over 5 days.

The risk to public health depends on the degree of potential human exposure, and this will be greatest if the receiving water is used for contact recreational purposes. Swimmers are therefore at greatest risk. Whilst some studies have demonstrated the relationship between gastroenteritis and FS levels (e.g., Wyer et al., 1995), Fewtrell and Kay (2015) found that although a higher risk of GI was generally reported in bathers compared with non-bathers, there was often no clear association with water quality as measured by FIOs. Given the increasing interest in bathing, paddle-boarding, and wild swimming in non-designated waters or outside the nominal bathing season, further research and policy updates in this area are urgently needed (Clary et al., 2022; NEPC, 2024; Sojobi and Zayed, 2022).

2.5.3.5 Aesthetics

Gross solids of obvious sanitary origin have a greater impact on the public's enjoyment of a visit to a river or beach than any other aesthetic pollution parameter (House et al., 1994). Clearly, such debris is both unsightly and offensive and there is evidence of its long-term buildup in certain locations. For example, 23,000 wet wipes were counted and removed from one stretch of the tidal River Thames foreshore in just a two-hour period. Follow-up studies indicate that the so-called "Thames Great Wet Wipe Reef" has continued to grow (Thames21, 2019).

2.5.4 Other causes of pollution

Intermittently discharging CSOs certainly have an impact on rivers and other waterbodies. However, not all river outfalls are CSOs and not all water pollution comes from CSOs. Other significant sources include the following:

- Contaminated runoff from agricultural areas (rural diffuse pollution)
- Final effluent from WTPs
- Contaminated runoff from roads entering rivers through storm sewers and highway drains (urban diffuse pollution)
- Illegal connections of wastewater into storm sewers discharging directly to watercourses (see Section 5.5)

In terms of reasons for waterbodies not achieving WFD Good Ecological Status (see Section 2.6.1.3) in the UK (expressed as a percentage of all reasons), agricultural pollution is the largest single polluter at 48% with the water industry second at 32% (CIWEM, 2021).

2.5.5 Scale and severity

There are approximately 15,000 permitted storm overflows in England, of which about 1/3rd are located at WTPs (Stantec, 2021). All of these are continuously monitored (see Section 18.4.2). In terms of their operation, headline findings for 2022 were (Defra, 2023):

- 3% of overflows spilled more than 100 times
- 18% of overflows did not spill
- the average number of spills per overflow was 23

Further analysis of the data shows:

- the average spill duration was 5.5 hours
- the average percentage of time spilling was 1.5% (or 98.5% not spilling)

Concerns over the scale of CSO discharges are not new. The Scottish Development Department (1977) report noted that CSOs with 6–8 DWF settings (see Chapter 13) produced annual spill volumes of 31 – 39% of all combined sewer flow. Mance (1981) estimated that 35% of the annual (wastewater) pollution discharged to rivers was a result of spillage from CSOs. Despite significant efforts to address this since then (notably the Urban Pollution Management (UPM) approach discussed in the next section), the Environment Agency estimates that approximately 400 inland river waterbodies (some 7% of the total number) fail to achieve Good Ecological Status because of intermittent storm overflow discharges (Stantec, 2021).

Less well known and understood is the impact of highway drainage of which there are around 1 million outfalls in England alone (CIWEM, 2021). These may have simple treatment processes, such as oil separators associated with them, or SuDS (Chapter 11) in newer systems. However, none are currently permitted or regularly monitored.

2.6 RECEIVING WATER STANDARDS

2.6.1 Legislation and regulatory regime

2.6.1.1 Urban Waste Water Treatment Directive

The EU Urban Waste Water Treatment Directive (CEC, 1991a) aims to protect the environment from the adverse effects of wastewater discharges from urban areas and biodegradable industrial wastewater by specifying uniform performance standards for wastewater treatment and collection across the European Union. The standards are more prescriptive for WTPs and less so for collection systems. Compliance dates for various aspects of the directive ranged from 31 December 1998 to 31 December 2005, with extended deadlines for post-1991 member states. However, by the end of 2018, the degree of compliance was still variable across the Union. For example, in terms of collecting systems, compliance averaged at 95% across all EU member states but varied from as high as 100% to as low as 15%. Compliance with required secondary treatment had reached 89% and tertiary treatment 88% (EC, 2022a).

The directive requires member states to ensure all urban areas with a population equivalent of 2000 or more are provided with collection systems. Further, the design, construction, and maintenance of collecting systems needs to be undertaken in accordance with *best technical knowledge not entailing excessive costs* with respect to

- Volume and characteristics of urban wastewater
- Prevention of leaks
- Limitation of pollution of receiving waters due to overflows

The directive acknowledges the difficulties that arise during periods of *unusually heavy rainfall* (but does not define this), and allows member states to decide on their own measures to limit pollution from overflows. These could be based on one of the following:

- Receiving water dilution rates
- Collection system capacity in relation to dry weather flow
- A certain number of overflows per year

Ambitious proposals for an update to the directive have been approved (April 2024) by the European Parliament. The updated directive includes new standards for micropollutants and MPs. Some 25 substances will be added to the hazardous/priority hazardous substances list, including PFAS, glyphosate (a herbicide), and some pharmaceuticals. The directive will require the systematic monitoring of wastewater for several viruses, including SARS-CoV-2 and other emerging pathogens. A "polluter pays" approach will be adopted. Obligations to treat wastewater will be extended to smaller municipalities with 1000 inhabitants. There is a requirement to establish integrated water management plans in larger cities to help manage more frequent and extreme rainfall events. There is a new requirement that stormwater overflows should not represent more than 1% of the annual volume and pollutant load of the collected dry weather flow. For urban areas with a population equivalent greater than 10,000, concentration and loads of pollutants discharged from overflows must be monitored. Finally, untreated discharges of urban runoff from separate collection networks will need to be progressively eliminated unless it can be demonstrated that those discharges do not cause adverse impacts on the quality of receiving waters (EC, 2022b).

2.6.1.2 Bathing Water Directive

The EU Bathing Water Directive (CEC, 2006) regulates the bacteriological water quality of coastal and inland bathing waters based on two faecal indicator parameters: intestinal *enterococci* and *E. coli*. Three levels or standards are set for inland waters and for coastal and transitional waters: excellent, good, and sufficient (see Table 2.6). There is also a "poor" classification for waters that do not reach these standards. All EU bathing waters were due to be classified by the end of the 2015 bathing season at which time all were to be at least "sufficient." In 2021, 95% of bathing waters were found to meet minimum water quality standards with more than 85% meeting the "excellent" standard (EEA, 2022). Figures for the UK are 95% and 71%, respectively (EA, 2022). Failure to comply with the standards, when and if it does happen, typically comes about by a combination of agricultural runoff, continuous wastewater discharges, and intermittent CSO discharges. Hence, control of overflow discharges becomes especially important where they discharge directly into bathing waters or to estuarine or coastal waters that lead to bathing waters.

2.6.1.3 Water Framework Directive

The Water Framework Directive (WFD) incorporates the main requirements for water management in Europe into one single, holistic system based on river basins (CEC, 2000). New or reorganised river basin district authorities were formed, each with a management plan and a "programme of measures" aimed at achieving the goals of the directive. The key guiding goal

Table 2.6 Bathing water quality standards

Waters	Parameter	Excellent quality[a]	Good quality[a]	Sufficient[b]
Inland	Intestinal *enterococci* (cfu/100 mL)	200	400	330
	Escherichia coli (cfu/100 mL)	500	1000	900
Coastal and transitional	Intestinal *enterococci* (cfu/100 mL)	100	200	185
	Escherichia coli (cfu/100 mL)	250	500	500

[a] Based on a 95 percentile evaluation.
[b] Based on a 90 percentile evaluation.

Source: CEC. 2006. *Directive Concerning the Quality of Bathing Water*, 2006/7/EC.

is to achieve "good status" of groundwaters and surface waters; "good" meaning that water meets the standards established in existing water directives and, in addition, new ecological quality standards. A surface water is considered to be of good ecological quality if there is only a slight departure from the biological community that would be expected in conditions of minimal anthropogenic impact.

Good chemical status is defined in terms of compliance with all of the quality standards established for chemical substances at the European level. A mechanism for controlling the discharge of dangerous *priority hazardous substances* (Section 2.3.12) is also provided. A combined approach to setting standards is used where both emission limit values and EQSs are legally binding. Derogations from good status are allowed in unforeseen or exceptional circumstances (e.g., droughts, floods). EQSs are specified as a maximum allowable concentration (MAC-EQS) to protect against short-term, acute exposure, and/or an Annual allowance (AA-EQS) to provide protection against long-term, chronic effects (Lundy et al., 2012).

The WFD was designed to provide an "umbrella" to existing directives as well as incorporating additional standards. River basin authorities designate specific protection zones within their area (i.e., bathing, shellfish, drinking water, or protected natural areas) where the standards in the respective existing EU directives apply, but zones with higher objectives are established where more stringent standards must be met. Good ecological and chemical status is the minimum for all waters. The Urban Waste Water Treatment Directive and the Nitrates Directive (CEC, 1991b) are considered tools to achieve the objectives of the river basin management plans.

The directive also sets rules for groundwater. All direct discharges to groundwater are prohibited and there is a requirement to monitor groundwater bodies so as to detect changes in composition due to non-point pollution and take measures to reverse them.

The initial period of implementation of the directive was 15 years to 2015 (9 years to prepare plans and 6 years more to achieve specific targets) plus an additional deferment of up to two periods of 6 years if justified on technical or economic grounds (Kallis and Butler, 2001). By 2015, only about 40% of surface waterbodies had achieved good status (EEA, 2021) with the 2021 UK figure being 36% (although only 16% in England) (JNCC, 2021). Voulvoulis et al. (2017) discuss why the initial great expectations have yet to be fully realised.

2.6.1.4 Marine Strategy Framework Directive

The Marine Strategy Framework Directive required member states to develop strategies to protect their marine waters with the aim of achieving "good environmental status" by 2020. This is being implemented by protecting the marine environment, preventing its deterioration, and restoring degraded marine ecosystems (CEC, 2008). The directive is complementary to the Water Framework Directive.

2.6.2 Urban Pollution Management (UPM) Manual

The UPM Manual (FWR, 2019) is widely accepted as the best current practice for the management of urban wet weather discharges and includes environmental standards ratified by the regulators for use in the UK. It sets out a methodical procedure for dealing with wet weather discharges from sewer systems. There are four recurring themes in the guidance. The first is that analysis should be holistic, covering all elements in the system that determine the pollution impact of a sewer system: rainfall, the sewer system itself, the WTP, and the receiving waterbody. The second is that the level of detail of any study – and, in particular, of the models used – should be appropriate and that, in the right circumstances, a holistic approach may also be simple. Third, the approach should be underpinned by relevant water quality standards of appropriate complexity and rigour (discussed below) with models able to demonstrate compliance with those standards. Finally, a "partnership approach" to planning and executing a UPM study should be adopted.

2.6.3 Environmental quality standards

Environmental quality standards (EQSs) are routinely used to control continuous discharges. They are based on a receiving water's ability to assimilate discharged pollutants without detriment to legitimate uses of the water – based on environmental quality objectives (EQOs). These standards are monitored by checking the compliance of routinely taken samples against water quality criteria (usually 90 or 95 percentiles). However, discharges from urban drainage systems should be infrequent and of short duration, although they can be of high pollutant concentration, resulting in a disproportionately high impact on receiving water "usage." The UPM Manual (FWR, 2019) offers two different types of EQS standards to regulate intermittent polluting discharges:

- High percentile standards based on extrapolation of the 90/95 percentile thresholds used for protecting ecosystems subject to continuous discharges
- Intermittent standards directly related to the characteristics of events that cause stress in river ecosystems, expressed in terms of concentration-duration thresholds at an allowable return period

These are discussed below.

2.6.3.1 Freshwater aquatic life standards

The *Manual* sets high, 99 percentile standards for BOD, total ammonia, and unionised ammonia to protect freshwater aquatic life based on the WFD waterbody type. Standards for BOD are given in Table 2.7.

Fundamental intermittent standards (FIS), based on the LC50 values mentioned in Section 2.5.3.3, are also given in the *Manual*. These standards consist of a relationship between three variables:

- Pollutant concentration
- Duration of the event
- Return period of an event in which that concentration is exceeded

Table 2.8 shows this three-way relationship for dissolved oxygen and unionised ammonia based on sustaining cyprinid fisheries. Thus, minimum river DO levels of 3.0–5.5 mg/L are allowed, depending on the duration and frequency of the storm event. Fish kills due to ammonia poisoning can be avoided if NH_3–N levels are limited to 0.03–0.25 mg/L.

2.6.3.2 Shellfish standards

Several EQSs have been developed for bacteria in shellfish waters to ensure compliance with the standards for shellfish flesh. In situations where a water is affected by both intermittent and continuous discharges, a standard of 1500 faecal coliforms per 100 mL for at least 97% of the time in the long term has been derived to ensure compliance with category B of the Food Hygiene Regulations. An alternative EQS is to determine the significant spill volume for a single or group of CSO discharges based on the shellfish water achieving a geometric

Table 2.7 Example 99 percentile river quality standards for biochemical oxygen demand derived from WFD waterbody ecological status 90 percentile standards

Type of river	BOD (mg/L)	
	90 percentile	99 percentile
High Status for Types 1, 2, 4 and 6 and Salmonid	3.0	7.0
Good Status for Types 1, 2, 4 and 6 and Salmonid High Status for Types 3, 5 and 7	4.0	9.0
Good Status for Types 3, 5 and 7	5.0	11.0
Moderate Status for Types 1, 2, 4 and 6 and Salmonid	6.0	14.0
Moderate Status for Types 1, 3, 5 and 7	6.5	14.0

Source: Adapted from FWR. 2019. *Urban Pollution Management Manual*, 3rd edn, Foundation for Water Research, www.fwr.org/UPM3.

Table 2.8 Fundamental intermittent standards for dissolved oxygen and ammonia concentration/duration thresholds for sustaining cyprinid fisheries

Return period (months)	DO concentration (mg/L)[a]		
	1 hour	6 hours	24 hours
1	4.0	5.0	5.5
3	3.5	4.5	5.0
12	3.0	4.0	4.5
	NH$_3$–N concentration (mg/L)[b]		
1	0.150	0.075	0.030
3	0.225	0.125	0.050
12	0.250	0.150	0.065

[a] Applicable when NH$_3$–N < 0.02 mg/L.
[b] Applicable when DO > 5 mg/L, pH > 7 and T > 5°C

Source: Adapted from FWR. 2019. *Urban Pollution Management Manual*, 3rd edn, Foundation for Water Research, www.fwr.org/UPM3.

mean of 110 *E. coli* cfu/100 mL. A geometric mean of 5 *E. coli* cfu/100 mL may be specified in order to comply with the WFD shellfish flesh standard (CEC, 2000; FWR, 2019).

2.6.3.3 Bathing standards

Bathing water EQSs are based on limiting the concentration of intestinal *enterococci* and *E. coli* to the levels given in Table 2.6 for at least 98.2% of the bathing season (May to September) as judged over an average period of at least 10 years (FWR, 2019). Compliance is based on up to 20 samples taken throughout the bathing season at each designated bathing water. The Environment Agency's *Swimfo* website provides detailed information on each of the 400+ bathing waters in England: https://environment.data.gov.uk/bwq/profiles/.

Undesignated waters are not subject to any microbiological monitoring, nor are they covered by any microbiological standards.

PROBLEMS

1 Sediment transported in a sewer as bed-load has been measured at a concentration of 20 ppm (by volume). What is the concentration in mg/L if the specific gravity of the sediment is 2.65?   [53 mg/L]

2 Plot the following hydrograph and pollutograph (concentration and load rate).

Time (hours)	Flow (L/s)	COD (mg/L)
0.5	80	50
1	170	160
1.5	320	380
2	610	400
2.5	670	230
3	590	130
3.5	380	70
4	220	40
4.5	100	20
5	50	0

Compute the average and flow-weighted (event mean) concentration of COD. [148 mg/L, 208 mg/L]

3 A wastewater sample has an organic nitrogen content of 15 mg org.N-N/L and an ammonium concentration of 35 mg NH_4^+ / L . If the nitrite and nitrate concentrations are negligible, what is the total nitrogen concentration of the sample?   [42 mg N/L].

4 Define the main types of solids found in urban drainage systems, and discuss their importance.

5 Compare and contrast the main methods for determining the organic content of a wastewater sample.

6 Describe the main forms of nitrogen found in wastewater. Why are they of interest?

7 What are priority and hazardous substances, and why are they important?

8 Many substances of anthropogenic origin have been found to be widespread in the environment. Explain what these substances are, why they are important and what the role of urban drainage is in their transmission.

9 What type of emissions into the environment can be expected from an urban drainage system? Explain how their impact may be acute, delayed, or chronic.

10 Discuss the main types of receiving water impact caused by intermittent discharges.

11 Discuss to what extent all waters should be safely accessible for contact recreational purposes and what impact that would have on urban drainage system design and operation.

12 What are the implications of the Water Framework Directive for urban drainage discharges? How effective has it been to date?

11 Explain the difference between Environmental Quality Objectives and Environmental Quality Standards.

12 How do intermittent standards differ from ordinary water quality standards? How are they derived?

13 The following table shows the number of times that dissolved oxygen fell below 4 mg/L for 6 hours or more in a river. Is this in compliance with the 4 mg/L–6 hour–1 year standard? [No]

Year	1	2	3	4	5	6	7	8	9	10
Number	0	1	1	3	0	0	2	1	0	3

KEY SOURCES

Botturi, A., Gozde Ozbayram, E., Tondera, K., Gilbert, N.I., Rouault, P., Caradot, N., Gutierrez, O., Daneshgar, S., Frison, N. Akyol, C., Foglia, A., Eusebi, A.L. and Fatone, F. 2021. Combined sewer overflows: A critical review on best practice and innovative solutions to mitigate impacts on environment and human health, *Critical Reviews in Environmental Science and Technology*, 51(15), 1585–1618.

Chave, P. 2001. *The EU Water Framework Directive. An Introduction*, IWA Publishing.

Ellis, J.B. & Hvitved-Jacobsen, T. 1996. Urban drainage impacts on receiving waters, *Journal of Hydraulic Research*, 34(6), 771–783.

FWR 2019. *Urban Pollution Management Manual*, 3rd edn, Foundation for Water esearch, http://www.fwr.org/UPM3

Horton, A.A. and Dixon, S.J. 2018. Microplastics: An introduction to environmental transport processes, *Wiley Interdisciplinary Reviews Water* 5(2), 1268.

Kallis, G. and Butler, D. 2001. The EU water framework directive: measures and implications. *Water Policy*, 3, 125–142.

Marsalek, J., Jimenez-Cineros, B., Karamouz, M., Malmqvist, P.-A., Goldenfum, J. and Cocat, B. 2008. *Urban Water Cycle Processes and Interactions*. Urban Water Series Volume 2, UNESCO Publishing—Taylor and Francis.

Wang, Z., DeWitt, J.C., Higgins C.P. and Cousins, I.T. 2017. A Never-Ending Story of Per- and Polyfluoroalkyl Substances (PFASs)? *Environmental Science & Technology 51*, 2508–2518.

REFERENCES

Aalderink, R.H. and Lijklema, L. 1985. Water quality effects in surface waters receiving stormwater discharges, in Water in Urban Areas, *TNO Committee on Hydrological Research, Proceedings and Information No. 33*, 143–159.

Ahrens, L. and Bundschuh, M. 2014. Fate and effects of poly- and perfluoroalkyl substances in the aquatic environment: A review, *Environmental Toxicology and Chemistry*, 33, 1921–1929.

American Water Works Association (AWWA). 2023. *Standard Methods for the Examination of Water and Wastewater*, 24th edition, AWWA.

CIWEM. 2021. *River Water Quality and Storm Overflows. A Systems Approach to Maximising Improvement*. Chartered Institution of Water and Environmental Management.

Clara, M., Windhofer, G., Hartl, W., Braun, K., Simon, M., Gans, O., Scheffknecht, C. and Chovanec, A. 2010. Occurrence of phthalates in surface runoff, untreated and treated wastewater and fate during wastewater treatment. *Chemosphere*, 78, 1078–1084

Clary, J., Ervin, J., Steets, B. and Olson, C. 2022. Pathogens in Urban Stormwater Systems: Where Are We Now? *Journal of Sustainable Water in the Built Environment*, 8(1), 02521004.

Council of the European Communities (CEC). 1978. *Directive Concerning Quality of Freshwater Supporting Fish*, 78/659/EEC.

Council of the European Communities (CEC). 1991a. *Directive Concerning Urban Waste Water Treatment*, 91/271/EEC.

Council of the European Communities (CEC). 1991b. *Directive Concerning Protection of Water against Pollution by Nitrates from Agriculture*, 91/276/EEC.

Council of the European Communities (CEC). 2000. *Directive Establishing a Framework for Community Action in the Field of Water Policy*, 2000/60/EC.

Council of the European Communities (CEC). 2006. *Directive Concerning the Quality of Bathing Water*, 2006/7/EC.

Council of the European Communities (CEC). 2008. *Directive Establishing a Framework for Community Action in the Field of Marine Environmental Policy (Marine Strategy Framework Directive)*, 2008/56/EC.

Council of the European Communities (CEC). 2013. *Amending Directives 2000/60/EC and 2008/105/EC as regards Priority Substances in the Field of Water Policy*, Directive 2013/39/EU.

Defra. 2023. *Event Duration Monitoring - Storm Overflows - Annual Returns.* Department for Environment, Food and Rural affairs.

Droste, R.L. 2018. *Theory and Practice of Water and Wastewater Treatment.* 2nd edition, John Wiley.

Environment Agency (EA). 2021. *Monitoring discharges to water: guidance on selecting a monitoring approach,* https://www.gov.uk/guidance/monitoring-discharges-to-water-guidance-on-selecting-a-monitoring-approach

Environment Agency (EA). 2022. *Bathing water classifications 2021,* https://deframedia.blog.gov.uk/2022/01/20/bathing-water-classifications-2021/

European Commission (EC). 2022a. *Eleventh Technical Assessment on UWWTD Implementation 2018.* European Overview & National Situation.

European Commission (EC). 2022b. *Proposal for a Directive of the European Parliament and of the Council Concerning Urban Wastewater Treatment (Recast),* COM/2022/551 final.

European Environment Agency (EEA). 2021. *Ecological status of surface waters in Europe,* https://www.eea.europa.eu/en/analysis/indicators/ecological-status-of-surface-waters

European Environment Agency (EEA). 2022. *European Bathing Water Quality in 2021, Report.*

Fewtrell, L. and Kay, D. 2015. Recreational Water and Infection: A Review of Recent Findings. *Current Environmental Health Reports, 2,* 85–94.

Garcia-Aljaro, C., Blanch, A.R., Campos, C., Jofre, J. and Lucena, F. 2018. Pathogens, faecal indicators and human-specific microbial source-tracking markers in sewage. *Journal of Applied Microbiology, 126,* 701–717.

Gillibert, R., Balakrishnan, G., Deshoules, Q., Tardivel, M., Magazzù, A., Donato, M.G., Maragò, O.M., Lamy De La Chapelle, M., Colas, F., Lagarde, F. and Gucciardi, P.G. 2019. Raman tweezers for small microplastics and nanoplastics identification in seawater. *Environmental Science & Technology, 53,* 9003–9013.

Gray, N.F. 2010. *Water Technology. An Introduction for Environmental Scientists and Engineers.* 3rd edition, Arnold.

House, M.A., Ellis, J.B., Herricks, E.E., Hvitved-Jacobsen, T., Seager, J. and Lijklema, L. 1993. Urban drainage – impacts on receiving water quality. *Water Science and Technology, 27*(12), 117–158.

House, M., Herring, M., Green, M. J. and Palfrey, E. A. 1994. *Public Perception of Aesthetic Pollution.* Report FR0439, Foundation for Water Research.

Inman, J.B. 1979. Sewage and its pretreatment in sewers. Chap. 7. in *Developments in Sewerage – 1* (ed. R.E. Bartlett), Applied Science.

Joakim Larsson, D.G. and Flach, C.-F. 2022. Antibiotic resistance in the environment. *Nature Reviews Microbiology, 20,* 257–269.

Joint Nature Conservation Committee (JNCC). 2021. B7. *Surface water status,* https://jncc.gov.uk/our-work/ukbi-b7-surface-water-status/

Kittipongvises, S., Phetrak, A., Hongprasith, N. and Lohwacharin, J., 2022. Unravelling capability of municipal wastewater treatment plant in Thailand for microplastics: Effects of seasonality on detection, fate and transport. *Journal of Environmental Management, 302,* 113990.

Lundy, L., Ellis, J.B. and Revitt D.M. 2012. Risk prioritisation of stormwater pollutant sources. *Water Research, 46,* 6589–6600.

Mance, G. 1981. *The Quality of Urban Storm Discharges: a review.* Report 192M, Water Research Centre, Stevenage.

National Engineering Policy Centre (NEPC). 2024. *Testing the waters. Priorities for mitigating health risks from wastewater pollution,* Royal Academy of Engineering. https://nepc.raeng.org.uk/testing-the-waters

Perkins, T.L., Perrow, K., Rajko-Nenow, P., Jago, C.F., Jones, D.L., Malham, S.K. and McDonald, J.E. 2016. Decay rates of faecal indicator bacteria from sewage and ovine faeces in brackish and freshwater microcosms with contrasting suspended particulate matter concentrations. *Science of the Total Environment, 572,* 1645–1652.

Polo, D., Quintela-Baluja, M., Corbishley, A., Jones, D., Singer, A., Graham, D. and Romalde, J. 2020. Making waves: wastewater-based epidemiology for COVID-19 – approaches and challenges for surveillance and prediction. *Water Research, 186,* 116404.

SCA. 2024. *Methods for the Examination of Waters and Associated Materials* (Separate documents for individual determinations), Standing Committee of Analysts. https://standingcommitteeofanalysts.co.uk/library/

Scottish Development Department. 1977. Storm sewage: Separation and disposal. Report of the Working Party on Storm Sewage (Scotland), HMSO, Edinburgh.

Stantec. 2021. *Storm Overflow Evidence Project*. Final Report, Water UK.

Sojobi, A.O. and Zayed, T. 2022. Impact of sewer overflow on public health: A comprehensive scientometric analysis and systematic review. *Environmental Research*, 203, 111609.

Stott, R., Tondera, K., Blecken, G.-T. and Schreiber, C. 2018. Microbial loads and removal efficiency under varying flows, in *Ecotechnologies for the treatment of variable stormwater and wastewater flows* (eds K. Tondera, G.-T. Blecken, F. Chazarenc and C. C. Tanner), Springer International Publishing, 57–74.

Sultana, N., Roddick, F., Gao, L., Guo, M. and Pramanik, B.K. 2023. Understanding the properties of fat, oil, and grease and their removal using grease interceptors. *Water Research* 225, 119141.

Sun, J., Dai, X., Wang, Q., van Loosdrecht, M. and Ni, B. 2019. Microplastics in wastewater treatment plants: Detection, occurrence and removal. *Water Research*, 152, 21–37.

Tavasoli, E. Luek, J.L., Malley Jr, J.P. and Mouser, P.J. 2021. Distribution and fate of per- and polyfluoroalkyl substances (PFAS) in wastewater treatment facilities. *Environmental Science Processes & Impacts*, 23, 903.

Thames21. 2019. *23 thousand wet wipes discovered on stretch of Thames river bank*, www.thames21.org.uk/2019/04/23-thousand-wet-wipes-discovered-stretch-thames-river-bank

Voulvoulis, N., Arpon, K.D. and Giakoumis, T. 2017. The EU Water Framework Directive: From great expectations to problems with implementation. *Science of the Total Environment*, 575, 358–366.

Wyer, M.D., Fleisher, J.M., Gough, J., Kay, D. and Merrett, H. 1995. An investigation into parametric relationships between enterovirus and faecal indicator organisms in the coastal waters of England and Wales. *Water Research*, 29(8), 1863–1868.

Young, S., Juhl, A. and O'Mullan, G. D. 2013. Antibiotic-resistant bacteria in the Hudson River Estuary linked to wet weather sewage contamination. *Journal of Water and Health*, 11(2), 297–310.

Zhang, J., Shao, J., Liu, G., Qi, L., Wang, H., Xu, X. and Liu, S. 2021. Wastewater COD characterization: RBCOD and SBCOD characterization analysis methods. *Scientific Reports*, 11, 691.

Zhou, Y., Li, Y., Yan, Z. Wang, H., Chen, H., Zhao, S., Zhong, N., Cheng, Y. and Acharya, K. 2023. Microplastics discharged from urban drainage system: Prominent contribution of sewer overflow pollution. *Water Research*, 236, 119976.

Chapter 3

Wastewater

3.1 INTRODUCTION

Wastewater, or sewage, is one of the two major urban water-based flows that form the basis of concern for the drainage engineer. The other, stormwater, is described in Chapter 5. Wastewater is the main liquid waste of the community. Safe and efficient drainage of wastewater is particularly important to maintain public health (particularly because of the high levels of potentially disease-forming microorganisms in wastewater) and to protect the receiving water environment (due to large amounts of oxygen-consuming organic material and many other pollutants in wastewater). This chapter provides background information and summary data on wastewater. The quantification for design purposes is dealt with in Chapter 8.

The basic sources of wastewater are summarised in Figure 3.1 and consist of

- Domestic
- Non-domestic (commercial and industrial)
- Infiltration/inflow

In practice, the relative importance of the components varies with a number of factors, including

- Location (climatic and weather conditions, the availability of water and its characteristics, and individual domestic water consumption)
- Diet of the population
- Presence of industrial and trade effluents

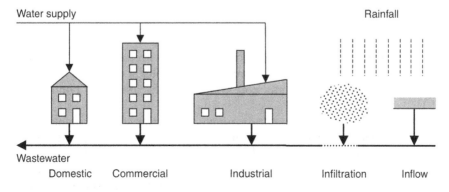

Figure 3.1 Sources of wastewater.

DOI: 10.1201/9781003408635-3

- Type of collection system (i.e., separate or combined)
- Condition of the collection system

This chapter is concerned with the generation and characteristics of wastewater. It collates quantity and quality information on the various sources of wastewater and discusses their relative importance. Section 3.2 covers domestic sources of wastewater, Section 3.3, non-domestic, and Section 3.4, the contribution of infiltration and inflow. Wastewater quality issues are dealt with in Section 3.5. Wastewater-based epidemiology (WBE) is discussed in Section 3.6.

3.2 DOMESTIC

In many (but not all) networks, the domestic component of wastewater is the most important. Domestic wastewater is generated primarily from residential properties but also includes contributions from institutions (e.g., schools and hospitals) and recreational facilities (such as leisure centres). In terms of flow quantity, the defining variable is domestic water *consumption*, which is linked to human behaviour and habits. In fact, very little water is actually consumed, or lost from the system. Instead, it is used intermittently (degrading its quality) and then discharged as wastewater. Hence, in this section, we look at the links between water usage and wastewater discharge and, in particular, how these vary with time.

3.2.1 Water use

Important factors affecting the magnitude of *per capita* water demand, referred to as PCC (*per capita* consumption) include the following.

3.2.1.1 Climate and weather

Climate and its weather effects such as temperature and rainfall can significantly affect water demand. Water use tends to be greatest when it is hot and dry, due largely to increased garden watering/sprinkling and landscape irrigation. The impact on wastewater is less pronounced, as this additional water will probably not find its way into the sewer. Downing et al. (2003) estimated that for South East England, climate change would increase household demand by 3.8 L/hd.d by 2025 of which 2.4 L/hd.d would become wastewater. Further discussion of the impact of climate change can be found in Chapter 4 (Rainfall).

3.2.1.2 Demography and user behaviour

It has been demonstrated that household occupancy levels are important, with larger families tending to have lower PCC (Willis et al., 2013). While, at the other end of the scale, retired people have been shown to use more water than the rest of the population (Russac et al., 1991).

User behaviour is clearly important, for example, showering for one minute less each day can save about 3000 litres of water per person per year (NIC, 2018). Campaigning and public engagement has been widely implemented and Lede and Meleady (2019) review the application and potential of various social influencing strategies. Mandatory water efficiency labelling on water using products also has a role in allowing consumers to make more informed decisions (Defra, 2023).

3.2.1.3 Socio-economic factors

The greater the affluence or economic capabilities of a community, the greater the water use tends to be. Work in the UK (Russac et al., 1991) and Australia (Willis et al., 2013) has confirmed the link between water demand and economic indicators such as dwelling type, dwelling value, and household income. This is probably due to greater ownership and use of water-using domestic appliances such as washing machines, dishwashers, and power showers.

3.2.1.4 Development type

Dwelling type is important. In particular, dwellings with gardens typically use more water than flats or apartments, but may not generate more wastewater.

3.2.1.5 Extent of and type of metering and water conservation measures

Water undertakers with metered supplies usually charge their customers based on the quantity of water used in a given period. Systems with unmetered services charge a flat rate for unlimited water use. In theory, at least, metered supplies should prevent waste of water by users, reduce actual water use, and therefore reduce wastewater flows. Water is not universally metered in the UK with about 56% (EA, 2022) of houses metered, albeit with a rising trend and wide regional variations (CCW, 2016). Ornaghi and Tonin (2019), in their study of the widespread rollout of meters to 150,000 customers in South East England, found that a water saving of 22% was achieved. Using expert consultation and averaging values in the literature, the National Infrastructure Commission (2018) estimated standard meters reduce average consumption by 15% and smart meters by 17%. Variable tariffs allied to smart meters could increase these savings further.

Water conservation measures such as low-flow taps/showers, low-flush WCs, and recycling/reuse systems also reduce PCC. The potential impacts on the urban drainage system of more widespread use of such measures are discussed further in Chapters 9 and 24.

3.2.1.6 COVID-19 pandemic

Many countries around the world were subject to compulsory lockdown periods during the COVID-19 pandemic. In England, these were in the period from March 2020 through to March 2021. This had a marked upward effect on PCC caused by the adoption of increased water-using practices such as the washing of groceries and more intensive personal hygiene, particularly early in the pandemic. In addition, widespread working from home relocated consumption from commercial spaces, such as offices and gyms, to domestic dwellings. Outdoor water consumption also increased, as gardens became more important for wellbeing and exercise (Bunney et al., 2021). Changes were also noted in diurnal water use patterns with a delayed morning peak (1–2 hours later than the norm) and higher water use throughout the day (Abu-Bakar et al., 2021).

In England, the average household PCC changed from 137.5 L/hd.d in 2020 to 148 litres L/hd.d in 2021 (EA, 2022); an increase of 7.6%. In the United States (US), increases of 11.8–13.7% were measured in six utilities in various states (Nemati and Tran, 2022).

3.2.1.7 Quantification

Water PCC can be extremely varied, as shown in Figure 3.2. However, average domestic water usage in England stood at 148 L/hd.d in 2021 (EA, 2022), a figure influenced by

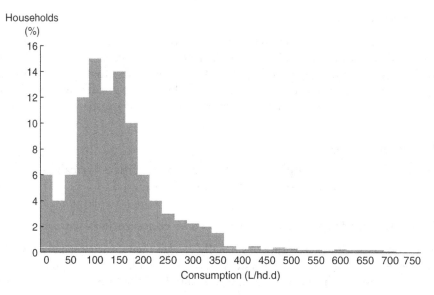

Figure 3.2 Variation of *per capita* water consumption. (Based on Russac et al., 1991. With permission of the Chartered Institution of Water and Environmental Management, London.)

Table 3.1 Percentage of water consumed for various purposes

Component	Water consumed (%)		
	Household	Commercial	Industrial and agricultural
WC flushing	27	35	5
Showering/bathing	28	26	1
Urinal flushing	–	15	2
Food preparation/drinking	15	9	13
Laundry (washing machine)	16	8	–
Washing-up (including dishwasher)	10	2	–
Car washing/garden use	5	4	17
Other	1	1	62

Source: Adapted from Friedler, E. et al. 2013. Wastewater composition. Chapter 17 in *Source Separation and Decentralization for Wastewater Management* (eds. T.A. Larsen, K.M. Udert, and J. Lienert), IWA Publishing.

COVID-19 restrictions (as discussed above). Usage ranges from 95 to 224 L/hd.d in the 11 countries cited by Friedler et al. (2013).

Water is used in three main areas in the home. Very approximately one-third of the water is used for WC flushing; one-third for personal washing via the wash basin, bath, and shower; and the final third for other uses such as washing up, laundry, and food/drink preparation (see Table 3.1). It is notable that only a very small percentage of this potable standard water is actually drunk (1–2 L/hd.d).

3.2.2 Water–wastewater relationship

As mentioned earlier, there is a strong link between water usage and wastewater disposal, with relatively little supplied water being "consumed" or taken out of the system. On a daily basis, we can simply say

Table 3.2 Percentage of water discharged as wastewater

Country	x (%)
United Kingdom	95
Middle East	
Poor housing	85
Good housing	75
United States	60

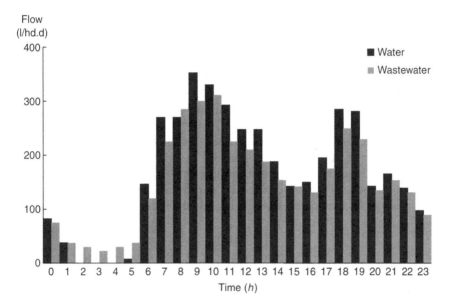

Figure 3.3 Typical diurnal plot of water consumption and wastewater flow.

$$G' = xG \tag{3.1}$$

where G is the water PCC (L/hd.d), G' is the wastewater generated per person (L/hd.d), and x is the return factor, given in Table 3.2 (–).

It is estimated that, in the UK, about 95% of water used is returned to the sewer network. The other 5% is made up of water used externally (watering the garden and washing the car, for example) and to miscellaneous losses within the household. In hotter climates with low rainfall, this proportion can be up to 40%.

Figure 3.3 shows a comparison made throughout the day between water use and wastewater flow in a household. In general, water use exceeds wastewater flow, especially in the early evening when gardens are being watered. At night, this situation is reversed due to sewer infiltration flows.

3.2.3 Temporal variability

It is emphasised that both wastewater quantity and quality vary widely from the very long term to the short term. Hence, any particular reported value should be related to the timescale over which it was measured. With the more widespread adoption of smart meters, ever-greater amounts of accurate and reliable data are becoming available to estimate the temporal dynamics of wastewater (Lund et al., 2021).

3.2.3.1 Long term

Until the turn of the last century, the long-term trend had been a steady increase in PCC on an annual basis, reflecting a number of factors such as increased ownership of water-using domestic appliances. However, that trend had reversed gradually downwards in the UK due to an increased emphasis on household water efficiency. It remains to be seen whether the COVID-19 uplift in consumption is merely a blip in this downward trend.

3.2.3.2 Annual

Variations within the year due to seasonal effects can be observed in water demand. Evidence (Thackray et al., 1978) suggests that WC flushing decreases in summer (probably due to an increased rate of body evaporation) and that bathing/showering increases. Outside water use increases significantly from gardening, and this can dominate the demand during summer months. For example, during the dry summer of 1995, increases in average demand of 50% or higher were observed in some areas. Figure 3.4 shows the monthly trends in the Anglian region for 3 years where the average consumption in July was up to 25% greater than in one of the winter months. The effect on wastewater flows is less clearly defined, but typically, summer dry weather flow discharges normally exceed winter flows by 10%–20%.

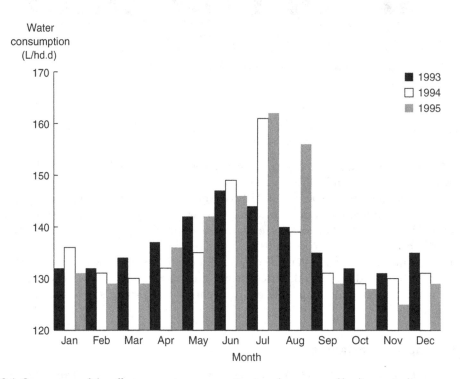

Figure 3.4 Comparison of the effects on water consumption in a dry summer (Anglian region).

3.2.3.3 Weekly

Variations in water demand and wastewater production can occur within the week, from day to day. Butler (1991) and Parker and Wilby (2013) found increased water consumption at weekends, probably due to increased WC flushing and bathing. This may also be due to a transfer of location rather than an increase *per se*.

3.2.3.4 Diurnal

A basic diurnal pattern showing variation from hour to hour of wastewater is given in Figure 3.3. Minimum flows occur during the early morning hours when activity is at its lowest. The first peak generally occurs during the morning, the exact timing of which is dependent on the social activities of the community, but in this example, it is between 09:00 and 10:00. A second flow peak occurs in the early evening between 18:00 and 19:00, and then a third can also be distinguished between 21:00 and 22:00, but this is less clearly defined in magnitude and timing. Detailed timing within the diurnal cycle is also affected by the day of the week, with some differences noted at weekends.

3.2.4 Appliances

Wastewater production is strongly linked to the widespread ownership and use of a wide range of domestic appliances, such as those in Table 3.3. The contribution of each individual appliance depends on both the volume of flow discharged after each operation and the frequency with which it is used.

Table 3.3 shows the typical discharge volumes of six different domestic appliances. Particularly large volumes are discharged by washing machines and during bathing, while relatively little is used during each use of the wash basin. The specific numbers shown in this table will change over time depending on the introduction of ever more water-efficient appliances. For example, in the UK, the mean WC flush volume in 2004 was approximately 9.4 L but by 2016 it had reduced to approximately 7.3 L. This was due to new water regulations, which prescribe the maximum flush volume for the installation of new WC cisterns (Artesia Consulting, 2018). Further information on water-efficient appliances and their effectiveness can be found on the Waterwise website (www.waterwise.org.uk).

Table 3.3 Average extant domestic appliance discharge volumes

Appliance	Volume (L)[a]
WC (per flush)	9.5
Bath (per use)	80
Shower (per use)	35
Washing machine (per cycle)	80
Dishwasher (per cycle)	35

Source: After Lallana, C. et al. 2001. *Sustainable Water Use in Europe, Part 2: Demand Management*, Environmental Issue Report 19, European Environment Agency.

Note:

[a] Introduction of low flow variants of these appliances will gradually reduce average volumes over time (see text)

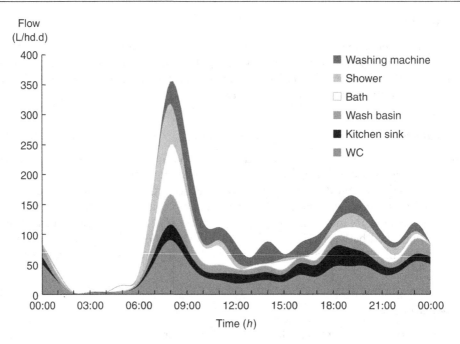

Figure 3.5 Appliance diurnal discharge patterns.

Figure 3.5 illustrates how the discharges from the individual appliances make up the general wastewater diurnal pattern. The most important contributor overall is the WC, which although only of modest volume is used very frequently throughout the day, particularly at peak periods. Further discussion of the implications of the diurnal wastewater pattern is given in Chapter 9.

3.3 NON-DOMESTIC

3.3.1 Commercial

This category includes businesses such as shops, offices, and light industrial units, and commercial establishments such as restaurants, laundries, public houses, and hotels.

Demand is generated by drinking, washing, and sanitary facilities, but patterns of use are inevitably different to those generated by domestic usage. For example, Table 3.1 shows how toilet/urinal usage is an even more dominant component of water use (50%) than in the domestic environment. Much less detailed information is available on commercial usage than on domestic usage.

3.3.2 Industrial

The component of wastewater generated by industrial processes can be important in specific situations but is more difficult to characterise in general because of the large variety of industries. Table 3.1 shows that many of the most important components of usage found in domestic and commercial premises are much less important in industry and agriculture.

In most cases, effluents result from the following water uses:

- Sanitary (e.g., washing, drinking, personal hygiene)
- Processing (e.g., manufacture, waste and by-product removal, transportation)
- Cleaning
- Cooling

The detailed rate of discharge will vary from industry to industry and will depend significantly on the actual processes used. Water consumption is often expressed in terms of volume used per mass of product. So, for example, papermaking consumes 50–150 m^3/t and dairy products 3–35 m^3/t.

Industrial effluents can be highly variable (in both quantity and quality) as a consequence of batch discharges, operation start-ups and shut-downs, working hours, and other factors. These may change significantly at weekends. Depending on the relative magnitude of the flows, industrial discharges can completely alter the normal diurnal patterns of flow. There may also be significant seasonal changes in demand, for example, due to agro-industrial practices responding to the needs of food production.

Other important factors include the size and structure of the organisation, the extent of process water recycling, and the availability and cost of water (Shang et al., 2017).

3.4 INFILTRATION AND INFLOW

Unlike the other sources of wastewater, infiltration and inflow are not deliberate discharges but occur as a consequence of the existence of a piped network. Infiltration and inflow have been briefly introduced in Chapter 1 and are defined as water that enters the sewer system, including laterals and private sewers, through indirect and direct means, respectively.

Infiltration is a *slow response* flow that enters the sewer system through defective drains and sewers (cracks and fissures), pipe joints, couplings, and manholes arising from

- Extraneous groundwater
- Spring water
- Seawater
- Water from other leaking pipes

Inflow is stormwater that enters separate foul sewers from illegal or misconnected yard gullies, through roof downpipes, or through manhole covers (see Chapter 5 for further discussion on misconnections). This is a common occurence, and a survey of one separate system (Inman, 1975) found that 40% of all houses had some arrangement whereby stormwater could enter the wastewater sewer. Inflow can also arise from adjacent storm sewers, land drainage, streams, and other watercourses.

Closely allied to inflow is rainfall-induced infiltration (RII). RRI is, in effect, the remaining portion of the flow hydrograph typically observed in combined sewers, which is not accounted for by the foul flows, slow response groundwater infiltration, and surface runoff. It is not a widely observed phenomenon but where it does occur can cause significant problems with flows remaining high for several days after rainfall (Allitt, 2002). This has particular implications for CSO "dry spills" (see Section 13.8.1).

3.4.1 Problems

The presence of excessive amounts of infiltration may cause one or more of the following problems:

- Reduced effective sewer capacity leading to possible surcharging and/or flooding preventing the flushing of toilets, and so on, or pollution for long periods of time
- Overloading of pumping stations and WTPs
- Higher frequency of CSO operation, and possibly in dry weather, during periods of high groundwater levels
- Increased entry of sediment (soil), resulting in higher maintenance requirements and possible surface subsidence

3.4.2 Quantification

The extent of infiltration is site-specific, but when excessive, it is usually a result of poor design and construction and will generally increase as the system physically degrades. Influencing factors include the following (Bishop et al., 1998; Karpf and Krebs, 2013):

- Age of the system
- Standard of materials and methods
- Standard of workmanship in laying pipes
- Settlement due to ground movement
- Height of groundwater level (varies seasonally)
- Type of soil (hydraulic conductivity)
- Properties of the sewer trench
- Aggressive chemicals in the ground
- Extent of the network – total length of sewer (including house connections); type of pipe joint, number of joints and pipe size; number and size of manholes and inspection chambers
- Proximity of other drainage networks to transfer flow (e.g., laid in the same trench)
- Frequency of surcharge

The amount of infiltration may range widely and can reach serious proportions in old systems. In the UK, rates up to 50% of average dry weather flow have been measured. The proportion of total infiltration arising from house connections may be up to 50% in places, but this is difficult to predict with any accuracy (Kohout et al., 2010). More complete details of the causes, costs, and control of infiltration can be found in White et al. (1997).

3.4.3 Exfiltration

Exfiltration is the opposite of infiltration. Under certain circumstances (e.g., when surcharged), wastewater (or stormwater) is able to leak out of the sewer into the surrounding soil and groundwater. This creates the potential for re-infiltration into the sewer or another system at a later stage or groundwater contamination, which could be critical in areas where an aquifer is used for drinking water supply.

Rutsch et al. (2008) argue that there are contradictory viewpoints on whether or not exfiltration is a serious problem. Values of exfiltration (as with infiltration) are variable, with published rates as low as 1% and as high as 13% of dry weather flow in UK case studies

(Rueedi et al., 2009; Wakida and Lerner, 2005), with even higher rates noted in unpublished studies. A study in Canada (Guériguneau et al., 2014) found exfiltration to vary between 0.6 and 15.7 m^3/d per kilometre during dry weather, and 1.1 and 19.5 m^3/d per kilometre during wet weather. This variability is a result of the inherent differences across systems and of the many alternative methods of quantification used, many of which are indirect.

Factors affecting the likelihood of exfiltration are similar to those discussed for infiltration. More complete details of the causes, costs, control, and implications of exfiltration can be found in Anderson et al. (1996) and Reynolds and Barrett (2003).

3.5 WASTEWATER QUALITY

Wastewater contains a complex mixture of natural organic and inorganic material present in various forms, from coarse grits, through fine suspended solids, to colloidal and soluble matter. Much is in the form of highly putrescible compounds. In addition, a small proportion of man-made substances, derived from household products and commercial and industrial practices, will be present.

In fact, wastewater is 99.9% water although the remaining 0.1% is significant, particularly if it is allowed to enter the environment. Fresh domestic wastewater is typically cloudy grey in colour with some recognisable solids and has a musty/soapy odour. With time (2–6 hours depending on ambient conditions), the waste "ages" and gradually changes in character as a result of physical and biochemical processes. Stale wastewater is dark grey/black with smaller and fewer recognisable solids, and "older" flows can have a pungent "rotten eggs" odour due to the presence of hydrogen sulphide.

Wastewater quality is variable with respect to both location and time. In addition, the techniques commonly used for sampling and analysis are subject to error (see Chapter 2). Therefore, caution is needed in interpreting standard or typical values. Such data should never be assumed to accurately represent the wastewater from a particular community – this can only be properly confirmed by a (possibly extensive) testing programme or access to historical data.

The impact of the COVID-19 pandemic on wastewater quality was less clear-cut than that on quantity (see Section 3.2.1.6), mainly depending on the number of inward and outward commuters in the urban areas under consideration. In a study by Pons et al. (2020), some catchments showed no change in terms of pollutant load whilst others saw significant reductions in the main pollutant determinands (e.g., COD, BOD, TKN, SS) of 20–45%.

3.5.1 Pollutant sources

Wastewater quality is influenced by the contaminants discharged into it derived mainly from human, household, and industrial activities. The quality of the carriage water (the original drinking water) or infiltrating groundwater can also be influential.

3.5.1.1 Human excreta

Human excreta are responsible for a large proportion of the pollutants in wastewater. Adults produce on average 200–300 g of faeces and 1–1.5 kg of urine per day. Faeces account for 10–15 g/hd.d of biochemical oxygen demand (BOD) and urine 6 g/hd.d of BOD, but together only contribute a small proportion of wastewater fats (Friedler et al., 2013).

Excreta are also an important source of nutrients. The bulk (94%) of the organic nitrogen in wastewater is derived from excreta. Of this percentage, 50% derives from urine (urea), which is most abundant in fresh wastewater as it is rapidly converted to ammonia under both aerobic and anaerobic conditions (see Chapter 2). Approximately 50% of the phosphorus discharged to sewer (1.5 g/hd.d) is derived from excreta. Excreta also contains about 1 g/hd.d of sulphur (Gilmour et al., 2008).

The bulk of the microorganisms in wastewater originate in faeces; urine is relatively microbe free.

3.5.1.2 Toilet/WC

A wide range of large (gross) solids is discharged, either deliberately or accidentally, via the toilet such as condoms, sanitary towels, panty liners, tampons, disposable nappies (diapers), toilet paper, paper towels, wet wipes, and cotton buds. Wet wipes and sanitary towels have also been shown to be a source of microplastic (MP) fibres (Briain et al., 2020).

Toilet paper is used in large quantities. Although this typically disintegrates in a matter of hours in the turbulent flow in sewers (Eren and Karadagli, 2012), it is only slowly biodegradable due to the presence of cellulose fibres. Spence et al. (2016) found by in-sewer sampling that 12–23 g/hd.d is disposed of influenced particularly by the demographics of the sampled catchments. Tests have shown that coloured papers contribute some 15% of the wastewater chemical oxygen demand (COD). In total, some 0.15 sanitary items/hd are disposed of each day (Friedler et al., 1996).

A number of cleaning, disinfecting, and descaling chemicals are also routinely discharged into the system via the toilet.

3.5.1.3 Food

Digested food is the source of many of the excreta-related pollutants mentioned above. However, undigested food is a major contributor of fats, oils, and grease (FOG) including butter, margarine, cooking oils, vegetable fats, meats, cereals, and nuts. The COD of food varies widely, for example, bread at 790 mg COD/g food, chicken at 437 mg/g and cabbage at 79 mg/g (Legge et al., 2022), and food residues are also a source of some organic nitrogen and phosphorus and of salt (NaCl).

Food waste disposers are not widely used in the UK but are common in Australia, New Zealand, and the US (Evans, 2012). Clearly, food waste load entering the sewer will be greater if this appliance is attached to the kitchen sink. Significant benefits (e.g., increased biogas production) are claimed on the basis of trials in Sweden, with very few drawbacks (e.g., increased water consumption, sewer deposition) in evidence (Evans et al., 2010; Mattsson et al., 2014).

3.5.1.4 Washing, laundry, and consumer products

Washing and laundry activities add soaps and detergents to the sewer (e.g., washing machine and dishwater detergents). The polyphosphate builders used in synthetic detergents contribute approximately 50% of the phosphorus load. Phosphorus concentrations have diminished significantly in countries where legislation has imposed significant reductions in the amounts of phosphorus used by manufacturers of detergents (Morse et al., 1993). Values for total phosphorus are approximately 0.3 g/hd.d from laundry activity and 0.2 g/hd.d from dishwashers (Gilmour et al., 2008). The washing of synthetic textiles is also a significant source of MP fibres. Further, large amounts are contributed by microbeads in cosmetics and toothpaste (Prata, 2018). A number of countries have prohibited the sale of rinse-off cosmetics and personal care products containing microbeads, including the UK (since June 2018).

3.5.1.5 Industry

The characteristics of industrial wastewaters, or trade effluents as they are often called, are similar to those of domestic wastewater in that they are likely to contain a very high proportion of water, and the impurities may be present as suspended, colloidal, or dissolved matter. But, in addition, a very large variety of pollutant types can be generated and industrial wastewater may contain

- Extremes of organic content
- A deficiency of nutrients
- Inhibiting chemicals (acids, toxins, bactericides)
- Resistant organic compounds
- Heavy metals and accumulative persistent organics
- Microplastics (MPs)
- PFAS

Processing liquors from the main industrial processes tend to be relatively strong, while wastewaters from rinsing, washing, and condensing are comparatively weak. Discharges may be seasonal and vary considerably from day to day both in volume and strength.

3.5.1.6 Carriage water and groundwater

The sulphate present in wastewater is derived principally from the mineral content of the municipal water supply or from saline groundwater infiltration (see Chapter 2).

In hard water areas, the use of softeners can result in significant increases in wastewater chloride concentrations. Infiltration of saltwater (if present) can contribute similarly.

3.5.2 Pollutant levels

Typical values and ranges of pollutant levels in UK wastewater are given in Table 3.4.

Table 3.4 Pollutant concentrations and unit loads for wastewater

Parameter type	Parameter	Unit load (g/hd.d)	Concentration (mg/L) mean (range)
Physical	Suspended solids		
	Volatile	48	240
	Fixed	12	60
	Total	60	300 (180–450)
	Microplastics (MPs)		(1–430) MP/L
	Gross (sanitary) solids		
	Sanitary refuse	0.15 items/hd.d	
	Toilet paper	7	
	Temperature		18 (15–20) °C: summer
			10°C: winter
Chemical	BOD_5		
	Soluble	20	100
	Particulate	40	200
	Total	60	300 (200–400)
	COD		

(Continued)

Table 3.4 (Continued)

Parameter type	Parameter	Unit load (g/hd.d)	Concentration (mg/L) mean (range)
	Soluble	35	175
	Particulate	75	375
	Total	110	550 (350–750)
	TOC	40	200 (100–300)
	Nitrogen		
	Organic N	4	20
	Ammonia	8	40
	Nitrites		0
	Nitrates		<1
	Total	12	60 (30–85)
	Phosphorus		
	Organic	1	5
	Inorganic	2	10
	Total	3	15
	pH		7.2 (6.7–7.5): hard water
			7.8 (7.6–8.2): soft water
	Sulphates	20	100: dependent on water supply
	DEHP		3.4–34 (ng/L)
	FOG	250	100
Microbiological	Total coliforms		10^7–10^8 MPN/100 mL
	Faecal coliforms		10^6–10^7 MPN/100 mL
	Viruses		10^2–10^3 infectious units/100 mL

Source: Adapted from Ainger, et al. 1997 and Williams et al., 2012.

3.6 WASTEWATER-BASED EPIDEMIOLOGY

The concept and practice of *wastewater-based epidemiology*, also known as *wastewater surveillance*, is a growing but relatively new approach for the surveillance of infectious diseases and substances of concern, and for the early warning of disease outbreaks. Applications include antimicrobial resistance (Hendriksen et al., 2019), and illicit drug and pharmaceutical use (Baker et al., 2014). More recently, WBE was widely adopted to monitor COVID-19 via detection of the SARS-CoV-2 coronavirus. Since a large proportion of infected individuals shed the virus ribonucleic acid (RNA) in their faeces, concentrations in wastewater can be used to provide an indicator of disease prevalence without relying on clinical testing data.

Wastewater samples are most commonly collected at WTPs giving a measure of SARS-CoV-2 prevalence across the entire WTP and its associated sewer catchment. However, some studies have focused on in-sewer monitoring to potentially provide more targeted actions to address and mitigate any outbreak detected. Sweetapple et al. (2022), for example, report on a study of an English university campus. Using wastewater biomarkers (ammoniacal nitrogen and orthophosphate – see Chapter 2), they found that population normalisation can reveal significant differences between days where SARS-CoV-2 concentrations are very similar. Confidence in the trends identified was strongest for samples collected during dry weather periods although wet weather samples still provided valuable information (see Figure 3.6, noting SARS-CoV-2 concentration is given in gene copies per litre (gc/L)). It was also shown

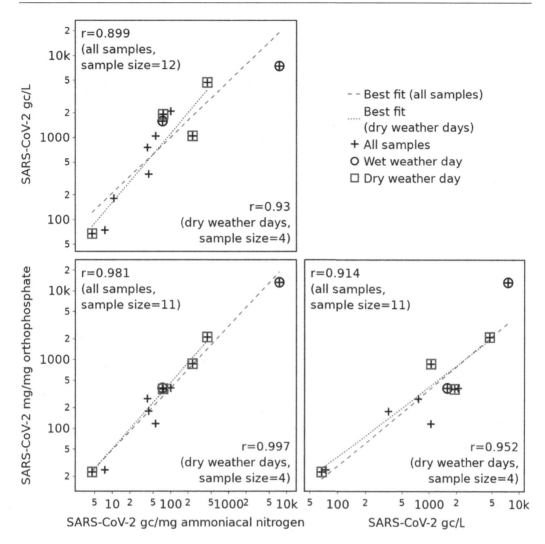

Figure 3.6 Correlations between wastewater SARS-CoV-2 concentrations and normalised values. (Reproduced from Sweetapple, C. et al., 2022. Building knowledge of university campus population dynamics to enhance near-to-source sewage surveillance for SARS-CoV-2 detection Science of the Total Environment, 806, 150406, with permission from Elsevier.)

that building-level occupancy estimates based on WC flush counters could aid local identification of potential sources of SARS-CoV-2 leading to more targeted infection control.

PROBLEMS

1 Classify the major sources of wastewater, and discuss the factors affecting their prevalence in practice.

2 Explain the quantitative link between water demand and wastewater generation. What are the major factors influencing domestic water use?

3 How did the COVID-19 epidemic affect the production of wastewater from various sources?

4 What are the main differences between domestic, commercial, and industrial water demand?

5 Describe how wastewater varies at various timescales, and explain the significance of this.

6 Compare and contrast the mechanisms, amounts, and implications of infiltration and exfiltration.

7 What are the main sources of pollutants in wastewater, and what is their importance?

8 Explain the difference between nano- and microplastics and discuss their sources and potential impact.

9 What are the benefits of monitoring wastewater quality for specific determinands? What additional benefits arise from in-sewer monitoring?

KEY SOURCES

Ainger, C.M., Armstrong, R.J., and Butler, D. 1997. *Dry Weather Flow in Sewers*, CIRIA R177.

Baker, D.R., Barron, L. and Kasprzyk-Hordern, B. 2014. Illicit and pharmaceutical drug consumption estimated via wastewater analysis. *Part A: Chemical analysis and drug use estimates. Science of the Total Environment*, 487, 629–641.

Friedler, E., Butler, D., and Alfiya, Y. 2013. Wastewater composition. Chapter 17 in *Source Separation and Decentralization for Wastewater Management* (eds. T.A. Larsen, K. P. Herrington). 1996. *Climate change and the Demand for Water*. HMSO, London.

Prata, J.C. 2018. Microplastics in wastewater: State of the knowledge on sources, fate and Solutions. *Marine Pollution Bulletin, 129*, 262–265.

Sims, N., Kasprzyk-Hordern, B. 2020. Future perspectives of wastewater-based epidemiology: Monitoring infectious disease spread and resistance to the community level. *Environment International, 139*, 105689. M. Udert, and J. Lienert), IWA Publishing.

REFERENCES

Abu-Bakar, H., Williams, L. and Hallett, S.H. 2021. Quantifying the impact of the COVID-19 lockdown on household water consumption patterns in England. *NPJ Clean Water, 4*, 13.

Allitt, R. 2002. Rainfall, runoff and infiltration re-visited. *WaPUG Spring Meeting*, Birmingham.

Anderson, G., Bishop, B., Misstear, B. and White, M. 1996. *Reliability of Sewers in Environmentally Sensitive Areas*, CIRIA PR44.

Artesia Consulting. 2018. *The Long Term Potential for Deep Reductions in Household Water Demand*, Commissioned Report for Ofwat, www.ofwat.gov.uk/publication/long-term-potential-deep-reductions-household-water-demand-report-artesia-consulting

Bishop, P.K., Misstear, B.D., White, M. and Harding, N.J. 1998. Impacts of sewers on groundwater quality. *Journal of the Chartered Institution of Water and Environmental Management, 12*(June), 216–223.

Briain O.O., Marques Mendes, A.R., McCarron, S., Healy, M.G. and Morrison, L. 2020. The role of wet wipes and sanitary towels as a source of white microplastic fibres in the marine environment, *Water Research, 182*, 116021.

Bunney, S., Lawson, E., Cotterill, S. and Butler, D. 2021. Water resource management: moving from single risk-based management to resilience to multiple stressors. *Sustainability, 13*, 8609.

Butler, D. 1991. The influence of dwelling occupancy and day of the week on domestic appliance wastewater discharge. *Building and Environment, 28*(1), 73–79.

Consumer Council for Water (CCW). 2016. *Delving into water 2016: Performance of the water companies in England and Wales, 2011–12–2015–16*, Birmingham, UK.

Department for Environment, Food & Rural Affairs (Defra). 2023. *Open consultation. UK mandatory water efficiency labelling,* www.gov.uk/government/consultations/uk-mandatory-water-efficiency-labelling

Downing, T.E., Butterfield R.E., Edmonds, B., Knox J.W., Moss, S., Piper, B.S., Weatherhead, E.K. (and the CCDeW project team). 2003. *Climate Change and the Demand for Water,* Research Report, Stockholm Environment Institute Oxford Office, Oxford.

Environment Agency (EA). 2022. *Water and Sewerage Companies in England: Environmental Performance Report 2021.* www.gov.uk/government/publications/water-and-sewerage-companies-in-england-environmental-performance-report-2021

Eren, B. and Karadagli, F. 2012. Physical disintegration of toilet papers in wastewater systems: Experimental analysis and mathematical modeling. *Environmental Science and Technology,* 46(5), 2870–2876.

Evans, T.D. 2012. Domestic food waste: The carbon and financial costs of the options. *Municipal Engineer, 165*(ME1), 3–10.

Evans, T.D., Andersson, P., Wievegg, A., and Carlsson, I. 2010. Surahammar: A case study of the impacts of installing food waste disposers in 50% of households. *Water and Environment Journal,* 24(4), 309–319.

Friedler, E., Brown, D.M., and Butler, D. 1996. A study of WC derived sewer solids. *Water Science and Technology,* 33(9), 17–24.

Gilmour, D., Blackwood, D., Comber, S. and Thornell, A. 2008. Identifying human waste contribution of phosphorus loads to domestic wastewater. *Proceedings of the 11th International Conference on Urban Drainage, Edinburgh, September, on CD-ROM.*

Guérineau, H., Dorner, S., Carrière, A., McQuaid, N., Sauvé, S. and Aboulfadl, K. 2014. Source tracking of leaky sewers: A novel approach combining fecal indicators in water and sediments. *Water Research,* 58(1), 50–61.

Hendriksen, R.S., Munk, P., Njage, P., van Bunnik, B., McNally, L. et al. 2019. Global monitoring of antimicrobial resistance based on metagenomics analyses of urban sewage. *Nature Communications,* 10, 1124.

Inman, J. 1975. Civil engineering aspects of sewage treatment works design. *Proceedings of the Institution of Civil Engineers, Part 1, 58(May),* 195–204, discussion, 669–672.

Karpf, C. and Krebs, P. 2013. Modelling of groundwater infiltration into sewer systems. *Urban Water Journal,* 10(4), 221–229.

Kohout, D., Princ, I. and Pollert, J. 2010. Measuring and scale-up of infiltration and exfiltration in house connections. In *Assessing Infiltration and Exfiltration on the Performance of Urban Sewer Systems (APUSS)* (eds. B. Ellis and J.L. Bertrand-Krajewski), IWA Publishing.

Lallana, C., Krinner, W., Estrela, T., Nixon, S., Leonard, J. and Berland J.M. 2001. *Sustainable Water Use in Europe, Part 2: Demand Management,* Environmental Issue Report 19, European Environment Agency, Copenhagen, Denmark.

Lede, E. and Meleady, R. 2019. Applying social influence insights to encourage climate resilient domestic water behavior: Bridging the theory-practice gap. *WIREs Climate Change,* 10, e562.

Legge, A., Jensen, H., Ashley, R., Tait, S. and Nichols, A. 2022. Food waste disposers: degradation of food waste in-sewer and the influence on energy recovery at wastewater treatment plants. *10th International Conference on Sewer Processes and Networks,* Graz, Austria, August.

Lund, N. S. V., Kirstei, J.K., Madsen, H., Mark, O. Mikkelsen, P.S. and Borup, M. 2021. Feasibility of using smart meter water consumption data and in-sewer flow observations for sewer system analysis: a case study. *Journal of Hydroinformatics,* 23(4), 795–812.

Mattsson, J., Hedstrom, A. and Viklander, M. 2014. Long-term impacts on sewers following food waste disposer installation in housing areas. *Environmental Technology,* 35(21), 2643–2651.

Morse, G.K., Lester, J.N. and Perry, R. 1993. *The Economic and Environmental Impact of Phosphorus Removal from Wastewater in the European Community,* Selper Publications.

Nemati, M. and Tran, D. 2022. The impact of COVID-19 on urban water consumption in the United States. *Water, 14,* 3096.

NIC 2018 *Analysis of Drought Resilience, Technical Annex*, National Infrastructure Assessment, National Infrastructure Commission.

Ornaghi, C. and Tonin, M. 2019. The effects of the universal metering programme on water consumption, welfare and equity. *Oxford Economic Papers*, 1–24.

Parker, J.M. and Wilby, R.L. 2013. Quantifying household water demand: A review of theory and practice in the UK. *Water Resources Management*, 27(4), 981–1011.

Pons, M.-N., Louis, P. and Vignati, D. 2020. Effect of lockdown on wastewater characteristics: a comparison of two large urban areas. *Water Science and Technology*, 82(12), 2813–2815.

Reynolds, J.H. and Barrett, M.H. 2003. A review of the effects of sewer leakage on groundwater quality. *Journal of the Chartered Institution of Water and Environmental Management*, 17(1), March, 34–39.

Rueedi, J., Cronin, A.A. and Morris, B.L. 2009. Estimation of sewer leakage to urban groundwater using depth-specific hydrochemistry. *Water and Environment Journal*, 23, 134–144.

Russac, D.A.V., Rushton, K.R., and Simpson, R.J. 1991. Insights into domestic demand from a metering trial. *Journal of the Institution of Water and Environmental Management*, 5(3), June, 342–351.

Rutsch, M., Rieckermann, J., Cullmann, J., Ellis, J.B., Vollertsen, J. and Krebs, P. 2008. Towards a better understanding of sewer exfiltration. *Water Research*, 42(10–11), 2385–2394.

Shang, Y., Wang, J., Ye, Y., Lei, X., Gong, J. and Shi, H. 2017. An analysis of the factors that influence industrial water use in Tianjin, China. *International Journal of Water Resources Development*, 33(1), 81–92.

Spence, K.J., Digman, C., Balmforth, D., Houldsworth, J., Saul, A. and Meadowcroft, J. 2016. Gross solids from combined sewers in dry weather and storms, elucidating production, storage and social factors. *Urban Water Journal*, 13(8), 773–789.

Sweetapple, C., Melville-Shreeve, P., Chen, A.S., Grimsley, J., Bunce, J.T., Gaze, W., Fielding, S. and Wade, M.J. 2022. Building knowledge of university campus population dynamics to enhance near-to-source sewage surveillance for SARS-CoV-2 detection. *Science of the Total Environment*, 806, 150406.

Thackray, J.E., Cocker, V., and Archibald, G. 1978. The Malvern and Mansfield studies of water usage. *Proceedings of the Institution of Civil Engineers, Part 1*, 64, 37–61.

Wakida, F.T. and Lerner, D.N. 2005. Non-agricultural sources of groundwater nitrate: A review and case study. *Water Research*, 39, 3–16.

White, M., Johnson, H., Anderson, G., and Misstear, B. 1997. *Control of Infiltration to Sewers*, CIRIA R175.

Williams, J.B., Clarkson, C., Mant, C., Drinkwater, A. and May, E. 2012. Fat, oil and grease deposits in sewers: Characterization of deposits and formation mechanisms. *Water Research*, 46, 6319–6328.

Willis, R.M., Stewart, R.A., Giurco, D.P., Talebpour, M.R. and Mousavinejad, A. 2013. End use water consumption in households: Impact of socio-demographic factors and efficient devices. *Journal of Cleaner Production*, 60(1), 107–115.

Chapter 4

Rainfall

4.1 INTRODUCTION

As already noted, urban drainage systems deal with both wastewater and stormwater. Most stormwater is the result of rainfall. Other forms of precipitation – snow, for example – are also contributors, but rainfall is the most significant in many places. Methods of representing and predicting rainfall are therefore crucial in the design, analysis, and operation of drainage systems.

The detailed study of rainfall is the work of meteorologists and hydrologists, and their work primarily entails interpreting and predicting nature – always a difficult task. They work primarily using observation, which is the origin of all our knowledge about rainfall. The more we observe, the more we learn.

Observation provides historical records and allows derivation of relationships between rainfall event properties (particularly intensity, duration, and frequency). Often, urban drainage engineers and modellers require long periods of rainfall (including different storms and the dry periods between) to input to models in order to see how a drainage system behaves. These may be real historical records, but in that case, it is hard to know exactly how representative a particular portion of history actually is. For this reason, it may be more appropriate to use specially created synthetic sets of data that represent the properties of actual rainfall.

This chapter describes the main methods of rainfall measurement and considers rainfall data requirements for different applications (Section 4.2). The methods of rainfall data analysis are presented in Section 4.3, with single events and multiple events described in Sections 4.4 and 4.5, respectively. Section 4.6 discusses climate change and its important implications for urban drainage. Data and techniques presented in this chapter are used in subsequent chapters on design and analysis.

4.2 MEASUREMENT

4.2.1 Rain gauges

Rain gauges are the most common device for measuring rainfall. A standard non-recording gauge (Figure 4.1a) collects rain falling on a standard area (a 127 mm diameter funnel with the rim placed 300 mm above ground level) over a known period of time. The volume of the stored rainfall is measured manually and, if necessary, converted to rainfall intensity (depth/time) by dividing by the collection area. Collection periods range from 6 hours to 1 month, but 1 day is typical. Since urban drainage systems can respond in less than 6 hours, the data from non-recording gauges is of limited value in this application.

DOI: 10.1201/9781003408635-4

Figure 4.1 Rain gauges: (a) standard; (b) tipping bucket. (Courtesy of Hydrokit, Poole.)

Recording gauges are able to provide a continuous record of rainfall. The tipping-bucket rain gauge collects rainfall over short periods of time (typically 2–15 minutes) in a balanced reservoir consisting of two miniature compartments. Rainwater enters the first compartment until the weight of the water makes it tilt. Water begins to enter the second compartment while the first empties (Figure 4.1b). Thus, the gauge produces a series of tips with a changing frequency depending on the rainfall intensity. The number of tips per unit time is therefore related to the rainfall intensity. A record is made either of the number of tips in a set time interval or the time of each tip. This is recorded electronically and stored in the memory of a data logger on site or transmitted over telephone lines or mobile-phone networks (typically using GSM or GPRS standards) to a central station. The range of rainfall depth resolution is 0.1–0.5 mm/tip. There are over 1500 online logging rain gauges across the UK (see environment.data. gov.uk/hydrology/explore for rainfall and other hydrological data in England).

A number of other types are also available including weighing, acoustic, and optical gauges. Their function as well as the pros and cons of all the gauge types are described in detail in a publication by the World Meteorological Organization (WMO, 2008).

4.2.1.1 Siting

The siting of gauges must be carefully planned in order to obtain representative data for the catchment. There are general rules (WMO, 2008) about the distances from obstacles, the level of the gauge orifice above ground level, and so on, but these rules cannot always be fully adhered to in the urban environment. In addition, in urban areas, some rain falls on roofs, so the siting of gauges on roofs is acceptable if the number of gauges on the different elevation levels corresponds approximately to the proportions of the different types of urban surface.

If more than one gauge is in use, careful synchronisation of records is crucial. This can be achieved manually via an accurate clock setting at each gauge or remotely via a grid of sensors. Settings should be regularly checked. As a rule, rain gauges should be visited and checked at least once a week.

4.2.2 Radar

Unlike the ground rain gauges that provide point estimates of rainfall values, radars can survey large areas, capturing the spatial and temporal variability of rainfall that is of high importance in the design and modelling of urban drainage systems. Radar stations sweep out short pulses (typically 2 microseconds) of electromagnetic waves at falling raindrops in a narrow 1° bandwidth around a vertical axis, and measure the time and strength of the return signal to infer the location and the intensity of rainfall accordingly (Collier, 1996). The sweep is repeated at various angles of elevation, usually between 0.5° and 4°, to take account of local topography and buildings that clutter the image.

Three main types of weather radar are available depending on the wavelength of the transmitted signal (from larger to shorter length): S-band, C-band, and X-band. The large wavelength radars, such as the S-band, are able to cover larger areas, given that they are not easily attenuated; however, there are problems associated with the interaction of the beam with the ground and the vertical variation of reflectivity. S-band radars are the biggest and most expensive given that larger dishes are needed to achieve a small beam width. On the contrary, C- and X-band radars require less power and possess smaller dishes, but sensitivity to attenuation is a critical issue. C-band radars balance power, size, and cost and are widely used in European countries, along with S-band units. X-band radars are more sensitive and can detect tiny droplets and, therefore, are often used to provide very short-term weather observations and forecasts that are of interest in urban drainage applications such as early warning systems. Table 4.1 summarises the main characteristics of each system.

In the British Isles, there are currently 18 weather (C-band) radar stations (Meteorological Office, 2013). These radars complete a series of scans every 5 minutes, providing detailed spatial information on rainfall intensities (approximately one measurement per 1 or 2 km) to a 75 km radius and the overall picture of the extent of precipitation (5 km resolution) to a range of 255 km.

The most commonly used equation to convert the power of the reflected signal, or equally reflectivity Z, to rainfall intensity i is (WMO, 2008)

$$Z = ai^b \tag{4.1}$$

where a and b are constants linked with drop size distribution (DSD).

Table 4.1 Typical operating characteristics for different types of weather radar used for hydrological applications

Characteristic	S-band	C-band	X-band
Spatial resolution (m)	1000–4000	250–2000	100–1000
Temporal resolution (m)	10–15	5–10	1–5
Maximum range (km)	100–200	100–130	30–60

Source: Adapted from Thorndahl et al., 2017.

Alternative values can be assigned to the two constants for different types of storms (e.g., Collier, 1996), while the most typical values for stratiform rainfall are $a = 200$ and $b = 1.60$.

Since radars provide an indirect measurement of rainfall, multiple sources of errors and uncertainties are inserted in the estimation (Harrison et al., 2000; Meteorological Office, 2013):

- Permanent echoes caused by obstacles or hills near the radar (ground clutter)
- Spurious echoes caused by aircraft and interference from other radars
- Attenuation of the signal by precipitation, especially for C- and X-band radars
- The radar beam may not detect lower-level clouds at long distances, under-predicting rainfall intensity.
- Evaporation of rainfall at lower levels beneath the beams may result in the over-prediction of rainfall intensity.
- Increase of rainfall intensity at lower levels upon hills where medium-level frontal clouds accumulate other small droplets as they pass through moist, cloudy layers at low levels. This may cause an underestimation of rainfall values.
- False hypothesis for DSD that may result in incorrect transformation of reflectivity to intensity. In general, radars underestimate rainfall from frontal clouds with small drops and over-estimate rainfall falling from convective shower clouds.
- Miscalibration

The typical pre-processing of the radar picture to correct some of the above errors comprises (Golz et al., 2005) (1) removal of ground clutter and spurious echoes, (2) correction of radial anomalies, and (3) correction of attenuation effects. Indeed, not all errors can be corrected, some can only be detected (Einfalt et al., 2002). Special attention has been given to the correction of errors inserted by the transformation of reflectivity into rainfall amounts (Goudenhoofdt and Delobbe, 2013; Smith et al., 2009). In general, the higher are the rainfall intensity and variability, the wider the error bandwidths in the reflectivity to rainfall conversion function become due to difficulty in accurately determining DSD.

Considerable reduction of some of these errors, as well as more accurate identification of DSD have been achieved with the deployment of Polarimetric Doppler Weather Radar, also referred to as *dual-polarization radar* (Bringi and Chandrasekar, 2001). This type of radar transmits and receives signals around both the horizontal and vertical axes, enabling a more accurate sizing of droplets and estimation of rainfall intensities (Bringi et al., 2011).

To improve the accuracy of radar estimates at local spatial scales, useful in urban drainage applications, ground measurements from rain gauge networks are often required. This process, referred to as *adjustment*, *combination*, or *merging*, dynamically corrects the error between ground observations and radar estimates. A great number of adjustment procedures have been proposed, which, can be classified into two main categories (Wang et al., 2013): (1) mean bias reduction techniques (e.g., Seo et al., 1999; Smith et al., 2007) and (2) error variance minimisation techniques (e.g., Gerstner and Heinemann, 2008; Todini, 2001). In our context of urban drainage applications, the dynamic geostatistical adjustment methods seem most suitable to reproduce the high variability of rainfall at fine spatio-temporal scales needed (Nielsen et al., 2014; Sinclair and Pegram, 2005; Wang et al., 2015).

Radar data are of particular value when radars are continuously operating and feed data online for flood warning (23.3), real-time control (23.2), weather forecasting, and water quality prediction (Dale and Stidson, 2007). Data can also be used offline for long-term model simulations (Chapter 19), and rainfall event assessment (e.g., rainfall extremes, high spatial variation) and can provide rapid and extensive urban water management information (Neale, 2008).

4.2.3 Satellites

Satellite imagery has also been used, usually by indirectly relating cloud-top temperature to rainfall (Rosenfeld and Collier, 1998). Despite the fact that rainfall estimates from satellites are less accurate than ground gauges and radars, they can scan much larger areas, sampling also oceans and remote regions. Two different techniques exist to estimate precipitation from space (WMO, 2008). The first method, named *cloud indexing*, analyses the structure of the clouds (i.e., the amount and type of cloudiness) to establish a relationship between a precipitation index and a function of the cloud surface area. The main limitation of this method is the bias from the presence of non-precipitating clouds, such as cirrus. The second method, called *life history* uses consecutive images from geostationary satellites, sampling every half hour, to take into consideration the stage of development of the cloud and estimate precipitation. Various techniques and methods have been developed to reduce the error in rainfall estimation from satellites, using observations from radars and rain gauges (Ebert et al., 2007; Krajewski et al., 2000).

Rising computational power availability, together with advances in remote sensing and data-driven algorithms, have enabled the fusion of multiple space-born products (both rainfall and other hydroclimatic variables, such as soil moisture) to provide gridded rainfall estimates of high accuracy and spatio-temporal coverage (Beck et al., 2017; Massari et al., 2020). Machine-learning algorithms (introduced in Chapter 21) have a key role to play in such data fusion approaches (also known as *data merging*, *blending* or *data integration*) due to their ability to efficiently combine individual data sources by handling large amounts of data and capturing complex relationships (Kumar et al., 2019; Zhang et al., 2021).

The unique advantage of satellites to cover much larger areas and provide weather forecasts with great lead times is of high importance in online applications of urban drainage such as early warning systems and real-time control (Bajracharya et al., 2017; Chiang et al., 2007; Fan et al., 2016; see also Chapter 22).

4.2.4 Microwave links

An emerging method for rainfall monitoring is the use of microwave links. These consist of two antennas, one sending and one receiving unit, typically between a few hundred metres and 15 km apart. Microwave links are used for mobile telephone communication. The information sent over these links travels through the air with any precipitation causing attenuation of the signal. The magnitude of the received signal power can be used to calculate the total integrated attenuation over the link path, from which the path-averaged rainfall intensity can be estimated (Willems and Einfalt, 2021). Research in this field started initially at small scale (e.g., Grum et al., 2005) but has progressed to using data from commercial cellular networks (e.g., Chwala et al., 2012).

These microwave links can be used as a standalone estimator of the rainfall or can be combined with rain gauges and/or radar measurements to give better rainfall estimations at ground level. The network of links is typically denser in urban areas, which could be an advantage for the correction of the radar estimates for urban drainage applications since there are normally only a limited number of rain gauges available in a city centre. Other advantages are that they are mostly clutter-free and very close to the ground compared to radar scans. A disadvantage is the limited availability of commercial microwave link data (Willems and Einfalt, 2021).

4.2.5 Other approaches

Zheng et al. (2023) report on an approach based on processing distributed surveillance camera network imagery. Machine learning algorithms are used to estimate rainfall intensities based on raindrop information extracted from the camera images. This could lead towards achieving real-time, high spatio-temporal resolution measurement of urban rainfall at a relatively low cost.

4.2.6 Data requirements

The appropriate level of detail in rainfall measurement depends on how the data will be used. Three broad categories can be identified.

In *design and planning*, the task is to produce the overall dimensions of the system. Examples include determining the peak flow rate in storm sewers (see Chapter 10) or the total volume of storm detention tanks (Chapters 12 and 13).

In *checking and evaluation*, the performance of the designed system is assessed under extreme or onerous conditions. This usually requires more effort than design and, consequently, more detailed rain data.

The third task, *analysis and operation*, is concerned with the evaluation of systems that already exist, and examples include the analysis of combined sewer overflows (CSO) spill frequencies (Chapter 13), verification of a flow simulation model with real flow data (Chapter 19) or operation of a system in real time (Chapter 22). This latter task has the most stringent requirements for rainfall data.

Table 4.2 lists the rainfall data requirements for each of the three main engineering tasks. The *rainfall record duration* is the length of historical data available for analysis, measured in years. This should be significantly longer than the return period (defined in the next section) of the storm event used in system design. The *gauge location* is ideally within the catchment, but this is less important in design and more important in analysis. *Temporal resolution* is

Table 4.2 Requirements for rainfall data in urban drainage applications

Engineering task	Rainfall record duration (years)	Rain gauge location (relative to catchment)	Temporal resolution (minutes)	Spatial resolution (km²/gauge)	Synchronisation error (minutes)
Design/planning					
Sewers	>10	Near	Block rain	Homogeneous	≤30
CSO volumes	>5	Vicinity	≤15	Homogeneous	≤30
Checking/evaluation					
Sewers	>20	Adjacent	1	Homogeneous	≤10
CSO volumes	>10	Adjacent	5	≤5	≤5
Analysis/operation					
Calibration	Several	Within	2	2	0.25
Verification	Events	Within			0.25
Real-time control	Online	Within	2	2*	0.25

Source: Adapted from Schilling, W. 1991. *Atmospheric Research*, 27, 5–21.
Note: * No less than 3 in total.

the desired time period between rainfall measurements, and *spatial resolution* indicates the desired distance between rain gauges. It is preferable to have several gauges in all catchments to provide data checks and detect spatial variations, including storm movement. Minimising *synchronisation errors* becomes important when multiple gauges are used.

In practice, urban drainage engineers often use single rainfall events (see Section 4.4) and multiple events (Section 4.5) in tandem. Single-design storms are used to analyse flooding (Chapter 12) and surcharge level (Section 7.4.5) performance and multiple event time series are used to analyse more frequent events such as CSO spills (Chapter 13). It is rare to use a time series length which exceeds the maximum design storm return period.

4.3 ANALYSIS

4.3.1 Basics

Rain data measured at an individual rain gauge are most commonly expressed either as *depth* in millimetres or *intensity* in millimetres per hour (mm/hour). This type of *point* rainfall data is therefore representative of one particular location in the catchment. Such data are of greater value if they can be related statistically to two other important rainfall variables: duration and frequency (or probability).

The rainfall *duration* refers to the time period D over which the rainfall falls. However, duration is not necessarily the time period for the whole storm, as any event can be sub-divided and analysed for a range of durations. It is common to represent the *frequency* or *probability* of the rainfall as a *return period*. An annual maximum rainfall event has a return period of T years if it is equalled or exceeded in magnitude once, on average, every T years. Thus, a rainfall event that occurs on average 20 times in 100 years has a probability of being equalled or exceeded of 0.2 and a return period of 5 years. Annual maximum storm events are often used to determine return period because it is assumed that the largest event in 1 year is statistically independent of the largest event in any other year. Alternatively, rainfall probabilities can be specified directly so, for example, a 100-year return period event has an annual exceedance probability (AEP) of 1%.

A partial duration series can also be used which considers all the events over a selected level of rainfall (sometimes called the *Peaks over Threshold* method). This typically generates more data points for analysis but there is a greater chance that the peaks are not statistically independent (Shaw et al., 2010). This method could find application in considering sub-annual return period events.

4.3.2 IDF relationships

4.3.2.1 Definition

The most commonly used tool for the design and planning of drainage systems is the intensity–duration–frequency (IDF) relationship of maximum rainfall amounts. A typical set of IDF curves is given in Figure 4.2 where it can be seen that (for a particular return period) rainfall intensity and duration are inversely related. As the duration increases, the intensity reduces. This confirms the commonsense observation that heavy storms only last a short time, but drizzle can go on for long periods. Also, frequency and intensity are related, as rarer events (greater return periods) tend to have higher intensities (for a given duration).

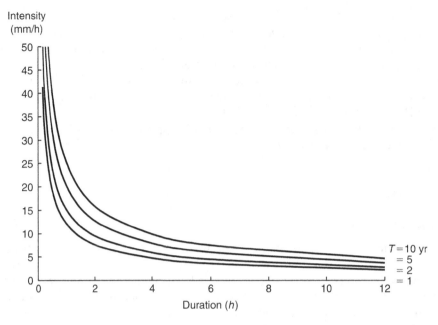

Figure 4.2 Typical intensity–duration–frequency curves.

4.3.2.2 Derivation

Several semi-empirical as well as statistically consistent methods have been developed to construct IDF curves (Chow et al., 1988; Shaw et al., 2010). A generalised IDF formula is given in Equation (4.2) (Koutsoyiannis et al., 1998):

$$i = \frac{a}{\left(D^{\alpha} + b^{\beta}\right)} \tag{4.2}$$

where a, b, α and β are non-negative coefficients with α, $\beta \leq 1$.

Estimation of the parameters is based on the frequency analysis of maximum rainfall depths and comprises the following steps:

- For each duration, the series of annual maximum values is retrieved (i.e., from 1 minute to 48 hours) and is fitted to a suitable statistical distribution (e.g., log-normal, Gumbel, Pareto).
- For each duration, the rainfall intensities are estimated from the fitted distribution, for a set of return periods, T (e.g., 5, 10, 20, 50, 100 years).
- For each return period, the intensities of the previous step are inserted in a numerical procedure to establish a relationship between i and D – that is, to estimate IDF parameters.

The last step results in different values of IDF parameters for each T. Alternative approaches enable the estimation of unique values for the parameters of the denominator, allowing only parameter a to vary with T (see Koutsoyiannis et al., 1998).

EXAMPLE 4.1

Using the data presented in Figure 4.2, determine the intensity of a 1-year return period 2-hour duration rainfall event. For a similar duration event of a 10-year return period, find the appropriate rainfall depth.

Solution

For $T = 1$ year, $D = 2$ hours $\therefore i = 7.5$ mm/hours
For $T = 10$ years, $D = 2$ hours $\therefore i = 16$ mm/hours $\therefore d = 16 \times 2 = 32$ mm

---

4.3.2.3 IDFs in practice

In most situations, it is not necessary to derive such a set of curves, since maps with IDF information are available for many countries. For example, the *NOAA Atlas 14* provides IDF curves as well as other information for rainfall extremes for the US (hdsc.nws.noaa.gov). The *Atlas* is divided into volumes based on geographic sections of the country, while the data are available for operational purposes via the *Precipitation Frequency Data Server* web application. Environment Canada provides short-duration IDF curves for each province and territory across Canada (Environment Canada, 2024).

In the UK, The *Flood Studies Report* (*FSR*; Natural Environment Research Council [NERC], 1975) gave point rainfall depth–duration–frequency (DDF) data for the whole of the UK for durations from 1 minute to 48 hours originally available as a set of paper maps. The *FSR* DDF model was superseded by the *Flood Estimation Handbook* (*FEH*; Institute of Hydrology, 1999) and its later updates, and this is discussed further in Section 4.3.6.

4.3.3 Wallingford Procedure

The *Wallingford Procedure* is a set of documents (DoE/NWC, 1981) that explain the basis for and development of an early sewer modelling package (WaSSP). Although the computer model is no longer in use, the procedure itself is, and still provides a useful reference document and set of maps where *FSR*-based methods and rainfall can be abstracted for any location in the UK and used for urban drainage design and analysis. Aspects of the procedure are returned to in Chapters 5 and 10, but the rainfall estimation approach is explained here.

The *FSR* and the *Wallingford Procedure* both use a standard notation when specifying rainfall information. Thus, *MT-D* represents the depth of rainfall (in millimetres) occurring for duration *D* with a return period of *T* years. Durations specified in minutes start at any minute in the hour, those in hours start "on the hour," and those in days begin at 9 a.m. GMT.

The method is based on working from standard M5–60 minute rainfall and the ratio (*r*) of M5–60 minutes/M5–2 day rainfall depth, both of which are mapped for the UK and given in Figures 4.3 and 4.4, respectively. The M5–60 minute rainfall effectively denotes the quantity of rainfall in an area, and the ratio *r* reflects the "type" of rainfall. Low values of *r* (< 0.2) represent rain that mostly falls as drizzle, whereas values > 0.4 indicate the prevalence of much higher-intensity storms.

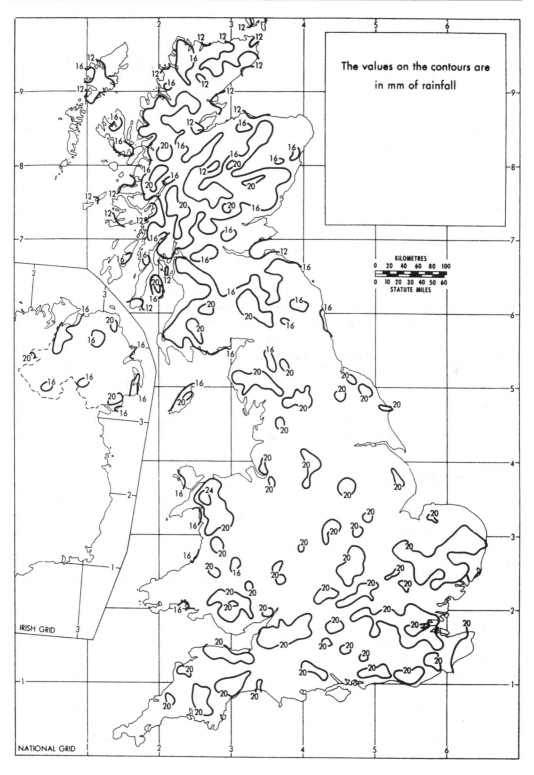

Figure 4.3 Rainfall depths of 5-year return period and 60-minute duration: M5–60 minutes. (Reproduced from "The Wallingford Procedure" with permission of HR Wallingford Ltd.)

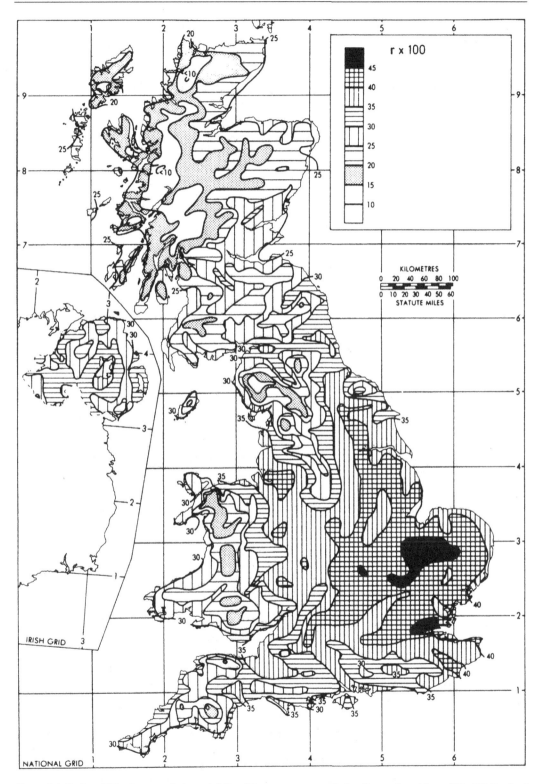

Figure 4.4 Ratio of 60-minute to 2-day rainfalls of 5-year return period: *r*. (Reproduced from "The Wallingford Procedure" with permission of HR Wallingford Ltd.)

By means of coefficients (or *growth factors*) Z_1 (Figure 4.5) and Z_2 (Table 4.3), the standard M5–60 minutes rainfall can be related to

- The 5-year rainfall depth for the required duration (M5–D)
- The rainfall depth for the required duration and return period (MT–D)

Example 4.2 shows how this approach can be used to produce an IDF relationship. IDF relationships of point rainfall are widely used in urban drainage applications. In particular, they are essential in the application of the Rational Method (Chapter 10).

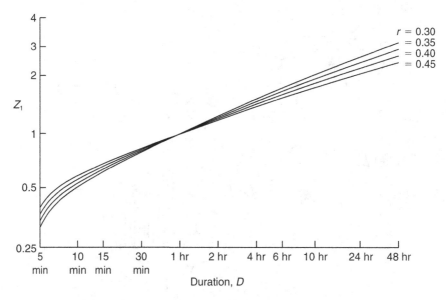

Figure 4.5 Relationship between Z_1 and D for different values of r (0.30 ≤ r ≤ 0.45). (Based on "The Wallingford Procedure" with permission of HR Wallingford Ltd.)

Table 4.3 Ratio Z_2 – relationship between rainfall of return period T(MT) and M5 for Englandand Wales

M5 (mm)	M1	M2	M3	M4	M5	M10	M20	M50
5	0.62	0.79	0.89	0.97	1.02	1.19	1.36	1.56
10	0.61	0.79	0.90	0.97	1.03	1.22	1.41	1.65
15	0.62	0.80	0.90	0.97	1.03	1.24	1.44	1.70
20	0.64	0.81	0.90	0.97	1.03	1.24	1.45	1.73
25	0.66	0.82	0.91	0.97	1.03	1.24	1.44	1.72
30	0.68	0.83	0.91	0.97	1.03	1.22	1.42	1.70
40	0.70	0.84	0.92	0.97	1.02	1.19	1.38	1.64
50	0.72	0.85	0.93	0.98	1.02	1.17	1.34	1.58
75	0.76	0.87	0.93	0.98	1.02	1.14	1.28	1.47
100	0.78	0.88	0.94	0.98	1.02	1.13	1.25	1.40

Source: After DoE/NWC. 1981. *Design and Analysis of Urban Storm Drainage. The Wallingford Procedure. Volume 1: Principles, Methods and Practice*, Department of the Environment, Standing Technical Committee Report No. 28.

EXAMPLE 4.2

Determine the relationship between intensity and duration for 10-year return period storms in the London area.

Solution

Calculate the range of M10–D rainfall intensities.

  Read from Figure 4.3: M5–60 minute = 20 mm
  Read from Figure 4.4: M5–60 minute/M5–2 day, r = 0.45
  Read from Figure 4.5: Z_1 values for various Ds
  Read from Table 4.2: Z_2 values for various Ms

(1) Storm duration (D)		(2) M5–60 minute rainfall total (mm)	(3) Z_1	(4) M5–D rainfall total (mm) (2) × (3)	(5) Z_2	(6) M10–D rainfall total (mm) (4) × (5)	(7) Intensity (mm/hour) (6) ÷ (1)
Hours	Minutes						
	5	20	0.39	7.8	1.21	9.4	112.8
	10	20	0.55	11.0	1.22	13.4	80.4
	15	20	0.63	12.6	1.23	15.5	62.0
	30	20	0.77	15.4	1.24	19.1	38.2
1		20	1.0	20	1.24	24.8	24.8
2		20	1.2	24	1.24	29.8	14.9
4		20	1.4	28	1.22	34.2	8.6
6		20	1.5	30	1.22	36.6	6.1
10		20	1.65	33	1.21	39.9	4.0
24		20	2.0	40	1.19	47.6	2.0
48		20	2.3	46	1.17	53.8	1.1

4.3.4 Annual rainfall

The Standard Annual Average Rainfall (*SAAR*) is the total annual rainfall at any location in the UK averaged over a 30-year period. It was first introduced as a parameter in the *FSR* and later included as a national map in the *Wallingford Procedure*, Volume 4. The most recent data available is for the period 1991–2020 and can be obtained for any UK climate station at: www.metoffice.gov.uk/research/climate/maps-and-data/uk-climate-averages.

4.3.5 Areal extent

Point rainfall is not necessarily representative of rainfall over a larger area, because average rainfall intensity decreases with increasing area. In order to deal with this problem, and avoid overestimating flows from larger catchments, *areal reduction factors* (ARFs) have been developed based on the comparison of point and areal data from areas where several gauges exist.

In the *Wallingford Procedure*, the ARF is calculated from

$$\mathrm{ARF} = 1 - f_1 D^{-f_2} \tag{4.3}$$

where f_1 is $0.0394A^{0.354}$, f_2 is $0.040 - 0.0208 \ln (4.6 - \ln A)$, A is the catchment area (km^2).

EXAMPLE 4.3

Adjust the point rainfall intensity of 25 mm/hour for a 15-minute storm falling over a 200-ha urban catchment.

Solution

D = 0.25 hour, A = 2 km² ∴ Equation (4.3) is valid.

f_1 = 0.0394 × $2^{0.354}$ = 0.050
f_2 = 0.040 − 0.0208 ln (4.6 − ln 2) = 0.012
ARF = 1 − 0.050 × $0.25^{-0.012}$ = 0.95

Areal intensity = 24 mm/hour.

The expression is valid for UK catchment areas < 20 km² and storm durations of 5 minutes to 48 hours (see Example 4.3).

The ARF is lower the larger the catchment area and the shorter event duration. For urban drainage applications, some specifications set a maximum size of catchment for ARF calculation such that ARF will exceed 0.9 for all event durations (CIWEM, 2016). For further information on ARFs in urban drainage applications, including the potential effects of climate change, see UKWIR (2022).

4.3.6 Flood Estimation Handbook

The FEH was originally published by the Institute of Hydrology (1999) as a major upgrade to supersede the *FSR* and its supplementary reports. Included within the *FEH* was a new rainfall DDF statistical model, which has been subject to continuous development since then as follows:

- *FEH99* released via the FEH CD-ROM in 1999
- *FEH13* released via the FEH Web Service in 2015 and based on rain gauge records up to 2005
- *FEH22* released via the FEH Web Service in 2022 and based on rain gauge records up to 2020

The *FEH99* DDF model covered a wide range of durations, from 1 hour up to 8 days, and return periods extending up to 1000 years or longer (Faulkner, 1999). It is valid for the estimation of rainfall-runoff on catchments with an area > 50 ha. It was subsequently revisited after concerns about the apparent high design rainfall depths that are provided for higher return periods (from 100 to 10,000 years) typically used for reservoir flood risk assessment. Its replacement, the *FEH13* DDF model (Stewart et al., 2013), was developed to correct this giving a better estimation of design rainfall over the full range of return periods from 2 to over 10,000 years. In general, the *FEH13* DDF model gives lower short-duration design rainfall values than *FEH99* for most areas of England, in contrast to Wales and Scotland where a considerable increase in the rainfall amounts is apparent. The *FEH22* DDF model maintains the methodology of *FEH13* while adding substantially to the period and spatial density of calibration data. Coverage for shorter durations, down to 1 hour, is particularly increased. Differences between *FEH22* and *FEH13* are greatest for shorter durations, where the quantity of data is most increased over *FEH13*, and for longer return periods, where the differences found at shorter return periods are often magnified. *FEH22*'s largest increases occur at the sites of extreme events that were not captured by the *FEH13* calibration dataset. Further detail is given by (Vesuviano, 2022). DDF point data is available on a pay-as-you-go basis for *FEH13* and *FEH22* via a web service (fehweb.ceh.ac.uk). Further advice on how to access this data and incorporate it into appropriate sewer modelling software is given in CIWEM (2016).

Some urban drainage engineers still use *FSR* rainfall depths, which is typically acceptable to local authority planners. However, the majority now favour using *FEH99* data and are in the process of transitioning to *FEH13*. As yet, there is little experience in using *FEH22* data, but this will probably become the de facto dataset in due course.

4.3.7 Seasonality

It is widely recognised that winter months are generally wetter than summer months in the UK. Rainfall depths generated by the FEH DDF model are based on annual data, with no means of deriving seasonal rainfall depths. This issue has been addressed by Kjeldsen et al. (2005), who generated seasonal correction factors (*SCF*) which can be applied to FEH design rainfall to obtain seasonal rainfall estimates. Table 4.4 shows how these are related to *SAAR* and rainfall duration. The data indicate that summer correction factors reduce as *SAAR* increases, and the opposite with winter correction factors. Some engineers recommend that a value of 1.00 should be used for all summer *SCF*s. The *SCF*s are now built into some software and can be applied automatically by users.

Table 4.4 Seasonal correction factors

SAAR(mm)	Rainfall duration(hr)	Seasonal correction factor (SCF)	
		Summer	Winter
500	1	1.00	0.51
800		0.98	0.60
1000		0.96	0.64
1500		0.92	0.73
500	2	1.00	0.55
800		0.98	0.65
1000		0.96	0.70
1500		0.93	0.79
500	6	1.00	0.62
800		0.98	0.73
1000		0.97	0.78
1500		0.95	0.87
500	24	1.00	0.63
800		0.97	0.75
1000		0.95	0.81
1500		0.90	0.89

Source: Adapted from CIWEM, 2016

4.4 SINGLE EVENTS

So far we have considered rainfall to consist of just a fixed rainfall depth for a given duration. Clearly, this is unrealistic as rainfall intensity varies with time throughout the storm. This is represented as a plot of rainfall intensity against time called a *hyetograph* (or "storm profile").

4.4.1 Synthetic design storms

A design storm is an idealised storm profile to which a statistically based return period has been attached. The defined pattern in time is designed to reproduce (albeit imperfectly) the "shape" of observed storms. The shape depends mainly on the type of event: a frontal storm usually has the highest intensities near the middle, and a convective storm has the highest intensities near the beginning.

The simplest (and least realistic) form of design storm is *block rainfall*, which may be simply derived from an IDF curve. Block rainfall has the same intensity over its duration and therefore has a rectangular time distribution. It is widely used in the Rational Method (given in Chapter 10) and has the advantage of being simple, quick to use, and easily understandable.

In order to facilitate more accurate design solutions, profiles that better represent observed rainfall profiles have been produced. This has become more important with the advent of more sophisticated surface routing methods and flow simulation models. A number of shapes have been proposed over the years, based successively on more comprehensive data sets.

Information on storm profiles can be found in the *FSR* based on the analysis of a wide range of storm events. A family of standard, symmetrical profiles was produced, with maximum rainfall intensity at the centre of the storm and varied in amplitude. The *peakedness* of a profile is defined as the ratio of maximum to mean intensity, and the *percentile peakedness* is the percentage of storms that are equally or less peaked. The profile shape was not found to vary significantly with storm duration, return period, or geographical region. However, on average, summer storms were found to be more peaked than winter ones. More recent evidence shows that storm profiles are approximately evenly distributed across three profile types: front-loaded, centred, and back-loaded. However, the implications of this have yet to be fully established (UKWIR, 2022).

The *Wallingford Procedure* recommends the 50 percentile summer profile (i.e., the storm that is more peaked than 50% of all summer storms) for the design of drainage systems (see Figure 4.6). A rainfall profile can be estimated by distributing the mean intensity over the storm duration as shown in Example 4.4. The *Procedure* also recommends a method of smoothing the point rainfall profile to allow for areal extent.

4.4.2 Historical single events

Historical single events are hyetographs of point rainfall constructed from measured data. Unlike design storms, they are not idealised and do not have an attached return period. Recorded data should be at intervals of 5 minutes, and preferably 1 minute. Their main use is in the verification of flow simulation models with measured hyetographs and simultaneous observations of flow. A model so verified is then assumed to give an accurate picture of catchment response (as considered in Chapter 20).

If the model allows consideration of spatially varying rainfall, the hyetographs used should adequately reflect the patterns of rainfall in time *and* space. The direction and movement (tracking) of a storm can be important in certain catchments (especially large ones) and can

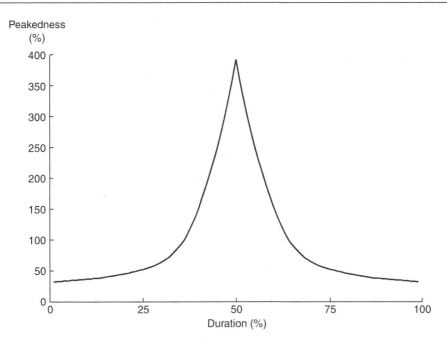

Figure 4.6 FSR 50 percentile summer storm profile.

EXAMPLE 4.4

Determine the intensity of a 50 percentile summer storm of mean intensity 25 mm/hour and duration 15 minutes at its one-third and one-half points.

Solution

   1/3 point = 33.3% duration. From Figure 4.6, % of mean intensity = 80
   i_{33} = 0.8 × 25 = 21 mm/hour @ t = 5 minutes
   1/2 point = 50% duration. From Figure 4.6:
   i_{50} = 3.9 × 25 = 98 mm/hour @ t = 7.5 minutes

---

be a source of error if neglected. Storms moving longitudinally downstream through the catchment have the greatest influence.

4.4.3 Critical input hyetograph (Superstorm)

To fully assess the performance of a sewer system, design storms of different return periods should be applied for a wide range of durations. Then, a *critical duration* can be determined by defining the event that has the greatest impact on the sewer system in terms of specific criteria, for example, risk of failure or cost of damages.

To alleviate the high computational burden imposed by time-consuming simulation models that need to run time and time again to find this critical event, the Critical Input Hyetograph (CIH), or Superstorm, method can be applied. This method substitutes a given set of rainfall

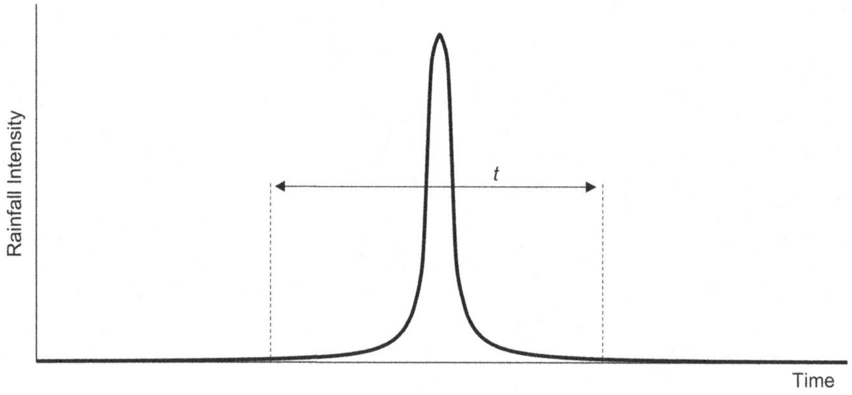

Figure 4.7 Typical critical input hyetograph.

hyetographs, of a specific return period, with a long-duration single synthetic hyetograph that encompasses the critical intensity and/or critical volume. In this case, only one simulation per return period is needed to characterise the behaviour of the sewer system under extreme conditions (CIWEM, 2016).

The CIH is composed so that, for any given duration, the middle section has the same average intensity as the critical, observed or synthetic, hyetograph with the same duration. A typical CIH is given in Figure 4.7, where *t* determines the section of the graph that has the same average intensity as the critical hyetograph with *t* duration, used to generate the CIH.

This method, also known as the Chicago Method, was studied first by Keifer and Chu (1957), while other CIH generators, based either on observed rainfall events or on synthetic *FEH* or *FSR* hyetographs, are also available (e.g., Newton et al., 2013).

One drawback of the CIH method is the computational effort that is required to analyse the observed rainfall series so as to determine the critical rainfall hyetographs of different durations. A further drawback is that the method tends to generate storm events with longer durations than are likely to be the critical durations for some parts of the drainage network.

4.5 MULTIPLE EVENTS

4.5.1 Historical time series

A historical series of rainfall events is the full set of all measured point rainfall for a particular time period (which would include all single historical events and the intervening dry periods) at a particular location. These are used in conjunction with pre-calibrated, continuous simulation models, for long-term catchment studies, evaluation of CSO spill volumes and frequencies, and flood risk assessment. The data can be sourced from specific rain gauges with an audited long-term record, a radar rainfall record, or as gridded data from the Met Office or environmental regulator. The rainfall record length needs to be as long as possible and certainly long enough to take account of the inherent variability of rainfall. This is typically 25 years for a pollution study and at least double that for flood risk assessments. In terms of time step, hourly or shorter data is required for deterministic modelling. Five-minute timestep data has been used, although some CSO and catchment-related studies have successfully used 15-minute timestep data (CIWEM, 2016). A typical daily time series is shown in Figure 4.8.

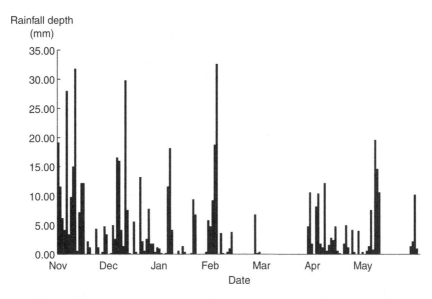

Figure 4.8 Time-series rainfall (6 months of daily data).

A key advantage of time series is that they can be used in the analysis of frequent events such as CSO spills which may be many per annum. Further, they should be expected to contain the conditions that are critical for the catchment being studied. Their main disadvantage is that large amounts of data are required but not always available for a particular site under consideration.

4.5.2 Synthetic time series

4.5.2.1 Stochastic rainfall generation

An alternative approach is the development and use of stochastic rainfall models. Output from these models is a synthetic rainfall time series, which reproduces the stochastic properties of observed data for the catchment. A widely known category of stochastic rainfall models for fine spatio-temporal scales is the "Poisson-cluster" models, introduced by Rodriguez-Iturbe et al. (1987). This type of model attempts to reproduce the characteristics of rainfall at multiple time scales, via a small number of parameters. These models have a plausible physical realism that assumes any rainfall event is triggered by arriving *storm origins* from which a sequence of *rain cells* is generated.

Clustering of the rain cells is accomplished using Bartlett–Lewis or Neyman–Scott models (Onof et al., 2000). Their structure (illustrated in Figure 4.9) is characterised by a Poisson process of cluster (or "storm") arrivals. In the Bartlett–Lewis process, it is assumed that a cell is located at the storm origin. The storm arrival is followed by a "Poisson process" of cell arrivals over a random duration (typically chosen as exponentially distributed, so that the number of cells has a geometric distribution). In the alternative Neyman–Scott process, cells arrive so that the times elapsing from the storm arrival (which is notional and not observed) are independently exponentially distributed, and the number of cells per cluster is randomly distributed (typically Poisson or geometrical). In both cases, the rectangular pulse is assumed to have a random intensity and duration. This defines models with five (if the random intensity is exponentially distributed) or, more generally, six parameters (Wheater et al., 2005).

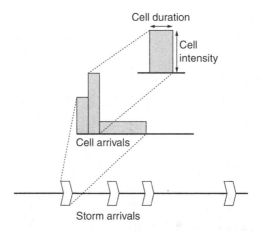

Figure 4.9 Poisson cluster rectangular pulse model schematic. (Adapted from Wheater, H.S., et al. 2005. *Stochastic Environmental Research and Risk Assessment*, 19, 403–416.)

The two basic models have been further modified to enable a better reproduction of different profiles of rainfall and their temporal properties, such as the proportion of dry intervals. Rodriguez-Iturbe et al. (1988) and Entekhabi et al. (1989) extended the Bartlett–Lewis and Neyman–Scott models, respectively, so that the parameter that specifies the duration of cells is randomly varied between storms according to a gamma distribution.

Greater diversity of rainfall types can also be achieved either by allowing different cell types within any storm or by superposing multiple independent processes for different storm types (Cowpertwait, 2004; Cowpertwait et al., 2007). In the former case, Cowpertwait (1994) formulated a two-cell Neyman–Scott model that requires eight parameters for exponentially distributed cell intensities, with the additional three being mean cell intensity and cell duration for the second cell type, and the probability of a cell being of this type. If two parameter distributions are used for the cell intensities, 10 parameters in all are needed. In the latter case, Cowpertwait (2004) superposed two independent Neyman–Scott processes to simulate the convective and stratified rainfall at a site in New Zealand. Despite the considerable flexibility that the above techniques provide in the simulation of different types of rainfall, the large number of parameters poses extra difficulties in the fitting of the models.

The performance of the models regarding the distribution of extremes is of importance for urban drainage applications. Empirical analyses have shown that the use of longer-tailed distribution for cell intensity (i.e., Gamma, Weibull, or Pareto), along with the use of the third- or higher-order moment in the fitting procedure, enables a better reproduction of extreme values, especially for lower time scales and high return periods (Verhoest et al., 2010; Wheater et al., 2005).

The rectangular intensity of the rain cell does not allow the original models to reproduce the high variability of sub-hourly rainfall (e.g., 5 minutes), which is important in the design and evaluation of urban drainage systems. To address this issue two different approaches have been proposed. The first concerns the replacement of the constant rectangular cell intensities by a cluster of instantaneous pulses that occur following a third Poisson process. Cowpertwait et al. (2007) applied this modification to the original Bartlett–Lewis model and Kaczmarska et al. (2014) to the randomised model version. The superposition of the two modified models to provide even greater flexibility has also been examined. The second approach concerns the re-parameterisation of the randomised Bartlett–Lewis model so as to introduce dependence between cell intensity and duration (Kaczmarska et al., 2014).

Specifically, the cell intensity parameter varies between storms, proportionally to the cell duration parameter. Both modifications enabled the reproduction of the main rainfall characteristics from 5-minute scale up to daily scale.

Further to pulse-based models, a different approach entails combining classical linear (auto-regressive) stochastic models (Box et al., 2008) with Nataf's joint distribution model (Nataf, 1962) to represent the peculiarities (e.g., non-Gaussianity, intermittency, dependence and periodicity) of physical and non-physical stochastic processes at fine spatio-temporal scales (Kossieris et al., 2019; Tsoukalas et al., 2019). Examples of the former include rainfall and the latter water demand and wastewater generation at different aggregation scales.

4.5.2.2 Stochastic disaggregation models

Observed data at fine time scales are often limited, and in some cases completely unavailable, in contrast to daily values, which have been routinely collected often for many decades. Taking advantage of the available coarser rainfall records, disaggregation techniques can be employed to produce possible, statistically consistent, realisations of rainfall events at finer time scales. Three "disaggregators" for sub-hourly time scales are "RainSim," "TSRsim," and "HyetosMinute," They are briefly presented below.

The first disaggregator is the RainSim family of models. In its initial version (Cowpertwait, 1994), the model simulated rainfall time series either at a single location or distributed across a region of up to 200 km in diameter. RainSim version 2 was developed to include third-moment properties, important for the modelling of extreme rainfall, providing point simulation with a single rain cell type. Version 3 includes improved model calibration and provides a spatial–temporal modelling capability (Burton et al., 2008).

The second approach, TSRSim, is coupled to a Bartlett–Lewis rectangular pulse model, BaLeReP (UKWIR, 2003). This utilises a multi-scaling random cascade (Lovejoy and Schertzer, 1995) that downscales the coarse-scale rainfall intensity in stages defined by a halving of the timescale. At each stage, the intensity is multiplied by identically distributed and independent random variables to obtain the finer-scale intensities. An update to TSRSim enables the generator to produce more realistic cells whose intensities are typically inversely proportional to their duration while avoiding the generation of unrealistically intense cells (Onof and Wang, 2020).

The third disaggregator, HyetosMinute (Kossieris et al., 2018) combines the use of the Bartlett–Lewis model, as a rainfall generator, along with accurate adjusting procedures (Koutsoyiannis and Onof, 2001) to disaggregate daily rainfall into any sub-hourly timescale, down to 1 minute.

Finally, an open-source software that implements the Nataf-based stochastic schemes, discussed in the previous section, is "anySim" (Tsoukalas et al., 2020), which supports the simulation (and disaggregation) of random variables, stochastic processes and random fields.

4.5.2.3 Comparison of historical and synthetic series

Historical time series are arguably the best option to use when available. In the UK, for example, long-term high-resolution historical rainfall data are available with good spatial coverage. This type of data is particularly useful for historical model verification such that measured rainfall can be run through the model to check events (e.g., CSO spills) in almost real time. It can also be merged to form a multi-profile series containing several rain gauges to pick up spatial variability. The main benefit and application of synthetic time series is in locations without long-term rainfall records.

4.6 CLIMATE CHANGE

4.6.1 Causes

The design of urban drainage systems that are able to last for a long time necessitates an understanding of phenomena and processes that could alter the characteristics of the various input parameters (such as precipitation, temperature, and evaporation) we used to design the systems in the first place. Having gone through the process of accurately reproducing historic rainfall time series to use in the design of future systems, using the methods described above, we must realise that the assumption of stationarity – the idea that natural systems fluctuate within an unchanging envelope of variability – is no longer valid (Milly et al., 2008). This is where it becomes important for the drainage engineer to understand and assess the impact of global climate change on the various input design variables of the drainage system.

Key to our increase in knowledge and understanding of global change is the Intergovernmental Panel on Climate Change (IPCC), set up in 1988. The IPCC prepares comprehensive assessment reports about climate change, its causes, potential impacts and response options. The first was published in 1990 and the most recent (sixth) in 2021–23. The first UK government scenarios and projections were published in 1991 informed by the IPCC work. The most recent are the UK Climate Projections 2018 (UKCP18).

The evidence for global climate change is compelling with each of the last four decades having been successively warmer than any decade that preceded it since 1850 (Figure 4.10). According to the sixth assessment (IPCC, 2021), the global surface temperature in the first

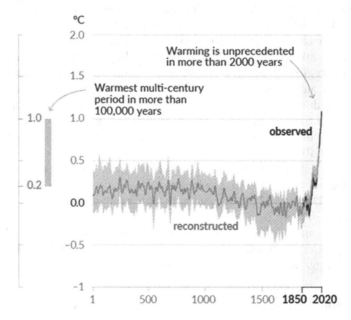

Figure 4.10 Change in global surface temperature (decadal average) as reconstructed (1–2000) and observed (1850–2020) (Reproduced with permission from Figure SPM.1 in IPCC, 2021: Summary for Policymakers. In: Climate Change 2021: The Physical Science Basis. Contribution of Working Group I to the Sixth Assessment Report of the Intergovernmental Panel on Climate Change [Masson-Delmotte, V., P. Zhai, A. Pirani, S.L. Connors, C. Péan, S. Berger, N. Caud, Y. Chen, L. Goldfarb, M.I. Gomis, M. Huang, K. Leitzell, E. Lonnoy, J.B.R. Matthews, T.K. Maycock, T. Waterfield, O. Yelekçi, R. Yu, and B. Zhou (eds.)]. Cambridge University Press, Cambridge, UK and New York, NY, USA, pp. 3–32, doi: 10.1017/9781009157896.001)

two decades of the twenty-first century (2001–2020) was 0.99°C higher than in 1850–1900. In the UK,

- all of the top ten warmest years since 1884 have occurred this century
- the most recent decade (2012–2021) has been on average 1.0°C warmer than 1961–1990
- 2022 was the hottest year on record with a mean temperature of 10.0°C

The main clue to the cause of these rises is the significant increase in the concentration of greenhouse gases (GHGs) observed over the past 200 years, attributed to the earth's climate system variability and emissions from human activities (mostly through the combustion of fossil fuel and industrial activities). In 2009, the anthropogenic GHG emissions (e.g., carbon dioxide, methane, nitrous oxide) were around 59 $GtCO_2$/year, increasing by a rate of 1.3% per year since 2010 (IPCC, 2022). This intensifies the natural greenhouse effect, trapping more energy in the lower atmosphere. Other human activities generate pollutants that actually cool the climate (sulphur dioxide that transforms into aerosols).

Unravelling the impact of various forces in the climate system and making predictions for the future require the combined use of different types of climate models, such as the Atmosphere–Ocean General Circulation Models, Earth System Models, Earth System Models of Intermediate Complexity, and Regional Complexity models (see Flato et al., 2013). A great variety of such models has been developed from leading climate centres, such as the GCM HadGEM3-GC3 and the ESM UKESM1 developed in the UK. The Coupled Model Intercomparison Project (CMIP) collates and compares the output from different models to provide multi-model ensemble projections, used in the assessments of the IPCC (Eyring, 2016).

CMIP6 projections show that only by incorporating natural *and* human factors can the model fit the data adequately, especially the warming since the 1970s (see Figure 4.11). This and other evidence have led the IPCC to conclude that "observed increases in well-mixed GHG concentrations since around 1750 are unequivocally caused by human activities" and that "it is very likely that well-mixed GHGs were the main driver of tropospheric warming since 1979" (IPCC, 2021).

4.6.2 Future trends

To obtain climate change projections, under uncertain driving forces of these changes, different scenarios have been developed and examined by the IPCC. The scenarios used in the *fifth* assessment are called Representative Concentration Pathways (RCPs) which describe different climate futures, all of which are considered possible depending on the volume of GHGs emitted in future years. The four pathways used - RCP2.6, RCP4.5, RCP6, and RCP8.5 - are based on a range of possible radiative forcing values in the year 2100 (IPCC, 2014). In the *sixth* assessment, five illustrative pathways are specified. They include scenarios with

- high and very high GHG emissions (SSP3-7.0 and SSP5-8.5) and CO_2 emissions that double from current levels by 2100 and 2050, respectively,
- intermediate GHG emissions (SSP2-4.5) and CO_2 emissions remaining at current levels until the middle of the century,
- very low and low GHG emissions and CO_2 emissions declining to net zero around or after 2050, followed by varying levels of net negative CO_2 emissions (SSP1-1.9 and SSP1-2.6)

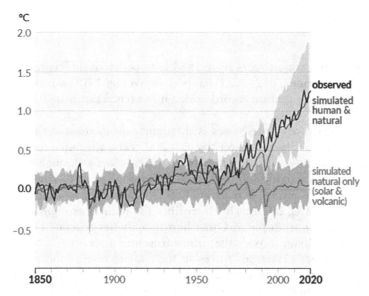

Figure 4.11 Change in global surface temperature (annual average) as observed and simulated using human & natural and only natural factors (both 1850–2020). (Reproduced with permission from Figure SPM.1 in IPCC, 2021: Summary for Policymakers. In: Climate Change 2021: The Physical Science Basis. Contribution of Working Group I to the Sixth Assessment Report of the Intergovernmental Panel on Climate Change [Masson-Delmotte, V., P. Zhai, A. Pirani, S.L. Connors, C. Péan, S. Berger, N. Caud, Y. Chen, L. Goldfarb, M.I. Gomis, M. Huang, K. Leitzell, E. Lonnoy, J.B.R. Matthews, T.K. Maycock, T. Waterfield, O. Yelekçi, R. Yu, and B. Zhou (eds.)]. Cambridge University Press, Cambridge, UK and New York, NY, USA, pp. 3–32, doi: 10.1017/9781009157896.001)

The pathway notation SSPx-y refers to the *shared socio-economic pathway* (SSP) x and the level of radiative forcing y (IPCC, 2021). The concept of socio-economic scenarios and pathways is discussed further in Chapter 24.

As part of UKCP18, two 5th IPCC RCPs – 8.5 and 2.6 – have been used as a basis for derived climate projections for the UK (Gohar et al., 2018). RCP8.5 represents a minimal mitigation policy with rising emissions, increasing global wealth and population. RCP2.6 assumes a strong global mitigation policy with falling emissions. For the RCP2.6 pathway global model projections, 15 realisations of the HadGEM3-GC3.05 global model were run from 1901 to 2100 plus 13 realisations of selected CMIP5 coupled ocean–atmosphere models. For the RCP2.6 pathway, an approximation approach was used based on the HadGEM3-GC3.05 RCP8.5 simulations and the CMIP5 simulations of RCP2.6. The derived data consists of UK-wide 60 km resolution time series for an ensemble of RCP2.6 as well as 50-year time slices, characteristic of the climate when global mean temperatures rise to either 2°C or 4°C above pre-industrial levels. Time series include internal variability and are available at monthly frequency for precipitation, temperature, relative humidity, solar radiation, surface wind, and daily frequency for temperature and precipitation.

In the next 50 years, all areas of the UK are expected to become warmer. At 2°C of global mean warming, the South East will see the largest warming with summer temperatures increasing 3–4°C relative to the present day. Warming of at least 1–2°C is expected across the rest of the country. Percentage changes in precipitation vary seasonally with wetter winters and drier summers. Precipitation increases are small in winter and broadly similar across the country. In summer, precipitation decreases toward the south. At 4°C of global warming, all seasons see increased temperatures, with up to 4–5°C in the south of England in the summer. Median summer precipitation decreases most in the south compared to current levels, with a median reduction of up to 20–30% across much of England and Wales. Winter precipitation increases by up to 20% across most of the country (Gohar et al., 2018).

Also as part of UKCP18, 12 much higher resolution (2.2 km spatial, hourly temporal) Local Projections were produced across the UK (Kendon et al., 2019), each representing a plausible future climate up to 2080 based on RCP8.5 (i.e., no curbs on GHG emissions). Due to the high resolution used, convection can be represented explicitly in the simulation and so these convection-permitting models (CPMs) are able to forecast localised extreme rainfall events. Results for the UK show there is a greater chance of warmer wetter winters and hotter drier summers in future, reinforcing the headline findings from coarser-resolution climate models. For summer rainfall, the average change predictions are broadly consistent with earlier regional model results, although underlying changes in rainfall on a day-to-day and an hour-to-hour basis are different. In particular, the CPM shows a greater shift to more intense rainfall in summer. Contrary to previous studies however, CPM results also show greater future increases in winter mean precipitation (Kendon et al., 2019).

In 2021, the Local Projections were updated to provide an additional ensemble of 12 simulations at 2.2 km resolution for three epochs or time periods (1981–2000, 2021–2040, and 2061–2080). The new 2.2 km reruns further reinforce the UKCP18 headline message of a greater chance of warmer wetter winters and hotter drier summers across the UK in future. These new projections are the currently recommended dataset for the UK (Kendon, et al., 2021).

Research has shown that in the UK not only has there been a rising trend in the yearly number of extreme rainfall events (20mm/hr or greater) over the last 40 years (at least), but that under RCP8.5 (minimal mitigation) these downpours could be four times or more as frequent by 2080 compared to the 1980s (Kendon et al., 2023).

4.6.3 Design rainfall under climate change

Regarding urban drainage applications, the impact of climate change on the intensity and frequency of extreme rainfall events is of great importance since an increase in these characteristics will result in more frequent and severe sewer overflows and flood events. The basic approach taken is to use the most up-to-date climate projections to represent predicted future changes via an "uplift" or "allowance" to existing rainfall data, individual design storms, or rainfall time series.

4.6.3.1 Design storms

BS EN 752: 2017 advises that historically derived rainfall rates should be multiplied by a Climate Change Allowance (CCA) to take account of climate change. A CCA of 1.4 should be used for the design of new drain and sewer systems (where the anticipated design life extends beyond 2070), and this should be considered to be the default value unless more detailed or more recent information is available.

The Future Drainage project (www.ukclimateresilience.org/projects/future-drainage-ensemble-climate-change-rainfall-estimates-for-sustainable-drainage) utilises the updated UKCP18 Local Projections (described in the previous section) to generate uplifts for design rainfall events of various durations (1, 3, 6, 12, and 24 hours) and return periods (2, 30, and 100 years) at two future dates (2050 and 2070) based on RCP8.5 and compared to the baseline of 1990.

An underlying statistical model (Youngman, 2018) was applied to each of the 12 UKCP local ensemble members individually and uplifts plus their uncertainties were combined using a method developed by Fosser et al. (2020). Two values of each uplift were calculated by estimating percentiles from the distribution obtained – central (50%) and high (95%). Values were rounded to the nearest 5%. The resulting uplifts are available for each 5 km grid point across the UK and are appropriate for use with *FEH13 DDF* data (Dale et al., 2021). Following free registration, the uplift values in the form of GIS shapefiles and CSV files can be obtained from: data.ceda.ac.uk/badc/ukcp18/data/land-cpm/derived/future-extremes.

Figure 4.12 shows 1-hour, 30-year uplift values for Scotland for both the central and high estimates. Uplifts range from 15% to 60%. The central estimate is the middle value across the ensemble but given the ensemble's small size, this cannot be associated with a given probability. The high estimate represents a "reasonable worst case" value. Drainage engineers need to be aware that there can be significant differences between the central and high estimates at any given location indicating considerable uncertainty in the uplift estimates.

When considering the upgrading of existing systems, BS EN752: 2017 recommends that designs be tested for increased peak rainfall intensity by using CCA values based on different design horizons for the central and upper (high) end climate projections. Peak rainfall allowances (based on the Future Drainage research) have been mapped for England by management catchment and are available here: https://environment.data.gov.uk/hydrology/climate-change-allowances/rainfall.

These allowances are appropriate for drainage design schemes and for stormwater flood mapping in small (less than 5 km²) catchments (EA, 2022).

4.6.3.2 Rainfall time series

UKWIR (2022) reports on the upgrade to their RED-UP software (to version 3) which provides rainfall "perturbation factors" for application on climate change-influenced rainfall time series. In this work, three separate perturbations are undertaken: number of dry days, annual rainfall, and sub-daily rainfall (hourly intensity). These are derived from UKCP18 Local ensemble data on the 2.2 km grid with standard values prepared for 13 regions across the UK at three future dates (2030, 2050, and 2070). Further details on the tool and its application in assessing flooding and CSO spill volumes are given in UKWIR (2022).

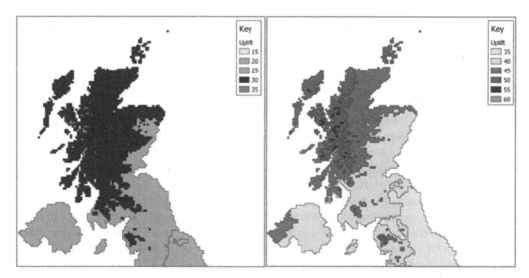

Figure 4.12 1-hour, 30-year uplifts (%) for Scotland for the central estimate (left) and high estimate (right) (Reproduced from Dale et al., 2021. Guidance for Water and Sewerage Companies and Flood Risk Management Authorities: Recommended Uplifts for Applying to Design Storms, courtesy of Murray Dale, JBA Consulting)

4.6.4 Implications

There are many studies, based either on global or regional models, that support the hypothesis of an increase in extreme precipitation events in a future climate over many regions (see Mailhot and Duchesne, 2010). The main implications for urban drainage systems are as follows:

- Increased volume and flow rate that may exceed the capacity of existing sewer systems leading to more frequent sewer surcharging, surface flooding, and property damage
- More frequent CSO spills
- Greater deterioration of sewers due to more frequent surcharging
- Greater buildup and mobilisation of surface pollutants in summer
- Poorer water quality in rivers due to extra stormwater outfall and CSO spills and reduced baseflows in summer

Increased temperature is also associated with more rapid chemical and biological reactions. This may lead to a greater number of sulphide-related problems (e.g., corrosion, odour – see Chapter 18) but may actually reduce wastewater organic concentrations due to increased degradation rates. The consequences of these effects on society and the environment are summarised in Table 4.5.

4.6.5 Solutions

A wide range of possible solutions is available to the urban drainage engineer to address the significant implications described in the previous section. The cost-effectiveness of each approach will vary and depends on the specific case under consideration. It is likely that a combination of measures will be required. The list below gives some suggested solutions and cross-refers to the chapter where they are covered in more detail.

- Increased belowground sewer capacity and/or storage (Chapter 14)
- Increased application of SuDS: infiltration devices, above-ground storage (Chapter 11)
- More widespread capture and reuse of rainwater (Chapter 24)
- Planned exceedance flow routing (major–minor systems) (Chapter 12)
- Enhanced flood protection of buildings (Chapter 12)
- Increased application of real-time control (Chapter 22)

Table 4.5 Societal and environmental consequences of urban drainage systems affected by climate change

Consequence	Impact	Severity
Health	Stress and psychological impacts caused by property flooding due to wastewater backup	High
	Contamination of bathing and shellfish waters caused by CSO spills	High
	Septicity and hydrogen sulphide impact on sewer operatives	Low
Receiving water quality	Reduced assimilative capacity of waterbodies during summer to cope with CSO spills	High
	Increased pollutant levels and lower dissolved oxygen concentrations in waterbodies	High

Source: Adapted from Campos and Darch (2015)

PROBLEMS

1 Describe the main types of rain gauges and assess their relative merits in urban drainage applications.
2 Calculate the average rainfall intensity for 5-year and 50-year return period storms of 30minutes duration using the generalised IDF formula with parameters $v = 1$, $\theta = 0.180$, $\eta = 0.90$, and $\omega(T) = 22.184\ T^{0.31}$, where T is the return period in years. [51.7, 105.5 mm/hour]
3 Derive a series of IDF curves (1, 2, and 5 years) for a location in Exeter.
4 Construct the 50 percentile rainfall intensity profile for a 20 mm/hour, 20-minute duration summer storm.
5 What is a synthetic design storm, and how can it be represented? What are the differences from a historical event?
6 What is the critical input hyetograph, and what are its main advantages?
7 Explain what you understand by a historical series of rainfall. How would you use one in practice?
8 What are the benefits of using synthetic rainfall time series?
9 Describe how stochastic rainfall generators work. What is the main advantage of Poisson-cluster models compared with simple Poisson models?
10 Describe the modifications that allow Poisson-cluster models to capture the high variability of sub-hourly rainfall.
11 Explain the main purpose of disaggregation techniques.
12 How can the impact of climate change be accounted for in design rainfall?
13 What are the main impacts of climate change, and how might urban drainage systems be adapted to take account of them?

KEY SOURCES

Arnbjerg-Nielsen, K., Willems, P., Olsson, J., Beecham, S., Pathirana, A., Bülow Gregersen, I., Madsen, H. and Nguyen, V.-T.-V. 2013. Impacts of climate change on rainfall extremes and urban drainage systems: a review. *Water Science & Technology*, 68.1, 16–28.

CIWEM 2016. *Rainfall Modelling Guide 2016*. The Chartered Institution of Water and Environmental Management, London.

DoE/NWC. 1981. *Design and Analysis of Urban Storm Drainage. The Wallingford Procedure. Volume 1: Principles, Methods and Practice*, Department of the Environment, Standing Technical Committee Report No. 28.

Onof, C., Chandler, R.E., Kakou, A., Northrop, P., Wheater, H.S., and Isham, V.S. 2000. Rainfall modelling using Poisson-cluster processes: A review of developments. *Stochastic Environmental Research and Risk Assessment*, 14, 384–411.

Shaw, E.M., Beven, K.J., Chappell, N.A. and Lamb, R. 2010. *Hydrology in Practice*, 4th edn, CRC Press, Boca Raton, FL.

Willems, P. and Einfalt, T. 2021. Sensors for Rain Measurements. Chapter 2 *in Metrology in Urban Drainage and Stormwater Management: Plug and Pray*, Jean-Luc Bertrand-Krajewski, Francois Clemens-Meyer, Mathieu Lepot (Eds.). IWA Publishing.

Willems, P., Olsson, J., Arnbjerg-Nielsen, K., Beecham, S., Pathirana, A., Bülow Gregersen, I., Madsen, H. and Nguyen, V.-T.-V. 2012 (Eds.). *Impacts of Climate Change on Rainfall Extremes and Urban Drainage Systems*. IWA Publishing.

REFERENCES

Bajracharya, S.R., Shrestha, M.S., and Shrestha, A.B. 2017. Assessment of high-resolution satellite rainfall estimation products in a streamflow model for flood prediction in the Bagmati basin, Nepal. *Journal of Flood Risk Management, 10(1)*, 5–16.

Beck, H.E., Vergopolan, N., Pan, M., Levizzani, V., van Dijk, A.I.J.M., Weedon, G.P., Brocca, L., Pappenberger, F., Huffman, G.J. and Wood, E.F. 2017. Global-scale evaluation of 22 precipitation datasets using gauge observations and hydrological modeling. *Hydrology and Earth System Sciences, 21*, 6201–6217.

Box, G.E.P., Jenkins, G.M. and Reinsel, G.C. 2008. *Time Series Analysis*. John Wiley & Sons, Inc.

Bringi, V.N. and Chandrasekar, V. 2001. *Polarimetric Doppler Weather Radar. Principles and Applications*. Cambridge University Press, New York.

Bringi, V.N., Rico-Ramirez, M.A., and Thurai, M. 2011. Rainfall Estimation with an Operational Polarimetric C-Band Radar in the United Kingdom: Comparison with a Gauge Network and Error Analysis. *Journal of Hydrometeorology, 12*, 935–954.

BS EN 752: 2017 *Drain and sewer systems outside buildings – Sewer system management*.

Burton, A., Kilsby, C.G., Fowler, H.G., Cowpertwait, P.S.P., and O'Connell, P.E. 2008. RainSim: A spatial–temporal stochastic rainfall modelling system. *Environmental Modelling and Software, 23*, 1356–1369.

Campos, L.C. and Darch, G. 2015. Adaptation of UK wastewater infrastructure to climate change. *Infrastructure Asset Management, 2(3)*, 97–106.

Chwala C., Gmeiner A., Qiu W., Hipp S., Nienaber D., Siart U., Eibert T., Pohl M., Seltmann J., Fritz J. and Kunstmann H. 2012. Precipitation observation using microwave backhaul links in the alpine and pre-alpine region of southern Germany. *Hydrology and Earth System Sciences Discussions, 9(1)*, 741–776.

Chiang, Y.-M., Hsu, K.-L., Chang, F.-J., Hong, Y., and Sorooshian, S. 2007. Merging multiple precipitation sources for flash flood forecasting. *Journal of Hydrology, 340(3–4)*, 183–196.

Chow, V.T., Maidment, D.R., and Mays, L.W. 1988. *Applied Hydrology*, McGraw-Hill, New York.

Collier, C.G. 1996. *Applications of Weather Radar Systems: A Guide to Uses of Radar Data in Meteorology and Hydrology*, 2nd edn, Praxis Publishers, John Wiley and Sons, Chichester/London.

Cowpertwait, P., Isham, V., and Onof, C. 2007. Point process models of rainfall: Developments for fine-scale structure. Proceedings of the *Royal Society: Mathematical Physical and Engineering Sciences, 463*, 2569–2587.

Cowpertwait, P.S.P. 1994. A generalized point process model for rainfall. *Proceedings of the Royal Society: Mathematical Physical and Engineering Sciences, A447*, 23–37.

Cowpertwait, P.S.P. 2004. Mixed rectangular pulses models of rainfall. *Hydrology and Earth System Sciences, 8*, 993–1000.

Dale, M. and Stidson, R. 2007. Weather radar for predicting beach bathing water quality. *WaPUG Autumn Conference*, Blackpool.

Dale, M., Fowler, H., Kendon, E. and Chan, S. 2021. *Guidance for Water and Sewerage Companies and Flood Risk Management Authorities: Recommended Uplifts for Applying to Design Storms*. Future Drainage, https://artefacts.ceda.ac.uk/badc_datadocs/future-drainage/FUTURE_DRAINAGE_Guidance_for_applying_rainfall_uplifts.pdf

EA, 2022 *Flood risk assessments: climate change allowances. Guidance*. Environment Agency, www.gov.uk/guidance/flood-risk-assessments-climate-change-allowances

Ebert, E.E., Janowiak, J.E., Kidd, C. 2007. Comparison of near-real-time precipitation estimates from satellite observations and numerical models. *Bulletin of the American Meteorological Society, 88*, 47–64.

Einfalt, T., Arnbjerg-Nielsen, K., and Spies, S. 2002. An enquiry into rainfall data measurement and processing for model use in urban hydrology. *Water Science and Technology, 45(2)*, 147–152.

Entekhabi, D., Rodriguez-Iturbe, I., and Eagleson, P.S. 1989. Probabilistic representation of the tempo-ral rainfall process by a modified Neyman-Scott Rectangular Pulses Model: Parameter estimation and validation. *Water Resources Research*, 25, 295–302.

Environment Canada. 2024. *Engineering Climate Datasets*. www.climate.weather.gc.ca/prods_servs/engineering_e.html

Eyring, V., Bony, S., Meehl, G. A., Senior, C. A., Stevens, B., Stouffer, R. J., and Taylor, K. E. 2016. Overview of the Coupled Model Intercomparison Project Phase 6 (CMIP6) experimental design and organization, *Geoscientific Model Development*, 9, 1937–1958.

Fan, F.M., Collischonn, W., Quiroz, K.J., Sorribas, M.V., Buarque, D.C. and Siqueira, V.A. 2016. Flood forecasting on the Tocantins River using ensemble rainfall forecasts and real-time satellite rainfall estimates. *Journal of Flood Risk Management*, 9, 278–288.

Faulkner, D. 1999. *Rainfall Frequency Estimation. Volume 2 of the Flood Estimation Handbook*, Institute of Hydrology, Wallingford.

Flato, G., Marotzke, J., Abiodun, B., Braconnot, P., Chou, S.C., Collins, W. 2013. *Evaluation of Climate Models. Climate Change 2013: The Physical Science Basis*. Contribution of Working Group I to the Fifth Assessment Report of the Intergovernmental Panel on Climate Change.

Fosser, G., Kendon, E.J., Stephenson, D. and Tucker, S. 2020. Convection-permitting models offer promise of more certain extreme rainfall projections, *Geophysical Research Letters* 47, e2020GL088151.

Gerstner, E.M. and Heinemann, G. 2008. Real-time areal precipitation determination from radar by means of statistical objective analysis. *Journal of Hydrology*, 352, 296–308.

Gohar, L., Bernie, D., Good, P. and Lowe, J.A. 2018. *UKCP18 Derived Projections of Future Climate over the UK*, Met Office, UK.

Golz, C., Einfalt, T., Gabella, M., and Germann, U. 2005. Quality control algorithms for rainfall mea-surements. *Atmospheric Research*, 77, 247–255.

Goudenhoofdt, E. and Delobbe, L. 2013. Statistical characteristics of convective storms in Belgium derived from volumetric weather radar observations. *Journal of Applied Meteorology and Climatology*, 52, 918–934.

Grum M., Krämer S., Verworn H.-R. and Redder A. 2005. Combined use of point rain gauges, radar, microwave link and level measurements in urban hydrological modelling. *Atmospheric Research*, 77(1–4), 313–321.

Harrison, D.L., Driscoll, S.J., and Kitchen, M. 2000. Improving precipitation estimates from weather radar using quality control and correction techniques. *Meteorological Applications*, 7, 135–144.

Institute of Hydrology. 1999. *Flood Estimation Handbook*, 5 volumes, Institute of Hydrology, Wallingford.

Intergovernmental Panel on Climate Change (IPCC). 2014. Climate Change 2014 Synthesis Report Summary Chapter for Policymakers. www.ipcc.ch/report/ar5/syr/

Intergovernmental Panel on Climate Change (IPCC). 2021. Summary for Policymakers. In: *Climate Change 2021: The Physical Science Basis. Contribution of Working Group I to the Sixth Assessment Report of the Intergovernmental Panel on Climate Change* [Masson-Delmotte, V., P. Zhai, A. Pirani, S.L. Connors, C. Péan, S. Berger, N. Caud, Y. Chen, L. Goldfarb, M.I. Gomis, M. Huang, K. Leitzell, E. Lonnoy, J.B.R. Matthews, T.K. Maycock, T. Waterfield, O. Yelekçi, R. Yu, and B. Zhou (eds.)]. Cambridge University Press.

Intergovernmental Panel on Climate Change (IPCC). 2022. Summary for Policymakers. In: *Climate Change 2022: Mitigation of Climate Change. Contribution of Working Group III to the Sixth Assessment Report of the Intergovernmental Panel on Climate Change* [P.R. Shukla, J. Skea, R. Slade, A. Al Khourdajie, R. van Diemen, D. McCollum, M. Pathak, S. Some, P. Vyas, R. Fradera, M. Belkacemi, A. Hasija, G. Lisboa, S. Luz, J. Malley, (eds.)]. Cambridge University Press.

Kaczmarska, J., Isham, V., and Onof, C. 2014. Point process models for fine-resolution rainfall. *Hydrological Sciences Journal*, 59, 1972–1991.

Keifer, C.J. and Chu, H.H. 1957. Synthetic storm pattern for drainage design. *ASCE Journal of the Hydraulics Division*, 83, 1–25.

Kendon, E.J., Fischer, E.M. and Short, C.J. 2023. Variability conceals emerging trend in 100yr projec-tions of UK local hourly rainfall extremes, *Nature Communications*, 14, 1133.

Kendon, E.J. Fosser, G., Murphy, J., Chan, S., Clark, R., Harris, G., Lock, A., Lowe, J., Martin, G., Pirret, J., Roberts, N., Sanderson, M and Tucker, S. 2019. *UKCP convection-permitting model projections: Science report*. Met Office Tech. Report.

Kendon, E. J., Short, C. Pope, J., Chan, S., Wilkinson, J., Tucker, S., Bett, P., Harris, G. 2021. *Update to UKCP Local (2.2km) projections*, Met Office, UK

Kjeldsen, T.R., Stewart, E.J. Packman, J.C., Folwell, S.S. and Bayliss, A.C. 2005. *Revitalisation of the FSR/FEH Rainfall-Runoff Method*, Defra, R&D Technical Report FD1913/TR.

Kossieris, P., Makropoulos, C., Onof, C. and Koutsoyiannis, D. 2018. A rainfall disaggregation scheme for sub-hourly time scales: Coupling a Bartlett-Lewis based model with adjusting procedures. *Journal of Hydrology*, 556, 980–992.

Kossieris, P., Tsoukalas, I., Makropoulos, C.and Savic, D. 2019. Simulating marginal and dependence behavior of water demand processes at any fine time Scale. *Water*, 11, 885.

Koutsoyiannis, D., Kozonis, D. and Manetas, A. 1998. A mathematical framework for studying rainfall intensity-duration-frequency relationships. *Journal of Hydrology*, 206, 118–135.

Koutsoyiannis, D. and Onof, C. 2001. Rainfall disaggregation using adjusting procedures on a Poisson cluster model. *Journal of Hydrology*, 246, 109–122.

Krajewski, W.F., Ciach, G.J., McCollum, J.R., and Bacotiu, C. 2000. Initial validation of the Global Precipitation Climatology Project monthly rainfall over the United States. *Journal of Applied Meteorology*, 39, 1071–1086.

Kumar, A., Ramsankaran, R., Brocca, L. and Munoz-Arriola, F. 2019. A Machine Learning Approach for Improving Near-Real-Time Satellite-Based Rainfall Estimates by Integrating Soil Moisture. *Remote Sensing*, 11, 2221

Lovejoy, S. and Schertzer, D. 1995. *Multifractals and Rain. New Uncertainty Concepts in Hydrology and Hydrological Modelling*. Cambridge University Press, Cambridge, 62–103.

Mailhot, A. and Duchesne, S. 2010. Design criteria of urban drainage infrastructures under climate change. *Journal of Water Resources Planning and Management.*, 136, 201–208.

Massari, C., Brocca, L., Pellarin, T., Abramowitz, G., Filippucci, P., Ciabatta, L., Maggioni, V., Kerr, Y. and Fernandez Prieto, D. 2020. A daily 25 km short-latency rainfall product for data-scarce regions based on the integration of the Global Precipitation Measurement mission rainfall and multiple-satellite soil moisture products. *Hydrology and Earth System Sciences*, 24, 2687–2710.

Meteorological Office. 2013. *National Meteorological Library and Archive Fact sheet 15—Weather radar*.

Milly, P. C. D., Betancourt, J., Falkenmark, M., Hirsch, R.M., Kundzewicz, Z.W., Lettenmaier, D.P. and Stouffer, R.J. 2008. Stationarity is dead: Whither water management? *Science*, 319, 573–574.

Nataf, A. 1962. Statistique mathematique-determination des distributions de probabilites dont les marges sont donnees. *Comptes Rendus l'Academie des Sci.*, 255, 42–43.

Natural Environment Research Council (NERC). 1975. *Flood Studies Report*, 5 vol., Institute of Hydrology, Wallingford.

Neale, W. 2008. Practical use of radar rainfall data at Thames Water. *WaPUG Autumn Conference*, Blackpool.

Newton, C., Jarman, D., Memon, F.A., Andoh, R. and Butler, D. 2013. Implementation and assessment of a critical input hyetograph generation for use in a decision support tool for the design of flood attenuation systems. *International Conference on Flood Resilience: Experiences in Asia and Europe, Exeter*, September.

Nielsen, J.E., Thorndahl, S., and Rasmussen, M.R. 2014. A numerical method to generate high temporal resolution precipitation time series by combining weather radar measurements with a nowcast model. *Atmospheric Research*, 138, 1–12.

Onof, C. and Wang, L.-P. 2020. Modelling rainfall with a Bartlett–Lewis process: new developments, *Hydrology and Earth System Science*, 24, 2791–2815,

Rodriguez-Iturbe, I., Cox, D.R., and Isham, V.S. 1987. Some models for rainfall based on stochastic point processes. *Proceedings of the Royal Society: Mathematical Physical and Engineering Sciences*, A410, 269–288.

Rodriguez-Iturbe, I., Cox, D.R., and Isham, V.S. 1988. A point process model for rainfall: Further developments. *Proceedings of the Royal Society: Mathematical Physical and Engineering Sciences*, 417, 283–298.

Rosenfeld, D. and Collier, C.G. 1998. Estimating surface precipitation. Chapter 14.1, in *Global Energy and Water Cycles* (eds. K.A. Browning and R. Gurney), Cambridge University Press, 124–133.

Schilling, W. 1991. Rainfall data for urban hydrology: What do we need? *Atmospheric Research*, 27, 5–21.

Seo, D.-J., Breidenbach, J., and Johnson, E. 1999. Real-time estimation of mean field bias in radar rainfall data. *Journal of Hydrology*, *223*, 131–147.

Sinclair, S. and Pegram, G. 2005. Combining radar and rain gauge rainfall estimates using conditional merging. *Atmospheric Science Letters*, *6*, 19–22.

Smith, J.A., Baeck, M.L., Meierdiercks, K.L., Miller, A.J. and Krajewski, W.F. 2007. Radar rainfall estimation for flash flood forecasting in small urban watersheds. *Advances in Water Resources*, *30*, 2087–2097.

Smith, J.A., Hui, E., Steiner, M., Baeck, M.L., Krajewski, W.F. and Ntelekos, A.A. 2009. Variability of rainfall rate and raindrop size distributions in heavy rain. *Water Resources Research*, *45*, 1–12.

Stewart, E.J., Jones, D.A., Svensson, C., Morris, D.G., Dempsey, P. and Dent, J.E. 2013. *Reservoir Safety—Long Return Period Rainfall*. R&D Technical Report WS 194/2/39/TR (two volumes), DEFRA/EA.

Thorndahl S., Einfalt T., Willems P., Nielsen J. E., ten Veldhuis M.-C., Arnbjerg-Nielsen K., Rasmussen M. R. and Molnar P. 2017. Weather radar rainfall data in urban hydrology. *Hydrology and Earth System Sciences Discussions*, *21*(3), 1359–1380.

Todini, E. 2001. A Bayesian technique for conditioning radar precipitation estimates to rain-gauge measurements. *Hydrology and Earth System Sciences*, *5*, 187–199.

Tsoukalas, I., Efstratiadis, A. and Makropoulos, C. 2019. Building a puzzle to solve a riddle: A multi scale disaggregation approach for multivariate stochastic processes with any marginal distribution and correlation structure. *Journal of Hydrology*. *2019*, *575*, 354–380.

Tsoukalas, I., Kossieris, P.and Makropoulos, C. 2020. Simulation of Non-Gaussian Correlated Random Variables, Stochastic Processes and Random Fields: Introducing the anySim R Package for Environmental Applications and Beyond. *Water*, *12*, 1645.

UK Water Industry Research (UKWIR). 2003. *Climate Change and the Hydraulic Design of Sewerage Systems*. Summary report no. 03/CL/10/0. https://ukwir.org/reports/03-CL-10-0/66635/Climate-Change-and-the-Hydraulic-Design-of-Sewerage-Systems-Summary-Report

UK Water Industry Research (UKWIR). 2022. *Climate Change Rainfall for use in Sewerage Design – Design Storm Profiles, Antecedent Conditions, Red-Up Tool Update and Seasonality Impacts*. Report no. 22/CL/10/19, London.

Verhoest, N.E.C., Vandenberghe, S., Cabus, P., Onof, C., Meca-Figueras, T., and Jameleddine, S. 2010. Are stochastic point rainfall models able to preserve extreme flood statistics? *Hydrological Processes*, *24*, 3439–3445.

Vesuviano, G. 2022. *The FEH22 Rainfall Depth-Duration-Frequency (DDF) Model*. UK Centre for Ecology & Hydrology.

Wang, L.P., Ochoa-Rodriguez, S., Onof, C. and Willems, P. 2015. Singularity-sensitive gauge-based radar rainfall adjustment methods for urban hydrological applications. *Hydrology and Earth System Sciences*, *19*, 4001–4021.

Wang, L.P., Ochoa-Rodríguez, S., Simões, N.E., Onof, C. and Maksimović, Č. 2013. Radar-raingauge data combination techniques: A revision and analysis of their suitability for urban hydrology. *Water Science and Technology*, *68*, 737–747.

Wheater, H.S., Chandler, R.E., Onof, C.J., Isham, V.S., Bellone, E. and Yang, C. 2005. Spatial-temporal rainfall modelling for flood risk estimation. *Stochastic Environmental Research and Risk Assessment*, *19*, 403–416.

World Meteorological Organization (WMO). 2008. *Guide to Meteorological Instruments and Methods of Observation*, 7th edn, Geneva.

Youngman, B.D. 2018. Generalized additive models for exceedances of high thresholds with an application to return level estimation for U.S. wind gusts, *Journal of the American Statistical Association*, *114* (528), 1865–1879.

Zhang, L., Li, X., Zheng, D., Zhang, K., Ma, Q., Zhao, Y. and Ge, Y., 2021. Merging multiple satellite-based precipitation products and gauge observations using a novel double machine learning approach. *Journal of Hydrology*, *594*, 125969.

Zheng, F., Yin, H., Ma, Y., Duan, H.-F., Gupta, H., Savic, D. and Kapelan, Z. 2023. Toward improved real-time rainfall intensity estimation using video surveillance cameras. *Water Resources Research*, *59*, e2023WR034831

Chapter 5

Stormwater

5.1 INTRODUCTION

Stormwater (or surface water runoff) is the second major urban flow of concern to the drainage engineer. Safe and efficient drainage of stormwater is particularly important to maintain public health and safety (due to the potential impact of flooding on life and property) and to protect the receiving water environment. Reliable data on the quantity and quality of existing and projected stormwater flows are a prerequisite for cost-effective urban drainage design and analysis.

Stormwater is generated by precipitation, typically rainfall, and consists of that proportion that runs off from urban surfaces (see Figure 5.1). Hence, the properties of stormwater, in terms of quantity and quality, are intrinsically linked to the nature and characteristics of both the rainfall and the catchment. This chapter focuses on the generation and characteristics of stormwater, with many of the concepts underlying the design and analysis techniques described in later chapters. Section 5.2 explains the mechanisms of runoff generation, while Section 5.3 discusses overland flow processes. Stormwater quality issues are dealt with in Section 5.4 and the chapter concludes with Section 5.5 concerning sewer misconnections.

5.2 RUNOFF GENERATION

The transformation of a rainfall hyetograph into a surface runoff hydrograph involves two principal parts. First, *losses* due to interception, depression storage, infiltration, and evapo-transpiration are deducted from the rainfall. Second, the resulting *effective rainfall* is transformed by *surface routing* into an *overland flow* hydrograph.

For most urban drainage applications, the runoff processes are at least as important as the pipe flow processes (discussed in Chapter 7) and of equal importance to the rainfall processes (discussed in Chapter 4).

Much of the rainfall that reaches the ground does not, in fact, run off. It is "lost" immediately or as it runs overland. The water may be completely lost from the catchment surface by processes such as evapo-transpiration, it may be temporarily retained in local depression storage, or it may eventually find its way to the drainage system via groundwater.

5.2.1 Initial losses

5.2.1.1 Interception and wetting losses

Interception consists of the collection and retention of rainfall by vegetation cover. There is an initial retention period, after which excess rain falls through the foliage or flows to the

DOI: 10.1201/9781003408635-5

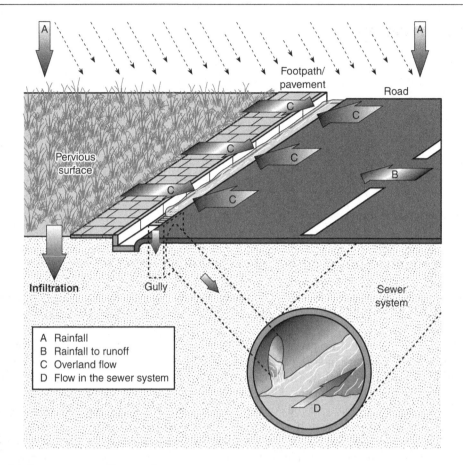

Figure 5.1 Stormwater runoff generation processes.

soil over the stems. The interception rate then rapidly approaches zero. The interception loss for impervious areas is small in magnitude (<1 mm) and is normally neglected or combined with depression storage.

5.2.1.2 Depression storage

Depression storage accounts for rainwater that has become trapped in small depressions on the catchment surface, preventing the water from running off. Infiltration, evaporation, or leakage will eventually remove the water that has been retained. Factors affecting the magnitude of depression storage are surface type, slope, and rainfall return period (Kidd and Lowring, 1979). Depression storage d (mm) can be represented as

$$d = \frac{k_1}{\sqrt{s}} \tag{5.1}$$

where k_1 is a coefficient depending on the surface type (0.07 for impervious surfaces and 0.28 for pervious surfaces) (mm), and s is the ground slope (–).

Typical values for d are 0.5–2 mm for impervious areas, 2.5–7.5 mm for flat roofs, and up to 10 mm for gardens.

5.2.1.3 Representation

For intense summer storms in urban areas, the initial losses are not important, but for less severe storms or for less urbanised catchments, they should not be neglected. For modelling purposes, the combined initial losses are usually subtracted from the rainfall at the beginning of the storm to leave the *net* rainfall. This is illustrated in Example 5.1.

5.2.2 Continuing losses

5.2.2.1 Evapo-transpiration

Evapo-transpiration is the vaporisation of water from plants and open waterbodies and therefore its removal from surface runoff. Potential evaporation (PET) values in the UK are typically 0 mm/day in winter and 2–4 mm/day in summer. Although it is a continuing, constant loss, its effect during short-duration rainfall events is treated as negligible. PET data is required for the variable and UK Water Industry Research (UKWIR) runoff equations (see Sections 5.2.4 and 5.2.5, respectively) and 1 km gridded observations are available across the UK for the years 1969–2021 (Brown et al., 2022).

5.2.2.2 Infiltration

Infiltration represents the process of rainfall passing through the ground surface into the pores of the soil. The infiltration capacity of a soil is defined as the rate at which water infiltrates into it. The magnitude depends on factors including soil type, structure and compaction, initial moisture content, surface cover, and the depth of water in the soil. The infiltration rate tends to be high initially but decreases exponentially to a final quasi-steady rate when the upper soil zone becomes saturated.

EXAMPLE 5.1

For an urban catchment of average slope of 1% with an estimated interception loss of 0.5 mm, calculate the net rainfall profile (based on initial losses only) of the following storm:

Time (minutes)	0–10	10–20	20–30	30–40
Rainfall intensity (mm/h)	6	12	18	6

Take k_i = 0.1 mm.

Solution

Interception loss = 0.5 mm.

Depression storage loss from Equation (5.1): $d = 0.1 / \sqrt{0.01} = 1\text{mm}$

Time (minutes)	0–10	10–20	20–30	30–40
Rainfall intensity (mm/h)	6	12	18	6
Rainfall depth (mm)	1	2	3	1
Net rainfall depth (mm)	0	1.5	3	1
Net rainfall intensity (mm/h)	0	9	18	6

Initial losses are deducted from rainfall *depth* at the beginning of the storm.

Table 5.1 Typical Horton parameters for various surface types

Surface type	f_o (mm/h)	f_c (mm/h)	k_2 (h^{-1})
Coarse textured soils	250	25	2
Medium textured soils	200	12	2
Fine textured soils	125	6	2
Clays/paved areas	75	3	2

A common empirical relationship used to represent infiltration is Horton's (1940) equation:

$$f_t = f_c + \left(f_o - f_c\right)e^{-k_2 t} \tag{5.2}$$

where f_t is the infiltration rate at time t (mm/h), f_c is the final (steady state) infiltration rate or capacity (mm/h), f_o is the initial rate (mm/h), and k_2 is the decay constant (h^{-1}).

The equation is valid when $i > f_c$. These parameters depend primarily on soil/surface type and initial moisture content of the soil. The range of values encountered for f_c, f_o, and k are given in Table 5.1. Careful adaptation of the equation is required to render it suitable for application in continuous simulation models.

Other, more physically based approaches have been formulated, such as Green and Ampt's (1911) equation and Richard's (1933) equation. However, these are less widely implemented and used in urban drainage models.

5.2.2.3 Representation

Continuing losses are always important in urban catchments, but are of most prominence in areas with relatively large open spaces. A simplified, but common, approach to representing them is by a constant proportional loss model applied after initial losses have been deducted to produce the *effective* rainfall:

$$i_e = C\, i_n \tag{5.3}$$

where i_e is the effective rainfall intensity (mm/h), C is the dimensionless runoff coefficient (–), and i_n is the net rainfall intensity (mm/h).

The runoff coefficient C depends primarily on land use, soil and vegetation type, and slope. It is also influenced by rainfall characteristics (e.g., intensity, duration) and antecedent conditions. Values of C range from 0.70 to 0.95 for impervious surfaces such as pavements and roofs, and from 0.05 to 0.35 for pervious surfaces. A more comprehensive listing of coefficients is given in Table 10.3. This model also forms the basis of the Rational Method used for estimating stormwater peak flow rates. This important method is described in more detail in Chapter 10.

5.2.3 Fixed (Wallingford) runoff equation

For urban catchments in the UK, the dimensionless runoff coefficient can be estimated from Equation (5.4) (where $C = PR/100$) originally prepared as part of the *Wallingford Procedure* (DoE/NWC, 1981). This is a regression equation derived from data obtained from 17 catchments and 510 (summer) events:

$$PR = 0.829PIMP + 25.0SOIL + 0.078UCWI - 20.7 \, [PR > 0.4PIMP]$$
$$PR = 0.4PIMP \, [PR \leq 0.4PIMP] \tag{5.4}$$

where PR is the percentage runoff (%), $PIMP$ is the percentage impervious area of the catchment (25–100), $SOIL$ is a soil index (0.15–0.50), and $UCWI$ is the urban catchment (antecedent) wetness index (30–300).

This equation is reasonably reliable provided it is used with variables that are within the range of those upon which it is based (shown in brackets). Since its development, it has been used successfully to represent many thousands of catchments (see Example 5.2). The principal variables are described in further detail in the following sections.

5.2.3.1 PIMP

The percentage imperviousness represents the degree of urban development of the catchment and is defined as

$$PIMP = \frac{A_i}{A} \times 100 \tag{5.5}$$

where A_i is the impervious (roofs and paved areas) area (ha), and A is the total catchment area (ha).

EXAMPLE 5.2

Calculate the effective rainfall profile for the storm specified in Example 5.1. The rain falls on a catchment that is 78% impervious, has a soil type index of 0.25, and has a standard average annual rainfall ($SAAR$) of 540 mm.

Solution

$SAAR$ = 540 mm.
Read from Figure 5.2: $UCWI$ = 40.
Equation (5.4):

$$PR = 0.829 \times 78 + 25.0 \times 0.25 + 0.078 \times 40 - 20.7 = 53\%$$

Equation is valid as [PR = 53] > [0.4 × 78 = 31].

Total net rainfall depth = 5.5 mm (from Example 5.1)
Runoff rainfall depth   = 0.53 × 5.5 = 2.9 mm
Runoff loss   = 5.5–2.9 = 2.6 mm
Continuing loss   = 2.6/0.5 = 5.2 mm/h (over 30 minutes)

The profile is therefore as follows:

Time (minutes)	0–10	10–20	20–30	30–40
Net rainfall profile (mm/h)	0	9	18	6
Effective rainfall profile (mm/h)	0	3.8	12.8	0.8

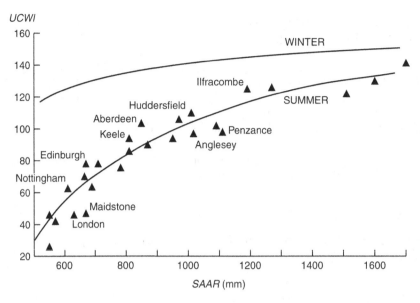

Figure 5.2 Relationship between UCWI and SAAR.

Source: Based on Packman, J.C. 1986. Runoff estimation in urbanising and mixed urban/rural catchments. *IPHE Seminar
          on Sewers for Adoption*, Imperial College, London, with permission of the Chartered Institution of Water and
          Environmental Management, London.

5.2.3.2 SOIL

The *SOIL* index is based on the winter rain acceptance parameter in the *Flood Studies Report*
(NERC, 1975) and is a measure of the infiltration potential of the soil. It can be obtained
from maps in the *Flood Studies Report* or the *Wallingford Procedure* (DoE/NWC, 1981)
although these are no longer commercially available.

5.2.3.3 UCWI

The *UCWI* represents the degree of wetness of the catchment at the start of a storm event. As
UCWI increases, so does the *PR* value reflecting the increased runoff expected from a wetter catchment. It can be estimated for design purposes from its relationship with the *SAAR*
(defined in Chapter 4) and given in Figure 5.2.

   When simulating historical events,

$$UCWI = 125 + 8 \, API5 - SMD \tag{5.6}$$

where *API5* is the 5-day antecedent precipitation index, and *SMD* is the soil moisture deficit.

   API5 is calculated according to a methodology described in the Wallingford Procedure
based on rainfall depths in the 5 days prior to the event. *SMD* is a measure of the amount of
water that can be retained within the soil matrix, values of which are available for UK locations from the Met Office.

5.2.3.4 Limitations

In specific circumstances, the fixed runoff equation has been found to have limitations:

  • For catchments with relatively low *PIMP* (i.e., with a large proportion of pervious sur-
    face), particularly those with light soils in dry conditions, the equation tends to under-
    predict the runoff volume.

- During long-duration storms, catchment surfaces can be significantly wetted, increasing the proportion of runoff. This expected increase in runoff is not properly represented.
- The equation was developed for use with discrete rainfall events and is not directly applicable for continuous simulation using rainfall time series now commonly applied (Section 4.5).

5.2.4 Variable (new) runoff equation

The variable (new) runoff equation (Packman, 1990) was developed to try and overcome some of the limitations mentioned in the previous section. It was based on data from 11 catchments and 112 events and has two major differences. The first is that runoff is calculated separately for impervious and pervious areas (not combined, as in the fixed equation). The second difference lies in the way *API* is allowed to vary during the storm rather than being a fixed value.

The model has three components: initial losses (see Section 5.2.1), runoff from impervious areas, and runoff from pervious areas. It has the following form:

$$PR = IF.PIMP + (100 - IF.PIMP) NAPI / PF \tag{5.7}$$

where *IF* is the effective impervious area factor, *NAPI* is the net antecedent precipitation index (mm), and *PF* is the porosity factor (mm).

5.2.4.1 Impervious area runoff

The model deals with continuous losses following deduction of the initial losses (depression storage $d = 0.5$ mm for paved surfaces). Impervious areas are dealt with, simply, in two parts:

- A proportion of the surface is assumed to be directly connected to the drainage network and to generate 100% runoff (*IF.PIMP*). The effective impervious area factor can be estimated from Table 5.2.
- The rest of the surface is assumed to be less effectively connected and to behave hydrologically as if it were pervious. This part is therefore added to the pervious area total (remainder of the right-hand side of Equation (5.7)).

5.2.4.2 Pervious area runoff

The pervious area and the less effectively connected impervious area are taken together, and the runoff is calculated using a soil moisture storage model, applied progressively throughout a storm (rather than once, before the storm).

Table 5.2 Effective impervious area factor

Surface type	IF
Normal urban paved surfaces	0.6
Roof surfaces	0.8
Well-drained roads	0.8
Very high-quality roads	1.0

Source: Osborne, M.P. 2009. *A New Runoff Volume Model*, WaPUG User Note No. 28, version 3.

NAPI in Equation (5.7) is defined as a 30-day *API* with evapo-transpiration and initial losses subtracted from the rainfall. This formulation takes account of evaporation, better represents the rate of drying out of different soil types, and can be continuously updated during the storm. Details of its calculation are given by Osborne (2009). The default value for the porosity factor *PF* (approximating soil moisture storage depth) is 200 mm, which notionally represents the soil depth that is wetting and drying. A detailed analysis of the performance of this model is given by Balmforth et al. (2006).

5.2.5 UK Water Industry Research (UKWIR) runoff equation

The UKWIR runoff equation (2014) is a further development of the variable runoff equation with the intention of overcoming some of the weaknesses of previous models.

$$
PR = \sum_{N}^{n=1}\left(IF_n \cdot PIMP_n + \left(1 - IF_n\right) \cdot PIMP_n \cdot \frac{PI_{pv}^{\beta}}{PF_{pv}} \right)
$$
$$
+ \left(\left(1 - PIMP_{\text{TOTAL}}\right) \cdot \frac{\left(NAPI_s + PI_s\right)^{\alpha} \cdot SPR}{PF_s} \right)
\tag{5.8}
$$

where n is the number of surface types, 1 to N, PI_{pv} is the precipitation index for paved surfaces, β is the empirical coefficient for paved surfaces, PF_{pv} is the porosity factor for paved surfaces, $NAPI_s$ is the *NAPI* for pervious surfaces, PI_s is the *PI* for pervious surfaces, α is the empirical coefficient for pervious surfaces, *SPR* is the standard percentage runoff, and PF_s is the porosity factor for pervious surfaces.

Further details of the model and an explanation of the various parameters and their values can be found in the SuDS Manual (Woods Ballard et al., 2016). Advice on calculating NAPI and NAPI$_s$ values is given in CIWEM (2016).

5.2.6 Direct runoff approach

With the advances in the modelling of surface flows to represent aboveground flooding (see Chapter 12), an increasingly popular approach is to apply rainfall directly to a "2D surface." Rather than representing runoff from sewer sub-catchments, runoff can be calculated from each segment of a DTM/DEM. This has the advantage of being able to represent in detail which way runoff is routed towards each sewer, and indeed each individual gully. The main disadvantage is the required level of detail of surface features. Also, currently, models only tend to represent a fixed PR (UKWIR, 2022).

5.3 OVERLAND FLOW

Once the losses from the catchment have been accounted for, the effective rainfall hyetograph can be transformed into a surface runoff hydrograph – a process known as *overland flow* or *surface routing*. In this process, the runoff moves across the surface of the sub-catchment to the nearest entry point to the sewerage system.

There are two general approaches currently used for routing overland flow. The most common utilises the *unit hydrograph* method, although this is actually implemented in a number of different ways. The second, more physically based approach, usually utilises a *kinematic wave* model.

5.3.1 Unit hydrographs

The unit hydrograph (UH) is a widely used concept in hydrology that has also found application in urban hydrology. It is based on the premise that a unique and time-invariant hydrograph results from effective rain falling over a particular catchment. Formally, it represents the out-flow hydrograph resulting from a unit depth (generally 10 mm) of effective rain falling uniformly over a catchment at a constant rate for a unit duration D: the D-h UH is shown in Figure 5.3. The ordinates of the D-h UH are given as $u(D, t)$, at any time t. D is typically 1 hour for natural catchments but could, in principle, be any time period.

Once derived, the UH can be used to construct the hydrograph response to any rainfall event based on three guiding principles:

- Constancy: The time base of the UH is constant, regardless of the intensity of the rain.
- Proportionality: The ordinates of the runoff hydrograph are directly proportional to the volume of effective rain – doubling the rainfall intensity doubles the runoff flow rates.
- Superposition: The response to successive blocks of effective rainfall, each starting at particular times, may be obtained by summing the individual runoff hydrographs starting at the corresponding times.

This approximate, linear approach (known as *convolution*) is stated succinctly in Equation (5.9). If a rainfall event has n blocks of rainfall of duration D, the runoff $Q(t)$ at time t is

$$Q(t) = \sum_{\omega=1}^{N} u(D, j) I_\omega \qquad (5.9)$$

$$Q(t) = u(D, t) I_1 + u(D, t - D) I_2 \ldots + (D, t - (N-1)D) I_n$$

where $Q(t)$ is the runoff hydrograph ordinate at time t (m³/s), $u(D,j)$ is the D-h UH ordinate at time j (m³/s), I_ω is the rainfall depth in the ωth of N blocks of duration D (m), and j is the $t - (\omega - 1)D$ (s).

Further details and examples of using UHs are given in Shaw et al. (2010).

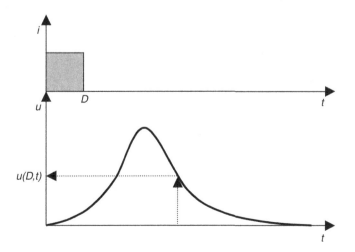

Figure 5.3 The unit hydrograph.

In order to use this concept in ungauged urban catchments for design purposes, some way of predicting UHs is required, based on catchment characteristics. Three methods of doing this are available: synthetic UHs, the time–area method, or reservoir models.

5.3.2 Synthetic unit hydrographs

The detailed shape of the UH reflects the characteristics of the catchment from which it has been derived. When converted into dimensionless form, it is found that similar shapes are observed in catchments in the same region.

Harms and Verworn (1984), for example, derived a dimensionless UH suitable for urban areas, as shown in Figure 5.4. This has a linear rise up to the peak flow, an exponential recession, and an endpoint at 1% of peak flow:

$$Q = \frac{t}{t_p} Q_p \quad 0 < t < t_p \tag{5.10}$$

$$Q = Q_p e^{-\left(t - t_p / k_3\right)} \quad t \geq t_p \tag{5.11}$$

where Q is the flow rate (m³/s), Q_p is the peak flow rate (m³/s), t is the time (s), t_p is the time to peak (s), and k_3 is the exponential decay constant (s⁻¹).

The three parameters, Q_p, t_p, and k can be related to catchment characteristics.

This is the most direct application of the UH approach, but it is the least common in practice.

5.3.3 Time–area diagrams

An alternative approach is to derive a time–area diagram, which is a special case of a UH. To do this, lines of equal flow "travel time" to the catchment outfall are delineated, called

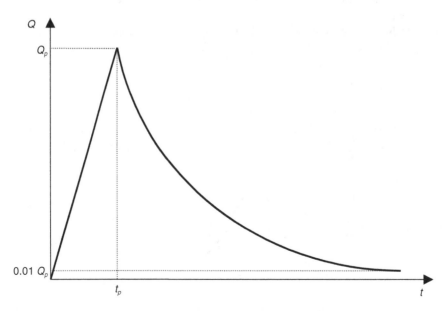

Figure 5.4 Synthetic unit hydrograph.

isochrones. The maximum travel time represents the time of concentration (t_c) of the catchment (considered in Chapter 10). The time–area diagram that is constructed by summing the areas between the isochrones (see Figure 5.5) defines the response of the catchment.

When combined with rainfall in depth increments of I_1, I_2, ..., I_N, flow at any time $Q(t)$ is

$$Q(t) = \sum_{\omega=1}^{N} \frac{dA(j)}{dt} I_\omega \tag{5.12}$$

where $dA(j)/dt$ is the slope of the time–area diagram at time j.

The time–area diagram can be used in design as an extension to the Rational Method (Chapter 10 has a description and examples) and is also implemented in some simulation models using standard time–area diagram profiles and empirical equations to estimate t_c.

5.3.4 Reservoir models

The third approach is to propose the analogy that the catchment surface acts on the flow generated by an effective rainfall profile as one or more reservoirs connected in series. Each reservoir then experiences inflows of rainfall (and/or inflows from upstream reservoirs) and outflows of runoff. The model is based on the two equations of continuity and storage:

$$\frac{dS}{dt} = I - O \tag{5.13}$$

$$S = KO^m \tag{5.14}$$

where I is the inflow rate (m³/s), O is the outflow rate (m³/s), S is the storage volume (m³), K is the reservoir time constant (s), and m is the exponent (–).

These equations are returned to again in Section 14.4 in the context of reservoir routing for storage design.

If m is taken to be equal to unity (a physical impossibility, but a conceptual convenience), then the reservoir is referred to as "linear." Nash (1957) proposed that the overland flow process could be represented as a series of identical linear reservoirs, where the output from

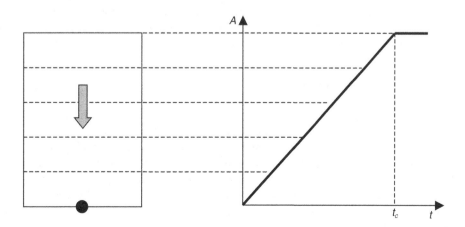

Figure 5.5 Linear time–area diagram.

one reservoir is considered as the input to a second, and so on. Assuming an instantaneous inflow of unit volume, the UH time to peak t_p and peak flow Q_p are

$$t_p = (\alpha - 1)K \tag{5.15}$$

and

$$Q_p = \frac{1}{K(\alpha - 1)!}\left(\frac{\alpha - 1}{e}\right)^{n-1} \tag{5.16}$$

This approach can be used by specifying the number of reservoirs (α) and the time constant K, where α and K can be related to catchment characteristics. Alternatively, α and K can be used as calibration constants in models. A linear reservoir cascade is also a special case of the UH approach.

The reservoirs are termed linear because, in Equation (5.14), the storage S is linearly related to outflow O, so from Equation (5.13) outflow must be linearly related to inflow. However, other non-unity values of m (e.g., 0.67) can be used, resulting in a non-linear response from a *single* conceptual reservoir. The parameters K and m become calibration constants.

5.3.5 Kinematic wave

A more physically based approach is to simplify and solve the equations of motion to give

$$\frac{\partial d}{\partial t} + \frac{\partial q}{\partial x} = i_e \tag{5.17}$$

$$q = \frac{1}{n}d^{5/3}s^{1/2} \tag{5.18}$$

where d is the depth of flow (m), q is the flow per unit width $= Q/b$ (m²/s), i_e is the effective rainfall intensity (m/s), t is the time (s), x is the longitudinal distance (m), n is the Manning's roughness coefficient $\left(\text{m}^{-(1/3)} / \text{s}\right)$, and s is the catchment slope (–).

Equation (5.17) is a continuity equation, and Equation (5.18) is a simplified momentum equation based on Manning's equation. The kinematic wave approximation is elaborated further in Section 20.8.3 and Manning's equation is explained in Section 7.5.1.1. Manning's n for urban surfaces is typically 0.05–0.15, an order of magnitude greater than pipe roughness.

5.4 STORMWATER QUALITY

Urban stormwater contains a complex mixture of natural organic and inorganic materials, with a small but important proportion of man-made substances derived from transport, commercial, and industrial practices. These materials find their way into the drainage system from atmospheric sources and as a result of being washed off or eroded from urban surfaces. In certain respects, stormwater can be as polluting as wastewater.

It was stressed in Chapter 3 that wastewater is variable in character. The quality of stormwater is even more variable from place to place and from time to time. As with wastewater, care should be taken in interpreting "standard" or "typical" values.

5.4.1 Pollutant sources

Stormwater quality is influenced by rainfall and, especially, by the catchment. The major catchment sources include vehicle emissions, corrosion, and abrasion; building and road corrosion and erosion; bird and animal faeces; street litter deposition, fallen leaves and grass residues; and spills.

5.4.1.1 Atmospheric deposition

Pollutants in the urban atmosphere with the potential for deposition result mainly from human activities: heating, vehicular traffic, industry, or waste incineration, for example. They may either be absorbed and dissolved by precipitation (known as wet fallout), to be carried directly into the drainage system with the stormwater; or they may settle on land surfaces (as dry fallout), and subsequently be washed off. Dry fallout particles can be transported by winds over long distances.

Although atmospheric sources are accepted as a major contributor to stormwater pollution, the importance of dry and wet fallout appears to be dependent on the site and the pollutant. In Gothenburg, for example, wet fallout has been identified as the dominant form of atmospheric pollutant (at 60%) for nitrogen, phosphorus, lead, zinc, and cadmium. Even higher percentages have been noted in some locations.

Dry fallout is thought to be of more importance in urban areas or areas with significant sources of solids. In Sweden, it was estimated that 20% of the organic matter, 25% of phosphorus, and 70% of the total nitrogen in stormwater can be attributed to atmospheric fallout (Malmqvist, 1979). Granier et al. (1990) found that approximately half the total loads of lead and chromium come from the atmosphere, with a significantly lower proportion of zinc.

Numerous studies have confirmed the importance of this pollutant source with the key pollutants identified being: solids, nitrogen, phosphorus, metals, and polyaromatic hydrocarbons (PAHs).

5.4.1.2 Vehicles

Vehicle emissions include volatile solids and PAHs derived from unburned fuel, exhaust gases and vapours, lead compounds (from petrol additives), and hydrocarbon losses from fuels, lubrication, and hydraulic systems. Catalytic converters are sources of platinum, palladium and rhodium (Rauch et al., 2005).

Pollutants are generated by the everyday passage of traffic. Tyre wear releases zinc, hydrocarbons, and microplastics (MPs) (Horton et al., 2017). Vehicle corrosion releases pollutants such as iron, chromium, lead, and zinc. Other pollutants include metal particles, especially copper and nickel, released by the wear of clutch and brake linings. Most metals are predominantly associated with the particulate phase.

Wear of the paved surface will release various substances: bitumen and aromatic hydrocarbons, tar and emulsifiers, carbonates, metals, and fine sediments, depending on the road construction technique and materials used. MPs have also been detected originating from asphalt bitumen and road marking paints (Vijayan et al., 2022).

Improvements in analytical techniques and a greater understanding of environmental pathways have unearthed an increasing range of micropollutants in road runoff including phthalates and PFAS (Müller et al., 2020). However, further research is still required to identify specific sources.

5.4.1.3 Buildings and roads

Urban erosion produces particles of brick, concrete, asphalt, and glass. These particles can form a significant constituent of sediment in stormwater. The extent of pollution depends on the condition of the buildings/roads. Roofs, gutters, and exterior paint can release varying amounts of particles, again depending on their condition. Metallic structures, such as street furniture (e.g., fences, benches), roofs, and facades corrode, releasing toxic substances such as chromium. Galvanised steel is an important source of zinc because of its ubiquity in the urban environment (Clark et al., 2008). Building surface materials are also significant sources of pesticides, which are added to renders, paints and roofing tiles to prevent weed and moss growth (Müller et al., 2020). Roads and pavements degrade over time, releasing particles of various sizes.

5.4.1.4 Animals

Urine and faeces deposited on roads and pavements by animals (pets or wild) are a source of bacterial pollution in the form of faecal coliforms and faecal streptococci. They are also a source of high oxygen demand. Hobbie et al. (2017) estimated that dog waste was the largest source of phosphorus and the third largest source of nitrogen in a US urban catchment.

5.4.1.5 De-icing

The most commonly used de-icing agent is salt (sodium chloride). Salt applications to roads cause the annual chloride loads in stormwater to be (on average) 50–500 times higher than would occur naturally (Stotz, 1987). Rock salt contains other impurities, including an insoluble fraction shown to contribute 25% of the winter suspended solids load in a motorway study (Colwill et al., 1984). The presence of salt also accelerates the corrosion of vehicles and metal structures.

5.4.1.6 Urban debris

Urban surfaces can contain large amounts of street debris, litter, and organic materials such as dead and decaying vegetation. Litter will generally result in elevated levels of solids and greater consequential oxygen demand, and is also a source of MPs. Fallen leaves and grass cuttings may lie on urban surfaces, particularly in road gutters, and decompose, or may be washed into gullies. Construction sites are also a major source of solids (Sajjad et al., 2019).

5.4.1.7 Spills/leaks

Household cleaners, motor fluids/lubricants, and bin washings are sometimes illegally discarded or spilt into gutters and gullies. The range and amounts of these pollutants vary considerably, depending on land use and public behaviour. However, domestic sources of chemical pollutants are usually minor when compared with industrial spills or illicit toxic waste disposal.

5.4.2 Surface pollutants

The bulk of the pollutants, derived from the sources mentioned, are attached to particles of sediment that deposit temporarily on the catchment surface. Analysis of this particulate material shows a large range of sizes from below 1 μm to above 10 mm. Larger, denser sediment causes particular problems in the drainage system itself, and this is addressed further in Chapter 17. Despite typically comprising less than 5% of the material present, particulates less than 50 μm in size have most of the pollutants associated with them: 25% of the COD, up to 50% of the nutrients, and 15% of total coliforms (Ellis, 1986). These, of course, are the particles that are most readily washed off by the stormwater.

The situation with pollutants washed off from the surface is somewhat different. The median particle size (d_{50}) is much smaller at around 50 μm (see Figure 5.6), although the d_{50} does vary to a certain extent depending on location (Charters et al., 2015). Butler and Memon found it to be positively correlated with the runoff rate (Figure 5.7).

5.4.3 Pollutant levels

Pollutants in stormwater include solids, oxygen-consuming materials, nutrients, hydrocarbons, heavy metals, trace organics, and bacteria. Typical values and ranges of pollutant discharges from stormwater systems are given in Table 5.3. Table 5.3 demonstrates the inherent variability of runoff quality. The average quality of "clean" and "dirty" catchments can vary by a factor of 10, and the variation in quality between stormwater events for any single catchment can vary by a factor of 3 (Ellis, 1986). Runoff quality will depend on a number of factors, including the following:

- Geographical location
- Road and traffic characteristics
- Building and roofing types
- Weather, particularly rainfall

As a specific example, for two catchments in France, the total atmospheric contribution to the pollutants studied (SS, COD, 7 metals, and 4 pesticides) was mainly less than 20%.

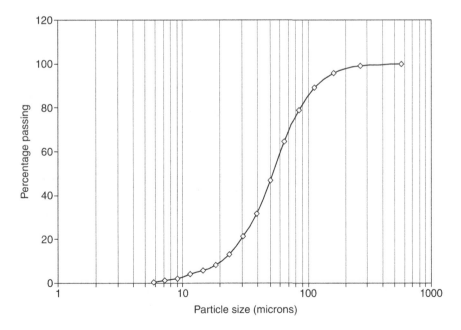

Figure 5.6 Particle size analysis for solids in urban road runoff. (Based on Memon, F.A. and Butler, D. 2005. *Urban Water Journal*, 2(3), 171–182.)

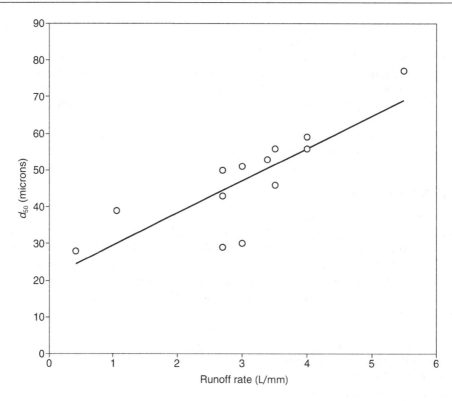

Figure 5.7 Relationship between median particle size (d₅₀) and runoff flow rate. (Based on Memon, F.A. and Butler, D. 2005. *Urban Water Journal*, 2(3), 171–182.)

Table 5.3 Pollutant event mean concentrations (range and mean) and unit loads for urbanstormwater

Quality parameter	EMC (mg/L)	Unit load (kg/imp ha.yr)[a]
Suspended solids (SS)	21–2582 (90)	347–2340 (487)
Microplastics (MPs)	0–8580 (MP/L)	
BOD₅	7–22 (9)	35–172 (59)
COD	20–365 (85)	22–703 (358)
Ammoniacal nitrogen	0.2–4.6 (0.56)	1.2–25.1 (1.76)
Total nitrogen	0.4–20.0 (3.2)	0.9–24.2 (9.9)
Total phosphorus	0.02–4.30 (0.34)	0.5–4.9 (1.8)
Total lead	0.01–3.1 (0.14)	0.09–1.91 (0.83)
Total zinc	0.01–3.68 (0.30)	0.21–2.68 (1.15)
Total hydrocarbons	0.04–25.9 (1.9)	0.01–43.3 (1.8)
Polyaromatic hydrocarbons (PAH)	0.01	0.02
DEHP	0.45–24 (µg/L)	
Faecal coliforms (*E. coli*)	400–50,000 (MPN/100 mL)	0.9–3.8 (2.1) (×10⁹ counts/ha)

[a] imp ha = impervious area measured in hectares.

Source: Adapted from Ellis and Mitchell, 2006, Clara et al. 2010, and Wang et al. 2022.

The biggest contribution (>50%) came from the catchment surface runoff (Becouze-Lareure et al., 2016).

5.4.4 Representation

The most common methods used to quantify stormwater quality are event mean concentrations (EMCs), regression equations/rating curves, and build-up/washoff models. These are described in the following sections.

5.4.4.1 Event mean concentrations

EMCs, such as those given in Table 5.3, can be used in relation to environmental quality standards (EQSs) expressed in terms of maximum allowable concentrations (MACs) or annual allowances (AAs). These were discussed in Chapter 2. They can also be readily integrated into standard flow simulation models. Lundy et al. (2012) have demonstrated how such data can be used to estimate the risks associated with pollutants from a variety of different land use types and activities.

5.4.4.2 Regression equations

In this approach, the quality of stormwater is statistically regressed against a number of describing variables, for example, catchment characteristics or land use. This is the quality equivalent to the fixed runoff equation (5.4) discussed earlier in the chapter. Regression equations can usually be relied on to give good representations of the catchment(s) on which they were based and perhaps similar ones. They will be less accurate on other catchments, but can often give a reasonable first approximation.

5.4.4.3 Buildup

The most common model-based approach to quality representation is by separately predicting pollutant buildup and washoff. In practice, the distinction between these two processes is not clearly defined. Factors affecting the buildup of pollutants on impervious surfaces include:

- Land use
- Population
- Traffic flow
- Effectiveness of street cleaning
- Season of the year
- Meteorological conditions
- Antecedent dry period
- Street surface type and condition

Buildup on the surface dM_s/dt can be assumed to be linear, so

$$\frac{dM_s}{dt} = aA \tag{5.19}$$

where M_s is the mass of pollutant on the surface (kg), a is the surface accumulation rate constant (kg/ha.d), A is the catchment area (ha), and t is the time since the last rainfall event or road sweeping (d).

Accumulation rate a values for solids in residential areas are approximately 5 kg/imp ha.d. but can be much higher in dense urban areas.

Detailed observation of sites in the US (Sartor and Boyd, 1972) revealed that, despite there being no rainfall or street cleaning, the pollutant deposition often has a reducing rate of increase rather than a uniform linear increase. The first-order removal concept can be used to represent this, which implies that equilibrium is reached when the supply rate of pollutants matches their removal:

$$\frac{dM_s}{dt} = aA - bM_s \qquad (5.20)$$

where b is the removal constant (d^{-1}).

The equilibrium mass on the catchment is therefore A (a/b). Novotny and Chesters (1981) reported values for b from a US medium-density residential area of 0.2–0.4 d^{-1}. Studies in London (Ellis, 1986) suggested equilibrium is reached within 4–5 days where vehicle-induced re-suspension is dominant. The levelling-off phenomenon is most profound in areas where:

- Adjacent pollution traps (pervious areas) are available.
- Vehicle-induced wind and vibration are high.

This will be the case on motorways and trunk roads and in busy commercial/industrial areas. Pollutants other than solids can be predicted using "potency" factors (defined in Chapter 20).

5.4.4.4 Washoff

Washoff occurs during rainfall/runoff by raindrop impact, erosion, or solution of the pollutants from the impervious surface. Important factors include the following:

- Rainfall characteristics
- Topography
- Solid particle characteristics
- Street surface type and condition

This extensive range of influences makes pollutant washoff difficult to quantify with any certainty. The simplest approach adopted is to assume there is effectively an infinite store of pollutants always available on the surface to be washed off, and hence no buildup. Experimental evidence suggests this assumption may be valid in British conditions (Mance and Harman, 1978). Washoff can then be modelled as a function of rainfall intensity:

$$W = z_1 i^{z_2} \qquad (5.21)$$

where W is the pollutant washoff rate (kg/h), i is the rainfall intensity (mm/h), and z_1, z_2 are the pollutant-specific constants.

The exponent z_2 usually has values between 1.5 and 3 for particulate pollutants and <1 for those in solution (Delleur, 1998). Price and Mance (1978) found z_2 to be 1.5 and z_1 0.02 in some British catchments. This is a convenient form for ready addition to flow models.

Alternatively, a first-order relationship, where the rate at which pollutants wash off is assumed to be directly proportional to the amount of pollutant remaining on the surface, can be used:

$$W = -\frac{dM_s}{dt} = k_4 i M_s(t) \tag{5.22}$$

where k_4 is the washoff constant (mm^{-1}).

Integrating Equation (5.21) gives

$$M_s(t) = M_s(0) e^{-k_4 i t} \tag{5.23}$$

where $M_s(0)$ is the initial amount of pollutant on the surface (kg), $M_s(t)$ is the amount of pollutant on the surface after time t (kg), and $M_w(t)$ is the amount of pollutant washed off after time t (kg).

Since $M_s(t) = M_s(0) - M_w$,

$$M_w(t) = M_s(0)\left[1 - e^{-k_4 i t}\right] \tag{5.24}$$

Typical values for k_4 are 0.1–0.2 mm^{-1}. As with the build-up parameters, this requires calibration for each individual catchment. Washoff concentration (c) can be obtained as

$$c = \frac{W}{Q} = \frac{k_4 M_s}{A_i} \tag{5.25}$$

where A_i is the catchment impervious area (ha).

A disadvantage of this formulation is that pollutant concentration will only decrease with time as M_s decreases. This can be remedied by introducing an exponent w for i in Equation (5.22), where i is in the range 1.4–1.8:

$$W = k_5 i^w M_s \tag{5.26}$$

where k_5 is the amended washoff constant (mm^{-1}).

5.5 SEWER MISCONNECTIONS

The issue of sewer misconnection has been discussed in Chapter 3 in the context of inflow of stormwater into separate foul sewers. However, an arguably more serious situation arises when wastewater from buildings gets into separate stormwater sewers (via soil and waste drainage systems) with the potential to directly impact the water quality and the amenity value of an area. The main pollutant loading contributions arise from misconnected toilets (BOD, NH_4-N, FIOs), kitchen sinks (BOD and PO_4-P), washing machines (PO_4-P and BOD), and, to a lesser extent, dishwashers (PO_4-P).

The scale of the issue is hard to establish with any accuracy, but Ellis and Butler (2015) report that the average misconnection rate in England and Wales is about 3% with a working range of 1–5%. However, outlier rates as high as 30% or more can occur arising from specific local conditions. The organic and nutrient loads from misconnections at this level have been shown to have a significant impact on receiving water quality and the achievement of river quality standards (Revitt and Ellis, 2016). Methods to detect and eliminate illicit discharges

---

EXAMPLE 5.3

A storm of duration 30 minutes and intensity 10 mm/h falls on a 1.5 ha impervious area urban catchment. If the initial pollutant mass on the surface is 12 kg/ha, calculate:

1. The mass of pollutant washed off during the storm (k_4 = 0.19 mm^{-1})
2. The average pollutant concentration

Solution

a

$$M_s(0) = 12 \times 1.5 = 18 \text{ kg}$$

From Equation (5.24),

$$M_w(0.5) = 18\left[1 - e^{-0.19 \times 10 \times 0.5}\right] = 11.0 \text{ kg}$$

b

$$c = \frac{M_w(0.5)}{Q} = \frac{11.0 (kg)}{0.01(m/h) \times 0.5(h) \times 15,000(m^2)} = 0.147 \text{ kg} / m^3$$

$$= 147 \text{ mg} / L$$

---

include visual reconnaissance to document flowing stormwater outfalls in dry weather, sampling to identify possible sources, and testing for key chemical/microbiological determinands (Irvine et al., 2011). Chandler and Lerner (2015) developed a technique based on the passive sampling of optical brighteners. These chemicals, commonly found in toilet paper, sanitary products, detergents, and cleaning products, do not occur naturally in the environment but are commonly associated with illicit discharges. Lepot et al. (2017) demonstrated the application of an in-sewer infrared camera to detect and locate active (leaking) lateral connections.

PROBLEMS

1. List and explain the initial rainfall losses. How are they represented mathematically?
2. List and explain the continuing rainfall losses. How are they represented mathematically?
3. What are the disadvantages of the fixed runoff equation? How are some of these overcome with the variable runoff equation?
4. Reassess the effective rainfall profile calculated in Example 5.2 using Horton's equation as the continuing loss model where f_0 = 10 mm/h, f_c = 1 mm/h, and k_2 = 1 h^{-1} [0, 0.4, 10.6, 0 mm/h]
5. A 1 mm, 10-minute UH for an urban catchment is given in the table. Derive the runoff hydrograph resulting from the effective rainfall hyetograph also given. [Peak = 2500 L/s @ 40 minutes]

Time (minutes)	0–10	10–20	20–30	30–40	40–50	50–60	60–70
UH flow rate (L/s)	0	250	500	375	250	125	0
Rainfall intensity (mm)	1	2	3	0	1	0	0

6 Compare and contrast the three main approaches to predicting UHs in ungauged urban catchments.

7 A three-reservoir Nash cascade (K = 12 minutes) is used to represent the runoff response of a 10 ha urban catchment. Calculate the 10mm 1-hour UH peak flow and time to peak. [278 L/s, 24 minutes]

8 What are the main sources of pollutants in stormwater, and what is their importance?

9 Explain the main ways in which stormwater quality is modelled. What are their relative merits?

10 For the conditions described in Example 5.3, reevaluate the stormwater pollutant concentration at 10-minute time intervals. [166, 122, 88 mg/L]

11 What are the main problems associated with sewer misconnections?

KEY SOURCES

Luker, M. and Montague, K.N. 1994. *Control of Pollution from Highway Drainage Discharges*, CIRIA R142, London.

Marsalek, J., Maksimovic, C., Zeman, E., and Price, R. (eds.). 1998. *Hydroinformatic Tools for Planning, Operation, Design, Operation and Rehabilitation of Sewer Systems*, NATO ASI Series 2: Environment – Vol. 44, Kluwer Academic Press.

Shaw, E.M., Beven, K.J., Chappell, N.A., and Lamb, R. 2010. *Hydrology in Practice*, 4th edn, CRC Press.

Torno, H.C., Marsalek, J., and Desbordes, M. (eds.). 1986. *Urban Runoff Pollution*, NATO ASI Series G: Ecological Sciences – Vol. 10, Springer-Verlag.

UK Water Industry Research (UKWIR). 2022. *Urban Runoff (including Road Runoff) and Atmospheric Deposition – How to Apportion Pollution Load Especially Chemicals of Emerging Concern*, Report 22/WW/02/15, London. https://ukwir.org/urban-runoff-including-road-runoff-and-atmospheric-deposition-how-to-apportion-pollution-load-especially-chemicals-of-emerging-concern-19

REFERENCES

Balmforth, D., Digman, C., Kellagher, R., and Butler, D. 2006. Designing for Exceedance in Urban Drainage—Good Practice, CIRIA C635, London.

Becouze-Lareure, C., Dembélé, A., Coquery, M., Cren-Olivé, C., Barillon, B., and Bertrand-Krajewski, J.-L. 2016. Source characterization and loads of metals and pesticides in urban wet weather discharges, *Urban Water Journal*, 13(6), 600–617.

Brown, M.J., Robinson, E.L., Kay, A.L., Chapman, R., Bell, V.A. and Blyth, E.M. 2022. *Potential evapotranspiration derived from HadUK-Grid 1km gridded climate observations 1969-2021 (Hydro-PE HadUK-Grid)*. https://doi.org/10.5285/9275ab7e-6e93-42bc-8e72-59c98d409deb

Chandler, D.M. and Lerner, D.N. 2015. A low cost method to detect polluted surface water outfalls and misconnected drainage, *Water and Environment Journal*, 29, 202–206.

Charters, F.J., Cochrane, T.A. and O'Sullivan, A.D. 2015. Particle size distribution variance in untreated urban runoff and its implication on treatment selection, *Water Research*, 85, 337–345.

CIWEM 2016. *Rainfall Modelling Guide 2016*. The Chartered Institution of Water and Environmental Management, London.

Clara, M., Windhofer, G., Hartl, W., Braun, K., Simon, M., Gans, O., Scheffknecht, C. and Chovanec, A. 2010. Occurrence of phthalates in surface runoff, untreated and treated wastewater and fate during wastewater treatment, *Chemosphere*, 78, 1078–1084.

Clark, S.E., Steele, K.A., Spichers, J., Siu, C.Y., Lalor, M.M., Pitt, R. and Kirby, J.T. 2008. Roof materials' contributions to storm-water runoff pollution, *Journal of Irrigation and Drainage*, *134*(5), 638–645.

Colwill, D.M. Peters, C.J., and Perry, R. 1984. *Water Quality in Motorway Runoff*. TRRL Supplementary Report 823, Transport and Road Research Laboratory.

Delleur, J.W. 1998. Modelling quality of urban runoff, in *Hydroinformatic Tools for Planning, Operation, Design, Operation and Rehabilitation of Sewer Systems* (eds. J. Marsalek, C. Maksimovic, E. Zeman, and R. Price), NATO ASI Series 2: Environment – Vol. 44, Kluwer Academic Press, 241–285.

DoE/NWC. 1981. *Design and Analysis of Urban Storm Drainage. The Wallingford Procedure. Volume 1: Principles, Methods and Practice*. Department of the Enviroment, Standing Technical Committee Report No. 28.

Ellis, J.B. 1986. Pollutional aspects of urban runoff, in *Urban Runoff Pollution* (eds. H.C. Torno, J. Marsalek, and M. Desbordes), NATO ASI Series G: Ecological Sciences – Vol. 10, Springer-Verlag, 1–38.

Ellis, J.B. and Butler, D. 2015. Surface water sewer misconnections in England and Wales: Pollution sources and impacts. *Science of the Total Environment*, *526*, 98–109.

Ellis, J.B. and Mitchell, G. 2006. Urban diffuse pollution: Key data information approaches for the water framework directive. *Water and Environment Journal*, *20*(1), 19–26.

Granier, L., Chevrevil, M., Carru, A., and Letolle, R. 1990. Urban runoff pollution by organochlorines (polychlorinated biphenyls and lindane) and heavy metals (lead, zinc and chromium). *Chemosphere*, *21*(9), 1101–1107.

Green, W.H. and Ampt, G.A. 1911. Studies of soil physics, 1: The flow of air and water through soils. *Journal of Agricultural Science*, *4*(1), 1–24.

Harms, R.W. and Verworn, H.-R. 1984. HYSTEM—ein hydrologisches Stadtent-waesser-ungsmodell. Teil I: Modellbeschreibung. *Korrespondenz Abwasser*, *31*(2), 112–117.

Hobbie, S.E. Finlay, J.C., Janke, B.D., Nidzgorski, D.A., Millet, D.B., and Baker, L.A. 2017. Contrasting nitrogen and phosphorus budgets in urban watersheds and implications for managing urban water pollution, *Proceedings of the National Academy of Sciences of the United States of America*, *114*(16), 4177–4182.

Horton, R.E. 1940. An approach towards a physical interpretation of infiltration capacity. *Proceedings of Soil Science Society of America*, *5*, 399–417.

Horton, A. A., Svendsen, C., Williams, R.J., Spurgeon, D.J., Lahive, E. 2017. Large microplastic particles in sediments of tributaries of the River Thames, UK – Abundance, sources and methods for effective quantification, *Marine Pollution Bulletin 114*, 218–226.

Irvine, K., Rossi, M.C., Vermette, S., Bakert, J., and Kleinfelder, K. 2011. Illicit discharge detection and elimination: Low cost options for source identification and trackdown in stormwater systems. *Urban Water Journal*, *8*(6), 379–395.

Kidd, C.H.R. and Lowring, M.J. 1979. *The Wallingford Urban Sub-catchment Model*. IoH Report No. 60, Institute of Hydrology.

Lepot, M., Makris, K.F. and Clemens, F.H.L.R. 2017 Detection and quantification of lateral, illicit connections and infiltration in sewers with Infra-Red camera: Conclusions after a wide experimental plan, *Water Research*, *122*, 678–691.

Lundy, L., Ellis, J.B., and Revitt, D.M. 2012. Risk prioritisation of stormwater pollutant sources. *Water Research*, *46*, 6589–6600.

Malmqvist, P.-A. 1979. Atmospheric fallout and street cleaning—Effect on urban storm water and snow. *Progress in Water Technology*, *10*, 417–431.

Mance, G. and Harman, M. 1978. The quality of urban stormwater runoff, in *Urban Storm Drainage* (ed. P.R. Helliwell), Pentech Press, 603–617.

Memon, F.A. and Butler, D. 2005. Characterisation of pollutants washed off from road surfaces during wet weather. *Urban Water Journal*, *2*(3), 171–182.

Müller, A., Österlund, H., Marsalek, J. and Viklander, M. 2020. The pollution conveyed by urban runoff: A review of sources, *Science of the Total Environment*, *709*, 136125.

Nash, J.E. 1957. The form of the instantaneous unit hydrograph. *International Association of Science Hydrology Publication*, 45(3), 114–121.

Natural Environment Research Council (NERC). 1975. *Flood Studies Report*, 5 volumes, Institute of Hydrology.

Novotny, V. and Chesters, G. 1981. *Handbook of Nonpoint Pollution, Sources and Management*, Van Nostrand Reinhold.

Osborne, M.P. 2009. *A New Runoff Volume Model*, WaPUG User Note No. 28, version 3.

Packman, J.C. 1986. Runoff estimation in urbanising and mixed urban/rural catchments. IPHE Seminar on Sewers for Adoption, 2/1-2/12, Imperial College, London.

Packman, J.C. 1990. New hydrology model. *Proceedings of WaPUG Conference*, Birmingham, UK, June.

Price, R. and Mance, G. 1978. A suspended solids model for storm water runoff, in *Urban Storm Drainage* (ed. P.R. Helliwell), Pentech Press, 546–555.

Rauch, S., Hemond, H.F., Barbante, C., Owari, M., Morrison, G.M. Peucker-Ehrenbrink, B. and Wass, U. 2005. Importance of automobile exhaust catalyst emissions for the deposition of platinum, palladium, and rhodium in the northern hemisphere, *Environmental Science & Technology*, 39, 8156–8162.

Revitt, D.M. and Ellis, J.B. 2016. Urban surface water pollution problems arising from misconnections. *Science of the Total Environment*, 551–552, 163–174.

Richards, L.A. 1933. Capillary conduction of liquids through porous mediums. *Physics*, 1, 318–333.

Sajjad, R.U., Paule-Mercado, M.C., Salim, I., Memon, S., Sukhbaater, C. and Lee, C. 2019. Temporal variability of suspended solids in construction runoff and evaluation of time-paced sampling strategies, *Environmental Monitoring and Assessment*, 191, 110.

Sartor, J.D. and Boyd, G.B. 1972. *Water Pollution Aspects of Street Surface Contaminants*. EPA Report No. EPA/R2/72/081.

Stotz, G. 1987. Investigations of the properties of the surface water runoff from federal highways in the FRG. *Science of the Total Environment*, 59, 329–337.

UK Water Industry Research (UKWIR). 2014. *Development of the UKWIR Runoff Model: Main Report*. Report 14/SW/01/6, London. https://ukwir.org/reports/14-SW-01-6/66925/Development-of-the-UKWIR-Runoff-Model-Main-Report

UK Water Industry Research (UKWIR). 2022. *Ground Water Infiltration Model Project Report and Guidance Manual*. Report 22/SW/01/22, London. https://ukwir.org/ground-water-infiltration-model-project-report-and-guidance-manual

Vijayan, A., Österlund, H. Magnusson, K. Marsalek, J. and Viklander, M. 2022 Microplastics (MPs) in urban roadside snowbanks: Quantities, size fractions and dynamics of release, *Science of the Total Environment*, 851, 158306.

Wang, C., O'Connor, D., Wang, L., Wu, W.-M., Luo, J. and Hou, D. 2022. Microplastics in urban runoff: Global occurrence and fate, *Water Research*, 225, 119129.

Woods Ballard, B., Wilson, S., Udale-Clark, H., Illman, S., Scott, T., Ashley, A., and Kellagher, R. 2016. *The SuDS Manual*, 5th edn, CIRIA C753.

Chapter 6

System components and layout

6.1 INTRODUCTION

This chapter gives an overview of the elements that make up any piped urban drainage system operating under gravity, including building drainage (Section 6.2) and other main system components (Section 6.3). The key stages in the design process are also described (Section 6.4). Gravity system "exceedance" components are described separately in Chapter 11. Piped systems operating under different pressure regimes are discussed in Chapter 14, and non-piped system components (SuDS) are covered in Chapter 11.

6.2 BUILDING DRAINAGE

Even though urban drainage engineers are not normally involved directly in the planning, design, and construction of building drainage, it is important that they are at least aware of the main components and layout of systems in and around buildings. This includes, in particular, an understanding of how building drains connect with the main sewer system. Building drainage systems are subject to BS EN 12056-2: 2000 and Approved Document H (2015), part of the 2010 Building Regulations.

6.2.1 Soil and waste drainage

6.2.1.1 Inside

A common arrangement for the soil (WC/toilet) and waste (other appliances) drainage of modern domestic properties is shown in Figure 6.1. This illustrates a 2-storey dwelling with appliances on both floors connected to a single vertical stack. Each appliance is protected by a trap (U-bend or S-bend) and water seal to prevent odours from reaching the house from the downstream drainage system.

The stack (typically 100–150 mm in diameter) has a top open to the atmosphere that should be at least 900 mm from the top of any adjacent opening into the property. The flow regime in a vertical stack is quite different from that in sloping pipes. Flow tends to adhere to the perimeter of the pipe, forming an annulus with a central air core. The pressure of the air in the core varies with height, depending on the appliances in use, and can be both positive and negative. Design rules and details have been devised to avoid the risk of water being siphoned from the traps.

The distance x from the lowest branch to the invert of the building drain typically exceeds 450 mm for dwellings up to three storeys high.

DOI: 10.1201/9781003408635-6

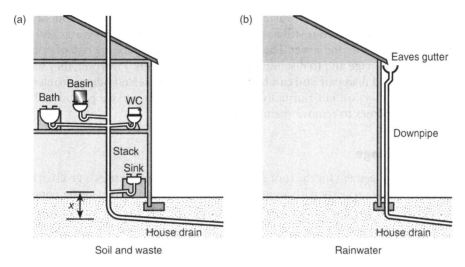

Figure 6.1 Typical building drainage arrangement in a 2-storey house.

6.2.1.2 Outside

The individual lengths of drains connecting each property to the public sewer tend to be short (usually <20 m) and small (diameter <150 mm). However, in terms of the total length of the whole piped system, they make up a surprisingly large fraction – perhaps as much as half.

6.2.1.3 Components

Outside building drainage systems have a number of common components, particularly associated with providing access for testing, inspection, and blockage clearance from the surface. A *rodding eye* permits rodding along the drain from the surface. It consists of a vertical or inclined riser pipe with a sealed, removable cover. Access can also be gained using an *access chamber* over a pipe fitting with a sealed, removable cover. *Inspection chambers* are also used and consist of shallow access points on the drain, and also have a sealed, removable cover. Woolley (1988) gives comprehensive coverage of building drainage details.

Building drains are typically designed using procedures similar to small foul sewers (described in Chapter 9). Gradients tend to be quite steep (>1:80 for 100 mm diameter pipes), although field evidence suggests that very flat drains are no more likely to block than steep ones (Lillywhite and Webster, 1979); good quality construction is more influential in reducing blockage potential. Drains and private sewers are relatively shallow with a minimum cover of 0.75 m under gardens and 1.25 m under roads and paths. The height x, plus the length and gradient of the drain, determine the minimum feasible depth of the public sewer.

6.2.1.4 Layout

The main aim of the layout of external building drainage is to minimise the length of pipework and associated components while ensuring that adequate accessibility is maintained. Generally, changes of direction should be minimised and appropriate access points provided where necessary.

Building drains carrying soil and waste should discharge only to a public foul or combined sewer. Some existing installations still feature an interceptor trap with a water seal in the last inspection chamber before the sewer. These were provided to reduce the risk of odour release into the building drainage and to discourage the entry of rodents. However, they have tended to fall into disuse and disrepair and can be a source of blockage and odour problems in their own right. Today, they are not normally specified, and water service providers have often undertaken programmes to remove them in problem areas.

6.2.2 Roof drainage

A conventional arrangement for the roof drainage of domestic properties is given in Figure 6.1. This shows a 2-storey property with a pitched roof, drained by an eaves gutter connected to a single, vertical downpipe, positioned at one end. A typical eaves gutter is a 75 mm half-round channel with a nominal fall. Its capacity can be estimated using the theory of spatially varied flow, and also depends on the position and spacing of the outlets.

Flow in the rainwater downpipe is annular, just like in the soil and waste stack. The type of inlet dictates capacity. For single-family dwellings, downpipes are 75–100 mm in diameter.

The downpipes can discharge directly to a separate storm sewer but will need a water seal trap if connected to a combined sewer. Roof drainage should *not* be discharged to separate foul sewers.

The design of roof drainage systems is similar to that of small storm sewer networks (Chapter 10 gives details). In this situation, the catchments are very small (<60 m^2), the time of concentration is low (1–2 minutes), and so short-duration, high-intensity rainfall events are critical for pipe capacity estimation. Approved Document H of The Building Regulations 2010 gives rainfall intensities appropriate for different parts of the UK, for example, 0.018–0.020 L/s/m^2 for Devon and Cornwall (equivalent to 65–72 mm/h). This figure is multiplied by roof area to determine runoff, with allowance for the slope of pitched roofs.

Readers are referred to Swaffield et al. (2015) and Wise and Swaffield (2002) for a fuller exposition of building drainage hydraulic design.

6.3 SYSTEM COMPONENTS

The design of sewer systems is covered in Chapters 9 and 10, but this section introduces the main "hardware" components of the sewerage.

6.3.1 Sewers

Sewer pipes have historically been made of a wide range of materials including vitrified clay, concrete, fibre cement, PVC-U and other polymers, pitch fibre, and brick. Information on current pipe materials and jointing methods is given in Chapter 16. Sewer cross-sectional shapes have also varied historically, and some of these are discussed in Chapter 7. Most new sewers are circular in cross section and range upwards in diameter from 150 mm.

6.3.1.1 Vertical alignment

Figure 6.2 illustrates how the vertical position of a sewer is defined by its invert level (IL). The invert of a pipe refers to the lowest point on the inside of the pipe. The IL is the vertical distance of the invert above some fixed level or *datum* (e.g., in the UK, above ordnance datum [AOD]). Other important levels shown in Figure 6.2 are the *soffit level*, which is the highest

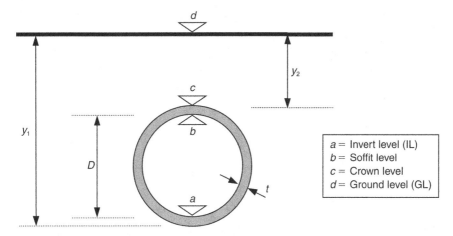

Figure 6.2 Level definitions associated with sewers.

EXAMPLE 6.1

A 375 mm diameter pipe with 15 mm walls has an invert level of 52.665 m. If the ground level is 54.930, calculate the pipe: (a) soffit level, (b) depth, and (c) cover.

Solution

a. Soffit level: $b = a + D = 52.665 + 0.375 = 53.040$ m
b. Depth (Equation 6.1): $y_1 = d - a + t = 54.930 - 52.665 + 0.015 = 2.280$ m
c. Cover (Equation 6.2): $y_2 = y_1 - D - 2t = 2.280 - 0.375 - 0.030 = 1.875$ m

---

point on the inside of the pipe, and the *crown*, which is the highest point on the outside of the pipe. Using the nomenclature defined in Figure 6.2, $b = a + D$ and $c = b + t = a + D + t$, where D is the internal diameter of the pipe (mm) and t is the pipe wall thickness (mm).

The depth of the pipe (y_1) is, therefore,

$$y_1 = d - a + t \tag{6.1}$$

and the cover of the pipe (y_2) is

$$y_2 = d - c = y_1 - D - 2t \tag{6.2}$$

Figure 6.3 shows a typical sewer vertical alignment plotted on a *longitudinal profile*. The profile contains the main information required in the vertical plane to construct the pipeline.

Two invert levels are given at two of the manholes since one refers to the exit and one to the entry level. At MH34, dissimilar diameter pipes meet, and good practice recommends (as shown) that soffit not invert levels are matched. "Chainage" refers to the plan (horizontal) distance along the pipe from a specific point. The top line of the profile box is fixed at a given level above datum (in this case 75 m). All other vertical levels can then be scaled from this line. The scale of the drawing is usually distorted to give more detail in the vertical plane.

Normal practice is to ensure that individual pipes between manholes have a constant gradient. Sewers are usually constructed under the highway with the storm sewer being on the

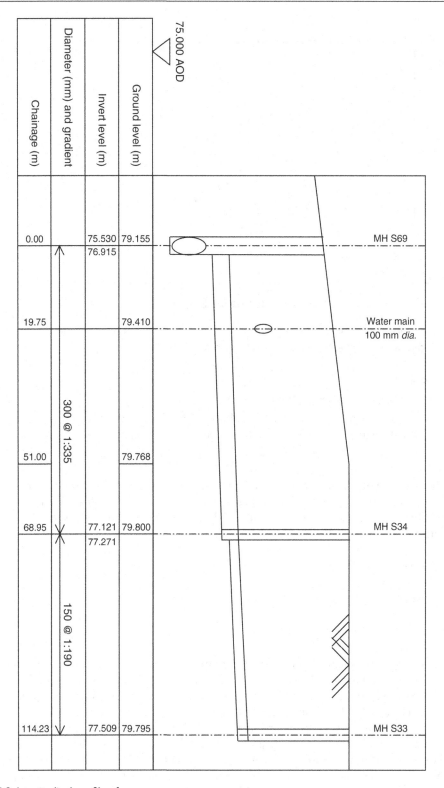

Figure 6.3 Longitudinal profile of a sewer.

centre line and the foul sewer being offset laterally and slightly lower. Should exfiltration occur, this will avoid pollution of the stormwater system. Sewers should be laid deep enough:

- To drain the lowest appliance in the premises served
- To withstand surface loads
- To prevent the contents from freezing

Typically, the minimum cover for rigid pipes is 0.9 m under gardens and fields and 1.2 m under roads.

6.3.1.2 Horizontal alignment

Figure 6.4 shows a typical sewer horizontal alignment with two ways of numbering the system. Figure 6.4 numbers the *pipes* in the form $(x.y)$, where x refers to the sewer branch and y refers to the individual pipe within the branch. Figure 6.4b numbers the manholes with a

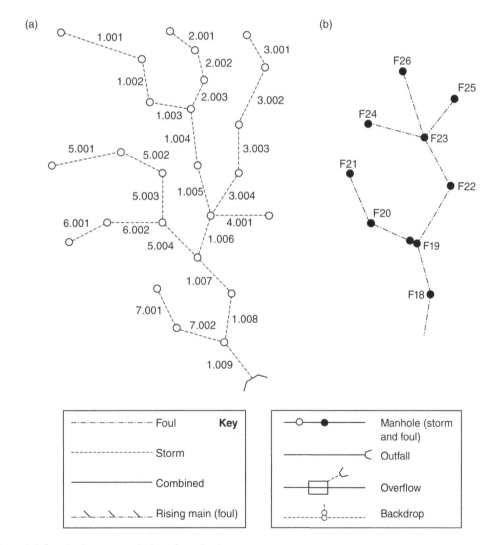

Figure 6.4 Standard sewer symbols and numbering systems.

unique code. The horizontal positions of manholes may be identified by their grid reference. Standard symbols are given in the key.

Good engineering practice is to ensure that individual pipe runs (between manholes) are straight in plan. However, larger (man-entry) sewers can be built with slight curves, if necessary. Further aspects of good practice in horizontal and vertical alignment are discussed later in the chapter under layout design.

Individual SuDS components (see Chapter 11) could also be included and mapped but although suggestions have been made for symbols, these have not yet been standardised (Kellagher and Woods-Ballard, 2019).

6.3.2 Manholes

As with building drainage systems, access points are required for testing, inspection, and cleaning. In sewer systems, access is usually by manholes that differ from inspection chambers in that they are deeper (>1 m) and can be entered if necessary. Manholes are provided at:

- Changes in direction
- Heads of runs
- Changes in gradient
- Changes in size
- Major junctions with other sewers
- Every 90–100 m

In larger pipes, where man-access is possible (although undesirable), the spacing of manholes may be increased up to 200 m. In tunnels, the lengths increase further and are dictated by operational requirements. Manholes are commonly constructed of precast concrete rings as specified in BS 5911-3: 2022. Figure 6.5 shows a detail of a precast concrete ring manhole. Smaller manholes may have precast benching. The diameter of the manhole depends on the size of the sewer and the orientation and number of inlets. Requirements for manhole

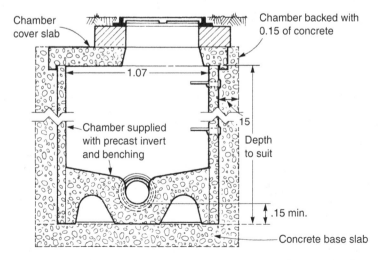

Figure 6.5 Precast concrete ring manhole. (Reproduced from Woolley, L. 1988. *Drainage Details*, 2nd edn, Taylor and Francis with permission of E&FN Spon.)

covers and frames are given in BS EN 124-1: 2015. More details on manholes can be found in Woolley (1988).

In situations where a high-level sewer is connected to one of a significantly lower level, a backdrop manhole can be used. These are typically used to bring the flow from higher-level laterals into a manhole rather than lowering the length of the last sewer lengths. Drops may be external or internal to the manhole, or sloping ramps may be used, depending on the drop height and the diameter of the pipe. Figure 6.6 shows an externally placed vertical backdrop manhole. Drop manholes can require additional maintenance.

Alternatives to straightforward drop arrangements, including vortex drop shafts, are considered in Chapter 8.

Historically, some separate sewer systems were provided with dual manholes consisting of a single chamber but giving access to both foul and storm sewers. However, the rodding eye stoppers have often become lost or broken resulting in flow transfer and pollution, so these are no longer specified.

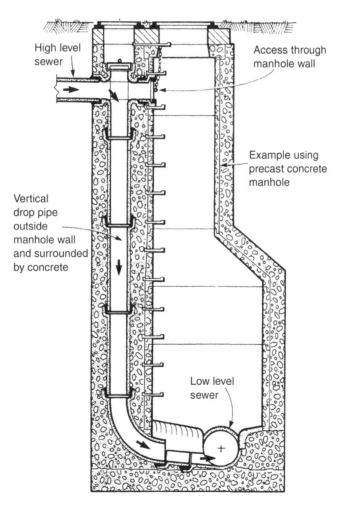

High level sewer

Access through manhole wall

Example using precast concrete manhole

Vertical drop pipe outside manhole wall and surrounded by concrete

Low level sewer

Figure 6.6 Backdrop manhole. (Reproduced from Woolley, L. 1988. *Drainage Details*, 2nd edn, Taylor and Francis with permission of E&FN Spon.)

6.3.3 Gully inlets

Surface runoff (stormwater) is admitted from roads and other paved areas via inlets known as *gullies* or *catch basins*. Gullies consist of a grating and usually an underlying sump (a *gully pot*) to collect heavy material in the flow. A water seal is incorporated to act as an odour trap for those gullies connected to combined sewers (see Figure 6.7). The gully is connected to the sewer by a lateral pipe. The relevant standard for gully gratings and frames is BS EN 124-1: 2015 and for precast concrete gully pots, BS 5911-6: 2021. Gully pots are also available in PVC-U and HDPE.

The size, number, and spacing of gullies determine the extent of surface ponding of runoff during storm events. Gullies are always placed at low points and, typically, are spaced along the road channel, adjacent to the curb. The simplest approach is to specify a standard of 50 m spacing or to require one gully per 200 m^2 of impervious area. More accurate (but more involved) gully spacing methods are explained in Chapter 8.

6.3.4 Ventilation

Ventilation is required in all urban drainage systems, particularly in foul and combined sewers and storage tanks. It is needed to ensure that aerobic conditions are maintained within the pipe and to avoid the possibility of the buildup of toxic or explosive gases. The implications of anaerobic conditions and health and safety issues are discussed in Chapter 18.

Nearly all sewer systems are ventilated passively, without air extraction equipment. However, some major pumping stations and wastewater treatment plants are mechanically ventilated. In larger and older schemes, above-ground ventilation shafts have been used to ensure good circulation of air. Care is needed in siting these structures to avoid odour nuisance. Some schemes use ventilated manhole covers. More modern practice is to utilise

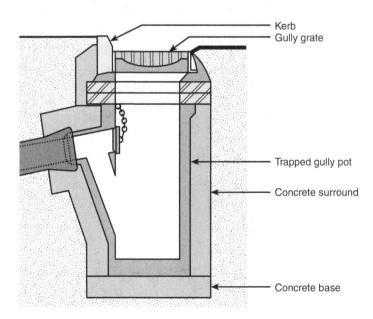

Figure 6.7 Trapped road gully.

the ventilation provided by the soil stacks on individual buildings (Figure 6.1) or individual vent pipes, such as on large storage chambers. Air is drawn through the system by the low pressure induced by the flow of air over the top of the stacks, and by the fall of wastewater. The water seals on domestic appliances avoid the backup of sewer gases into the building interior.

6.4 DESIGN

6.4.1 Stages

A number of fundamental stages need to be followed to design a rational and cost-effective urban drainage system. These are illustrated in Figure 6.8 and are valid for any type of system. Further details on sewer sizing are given in Chapters 9, 10 and 13.

The first stage is to define the *contributing area* (catchment area and population) and display it on a (digital) topographical map. Sewer software packages can often do this automatically. The map should already include contours, but other pertinent natural (e.g., rivers) and man-made (e.g., buildings, roads, services) features should also be marked up. Possible outfall or overflow points should be identified and investigations made as to the capacity of the receiving waterbody.

The next stage is to produce a preliminary *horizontal alignment* aiming to achieve a balance between the requirement to drain the whole contributing area and the need to minimise pipe run lengths. Least-cost designs tend to result when the pipe network broadly follows the natural drainage patterns and is branched, converging to a single major outfall.

Having located the pipes horizontally, the pipe sizes and gradients can now be calculated based on estimated flows from the contributing area as described in Chapters 9 and 10. Generally, sewers should follow the slope of the ground as far as possible to minimise excavation. However, gradients flatter than 1:500 should be avoided as they are difficult to construct accurately. A preliminary *vertical alignment* can then be produced, again bearing in mind the balance between coverage of the area and depths of pipes. The alignment can be plotted on longitudinal profiles as shown in Figure 6.3. The final stage involves revising both the horizontal and vertical alignment to minimise cost by reducing pipe lengths, sizes, and depths while meeting the hydraulic design criteria. Longer sewer runs may be cost-effective if shorter runs would require costlier excavation and/or pumping.

6.4.2 Sewers for adoption

When new residential areas are built, it is common for developers to construct the building drainage and some of the "communal" sewers. The aim of the developers and of the subsequent house owners is for the drainage network to be accepted by the local sewerage undertaker and be adopted as part of the network under its control. To aid this process, a set of design and construction guidelines has been produced, which is generally agreed upon and accepted (Water UK, 2023). The main areas covered by this document are:

- Design and construction advice
- Standard details
- Model civil, mechanical, and electrical engineering specifications

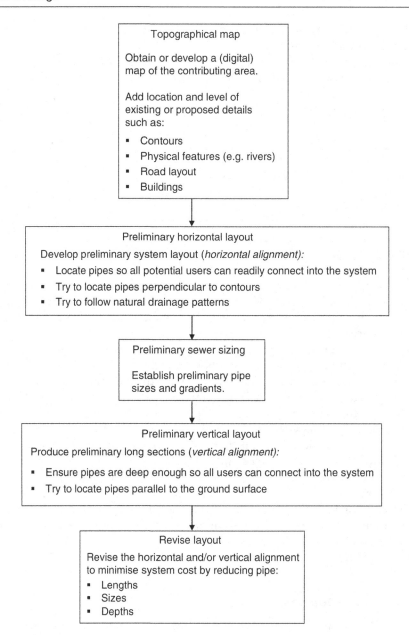

Figure 6.8 Urban drainage design procedure.

PROBLEMS

1 Describe the main components of building drainage. Compare and contrast them with those used in main sewer systems.

2 A 375 mm internal diameter, 1:258 gradient sewer connects two manholes A and B. The upstream manhole A has coordinates E 274.698, N 842.393, and the soffit level of the sewer leaving it is 16.438 m. Assuming negligible fall across manhole B (E 342.812, N 864.844), what is the invert level of the 450 mm diameter exiting pipe? If the cover level at manhole B is 18.590 m, what is its depth?   [15.710 m, 2.880 m]

3 Explain how manholes differ from inspection chambers. Why and where would you locate manholes? Where are backdrop manholes used and why?
4 Explain the function of road gullies. Why, where, and how would you locate them?
5 How are sewer systems ventilated?
6 List the main stages in urban drainage design. Assess the main factors affecting the horizontal and vertical alignment of the system.

KEY SOURCES

Swaffield, J., Gormley, M., Wright, G., and McDougall, I. 2015. *Transient Free Surface Flows in Building Drainage Systems*, Routledge, Oxon.

Wise, A.F.E. and Swaffield, J.A. 2002. *Water, Sanitary and Waste Services for Buildings*, 5th edn, Taylor and Francis.

Woolley, L. 1988. *Drainage Details*, 2nd edn, Taylor and Francis.

Water UK. 2023. *Sewerage Sector Guidance.* https://www.water.org.uk/sewerage-sector-guidance-approved-documents

REFERENCES

Approved Document H. 2015. *Drainage and Waste Disposal*, The Building Regulations 2010, HM Government.

BS 5911-3: 2022. *Concrete pipes and ancillary concrete products. Concrete manholes, inspection chambers and soakaways (complementary to BS EN 1917:2002). Specification.*

BS 5911-6: 2021. *Concrete Pipes and Ancillary Concrete Products - Road Gullies and Gully Cover Slabs. Specification.*

BS EN 124-1: 2015. *Gully Tops and Manhole Tops for Vehicular and Pedestrian Areas. Definitions, Classification, General Principles of Design, Performance Requirements and Test Methods.*

BS EN 12056-2: 2000. *Gravity drainage systems inside buildings. Sanitary pipework, layout and calculation.*

Kellagher, R. and Woods-Ballard, B. 2019. *SuDS Asset Register and Mapping. Review of Current Status and Recommendations.* Report MAR5930-RT001-R04-00, Release 04-00, HR Wallingford.

Lillywhite, M.S.T. and Webster, C.J.D. 1979. Investigations of drain blockages and their implications for design. *The Public Health Engineer*, 7(2), 53–60.

Chapter 7

Hydraulics

7.1 INTRODUCTION

An understanding of hydraulics is needed in the design of new drainage systems in order to specify the appropriate size of system components, especially pipes, channels, and tanks. It is also needed in the analysis and modelling of existing systems in order to predict the relationship between flow rate and depth for varying inflows and conditions.

The study of civil engineering hydraulics tends to concentrate on two main types of flow. The first is *pipe flow* in which a liquid flows in a pipe under pressure. The liquid always fills the whole cross section, and the pipe may be horizontal or inclined up or down in the direction of flow. The second is *open-channel flow*, in which a liquid flows in a channel by gravity, with a free surface at atmospheric pressure. The liquid only fills the channel when the flow rate equals or exceeds the designed capacity, and the bed of the channel slopes down in the direction of flow.

The most common type of flow in sewer systems is a hybrid of these two: *part-full pipe flow*, in which a liquid flows in a pipe by gravity, with a free surface. The liquid only fills the pipe area when the flow rate equals or exceeds the designed capacity, and the bed of the pipe slopes down in the direction of flow. Traditionally the theories used are most closely related to those for full pipes, though part-full pipe flow is actually a special case of open-channel flow.

In this chapter, we cover basic hydraulic principles (Section 7.2), pipe flow (Section 7.3), part-full pipe flow (Section 7.4), and open-channel flow (Section 7.5). Aspects of hydraulics are developed where appropriate in other chapters, for example, those relating to special features in Chapter 8; to the design of sewers in Chapters 9, 10, and 13; storage in Chapter 14; pumped systems in Chapter 15; and flow models in Chapter 20.

The content is intended as an introduction to the subject for those who have not studied it before, or as a refresher course for those who have. For more information, a number of sources are referred to in the text. For general reference, Chadwick et al. (2021) give a practical engineering treatment of all aspects. Hamill (2011) provides helpful explanations, and Kay (2016) uses a more descriptive approach.

7.2 BASIC PRINCIPLES

7.2.1 Pressure

Pressure is defined as force per unit area. The common units for pressure are kN/m^2 or bars (1 bar = 100 kN/m^2). *Absolute pressure* is the pressure relative to a vacuum, and *gauge*

DOI: 10.1201/9781003408635-7

pressure is the pressure relative to atmospheric pressure. Gauge pressure is used in most hydraulic calculations. Atmospheric pressure varies but is approximately 1 bar.

In a still liquid, pressure increases with vertical depth:

$$\Delta p = \rho g \Delta y \tag{7.1}$$

where Δp is the increase in pressure (N/m^2), ρ is the density of the liquid (for water, 1000 kg/m^3), g is the gravitational acceleration (9.81 m/s^2), and Δy is the increase in depth (m).

Pressure at a point is equal in all directions.

7.2.2 Continuity of flow

In a section of pipe with constant diameter and no side connections (Figure 7.1), in any period of time, the mass of liquid entering (at 1) must equal the mass leaving (at 2). Assuming that the liquid has a constant density (mass per unit volume), the volume entering (at 1) must equal the volume leaving (at 2).

In terms of flow rate (volume per unit time, Q),

$$Q_1 = Q_2 \tag{7.2}$$

The common units for flow rate are m^3/s or L/s (1 L [Litre] = 10^{-3} m^3).

The velocity of the liquid varies across the flow cross section, with the maximum for a full pipe in the centre. "Mean velocity" (v) is defined as the flow rate per unit area (A) through which the flow passes:

$$v = \frac{Q}{A} \tag{7.3}$$

The common units for velocity are m/s (see Example 7.1). Equation (7.2) can be rewritten as

$$v_1 A_1 = v_2 A_2$$

7.2.3 Flow classification

In hydraulics there are two terms for "constant": *uniform* and *steady*. Uniform means constant with distance, and steady means constant with time. The words have negative forms: *nonuniform* means not constant with distance; *unsteady* means not constant with time.

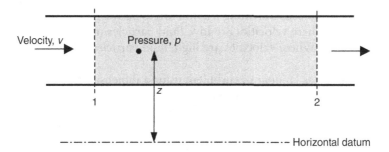

Figure 7.1 Continuity and definition of symbols for pipe flow.

EXAMPLE 7.1

The diameter of a pipe flowing full increases from 150 to 200 mm. The flow rate is 20 L/s (0.020 m³/s). Determine the mean velocity upstream and downstream of the expansion.

Solution

$$\text{Upstream}\,(1): A_1 = \frac{\pi 0.15^2}{4} \text{ so } \upsilon_1 = \frac{Q}{A_1} = \frac{0.02 \times 4}{\pi 0.15^2} = 1.13\,\text{m/s}$$

$$\text{Downstream}\,(2): A_2 = \frac{\pi 0.2^2}{4} \text{ so } \upsilon_2 = \frac{Q}{A_2} = \frac{0.02 \times 4}{\pi 0.2^2} = 0.64\,\text{m/s}$$

Hydraulic conditions in urban drainage systems can be

- Uniform steady: The flow cross-sectional area is constant with distance, and the flow rate is constant with time.
- Nonuniform steady: The flow area varies with distance, but the flow rate is constant with time.
- Uniform unsteady: The flow area is constant with distance, but the flow rate varies with time.
- Nonuniform unsteady: The flow area varies with distance, and the flow rate varies with time.

Flow in sewers is generally unsteady to some extent: wastewater varies with the time of day, and storm flow varies during a storm. However, in many hydraulic calculations, it is not necessary to take this into account, and conditions are treated as steady for the sake of simplicity. Most of the rest of this chapter considers only steady flow. In some cases, unsteady effects are significant and must be considered: for example, storage effects (considered in Chapter 14), sudden changes in pumping systems (referred to in 15.4.3), and storm waves in sewers (see Sections 10.6 and 20.8).

7.2.4 Laminar and turbulent flow

The property of a fluid that opposes its motion is called *viscosity*. Viscosity is caused by the interaction of the fluid molecules creating friction forces between the layers of fluid travelling at different velocities. Where velocities are low, fluid particles move in straight, parallel trajectories – *laminar* flow. Where velocities are high, fluid particles follow more chaotic paths, and the flow is *turbulent*.

Flow can be identified as laminar or turbulent using a dimensionless number, the Reynolds number (R_e), defined for a pipe flowing full as

$$R_e = \frac{\upsilon D}{\nu} \tag{7.4}$$

where υ is the mean velocity (m/s), D is the pipe diameter (m), and ν is the kinematic viscosity of liquid (for water, typically 1.1×10^{-6} m²/s, dependent on temperature).

EXAMPLE 7.2

For the pipe in Example 7.1, determine the Reynolds number for both sides of the change in diameter. Is the flow laminar or turbulent? Assume that the pipe carries water with kinematic viscosity of $1.1 \times 10^{-6} m^2/s$.

Solution

$$\text{Reynolds number upstream} \left(1\right) = \frac{\upsilon_1 D_1}{\nu} = \frac{1.13 \times 0.15}{1.1 \times 10^{-6}} = 154,100$$

$$\text{Reynolds number downstream} \left(2\right) = \frac{\upsilon_2 D_2}{\nu} = \frac{0.64 \times 0.2}{1.1 \times 10^{-6}} = 116,400$$

Both are well into the turbulent zone.

When $R_e < 2000$, the flow in the pipe is laminar; when $R_e > 4000$, the flow is turbulent. In most urban drainage applications, flow is firmly in the turbulent region (see Example 7.2).

7.2.5 Energy and head

A flowing liquid has three main types of energy: pressure, velocity, and potential. In hydraulics, the most common way of expressing energy is in terms of *head* which can be defined as energy per unit weight (common units, m).

The three types of energy expressed as head are

$$\text{pressure head} \left(\frac{p}{\rho g} \right) \quad \text{velocity head} \left(\frac{\upsilon^2}{2g} \right) \quad \text{potential head } z$$

The symbols p, υ, and z are illustrated in Figure 7.1.

Total head (H) is the sum of the three types of head, given by the *Bernoulli equation*:

$$H = \frac{p_1}{\rho g} + \frac{\upsilon^2}{2g} + z \tag{7.5}$$

When a liquid flows in a pipe or a channel, some head (h_L) is "lost" from the liquid. So, for water flowing between sections 1 and 2 in the full pipe in Figure 7.1:

$$H_1 - h_L = H_2 \tag{7.6}$$

or

$$\frac{p_1}{\rho g} + \frac{\upsilon_1^2}{2g} + z_1 - h_L = \frac{p_2}{\rho g} + \frac{\upsilon_2^2}{2g} + z_2 \tag{7.7}$$

If the flow in Figure 7.1 is uniform, and the pipe is horizontal, then

$$\upsilon_1 = \upsilon_2 \text{ and } z_1 = z_2$$

Therefore, Equation (7.7) becomes

$$h_L = \frac{p_1}{\rho g} - \frac{p_2}{\rho g}$$

The head loss is equal to the difference in pressure head.

7.3 PIPE FLOW

7.3.1 Head (energy) losses

The head or energy losses in flow in a pipe are made up of *friction losses* (h_f) and *local losses* (h_{local}). Friction losses are caused by forces between the liquid and the solid boundary (distributed along the length of the pipe), and local losses are caused by disruptions to the flow at local features like bends and changes in cross section. Total head loss h_L is the sum of the two components.

The distribution of losses, and the other components in Equation (7.7), can be shown by two imaginary lines:

- The *energy grade line* (EGL) is drawn at a vertical distance from the datum equal to the total head.
- The *hydraulic grade line* (HGL) is drawn at a vertical distance below the EGL equal to the velocity head.

The two lines are drawn for a pipe flowing full in Figure 7.2. The lines allow all the terms in Equation (7.7) to be identified.

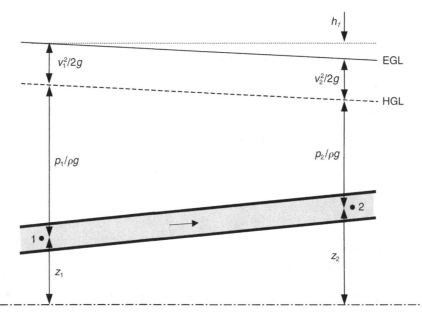

Figure 7.2 EGL and HGL for a pipe flowing full.

7.3.2 Friction losses

A fundamental requirement in the hydraulic design and analysis of urban drainage systems is the estimation of friction loss. The basic representation of friction losses, valid for both laminar and turbulent flow, is the Darcy–Weisbach equation:

$$h_f = \frac{\lambda L}{D} \cdot \frac{v^2}{2g} \qquad (7.8)$$

where h_f is the head loss due to friction (m) (Figure 7.2), λ is the friction factor (no units), L is the pipe length (m), and D is the pipe diameter (m).

The term h_f/L is the gradient of the EGL, and (for uniform flow) of the HGL, and is often referred to as the *hydraulic gradient* or *friction slope*.

7.3.3 Friction factor

The friction factor λ is one of the most interesting aspects of pipe hydraulics. Is it constant for a particular pipe or does it vary? Does it depend on pipe roughness or not?

As mentioned, velocity varies over the flow area. In turbulent flow (the type of flow of most significance in urban drainage), velocity levels are similar across most of the cross section but fall rapidly near the pipe wall (Figure 7.3). Very near to the wall, a boundary layer exists, where the velocity is low and laminar conditions occur: a *laminar sub-layer*.

Frictional losses are affected by the thickness of the laminar sub-layer relative to the "size" of the roughness of the pipe wall. In commercial pipes, the wall roughness is measured in terms of an equivalent sand roughness size (k_s) and can be thought of as the mean projection height of the roughness from the pipe wall.

Figure 7.4 presents the Moody diagram. This is a plot of the friction factor λ, against Reynolds number R_e for a range of values of relative roughness k_s/D. The Moody diagram demonstrates the relative effects of the thickness of the laminar sub-layer and the size of the roughness.

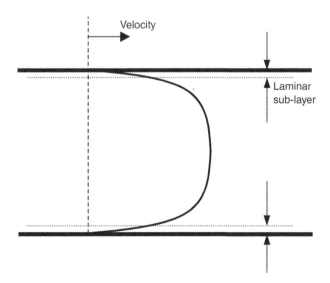

Figure 7.3 Velocity profile in turbulent flow, with laminar sub-layer.

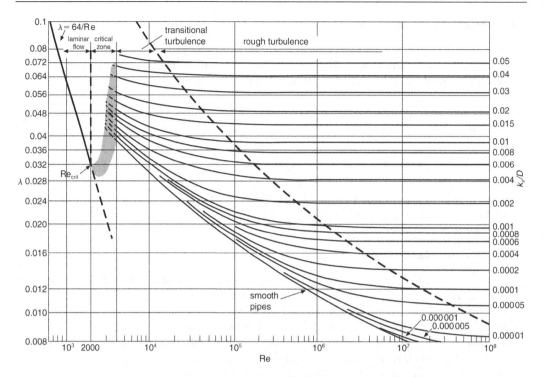

Figure 7.4 Moody diagram. (Reproduced from Chadwick A. et al. 2021. *Hydraulics in Civil and Environmental Engineering*, 6th edn, CRC Press. With permission of E & FN Spon.)

When the roughness projections are small compared with the thickness of the laminar sub-layer, the friction losses are independent of pipe roughness and dependent on Reynolds number. Flow is said to be "smooth turbulent." And λ is a function of R_e, but not of k_s/D.

When the roughness projections are much greater in height, losses are linked to pipe roughness only (not Reynolds number), and conditions are known as "rough turbulent." Then λ is a function of k_s/D, but not of R_e.

Between these two conditions lies a transitional turbulent region where conditions are dependent on roughness height and Reynolds number. Most urban drainage flows are rough or transitionally turbulent flows.

Whichever is the greater, the thickness of the laminar sub-layer or the size of the roughness, the friction factor can be determined mathematically from the Colebrook–White equation:

$$\frac{1}{\sqrt{\lambda}} = -2log_{10}\left(\frac{k_s}{3.7D} + \frac{2.51}{R_e\sqrt{\lambda}}\right) \tag{7.9}$$

The Moody diagram illustrates the relationship between the basic hydraulic variables and their relative importance, but it does not directly represent the variables routinely used in engineering practice: flow rate, velocity, pipe diameter, roughness, and gradient.

The Colebrook–White equation can provide an explicit expression for velocity by substitution of λ using Equation (7.8) and R_e using Equation (7.4), giving

$$v = -2\sqrt{2gS_fD}\,log_{10}\left(\frac{k_s}{3.7D} + \frac{2.51v}{D\sqrt{2gS_fD}}\right) \tag{7.10}$$

where k_s is the pipe roughness (m), S_f is the hydraulic gradient or friction slope, h_f/L (–), and ν is the kinematic viscosity (m²/s).

As illustrated by Example 7.3, this equation can be easily solved for ν. However, it is rather intractable if the variable of concern is the hydraulic gradient (S_f) or pipe diameter (D). For these cases solutions can be obtained using a spreadsheet or equivalent, or by using design tables or charts. Any software that is being used to design a pipe system will include this.

EXAMPLE 7.3

1. A 600 mm diameter circular pipe is used to pump stormwater. Calculate the mean velocity and flow rate in the pipe if the hydraulic gradient is 0.01 by solving the Colebrook–White equation. Assume the roughness height k_s is 0.6 mm and kinematic viscosity is 1.14×10^{-6} m²/s.

2. What value of hydraulic gradient is suggested by the Moody diagram in conjunction with the Darcy–Weisbach equation? (Use the value of v already determined, together with the remaining data given.) Are conditions smooth, transitional, or rough turbulent?

3. Check v and Q using the chart in Figure 7.5.

Solution

1. Velocity can be calculated by direct substitution into the Colebrook–White Equation (7.10) given

$$D = 0.6 \text{ m}, S_f = 0.01, k_s = 0.6 \text{ mm and } v = 1.14 \times 10^{-6} \text{ m}^2/\text{s}$$

$$\upsilon = -2\sqrt{2g0.01 \times 0.6} \log_{10} \left(\frac{0.6 \times 10^{-3}}{3.7 \times 0.6} + \frac{2.51 \times 1.14 \times 10^{-6}}{0.6\sqrt{2g0.01 \times 0.6}} \right)$$

$$= 2.43 \text{ m/s}$$

and flow rate by continuity (Section 7.3):

$$Q = \upsilon A = 2.43 \times \frac{\pi 0.6^2}{4} = 688\text{L/s}$$

2. $\dfrac{k_s}{D} = \dfrac{0.6}{600} = 0.001 \, R_e = \dfrac{2.43 \times 0.6}{1.14 \times 10^{-6}} = 1,279,000$

From the Moody diagram (Figure 7.4), $\lambda = 0.019$ (rough), so from Equation (7.8):

$$\frac{h_f}{L} = \frac{\lambda}{D}\frac{\upsilon^2}{2g} = \frac{0.019}{0.6}\frac{2.43^2}{2g} = 0.01$$

3. Figure 7.5 can be used by finding the point of intersection of the hydraulic gradient line (read downwards sloping right to left) with the diameter line (read vertically upwards).

Hydraulic gradient = 0.01, or 1 in 100, $D = 0.6$ m.
This gives

$$v \simeq 2.4 \text{ m/s} \quad Q \simeq 700 \text{ L/s}$$

7.3.4 Tables and charts

7.3.4.1 Tables

Output from the Colebrook–White equation can be presented in tables. Table 7.1 is an example, giving Q and v for a variety of values of D and S_f, for k_s of 1.5 mm and kinematic viscosity of 1.14×10^{-6} m²/s. Comprehensive sets of tables that fulfil this function are commercially available (Wallingford and Barr, 2006).

7.3.4.2 Charts

Output from the Colebrook–White equation can also be presented on charts. These give a good visual summary for use when carrying out calculations manually (as with the examples and problems in this book). Figure 7.5 is an example, giving the relationship between Q, v, D, and S_f for k_s of 0.6 mm and kinematic viscosity of 1.14×10^{-6} m²/s.

7.3.5 Roughness

Typical values of roughness k_s for use in the Colebrook–White equation, related to sewer type and age, are given in Table 7.2. The values described as "new" are appropriate for new, clean, and well-aligned pipes. They are appropriate in the design of stormwater pipes (where excessive sediment content is not expected) or in establishing the initial flow conditions in newly laid foul or combined sewers. "Old" values are generally preferred in the design or analysis of foul and combined sewers, where roughness is related more to the effect of biological slime than the pipe material.

For preliminary design purposes, or where existing pipe conditions are unknown, a value of $k_s = 0.6$ mm is suggested for storm sewers and $k_s = 1.5$ mm for foul sewers (irrespective of pipe material). Sewers subject to sediment deposition can have k_s values in the order of 30–60 mm (see Chapter 16 for further discussion). As an example of the effect of roughness height, for a 150 mm diameter pipe conveying 10 L/s, a change in the value of k_s from 0.6 to 1.5 mm would result in an increase in flow depth and a decrease in flow velocity of about 10%.

For rising mains, roughness can be empirically related to velocity (Forty et al., 2004; Lauchlan et al., 2005):

$$k_s \approx av^{-2.34} \tag{7.11}$$

where $a = 0.45$ for $v = 1.0$ m/s, $a = 0.17$ for $v = 1.5$ m/s, and $a = 0.09$ for $v = 2.0$ m/s.

Table 7.1 Table of output from the Colebrook–White equation, for k_s = 1.5 mm Q (L/s) in **bold**; v (m/s) in *italic*

Diameter (m)	$S_f = 0.001$		$S_f = 0.002$		$S_f = 0.005$		$S_f = 0.01$		$S_f = 0.02$		$S_f = 0.05$	
0.2	**10**	*0.33*	**15**	*0.47*	**23**	*0.75*	**33**	*1.06*	**47**	*1.50*	**75**	*2.38*
0.3	**31**	*0.43*	**43**	*0.62*	**69**	*0.98*	**98**	*1.38*	**139**	*1.96*	**220**	*3.11*
0.375	**55**	*0.50*	**79**	*0.71*	**125**	*1.13*	**177**	*1.60*	**250**	*2.27*	**396**	*3.59*
0.75	**347**	*0.79*	**492**	*1.11*	**780**	*1.76*	**1104**	*2.50*	**1562**	*3.54*	**2472**	*5.60*

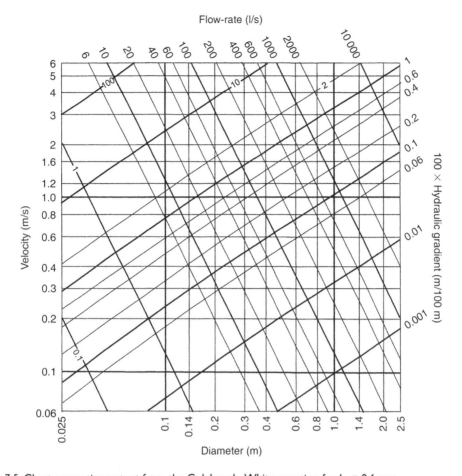

Figure 7.5 Chart presenting output from the Colebrook–White equation for k_s = 0.6 mm.

Table 7.2 Typical values of roughness (k_s)

Pipe material	k_s range (mm)	
	New	Old
Clay	0.03–0.15	0.3–3.0
PVC–U (and other polymers)	0.03–0.06	0.15–1.50
Concrete	0.06–1.50	1.5–6.0
Fibre cement	0.015–0.030	0.6–6.0
Brickwork – good condition	0.6–6.0	3.0–15
Brickwork – poor condition	–	15–30
Rising mains	0.03–0.60	

EXAMPLE 7.4

A circular storm sewer is to be designed to run just full when conveying a flow rate of 0.07 m³/s. If the mean velocity is specified at 1 m/s, calculate the required pipe diameter and gradient using an appropriate method. Assume the roughness height is 0.6 mm and kinematic viscosity is 1.141×10^{-6} m²/s.

Solution

In principle, this problem could be solved using the Colebrook–White equation. However, this would require an iterative solution, best suited to a computational method. A direct solution can be achieved by reading from the chart for $k_s = 0.6$ mm (Figure 7.5):

Q = 70 L/s, v = 1.0 m/s

This gives.

D = 0.3 m, and gradient = 0.42 in 100, or 1 in 240.

The Wallingford tables (Wallingford and Barr, 2006) recommend that for rising mains in "normal" condition, k_s should be taken as 0.3 mm for mean velocities up to 1.1 m/s, and 0.15 mm for velocities between 1.1 and 1.8 m/s.

7.3.6 Local losses

Local losses occur at points where the flow is disrupted, such as bends, valves, and changes of area. In certain circumstances (e.g., where there are many fittings in a short length of pipe), these can be equal to or greater than the friction losses. Local losses are usually expressed in terms of velocity head as follows:

$$h_{\text{local}} = k_L \frac{v^2}{2g} \tag{7.12}$$

where h_{local} is the local head loss (m) and k_L is a constant for the particular type of fitting (–).

In gravity sewers, local losses occur at manholes, but these are only usually significant when the system is surcharged.

Examples of the value of k_L used in design practice are given in Table 7.3.

For design cases when velocity is unknown and calculations are based on the Wallingford tables, a useful alternative to Equation (7.12) is to express local losses as an *equivalent length* of pipe. The sum of this equivalent length and the actual pipe length is then multiplied by the hydraulic gradient (taken from the table) to give the total (friction + local) losses.

It is helpful to relate this to the friction factor λ. Combining Equations (7.8) and (7.12) gives

$$\text{Total losses} = \frac{\lambda L}{D} \frac{v^2}{2g} + k_L \frac{v^2}{2g}$$

Table 7.3 Local head loss constants

Fitting	k_L
Pipe entry (sharp-edged)	0.50
Pipe entry (slightly rounded)	0.25
Pipe entry (bell-mouthed)	0.05
Pipe exit (sudden)	1.0
90° pipe bend ("elbow" – sharp bend)	1.0
90° pipe bend (long)	0.2
Straight manhole on gravity sewer (part-full)	< 0.1
Straight manhole on gravity sewer (surcharged)	0.15
Manhole with 30° bend (surcharged)	0.5
Manhole with 60° bend (surcharged)	1.0

So, if an equivalent length L_E is to be added to L in order to replace the separate term for local losses,

$$\frac{L_E}{D} = \frac{k_L}{\lambda} \tag{7.13}$$

In rough turbulence (Figure 7.4) L_E is independent of v, but in transitional or smooth turbulence it is not (because λ is affected by R_e). In addition to being available from the Moody diagram, the value of λ can be inferred from the Wallingford tables, since

$$\lambda = \frac{S_f D 2g}{v^2}$$

we have

$$L_E = \frac{k_L}{S_f} \frac{v^2}{2g}$$

The determination of equivalent length is shown in Example 7.5. Its use will be demonstrated in Example 8.2.

EXAMPLE 7.5

A surcharged manhole with a bend has local loss constant $k_L = 1$. Determine L_E/D (assuming that it is independent of velocity) if this feature occurs in a pipe with a diameter of (1) 300 mm or (2) 600 mm (k_s 1.5 mm for both). For both cases determine the conditions for which the equivalent length is independent of velocity. (Assume kinematic viscosity $= 1.14 \times 10^{-6}$ m²/s.)

· **Solution**

1. $\dfrac{k_s}{D} = \dfrac{1.5}{300} = 0.005$

Equivalent length is independent of velocity in rough turbulence. From the Moody diagram (Figure 7.4), $\lambda = 0.03$, so (Equation (7.13)

$$\frac{L_E}{D} = \frac{k_L}{\lambda} = \frac{1.0}{0.03} = 33$$

This is valid if R_e is greater than 200,000 – that is, velocity greater than 0.76 m/s.

2. $\dfrac{k_s}{D} = \dfrac{1.5}{600} = 0.0025$

Equivalent length is independent of velocity in rough turbulence. From the Moody diagram (Figure 7.4),

$$\lambda = 0.025, \text{so} \frac{L_E}{D} = \frac{k_L}{\lambda} = \frac{1.0}{0.025} = 40$$

This is valid if R_e is greater than 500,000 – that is, velocity greater than 0.95 m/s.

7.4 PART-FULL PIPE FLOW

The common flow condition in urban drainage pipes is part-full pipe flow. The presence of the free surface must be taken into account in hydraulic computations.

7.4.1 Normal depth

In uniform steady gravity flow, an equilibrium exists along a part-full pipe or channel. The energy consumed by friction between the liquid and the pipe wall is in balance with the fall along the pipe length. If the pipe slope could be increased for the same flow rate, additional energy would be available to the flow, resulting in higher velocity and lower depth. The equilibrium depth is referred to as the *normal depth*.

Since the depth of flow and velocity are constant when conditions are uniform, and pressure at the surface is atmospheric, the EGL and HGL are parallel to the bed, and the HGL coincides with the water surface (Figure 7.6).

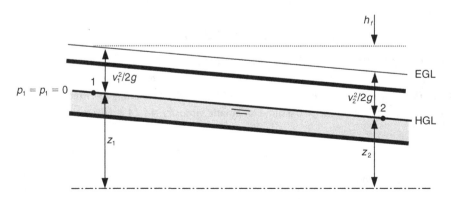

Figure 7.6 EGL and HGL for a pipe flowing part-full.

7.4.2 Geometric and hydraulic elements

When water flows along a part-full pipe, a number of properties (geometric elements), shown in Figure 7.7, can be defined as in Table 7.4. Geometric elements are defined only in terms of the geometry of the pipe cross section and the depth of flow.

Hydraulic radius can be related to geometrical properties and, for a circular pipe running full (for example),

$$R = \frac{A}{P} = \frac{\pi D^2 / 4}{\pi D} = \frac{D}{4} \qquad (7.14)$$

where D is the pipe diameter (m) (see Example 7.6).

A circular cross-sectional shape is most common for sewers and drains. Figure 7.7 shows the cross section of a pipe of diameter D with the flow of depth d. The angle subtended at the pipe centre by the free surface is θ. From geometrical considerations, θ is related to the proportional depth of flow d/D as follows:

$$\theta = 2\cos^{-1}\left[1 - \frac{2d}{D}\right] \qquad (7.15)$$

Expressions for area (A), wetted perimeter (P), hydraulic radius (R), top width (B), and hydraulic mean depth (d_m), based on D and θ, are given in Table 7.5.

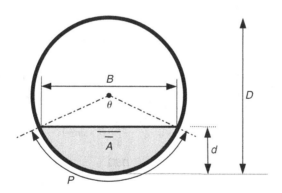

Figure 7.7 Definition of geometric elements for a circular pipe.

Table 7.4 Geometric elements

Property	Symbol	Definition	Common units
Depth	d	Height of water above the channel invert	m
Area	A	Cross-sectional area of flow	m²
Wetted perimeter	P	Portion of the flow area's perimeter that is in contact with the channel	m
Hydraulic radius	R	A per unit P	m
Top width	B	Flow width at the water surface	m
Hydraulic mean depth	d_m	A per unit B	m

EXAMPLE 7.6

A circular sewer of diameter 300 mm is flowing half full. Determine $d, A, P, R, B,$ and d_m.

Solution

Calculations are simply based on the property definitions (Table 7.4) and the geometry of a circle.

$$D = 0.3\,\text{m}$$

So,

$$d = 0.15\,\text{m}$$

$$A = \frac{1}{2} \cdot \frac{\pi D^2}{4} = \frac{1}{2} \cdot \frac{\pi 0.3^2}{4} = 0.0353\,\text{m}^2$$

$$P = \frac{\pi D}{2} = \frac{\pi 0.3}{2} = 0.471\,\text{m}$$

$$R = \frac{A}{P} = \frac{0.0353}{0.471} = 0.075\,\text{m}$$

$$B = D = 0.3\,\text{m}$$

$$d_m = \frac{A}{B} = \frac{0.0353}{0.3} = 0.118\,\text{m}$$

Table 7.5 Expressions for geometric elements in a part-full circular pipe

Parameter	Expression (θ in radians)
A	$\dfrac{D^2}{8}\left(\theta - \sin\theta\right)$
P	$D\theta/2$
R	$\dfrac{A}{P} = \dfrac{D}{4}\left[\dfrac{\theta - \sin\theta}{\theta}\right]$
B	$D\sin(\theta/2)$
d_m	$\dfrac{A}{B} = \dfrac{D(\theta - \sin\theta)}{8\sin\theta/2}$

Using these relationships, Figure 7.8 gives dimensionless relationships for the geometric elements of a part-full pipe, cross-sectional area, wetted perimeter, and hydraulic radius as a proportion of the full-depth value (i.e., A/A_f, P/P_f, R/R_f, and d/D). Also shown are lines representing the hydraulic elements (that is properties related to flow behaviour within the pipe), velocity ratio v/v_f and flow ratio Q/Q_f (where v and Q are part-full velocity and flow rate, and v_f and Q_f are the full-depth values). These latter lines are slightly dependent on the friction loss equation; the Colebrook–White equation is used here.

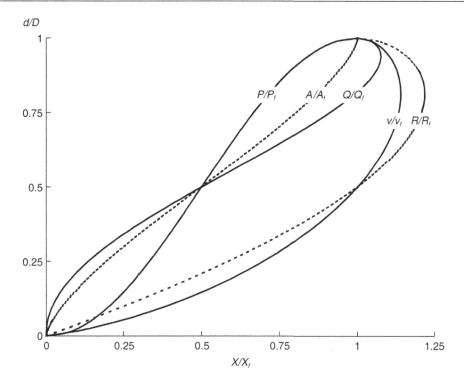

Figure 7.8 Values of geometric and hydraulic elements for part-full pipe flow.

Table 7.6 Proportional flow rate and velocity at various depths

	$d/D = 0.5$	$d/D = 1.0$	Maximum
Q/Q_f	0.5	1.0	1.08 at $d/D = 0.94$
v/v_f	1.0	1.0	1.14 at $d/D = 0.81$

In part-full pipes, maximum flow velocity and flow rate do not occur when the pipe is running full; they occur when it is slightly less than full. This is because the circular shape affects the relative magnitudes of the flow area and the wetted perimeter (which determines the magnitude of the frictional resistance). At low flows, the wetted perimeter is high compared with the flow area, resulting in low velocities. Velocity increases with depth (see Figure 7.8) until at the highest depths the increase in the wetted perimeter is again high compared with the flow area, and this results in a fall in velocity. It follows that, eventually, the flow rate will also fall, since the flow rate is the product of cross-sectional area and velocity. Table 7.6 summarises these effects.

The effect of a free surface within a pipe is largely understood, and conventionally the hydraulic radius (R) is substituted for pipe diameter (D) in the pipe flow equations to model the varying depth of flow (using Equation 7.14). The substitution appears to predict the effects quite accurately, but there is evidence to suggest that it overestimates flow rate and velocity in rough pipes by a few per cent.

Graphs of the velocity ratio v/v_f and flow rate ratio Q/Q_f in the form of Figure 7.8 can be used in conjunction with the Colebrook–White equation (using a chart like Figure 7.5,

---

EXAMPLE 7.7

Given Q = 70 L/s, S_f = 1:200, d/D = 0.25, and k_s = 0.6 mm, find the necessary diameter D and corresponding velocity.

Solution

From Figure 7.8, Q/Q_f = 0.14 (for d/D = 0.25), so Q_f = 500 L/s. From Figure 7.5, D = 600 mm and v_f = 1.7 m/s. From Figure 7.8 again, v/v_f = 0.7 so v = 1.2 m/s.

---

for example) to determine v and Q for any part-full case. This approach is appropriate for problems such as

- Given Q, S_f, and d/D, find D and v.
- Given Q, D, and S_f, determine d/D and v.

Example 7.7 gives an illustration.

7.4.3 Butler–Pinkerton charts

An alternative, more direct method for part-full pipe problems is provided by the Butler–Pinkerton charts (Butler and Pinkerton, 1987). These are based on a shape correction factor ψ, which can be used to modify pipe diameter rather than replace it. Usually, D is replaced by $4R$ (from Equation (7.14)), so incorporating the shape correction factor gives

$$\psi D \equiv 4R$$

$$\psi = \frac{4R}{D}$$

$$\psi = \frac{(\theta - \sin\theta)}{\theta} \tag{7.16}$$

By substituting the shape correction factor ψ into the Colebrook–White equation, an expression valid for any proportional depth of flow is obtained:

$$v = -2\sqrt{2gS_f\psi D}\,\log_{10}\left(\frac{k_s}{3.7\psi D} + \frac{2.51v}{\psi D\sqrt{2gS_f\psi D}}\right) \tag{7.17}$$

An example of a Butler–Pinkerton chart is given in Figure 7.9. A separate chart is required for each roughness value, and for each pipe diameter. Once the correct chart has been chosen, the definition of any two of the remaining variables allows the determination of the other two. This is particularly useful to cope with questions such as

- Given Q, and constraints for minimum v and maximum d/D, find minimum S_f and D (the design case).
- Given D and S_f, find Q and v at various values of d/D (the analysis case).

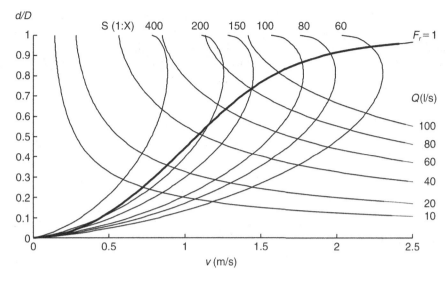

Figure 7.9 Butler–Pinkerton chart (D 300 mm, k_s 0.6 mm).

EXAMPLE 7.8

Given Q = 60 L/s, d/D = 0.75, minimum v = 0.75 m/s, and k_s = 0.6 mm, find the minimum gradient of a 300 mm diameter pipe.

Solution

The Butler–Pinkerton chart (Figure 7.9) can be used by estimating the point of intersection of the Q-curve (read from the right sloping upwards) with the hydraulic gradient S-curve (read downwards sloping first right then left).

Thus, at d/D = 0.75, S_f = 1:280 (v = 1.05 m/s).

The charts consist of two families of curves: the S-curves represent the modified Colebrook–White Equation (7.17) and the Q-curves represent the continuity equation:

$$v = \frac{8Q}{\psi\theta D^2} \tag{7.18}$$

The intersection of each S- and Q-curve gives the resulting normal flow depth and steady-state velocity (see Example 7.8). A curve labelled F_r = 1 is also shown and is explained in Section 7.5.4.

7.4.4 Non-circular sections

Circular pipes are by far the most common in shape. They have the shortest circumference per unit of cross-sectional area and so require the least wall material to resist internal and external pressures. They are also easy to manufacture. However, other shapes have been adopted in the past for sewers and drains – most commonly egg-shaped pipes, but also rectangular, trapezoidal, U-shaped, oval, horseshoe (arch), and compound types. Figure 7.10 shows a range of shapes, their geometries, and geometric elements.

Provided the cross-sectional shape does not differ much from the circular, the basic equations for pipe flow can be utilised by substituting $4R$ for D.

7.4.5 Surcharge

Surcharging refers to pipes designed to run full or part full, conveying flow under pressure. This can occur, for example, when flood flows exceed the design capacity, and it is therefore likely that all storm sewers will become surcharged at some time during their operational life.

A sewer pipe can surcharge in one of two ways, normally referred to as "pipe surcharge" and "manhole surcharge."

Figure 7.11a shows a longitudinal vertical section along a length of sewer running part-full (without surcharge). As explained in Section 7.4.1, the hydraulic gradient coincides with

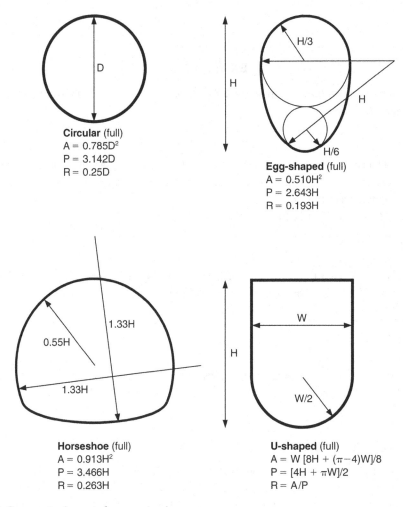

Figure 7.10 Geometric elements for non-circular sewers.

the water surface (and is parallel to the pipe bed). If there is an increase in flow entering the sewer, the consequence will be that the depth of flow in the pipe will increase.

Now imagine the sewer carrying the maximum flow rate (just less than full). If there is an increase in flow entering the sewer, the carrying capacity of the pipe can no longer be increased by a simple increase in depth. The capacity of a pipe is a function of diameter, roughness, and hydraulic gradient. To increase capacity, the only one of these that can change automatically (in response to "natural forces") is the hydraulic gradient. It follows that the new hydraulic gradient must be greater than the old (equal to the gradient of the pipe), and the result – pipe surcharge – is shown in Figure 7.11b. This increases local losses at manholes (Table 7.3) and so increases energy loss overall.

If inflow continues to increase, the hydraulic gradient will increase. The obvious danger is that the hydraulic gradient will rise above ground level. This may cause manhole covers to lift and the flow to flood onto the surface – "manhole surcharge". For these conditions, Rubinato et al. (2018) present the results of laboratory investigations of energy losses at manholes, including the impact of the manhole cover and the interaction of the flow surcharging from the manhole with the flow on the catchment surface. They found that losses increased when there was sewer-to-surface flow compared with flow through the straight manhole, especially when the effect of the cover was included.

The transition from conditions in Figure 7.11a to those in Figure 7.11b is sudden. As shown in Table 7.6, maximum flow is carried when the pipe is less than full. If the pipe is running at this maximum level, a further slight increase in flow rate or small disturbance will result in a sudden increase in pipe flow depth, not only filling the pipe completely but also establishing a hydraulic gradient in excess of S_o.

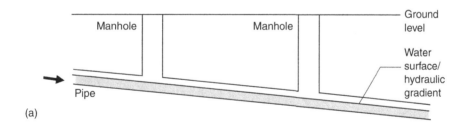

(a)

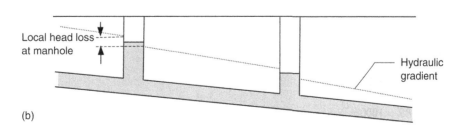

(b)

Figure 7.11 (a) Part-full pipe flow without surcharge. (b) Pipe flow with surcharge.

7.4.6 Velocity profiles

As discussed briefly in Sections 7.2.2 and 7.3.3, velocity varies over the cross section of a pipe. Velocity is at a minimum at the boundary and increases towards the centre. Maximum velocity may be at the surface when the flow depth is low, or a little below it when the flow depth is higher (Figure 7.12). The presence of a sediment bed in the invert of the pipe also affects the profile.

These profiles are significant when considering the transport of types of solids that are found only in specific parts of the cross section (floating solids close to the surface, or heavy solids close to the bed). This is discussed further in Chapters 9 and 17.

7.4.7 Minimum velocity

It is important for sewers to be able to convey wastewater or stormwater without long-term deposition of solid material. This is normally achieved by specifying a minimum mean velocity, a "self-cleansing" velocity, at a particular flow condition (e.g., pipe-full capacity) or for a particular frequency of occurrence (e.g., daily).

A common design criterion is to specify a minimum velocity when the pipe runs full. A value of 1 m/s is typical. The basis of this is that, although the pipe may never flow precisely full, mean velocities exceed the pipe-full velocity for flow rates greater than $0.5Q_f$ (see Figure 7.8). This method has the advantage of simplicity in computation but lacks precision. The other common approach is to specify self-cleansing velocities at some specified depth of flow (e.g., 0.75 m/s at $d/D = 0.75$).

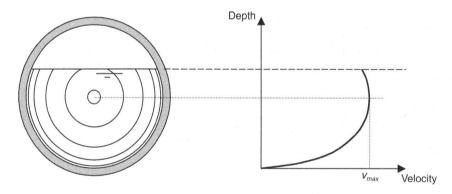

Figure 7.12 Velocity profile in a part-full pipe.

EXAMPLE 7.9

The metric equivalent of "Maguire's Rule" states that an appropriate self-cleansing pipe slope is given by $S_o = 1/D$ (D in mm). Interpret this in terms of a minimum shear stress standard.

Solution

Assuming $\rho = 1000$ kg/m³

$$\tau_o = \rho g R S_o = 1000g \cdot \frac{D}{4} \cdot \frac{1}{1000D} \approx 2.5\,N/m^2$$

In the past, the velocity criterion has sometimes been relaxed for larger-diameter sewers (say > 750 mm). However, research has shown this to be a mistake, and there is even evidence to suggest that higher self-cleansing velocities should be specified for larger diameters (see Chapter 17).

7.4.8 Minimum shear stress

Another potentially important parameter related to solid deposition/erosion is boundary shear stress. As water flows over the rigid boundary of the pipe channel, it exerts an average shear stress or drag τ_o (N/m^2) in the direction of flow, given by

$$\tau_o = \rho g R S_o \tag{7.19}$$

Substituting Equation (7.8) and assuming $D = 4R$ gives

$$\tau_o = \frac{\rho \lambda \upsilon^2}{8} \tag{7.20}$$

indicating that the applied shear stress varies linearly with the friction factor and the square of the velocity of flow. Shear stress is not uniform around the boundary because of the variations in velocity.

7.4.9 Maximum velocity

Historically, many sewerage systems were designed so that velocity would not exceed a specified maximum. This was no doubt a sensible criterion for early brick sewers with relatively weak lime-mortar joints. However, research has shown that abrasion is not normally a problem with modern pipe materials. Perkins (1977) has suggested that no fixed maximum limit is required, but where velocities are high (>3 m/s) careful attention needs to be given to

- Energy losses at bends and junctions
- Formation of hydraulic jumps leading to intermittent pipe choking
- Cavitation (see Section 14.3.5) causing structural damage
- Air entrainment (significant when $\upsilon = \sqrt{5gR}$)
- The possible need for energy dissipation or scour prevention
- Safety provisions

7.5 OPEN-CHANNEL FLOW

7.5.1 Uniform flow

Part-full pipe flow (covered in the last section) is the most common condition in sewer systems. Design methods, as we have seen, tend to be related to those for pipes flowing full. However, in hydraulic terms, part-full pipe flow is a special case of open-channel flow, the basic principles of which are considered in this section.

The concept of normal depth, and the nature of the EGL and HGL, explained in Section 7.4.1 for part-full pipe flow, apply to all cases of open-channel flow.

Table 7.7 Typical values of Manning's n

Channel material	n range
Glass	0.009–0.013
Cement	0.010–0.015
Concrete	0.010–0.020
Brickwork	0.011–0.018

7.5.1.1 Manning's equation

A number of purely empirical formulas for uniform flow in open channels have been developed over the years, a common example of which is Manning's equation:

$$v = \frac{1}{n} R^{2/3} S_o^{1/2} \tag{7.21}$$

where n = Manning's roughness coefficient; typical values are given in Table 7.7 (units are not usually given, but to balance Equation (7.21) the units of $1/n$ must be $m^{1/3}\ s^{-1}$), and S_o = bed slope (–).

If Manning's equation is plotted on the Moody diagram, it gives a horizontal line indicating the equation is only applicable to rough turbulent flow. Koutsoyiannis (2008) has developed a generalised form of the equation valid over a greater range of flow conditions.

Ackers (1958) showed that if k_s/D is in the typical range of 0.001–0.01, the values of k_s and n are related (to within 5%) by the following relationship:

$$n = 0.012 k_s^{1/6} \tag{7.22}$$

where k_s is in mm, and n is as defined for Equation (7.21).

7.5.2 Nonuniform flow

As stated, in uniform free surface flow, when the flow depth is normal, the total energy line, HGL, and channel bed (or pipe invert) are all parallel. In many situations, however, such as changes in pipe slope, diameter, or roughness, nonuniform flow conditions prevail, and these lines are not parallel. In sewer systems, it is likely that there will be regions of uniform flow interconnected with zones of nonuniform flow. Methods of predicting conditions in nonuniform flow are presented in the following sections.

7.5.3 Specific energy

If the Bernoulli Equation (7.5) is redefined so that the channel bed is used as the datum (in place of a horizontal plane), we have, with reference to Figure 7.13

$$\text{Total head} = \frac{p}{\rho g} + \frac{v^2}{2g} + z = \frac{\rho g h}{\rho g} + \frac{v^2}{2g} + x = h + x + \frac{v^2}{2g}$$

This gives *specific energy, E*:

$$E = d + \frac{v^2}{2g} \tag{7.23}$$

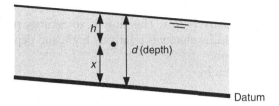

Figure 7.13 Terms for derivation of specific energy.

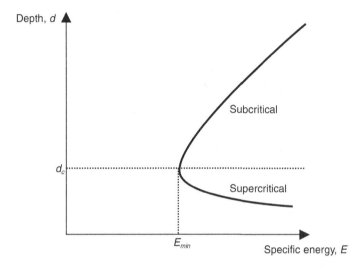

Figure 7.14 Depth against specific energy, for constant flow.

or

$$E = d + \frac{Q^2}{2gA^2}$$

Thus, for a given flow rate, E is a function of depth only (as A is a function of d). Depth can be plotted against specific energy (Figure 7.14) showing that there are two possible depths at which flow may occur with the same specific energy. The depth that actually occurs depends on the channel slope and friction, and on any special physical conditions in the channel. At the critical depth, d_c, the specific energy is a minimum for a given Q.

7.5.4 Critical, subcritical, and supercritical flow

The non-dimensional Froude number (F_r) is given by

$$F_r = \frac{\upsilon}{\sqrt{gd_m}} \tag{7.24}$$

where d_m is the hydraulic mean depth, as defined in Section 7.4.2.

It can be shown that $F_r = 1$ at critical depth. If $F_r < 1$, flow is *subcritical*; the depth is relatively high and the velocity relatively low. This flow is sometimes referred to as "tranquil" flow. If $F_r > 1$, flow is *supercritical*; velocity is relatively high, and depth low. This flow is also called "rapid" or "shooting" flow.

The critical velocity v_c is given by

$$v_c = \sqrt{gd_m} \qquad (7.25)$$

For rectangular channels, this leads directly to Equation (7.28) for critical depth

$$d_c = \sqrt[3]{\frac{q^2}{g}} \qquad (7.26)$$

where q is the flow rate per unit width (flow rate divided by channel width).

For circular channels, there is no simple analytical solution. A solution can be achieved computationally, graphically, or by approximation.

Critical conditions ($F_r = 1$) have been plotted on the Butler–Pinkerton chart given (see Figure 7.9), giving critical depth and critical slope for each flow rate. Subcritical conditions exist in the region above this line and supercritical below it.

As a good approximation, the critical depth in a circular pipe (d_c) can be determined from the following empirical equation (Straub, 1978):

$$\frac{d_c}{D} = 0.567 \frac{Q^{0.506}}{D^{1.264}} \qquad (7.27)$$

where $0.02 < d_c/D \leq 0.85$ (units for Q, m³/s). See Example 7.10.

EXAMPLE 7.10

What are the critical depth, velocity, and gradient in a 0.3 m circular sewer if the critical flow rate is 50 L/s? If the pipe is actually discharging 80 L/s, determine the depth of flow (assuming it to be uniform) and comment on the flow conditions ($k_s = 0.6$ mm).

Solution

The Butler–Pinkerton chart (Figure 7.9) can be used by estimating the point of intersection of the Q-curve (read from the right sloping upwards) with the $F_r = 1$ curve.

$Q = 50$ L/s, $D = 300$ m

This gives

$d_c / D = 0.57, v_c = 1.2$ m/s, $S_c = 1:200$

The same charts can be used to find the proportional depth of flow that is read at the intersection of the Q-curve and the relevant S-curve (read downwards sloping first right then left):

$Q = 80$ L/s, $S_f = 1:200$

This gives

$d / D = 0.84$

As the intersection is above the F_r curve, the flow must be subcritical.

Critical proportional depth can also be found using Straub's empirical equation (7.27):

$$\frac{d_c}{D} = 0.567 \frac{0.05^{0.506}}{0.3^{1.264}} = 0.57$$

Normal depth in an open channel may be subcritical or supercritical. A *mild* slope is defined as one in which normal depth is greater than critical depth (so uniform flow is subcritical), and a *steep* slope is defined as one in which normal depth is less than critical depth (so uniform flow is supercritical).

Most sewer designs are for subcritical flow. Flow in the supercritical state is acceptable but has the disadvantage that if downstream conditions dictate the formation of subcritical conditions, a hydraulic jump will form. This effect is described later in the chapter. Close to critical depth ($0.7 < F_r < 1.5$), the flow tends to be somewhat unstable with surface waves and would not be an appropriate position for a flow monitor (Section 18.4.1), for example (Hager, 2010).

7.5.5 Gradually varied flow

When variations of depth with distance must be taken into account, detailed analysis is required. This is done by splitting the channel length into smaller segments and assuming that the friction losses can still be accurately calculated using one of the standard equations such as the Colebrook–White equation.

The general equation of gradually varied flow can be derived as

$$\frac{d(d)}{dx} = \frac{S_o - S_f}{1 - F_r^2} \qquad (7.28)$$

where d is the depth of flow (m), x is the longitudinal distance (m), S_o is the bed slope (–), S_f is the friction slope, h_f/L as defined in Section 7.3.2(–), and F_r is the Froude number (–).

Examples of gradually varied flow in sewer systems are shown in Figure 7.15. Figure 7.15a shows flow ending at a "free overall" – a sudden drop at the end of the pipe or channel such as the inflow to a pumping station. Close to the end of the pipe, conditions are critical, and for a long distance upstream the depth will be subject to a "drawdown" effect (provided flow is subcritical). The effect is most pronounced for flatter pipes.

Figure 7.15b shows subcritical flow backing up behind an obstruction. As flow approaches the obstruction, the depth increases: a "backwater" effect.

7.5.6 Rapidly varied flow

When supercritical flow meets subcritical flow, a discontinuity called a *hydraulic jump* is formed (Figure 7.16) at which there may be considerable energy loss.

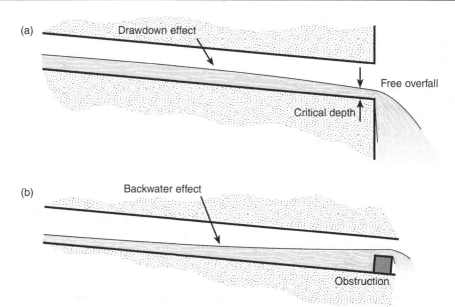

Figure 7.15 Drawdown and backwater effects (in a pipe).

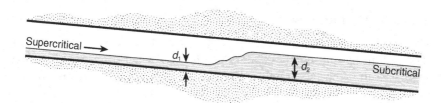

Figure 7.16 Hydraulic jump (in a pipe).

There is no convenient analytical expression for the relationship between d_1 and d_2 in Figure 7.16 in a part-full pipe. Straub (1978), however, has developed an empirical approach using an approximate value for the Froude number:

$$F_{r1} = \left(\frac{d_c}{d_1}\right)^{1.93} \tag{7.29}$$

where F_{r1} is the upstream Froude number. For cases where $F_{r1} < 1.7$, the depth d_2 is given by

$$d_2 = \frac{d_c^{\,2}}{d_1} \tag{7.30}$$

and for $F_{r1} > 1.7$.

$$d_2 = \frac{d_c^{\,1.8}}{d_1^{\,0.73}} \tag{7.31}$$

EXAMPLE 7.11

A 600 mm pipe flowing part-full has a slope of 1.8 in 100 (k_s = 0.6 mm). Flow depth (in uniform conditions) is 0.12 m. Confirm that flow is supercritical using Equation (7.27). An obstruction causes the flow downstream to become subcritical; therefore, a hydraulic jump forms. Determine the depth immediately downstream of the jump.

Solution

$$\frac{d}{D} = \frac{0.12}{0.6} = 0.2$$

From Figure 7.8,

$$\frac{Q}{Q_f} = 0.1$$

From Figure 7.5, Q_f = 900 L/s, so Q = 90 L/s = 0.09 m³/s:

$$\frac{d_c}{D} = 0.567\frac{Q^{0.506}}{D^{1.264}} = 0.567\frac{0.09^{0.506}}{0.6^{1.264}} = 0.32$$

so d_c = 0.19 m and $d < d_c$, so flow is supercritical. From Equation (7.29),

$$F_{rl} = \left(\frac{d_c}{d_l}\right)^{1.93} = \left(\frac{0.19}{0.12}\right)^{1.93} = 2.43$$

so from Equation (7.30),

$$d_2 = \frac{d_c^{1.8}}{d_l^{0.73}} = \frac{0.19^{1.8}}{0.12^{0.73}} = 0.24\,\text{m}$$

In drainage systems, hydraulic jumps have the potential to cause erosion of sewer materials due to turbulence and to cause the release of sewer gases (Section 18.10). It may be necessary to investigate their position so that suitable scour protection can be provided.

PROBLEMS

1 A pipe flowing full, under pressure, has a diameter of 300 mm and roughness k_s of 0.6 mm. The flow rate is 100 L/s. Use the Moody diagram to determine the friction factor λ and the nature of the turbulence (smooth, transitional, or rough). Determine the friction losses in a 100 m length. Check this by determining the hydraulic gradient using Figure 7.5. (Assume kinematic viscosity of 1.14 × 10⁻⁶ m²/s.) [0.024, transitional, 0.81 m]

2 A pipe is being designed to flow by gravity. When it is full, the flow rate should be at least 200 L/s and the velocity no less than 1 m/s. Use Figure 7.5 to determine the minimum gradient for a 600 mm diameter pipe (k_s 0.6 mm). What will the pipe-full flow rate actually be? At what part-full depth would velocity go below 0.8 m/s? [0.18 in 100, 300 L/s, 180 mm]

3 A surcharged manhole with a 30° bend has a local loss constant $k_L = 0.5$. Determine the pipe length, L_E, equivalent to this local loss (assuming that it is independent of velocity) for a pipe with a diameter of 450 mm and k_s of 1.5 mm. If velocity is 1.3 m/s, is the assumption above valid? (Assume kinematic viscosity = 1.14×10^{-6} m²/s.) [8.7 m, yes]

4 A gravity pipe has a diameter of 600 mm, a slope of 1 in 200, and when flowing full has a flow rate of 610 L/s and velocity of 2.2 m/s. Flowing part-full at a depth of 150 mm, what are the velocity, flow rate, area of flow, wetted perimeter, hydraulic radius, and applied shear stress? [1.5 m/s, 80 L/s, 0.055 m², 0.63 m, 0.09 m, 4.4 N/m²]

5 A 300 mm diameter pipe is being designed for the following: maximum flow rate 80 L/s, minimum allowable velocity 1 m/s, roughness k_s 0.6 mm. Determine the following, using the Butler–Pinkerton chart:
   a. The gradient required based on the pipe running full
   b. The depth at which it will actually flow at that gradient
   c. The minimum velocity that will be achieved if the working flow rate is 10 L/s
   d. The gradient at which the sewer would need to be constructed to just ensure that the minimum velocity is achieved at that flow rate. [1:190, 250 mm, 0.78 m/s, 1:95]

6 A pipe, diameter 450 mm, k_s 0.6 mm, slope 1.5 in 100, is flowing part-full with a water depth of 100 mm. Are conditions subcritical or supercritical? [supercritical]

7 If a hydraulic jump takes place in the pipe in Problem 6, such that conditions upstream of the jump are as in Problem 6, what would be the depth downstream of the jump? [0.18 m]

KEY SOURCE

Chadwick, A., Morfett, J., and Borthwick M. 2021. *Hydraulics in Civil and Environmental Engineering*, 6th edn, CRC Press.

REFERENCES

Ackers, P. 1958. *Resistance of Fluids in Channels and Pipes*, Hydraulics research paper No. 2, HMSO.
Butler, D. and Pinkerton, B.R.C. 1987. *Gravity Flow Pipe Design Charts*, Thomas Telford.
Forty, E.J., Lauchlan, C., and May, R.W.P. 2004. *Flow Resistance of Wastewater Pumping Mains*. Report SR641, H.R. Wallingford.
H.R. Wallingford and Barr, D.I.H. 2006. *Tables for the Hydraulic Design of Pipes, Sewers and Channels*, 8th edn, Volume 1, Thomas Telford.
Hager, W.H. 2010. *Wastewater Hydraulics. Theory and Practice*. 2nd edn, Springer.
Hamill, L. 2011. *Understanding Hydraulics*, 3rd edn, Palgrave Macmillan.
Kay, M. 2016. *Practical Hydraulics and Water Resources Engineering*, 3rd edn, CRC Press.
Koutsoyiannis, D. 2008. A power-law approximation of the turbulent flow friction factor useful for the design and simulation of urban water networks. *Urban Water Journal*, 5(2), 107–115.
Lauchlan, C., Forty, J., and May, R. 2005. Flow resistance of wastewater pumping mains. *Proceedings of the Institution of Civil Engineers, Water Management*, 158(WM2), 81–88.
Perkins, J.A. 1977. *High Velocities in Sewers*, Report No. IT165, Hydraulics Research Station, Wallingford.
Rubinato, M., Martins, R. and Shucksmith, J.D. 2018. Quantification of energy losses at a surcharged manhole. *Urban Water Journal*, 15(3), 234–241.
Straub, W.O. 1978. A quick and easy way to calculate critical and conjugate depths in circular open channels. *Civil Engineering (ASCE)*, December, 70–71.

Chapter 8

Hydraulic features

Urban drainage systems contain a wide range of different hydraulic features. In this chapter, we explore these elements in detail. Section 8.1 covers flow controls of various types, and Section 8.2 deals with weirs. Sewer drops and inverted siphons are explained in Sections 8.3 and 8.4, respectively. The last two sections (8.5 and 8.6) discuss gully spacing methods and culvert design.

8.1 FLOW CONTROLS

Flow controls can be used to limit the inflow to, or outflow from, elements in an urban drainage system. Typical uses include restricting the continuation flow at a combined sewer overflow (CSO) to the intended setting (Chapter 13) and controlling water level in tanks to ensure that the storage volume is fully exploited (Chapter 14). Flow controls can also be used to limit the rate at which stormwater actually enters the sewer system in the first place, deliberately backing up water in planned areas like car parks to prevent more damaging floods downstream in a city centre (Chapter 12).

Flow controls can be fixed, always imposing the same relationship between flow rate and water level, or adjustable, where the relationship can be changed by adjustment of the device.

8.1.1 Orifice plate

The simplest way of controlling inflow to a pipe is by an orifice plate. This forces the flow to pass through an area less than that of the pipe (Figure 8.1).

An orifice plate is fixed to the wall of the chamber where the inlet to the pipe is formed, and it usually either creates a smaller circular area (Figure 8.2a) or covers the upper part of the pipe area (Figure 8.2b). The area of the opening can only be changed by physically detaching and replacing or repositioning the plate.

Hydraulic analysis of an orifice is a simple application of the Bernoulli equation (Chapter 7). Comparing the total head at points 1 and 2 in Figure 8.1a, and assuming there is no loss of energy, we can write

$$\frac{p_1}{\rho g} + \frac{v_1^2}{2g} + z_1 = \frac{p_2}{\rho g} + \frac{v_2^2}{2g} + z_2$$

DOI: 10.1201/9781003408635-8

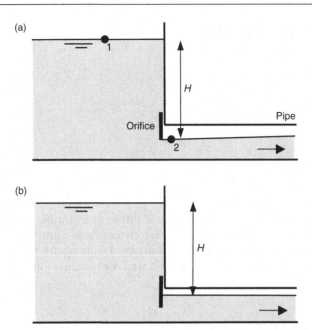

Figure 8.1 Orifice plate (vertical section): (a) free outfall and (b) drowned.

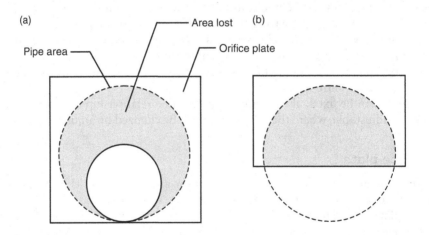

Figure 8.2 Orifice plate arrangements.

Now $p_1 = p_2 = 0$ (gauge pressure) and $z_1 - z_2 = H$, so, assuming that the velocity at 1 is negligible, we have

$$H = \frac{v_2^2}{2g}$$

or

$$v_2 = \sqrt{2gH}$$

So, the flow rate (Q) is as follows:

$$Q = A_o = \sqrt{2gH}$$

where A_o is the area of the orifice (m²).

The assumptions made above affect the accuracy of the answer, and this is compensated for by an "orifice coefficient," C_d giving

$$Q_{actual} = C_d A_o \sqrt{2gH} \tag{8.1}$$

This is alternatively written as $Q = CA_o\sqrt{gH}$, where C includes the $\sqrt{2}$.

Conditions downstream may cause the orifice to be "drowned" – the downstream water level to be above the top of the orifice opening. H in Equation (8.1) should now be taken as the difference in the water levels, as in Figure 8.1b. The minimum value of H for which the orifice will be not drowned (H_{min}) can be determined from Figure 8.3 (in which D_o is the

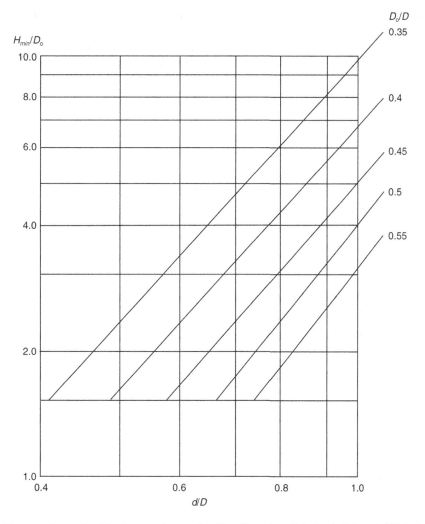

Figure 8.3 Chart to determine H_{min} for non-drowned orifice. (Based on Balmforth, D.J. et al. 1994. Guide to the Design of Combined Sewer Overflow Structures, Report FR 0488, Foundation for Water Research. With permission of Foundation for Water Research, Marlow.)

diameter of the orifice, and D is the diameter of the pipe). The use of Figure 8.3 requires the value of the water level in the pipe downstream (d), which can be calculated using the properties of part-full pipe flow described in Section 7.4. Example 8.1 demonstrates the calculation. For $H < H_{min}$, the orifice will be drowned.

For an orifice that is not drowned, C_d in Equation (8.1) generally has a value between 0.57 and 0.6. For a drowned orifice, C_d can be estimated from

$$C_d = \frac{1}{1.7 - \left(A_o / A\right)} \tag{8.2}$$

where A is the flow area in the pipe (m²).

8.1.2 Penstock

A penstock is an adjustable gate that creates a reduction in area at the inlet to a pipe in the manner of Figure 8.2b. The position of the penstock can be raised or lowered either manually or mechanically, by means of a wheel or a motorised actuator turning a spindle.

A penstock is more elaborate than an orifice plate. The advantage is that it can be adjusted to suit conditions – either, in the case of manual adjustment, to an optimum position to suit operational requirements, or, in the case of mechanical adjustment with remote control, to respond to changing requirements, perhaps as part of a real-time control system (described in Chapter 22).

EXAMPLE 8.1

The following arrangement is proposed. A tank will have an outlet pipe with a diameter of 750 mm, slope of 0.002, and k_s of 1.5 mm. Flow to the outlet pipe will be controlled by a circular orifice plate with a diameter of 300 mm. Determine the flow rate when the water level in the tank is 2 m above the invert of the outlet. Use Table 7.1 for pipe flow calculations.

Solution

First, assume that the orifice is not drowned.

So $H = 2.0 - 0.3 = 1.7$ m (see Figure 8.1a).

Equation (8.1) gives $Q_{actual} = C_d A_o \sqrt{2gH}$. Assuming $C_d = 0.6$,

$$Q_{actual} = 0.6\pi \frac{0.3^2}{4} \sqrt{2g1.7} = 0.244 \text{ m}^3/\text{s or } 244 \text{ L/s}$$

Now check the assumption that the orifice is not drowned, using Figure 8.4.

What is the flow rate in the outlet pipe flowing full?

Use Table 7.1: $Q_f = 492$ L/s. This gives $Q/Q_f = 0.5$, so (from Figure 7.8) $d/D = 0.5$. For the orifice,

$$\frac{D_o}{D} = \frac{0.3}{0.75} = 0.4$$

Note that the depth of uniform part-full flow in the outlet pipe would be above the top of the orifice. This does not mean that the orifice is necessarily drowned since conditions are nonuniform. For

$$\frac{d}{D} = 0.5 \text{ and } \frac{D_o}{D} = 0.4, \text{ Figure 8.3 gives } \frac{H_{min}}{D_o} = 1.7, \text{ so } H_{min} \text{ is } 0.51 \text{m},$$

which is less than the actual H of 1.7 m, and so the orifice is not drowned. Therefore, the flow rate calculated above (244 L/s) applies.

Blockage is a potential problem with both an orifice plate and a penstock, and both should be designed to allow a 200 mm diameter sphere to pass.

8.1.3 Vortex regulator

In a similar way to an orifice plate or penstock, a vortex regulator constricts flow, usually with the purpose of exploiting a storage volume; the magnitude of the flow rate passing through the device depends on the upstream water depth. The regulator consists of a unit (see Figure 8.4) into which flow is guided tangentially, creating (at sufficiently high flow rates) a rotation of liquid inside the chamber. This creates a vortex with high peripheral velocities and large centrifugal forces near the outlet. These forces increase with the upstream head until an air core occupies most of the outlet orifice creating a back-pressure opposing the flow.

(a)   (b)

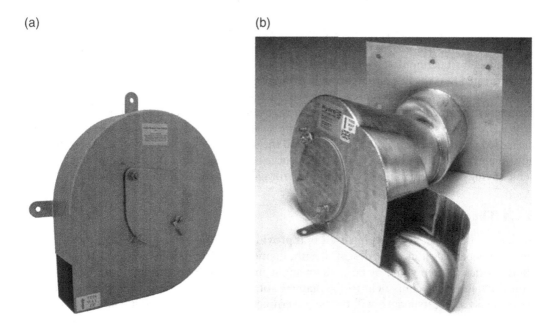

Figure 8.4 Vortex regulator for (a) stormwater and (b) wastewater. (Courtesy of Hydro International.)

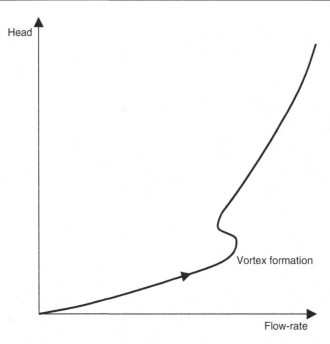

Figure 8.5 Head–discharge relationship for vortex regulator.

This type of device has a distinctive head–discharge curve as shown in Figure 8.5. The "kickback" occurs during the formation of a stable vortex in rising flow. The shape of the curve depends on the detailed geometry of the regulator and the downstream conditions. During falling flow, there can be a slight kickback, but not as pronounced as for rising flow.

The main advantage of the vortex regulator is that it provides a degree of throttling only possible with an orifice of a much smaller opening. Hence, regulators can avoid problems of blockage or ragging that would occur on small-diameter orifices. In addition, it has been demonstrated that the discharge through a vortex regulator is not directly related either to its inlet or outlet cross-sectional area (Butler and Parsian, 1993). Therefore, the impact of any ragging of the openings is less pronounced than might be expected in comparison with an orifice. The same study also showed that, in all cases, the retention of single solids within the device led to an increase in the discharge until the solid was eventually ejected. An additional advantage is that, since the head–discharge curve is initially flatter than an equivalent orifice, some savings can be made in the volume of storage required for flow balancing.

8.1.4 Throttle pipe

With a throttle pipe, it is the pipe itself that provides the flow control. The flow rate through the pipe depends on its inlet design, length, diameter, and hydraulic gradient. If the pipe is short, or has a steep slope or large diameter, it may be "inlet controlled"; the flow is controlled by an orifice equivalent to the diameter of the pipe. However, if the pipe is long, the friction loss along its length will be the governing factor. This condition is known as outlet control.

A common throttle pipe application is the continuation pipe of a stilling pond CSO (to be described in detail in Chapter 13). Figure 8.6 shows that, when the weir is operating, the throttle pipe will be surcharged and thus flow rate will be related to the hydraulic gradient (not the pipe gradient). There will also be local losses (not shown in Figure 8.6) that may be significant. So, with reference to Figure 8.6,

$$H = S_f L + k_L \frac{v^2}{2g}$$
(8.3)

where S_f is the friction slope, given by pipe design chart/table (–), and $k_L\left(v^2 / 2g\right)$ is the local losses (as defined in Section 7.3.6 and Table 7.3)

In throttle pipe calculations, it is sometimes convenient to represent local losses by an equivalent pipe length, as explained in Section 7.3.6 (and demonstrated in Example 8.2).

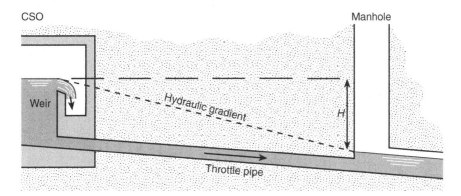

Figure 8.6 Throttle pipe (vertical section).

EXAMPLE 8.2

A throttle pipe carrying the continuation flow from a stilling pond CSO will have a length of 28 m. When the weir comes into operation, the water level in the CSO will be 1.8 m above the water level at the downstream end of the throttle pipe and 1.4 m above the soffit at the pipe inlet. Under these conditions, the continuation flow (in the throttle pipe) should be as close as possible to 72 L/s. The roughness, k_s, of the pipe material is assumed to be 1.5 mm, and local losses are taken as $1.4(v^2/2g)$. Determine an appropriate diameter for the throttle pipe. Confirm that this throttle pipe is not "inlet controlled."

If, as an alternative, there were no throttle pipe and flow control was achieved by an orifice, what would be its diameter (assuming that it would not be drowned)?

Solution

Solve by trial and error – a 200 mm pipe gives the following.

Represent local losses by equivalent length:

$$\frac{k_s}{D} = \frac{1.5}{200} = 0.0075$$

Assume rough turbulent, the Moody diagram (Figure 7.4) gives $\lambda = 0.034$

$$\frac{L_E}{D} = \frac{k_L}{\lambda} = \frac{1.4}{0.034} = 41$$

therefore $L_E = 8$ m

total length = 28 + 8 = 36 m
hydraulic gradient = 1.8/36 = 0.05

The chart for $k_s = 1.5$ mm, or see Table 7.1, gives a flow rate (for 200 mm diameter) of 75 L/s. So 200 mm diameter is suitable.

If the throttle pipe is inlet controlled, control comes solely from the inlet acting as an orifice. Apply orifice formula (assuming $C_d = 0.59$):

$$Q_{actual} = C_d A_o \sqrt{2gH} = 0.59 \times \pi \frac{0.2^2}{4} \sqrt{2g1.4} = 97 \text{ L/s}$$

so control does not solely come from the inlet: the pipe is not inlet controlled. Consider the use of an orifice plate

$$Q_{actual} = C_d A_o \sqrt{2gH}$$

What orifice diameter would give the same control as the throttle pipe?

$$0.075 = 0.59 \times \pi \frac{D_o^2}{4} \sqrt{2g1.4} \text{ giving } D_o = 0.175 \text{ m (unacceptably small)}$$

To prevent blockage, the diameter of the throttle pipe should not be less than 200 mm. Clearly, the length of the throttle pipe plays an important part in creating the flow control, and in design cases where it is inappropriate to reduce the diameter, the desired hydraulic control may be achieved by increasing the length (subject to restrictions in site layout). The diameter of an outlet-controlled throttle pipe will certainly be larger than that of the orifice plate giving equivalent flow control.

8.1.5 Flap valve

A flap valve is a hinged plate at a pipe outlet that restricts flow to one direction only. A typical application is at an outfall to receiving water with tidal variation in level. When the level of the receiving water is below the outlet, the outflow discharges by lifting the flap (Figure 8.7a). When the outlet is flooded, the flap valve prevents tidal water from entering the sewer (Figure 8.7b). In these circumstances, any flow in the sewer will back up in the pipe, and if the energy grade line rises above the tidal water level, there will be outflow. The flap (which may have considerable weight) will then create a local head loss. Methods of estimating this loss are proposed by Burrows and Emmonds (1988).

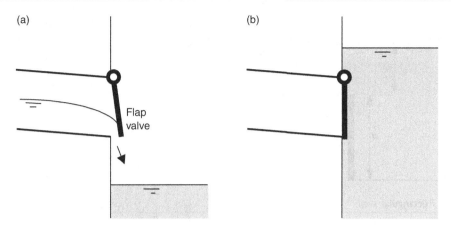

(a)   (b)

Flap
valve

Figure 8.7 Flap valve operation.

8.1.6 Summary of characteristics of flow control devices

Table 8.1 gives a summary of the characteristics of the flow control devices considered above (excluding the flap valve, which has a different function from the other devices).

8.2 WEIRS

8.2.1 Transverse weirs

Standard analysis, using the Bernoulli equation, of flow over a rectangular weir gives the theoretical equation for the relationship between flow rate and depth as

$$Q_{theor} = \frac{2}{3} b \sqrt{2g} H^{3/2}$$

where Q_{theor} is the flow rate (m^3/s), b is the width of the weir (m), and H is the height of water above the weir crest (m) (Figure 8.8).

Several assumptions are made in the analysis, and it is necessary to introduce a discharge coefficient to relate the theoretical result to the actual flow rate:

$$Q = C_d \frac{2}{3} b \sqrt{2g} H^{3/2} \tag{8.4}$$

Table 8.1 Summary of characteristics of flow control devices

Device	Characteristic
Orifice plate	Simple, cheap. Flow control can only be adjusted by physically detaching and replacing or repositioning the plate.
Penstock	Easily adjusted. When automated, can be used for real-time control.
Vortex regulator	Controls flow with a larger opening than the equivalent orifice.
Throttle pipe	Larger opening than an equivalent orifice.
	Significant construction costs.

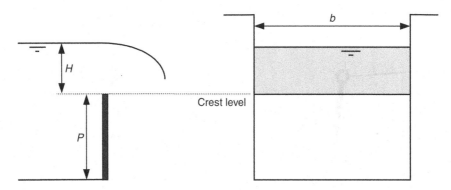

Figure 8.8 Rectangular weir.

where C_d = discharge coefficient. With this form of equation, C_d has a value between 0.6 and 0.7; C_d is sometimes written so that it incorporates some of the other constants in the equation.

The value of C_d and the accuracy of the equation depend partly on whether the weir crest fills the whole width of a channel or chamber, or is a rectangular notch that forces the flow to converge horizontally. For the former, an empirical relationship by Rehbock can be used:

$$Q = C_d \frac{2}{3} b \sqrt{2g} \left[H + 0.0012 \right]^{3/2} \tag{8.5}$$

in which

$$C_d = 0.602 + 0.0832 \frac{H}{P}$$

and where P is the height of the weir crest above the channel bed (m) (Figure 8.8).

A summary of discharge coefficients for a further range of cases is given by Balmforth (2009).

8.2.2 Side weirs

The flow arrangements for side weirs are more complex than for transverse weirs because the flow rate in the main channel decreases with length (as some flow passes over the weir) and conditions are nonuniform.

The possible flow conditions at a side weir can be classified into five types as illustrated in Figure 8.9. These conditions can be analysed by assuming that specific energy is constant along the main channel. The standard curve of depth against specific energy (for constant flow rate), introduced as Figure 7.14, is reproduced as Figure 8.10 with the curve for a slightly decreased flow rate added.

The classification of flow types is based partly on the slope of the channel. *Mild* and *steep* slopes are defined in Section 7.5.4.

Type I
          Channel slope: mild   Weir crest below critical depth.

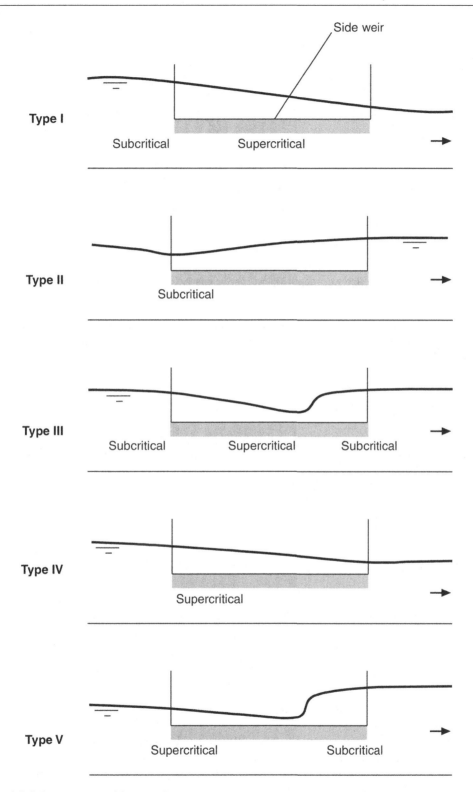

Figure 8.9 Side weir: types of flow condition.

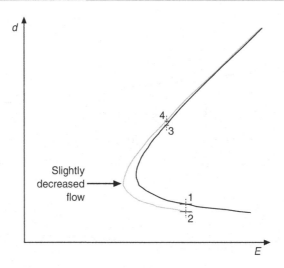

Figure 8.10 Flow parallel to side weir: depth against specific energy.

Depth along the weir is supercritical as a result of the fact that the weir crest is below critical depth. As the flow rate decreases, we move from point 1 to 2 in Figure 8.10, and the depth (d) decreases.

Type II
    Channel slope: mild   Weir crest above critical depth.

Depth along the weir is subcritical as a result of the fact that the weir crest is above critical depth. As the flow rate decreases, we move from point 3 to 4 in Figure 8.10, and the depth (d) increases.

Type III
    Channel slope: mild   Weir crest below critical depth.

At the start of the weir, conditions are as Type I. However, conditions downstream are such that a hydraulic jump forms before the end of the weir.

Type IV
    Channel slope: steep   Weir crest below critical depth.

Conditions are similar to those for Type I, except that supercritical conditions would prevail in the main channel in any case because it is steep.

Type V
    Channel slope: steep   Weir crest below critical depth.

Conditions are similar to those for Type III, except that supercritical conditions prevail before the start of the weir because the channel is steep.

For all types, the variation of water depth with distance, derived from standard expressions for spatially varied flow, is given by

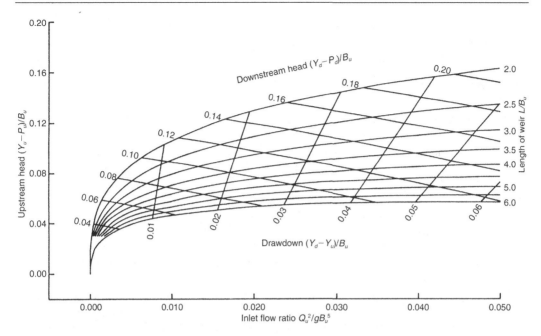

Figure 8.11 Chart for side weir design: double-side weir (horizontal weirs and channel bed), $Q_d/Q_u = 0.1$, $B_d/B_u = 1, P_u/B_u = P_d/B_u = 0.6, n = 0.010$. (Reproduced from Delo, E.A. and Saul, A.J. 1989. *Proceedings of the Institution of Civil Engineers*, Part 2, 87, June, 175–193. With permission of Thomas Telford Publishing.)

$$\frac{d(d)}{dx} = \frac{Q_c d\left[-\left(dQ_c/dx\right)\right]}{gB^2 d^3 - Q_c^2} \tag{8.6}$$

where d is the depth of flow (m), x is the longitudinal distance (m), Q_c is the flow rate in the main channel (m³/s), and B is the width of the main channel (m).

$[-(dQ_c/dx)]$ is the rate at which the flow rate in the main channel is decreasing – that is, the rate at which flow passes over the weir per unit length of the weir. Therefore, Equation (8.4) for flow over a weir can be adapted to give

$$-\frac{dQ_c}{dx} = C_d \frac{2}{3}\sqrt{2g}H^{3/2} \tag{8.7}$$

Note that H is water depth relative to the weir crest, whereas d is water depth relative to the channel bed. In a double-side weir arrangement (one weir on either side of the main channel), the right-hand side of Equation (8.7) is doubled. It has been found that the Rehbock expression, Equation (8.5), gives appropriate values of C_d for side weirs, even though it was originally proposed for transverse weirs.

For methods of solution of these equations, see Chow (2009). May et al. (2003) present a simple formula for total flow discharged over a side weir, backed up by charts for determining coefficients. They also provide general guidance on design and construction. For high side weir overflows (Type II flow conditions), design calculations can be based on charts presented by Delo and Saul (1989). One of these is given as Figure 8.11 where Q_u is the inflow (m³/s), Q_d is the continuation flow rate (m³/s), B_u is the upstream chamber width (m), B_d is the downstream chamber width (m), P_u is the upstream weir height (m), P_d is the downstream weir height (m), Y_u is the upstream water depth (m), Y_d is the downstream water depth (m), and L is the length of weir (m).

8.3 SEWER DROPS

Simple drop manholes were introduced in Chapter 6. We now consider more significant cases of sewer drops, arising for two main reasons. The first is that the vertical drop may have been specified as an alternative to using much steeper slopes in the approach sewer; this would avoid high sewer velocities, and the risks outlined in Chapter 7. The second is that topography, or the use of a deep tunnel for collection, means that significant vertical drops are unavoidable. A number of alternative drop arrangements can be used.

In a *plunge-flow* drop shaft, the flow falls down freely as a jet, plunging into a pool at the base of the shaft, or hitting the opposite wall of the shaft, or landing at the entry to the downstream sewer itself (Chanson, 2004a). Air entrainment can be a significant issue. Air may be entrained by the free-falling jet, or when the jet lands in the pool, or hits the shaft wall, or if a hydraulic jump forms in the downstream sewer. If there is not sufficient ventilation to compensate for the air entrainment, sub-atmospheric conditions may occur in some parts of the system, causing backing up in manholes and sewers. Air entrainment may also increase the risk of downstream choking caused by a sudden transition to pressurised flow, or cause a reduction in pipe discharge capacity due to bulking of the flow by the entrained air (Granata et al., 2011).

The main alternative to the plunge-flow drop shaft, particularly for drops greater than 7–10 m, is the vortex drop shaft.

8.3.1 Vortex drop shafts

In a vortex drop shaft, fluid flows helically downwards in contact with the inside of a circular wall, with a core of air at the centre. The intake must induce vortex flow in the drop shaft. A *scroll intake* is a common arrangement, with a design traditionally based on the analysis by Ackers and Crump (1960). The inlet channel is volute-shaped in plan (Figure 8.12), inducing a vortex in the vertically downward flow.

Del Guidice and Gisonni (2010), in the context of the historical need for drop shafts to cope with level differences in the city of Naples, discuss alternative inlet arrangements, and present hydraulic analysis and experimental results relating to scroll intakes. Similarly, Echavez and Ruiz (2008), this time based on experiences in Mexico City, give an overview of drop shaft analysis and also present a hydraulic study of scroll intakes.

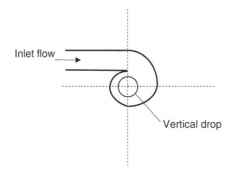

Figure 8.12 Scroll intake to vortex drop.

An alternative arrangement, common in the US and being used increasingly in the UK, is the *tangential intake*. This is more compact and simpler to construct than the scroll intake. Riisnaes et al. (2014) compare the dimensions and requirements for a tangential intake with those for a scroll intake, for a vortex drop located at a very restricted site in London. They explain the advantages of a tangential intake for this case.

We now consider the technical aspects of the tangential vortex intake. The typical layout is shown in plan in Figure 8.13a and in vertical section in Figure 8.13b. Hydraulic analysis, based on theory and physical modelling, is set out by Yu and Lee (2009), and is summarised here.

As can be seen from Figure 8.13, flow enters via a tapering and steeply sloping channel. At relatively small flow rates, water depth in the approach channel and the sloping intake is determined by the fact that critical depth (introduced in Chapter 7) occurs at section 1 (the start of the steeply sloping section). Conditions within the inlet upstream of this point are subcritical, and conditions downstream are supercritical. This depth can be determined from the standard expression for critical depth in a rectangular channel (Equation (7.26)), given here as Equation (8.8):

$$d_{c1} = \sqrt[3]{\frac{q_1^2}{g}} \tag{8.8}$$

where d_{c1} is the critical depth at section 1 (m), q_1 is the flow rate per unit width at section 1 = Q/B_1 (m²/s), Q is the flow rate (m³/s), and B_1 is the channel width at section 1 (m).

At much larger flow rates, critical depth occurs at point 2. This drowns the control at point 1, and conditions throughout are subcritical. The expression for this depth is similar

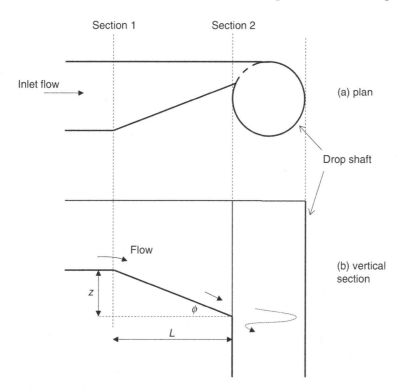

Figure 8.13 Tangential vortex intake.

(Equation (8.9)), but the vertical angle of the inlet channel must be included because depth is measured perpendicular to the bed.

$$d_{c2} = \sqrt[3]{\frac{q_2^2}{g\cos\varphi}} \tag{8.9}$$

where d_{c2} is the critical depth at section 2 (m), q_2 is the flow rate per unit width at section 2 = Q/B_2 (m²/s), B_2 is the channel width at section 2 (m), and φ is the vertical slope angle of intake channel (see Figure 8.13).

Based on specific energy in the approach channel (see Section 7.5.3) and assuming no overall loss of energy, the flow rate at which this shift in control occurs (Yu and Lee, 2009) can be determined from Equation (8.10):

$$Q_c = \frac{\sqrt{g}B_2\left[2z/3\right]^{3/2}}{\left(\left(\cos\varphi\right)^{2/3} - \left[B_2/B_1\right]^{2/3}\right)^{3/2}} \tag{8.10}$$

where Q_c is the flow rate at which the shift in control occurs (m³/s), and z is defined in Figure 8.13.

Conditions at the top of the vortex may cause disruption to the flow in the inlet channel as a result of re-entering the inlet or disturbing the inflow jet. Based on the geometry of the inlet, Yu and Lee (2009) give the maximum flow rate for which "free drainage" (no disruption) can be assumed, as

$$Q_f = \left[\tan\varphi\frac{\pi D}{1 - B_2/D}\right]^{3/2}\sqrt{g}B_2\left(\cos\varphi\right)^2 \tag{8.11}$$

where Q_f is the maximum flow rate for which free discharge can be assumed (m³/s) and D is the diameter of the drop shaft (m).

In design, Q_c should be less than Q_f so that the shift in control can occur when there is free discharge (without disruption to flow in the inlet channel caused by conditions at the top of the drop shaft), and Q_f should exceed the design maximum flow rate (Q_{max}).

For sizing of the drop shaft, Yu and Lee (2009) propose using $k = 1.2$ within Equation (8.12):

$$D = k\left[\frac{Q_{max}^2}{g}\right]^{1/5} \tag{8.12}$$

A suitable value for B_2 is given as $0.25D$.

Although air entrainment is not considered to be as significant within a vortex drop shaft as it is in a plunge-flow shaft, it can still be significant, and sufficient airflow is clearly an important element in maintaining a stable air core.

The outlet arrangement for a vortex drop shaft must fulfil a number of functions. It must direct the outflow into a horizontal conduit, it must facilitate the dissipation of energy, and it must allow de-aeration to take place as flow enters the outlet pipe (Zhao et al., 2006). Crispino et al. (2019) present laboratory findings relevant to the design of the dissipation chamber including estimation of chamber height and pressure profile (within a comprehensive study of the hydraulic behaviour of the inlet, vertical shaft, and outlet). Outlet arrangements

EXAMPLE 8.3

a. A vortex drop has a tangential intake with an inlet width (B_1) of 2.2 m. The design maximum flow rate is to be taken as 11 m³/s. The length available for the intake channel is 8 m. Propose acceptable dimensions for the intake.

b. For your selected dimensions, determine a known depth in the intake channel at the design flow rate. For a flow rate 20% greater than the design flow rate, determine a known depth.

Solution

a. Basing the drop shaft diameter on Equation (8.12), with $k = 1.2$, we have

$$D = 1.2\left[\frac{11^2}{g}\right]^{1/5} = 1.98 \text{ m, so make } D = 2 \text{ m}$$

Base the value of B_2 on 0.25D, so B_2 should be 0.5 m.
Trying a value for z of 4 m, and using the whole of the 8 m for L, gives φ = tan⁻¹(4/8) = 26.6°.
Equation (8.10) gives the value of Q_c as 16.5 m³/s.
And Equation (8.11) gives the value of Q_f as 10.8 m³/s.
This is not a good design because Q_f is not only less than Q_c but it is also less than Q_{max}. To increase the value of Q_f, we could increase the value of D, but that would be a significant and costly change to the design. Let's see first if increasing φ could be effective. We do this by decreasing L; as we do not need an increase in Q_c we also reduce z.
Try L = 4.25 m with z = 3 m. This gives φ = 35.2°.
Equation (8.10) now gives the value of Q_c as 12.5 m³/s.
And Equation (8.11) gives the value of Q_f as 15 m³/s.
This is a satisfactory design because Q_c is less than Q_f and Q_f exceeds Q_{max}.

b. The design flow rate of 11 m³/s is less than Q_c, so the depth control is at section 1 in Figure 8.13.

$$q_1 = \frac{Q}{B_1} = \frac{11}{2.2} = 5 \text{ m}^2 / \text{s}$$

$$d_{c1} = \sqrt[3]{\frac{5^2}{g}} = 1.37 \text{ m}$$

The known depth for a flow rate of 11 m³/s is at section 1: 1.37 m. A 20% increase gives Q = 13.2 m³/s. This is greater than Q_c, so the depth control moves to section 2:

$$q_2 = \frac{Q}{B_2} = \frac{13.2}{0.5} = 26.4 \text{ m}^2 / \text{s}$$

$$d_{c2} = \sqrt[3]{\frac{26.4^2}{g \cos 35.2}} = 4.43 \text{ m}$$

The known depth for a flow rate of 13.2 m³/s is at section 2: 4.43 m (perpendicular to the bed).

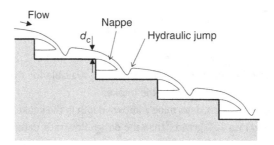

Figure 8.14 Stepped cascade (nappe flow regime).

are heavily dependent on the context of the sewer drop, and, like other aspects of the design of a specific vortex drop shaft of significant size, are often best determined through physical modelling (Riisnaes et al., 2014).

There are other hydraulic considerations in vortex drop shaft design. The risk of blockage must be considered, and this may define a minimum channel width (B_2, the width at section 2 in Figure 8.13a, in the case of a tangential intake). This, in turn, may define the minimum drop shaft diameter. Also, the approach conditions in the channel leading to the intake should be as stable as possible with no bends, steps, or other obstacles for a significant length upstream of the intake.

8.3.2 Other sewer drop arrangements

A variation on the concept of the vortex drop shaft is a hybrid arrangement described by Andoh et al. (2008), which uses an air intake control system to allow both vortex and vertical pipe-full flow, and removes the need for a vortex-inducing inlet.

Granata (2016) discusses the use of a *drop shaft cascade*: a series of plunge-flow drop shafts connected by lengths of sewer. The design is presented as an optimisation problem involving the selection of the configuration and number of drop shafts that minimise construction cost. This can be a more efficient solution than a single drop shaft for the same total drop height, provided the sewer drop does not need to be at a single location.

Another arrangement for a sewer drop made up of a number of vertical drops is a stepped cascade. This is a simple approach to energy dissipation but requires significantly more plan area than the use of drop shafts. Flow passes down a series of simple connected steps, each providing a small vertical drop. The nature of the flow conditions is dependent on the value of the flow rate. At relatively low flows, a nappe flow regime exists in which there is a free overfall from each step to the next (Figure 8.14). Assuming subcritical conditions upstream, the depth at the edge of each step is equal to the critical depth (Section 7.5.4). In the next step, the impact of the free-falling jet (or *nappe*) is followed by a hydraulic jump, then subcritical flow, then critical flow again at the edge of that step, and so on. At larger flow rates, a skimming flow regime is established where flow hits the edge of each step but does not form a particular flow pattern on each individual step. Principles for sizing cascade steps are given by Chanson (2004b).

8.4 INVERTED SIPHONS

Inverted siphons carry flows under rivers, canals, roads, and so on (e.g., Figure 8.15). They are necessary when this crossing cannot be made by means of a pipe-bridge, or by having the

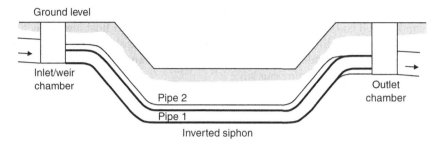

Figure 8.15 Inverted siphon for wastewater, vertical section (schematic).

whole sewer length at a lower level. Unlike normal siphons, inverted siphons do not require special arrangements for filling; they simply fill by gravity. However, they do present some problems and are avoided where possible.

Hydraulically, inverted siphons are an interesting case compared with other flowing sewers. As discussed in Chapter 7, the majority of sewers flow part-full, and when the flow rate is low, the depth is low. When sewer flows are pumped, the pipe flows full and the pumps tend to deliver the flow at a fairly constant rate, but not continuously (as described in Chapter 15). In contrast, inverted siphons flow full, and they flow continuously. At low flow, the velocity can be very low, which, unfortunately, creates the ideal conditions for sediment deposition.

The most important aim in design is to minimise silting. Some silting is virtually inevitable at low flows, but at higher flows, the system should be self-cleansing. It is normally assumed that this will be achieved if the velocity is greater than 1 m/s. (This subject is considered further in Chapter 17.) The higher the velocity, the lower the danger of silting.

Many siphons consist of multiple pipes as a means of minimising siltation. The low flows are carried by one pipe, smaller than the sewers on either side of the siphon. At higher flows, this pipe is self-cleansing, and an arrangement of weirs allows overflow into other pipes. In separate systems, two pipes for wastewater are usually enough (Figure 8.15); in combined sewers, a third much larger pipe is usually needed.

EXAMPLE 8.4

An existing single-pipe inverted siphon, carrying wastewater only, is to be replaced with a twin-barrelled siphon because of operational problems caused by sedimentation. The required length is 70 m; the available fall (invert to invert) is 0.85 m. Determine the pipe sizes required for an average dry weather flow (DWF) of 90 L/s and a peak flow of 3 DWF. Assume the inlet head loss is 150 mm, the self-cleansing velocity is 1 m/s, and k_s = 1.5 mm. Use Table 7.1 for pipe flow calculations.

Solution

One approach: use one pipe to carry DWF and a second pipe to carry excess.

Available hydraulic gradient = (0.85–0.15)/70 = 0.01.

Using Table 7.1 for pipe calculations

A 300 mm pipe carries 98 L/s at a velocity of 1.38 m/s. Velocity is sufficient.

Excess flow = (3 × 90) − 98 = 172 L/s.

A 375 mm pipe carries 177 L/s at a velocity of 1.6 m/s. Velocity is sufficient.

So, use pipe diameters of 300 mm and 375 mm.

Other devices for avoiding siltation are sometimes needed. On small systems, a penstock upstream can be used to back up the flow and create an artificial flushing wave. Silt can be removed directly by providing penstocks or stop boards for isolating sections of pipe, and access for removing silt. An independent washout chamber can be provided and used in conjunction with a system for pumping out silt.

8.5 GULLY SPACING

Several approaches to establishing the required spacing of road gullies have been proposed. The simplest are mentioned in Chapter 6, but in this section, more accurate methods are outlined.

8.5.1 Road channel flow

The typical geometry of flow in a road channel is as given in Figure 8.16.

For channels of shallow triangular section, Manning's equation (Equation (7.23)) can be simplified by assuming the top width of the channel flow (B) equals the wetted perimeter (P), to give

$$Q = 0.31Cy^{8/3} \tag{8.13}$$

where Q is the channel flow rate with "channel criterion" C (fixed for the road):

$$C = \frac{zS_o^{1/2}}{n} \tag{8.14}$$

where y is the flow depth (m), z is the side slope (1:z), S_o is the longitudinal slope (–), and n is the Manning's roughness coefficient (m$^{-1/3}$s).

Manning's n for roads ranges from 0.011 for smooth concrete to 0.018 for asphalt with grit. Example 8.5 demonstrates the use of these equations.

8.5.2 Gully hydraulic efficiency

The hydraulic efficiency of a gully depends on the depth of water in the channel immediately upstream, the width of flow arriving, and the geometry of the grating. A typical efficiency curve is given in Figure 8.17. This shows that at low flows, gullies are approximately 100% efficient and all flow is captured. Once the approach flow Q exceeds $\overline{Q}_l$, efficiency drops off

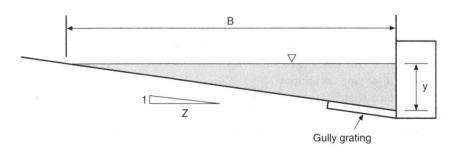

Figure 8.16 Geometry of road channel flow (exaggerated vertical scale).

EXAMPLE 8.5

Determine the flooded width of a concrete road ($n = 0.012$) when the flow rate is 20 L/s. The road has a longitudinal gradient of 1% and a crossfall of 1:40.

Solution

From Equation (8.14), calculate the channel criterion:

$$C = \frac{40 \times 0.01^{\frac{1}{2}}}{0.012} = 333.3$$

Rearranging Equation (8.13) gives

$$y = \left(\frac{Q}{0.3\,IC}\right)^{3/8} = \left(\frac{0.02}{0.31 \times 333.3}\right)^{3/8} = 0.041\,m$$

Thus, the depth of flow is 41 mm leading to a width of flow $B = yz = 0.041 \times 40 = 1.62$ m.

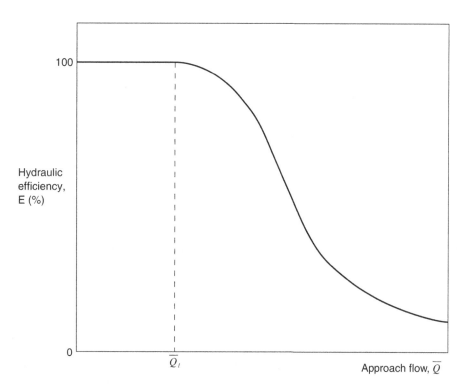

Figure 8.17 Typical gully efficiency curve. (After Davis, A. et al. 1996. *Journal of Chartered Institution of Water and Environmental Management*, 10, April, 118–122.)

rapidly. When the approach flow is plotted against the captured flow (as in Figure 8.18), it is clear that the captured flow $\overline{Q}_l$ corresponding to 100% efficiency is not the maximum flow that the gully can capture. Higher approach flows result in an increase in captured flow due to the greater flow depths over the grating. Thus, the capacity of a gully Q_c can be increased by allowing a small bypass flow. May (1994) suggests that an optimum value is 20%.

Thus, the hydraulic capture efficiency E for an individual gully grating is

$$E = \frac{Q_c}{\overline{Q}} \qquad (8.15)$$

where E is a function of grating type, water flow width, road gradient, and crossfall.

Data on the efficiency of a number of grating types can be found in CD 526 Spacing of road gullies (Highways England, 2020). An example is given in Table 8.2 for a particular gully type.

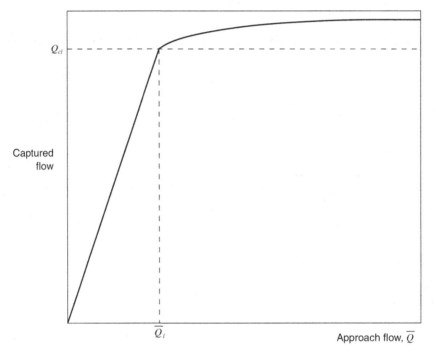

Figure 8.18 Relationship between approaching and captured flow for a typical gully. (After Davis, A. et al. 1996. *Journal of Chartered Institution of Water and Environmental Management*, 10, April, 118–122.)

Table 8.2 Example gully efficiencies (E) at 1:20 crossfall

Flow width	Longitudinal gradient (1:X)				
B (m)	20	30	50	100	300
0.5	92	94	95	97	98
0.75	85	88	90	93	96
1.0	76	80	85	89	94
1.5	52	61	70	78	88

Source: Adapted from Highways England (2020).

8.5.3 Gully inlet hydraulics

A number of studies treat a gully inlet as hydraulically equivalent to either an orifice or a weir. For example, laboratory work by Cosco et al. (2020) looks at representing the grated inlet to a gully, receiving inflow from a road surface, both as an orifice *and* as a weir. A range of values for the orifice coefficient (as in Equation (8.1)) and the weir discharge coefficient (Equation (8.4)) is presented. These values depend on hydraulic conditions of the surface flow, including the Froude number, and the configuration of the grated inlet. Values of orifice coefficient vary widely but are always less than the "standard" value for an orifice plate of around 0.6 (which is sometimes recommended for modelling purposes). The data is for cases where the whole inlet is completely covered by water. The term H (as contained in both Equations ((8.1)) and (8.4)) is given as the depth of water plus the velocity head immediately upstream of the grate.

When the system is surcharged such that flow from the drainage system is flooding onto the surface (described as "manhole surcharge" in Section 7.4.5) there can be upward flow through gully "inlets", in the opposite direction to usual. Gómez et al. (2019) describe a laboratory study of this "reverse flow" effect in which the gully inlet is treated as an orifice. They find that the orifice coefficient (in Equation (8.1)) is not constant, but increases with outflow discharge. As in the study above it is certainly not equal to the "standard" value of 0.6 (see Section 8.1.1), always being lower in this study.

Lopes et al. (2015) also studied surcharge flow upwards through a gully using laboratory experiments and numerical analysis. They present data on the jet created by the upward flow and the height it reaches above the gully. They also consider the behaviour of the flow inside the gully pot, with good agreement between numerical and experimental results.

Cárdenas-Quintero and Carvajal-Serna (2021) present a comprehensive overview of a range of published work on gully inlet hydraulics.

8.5.4 Spacing

The basic approach to gully hydraulic design is to make sure that they are sufficiently closely spaced to ensure that the flow spread in the road channel is lower than the allowable width (B). Figure 8.19 shows a schematic of the flow conditions along a road of constant longitudinal gradient and crossfall subject to constant inflow. Gullies are spaced at a distance L apart, except the first gully which is at a distance of L_1. The inflow per unit length q is generated by constant intensity rainfall. The flow bypassing each gully must be included in the flow arriving at the next inlet.

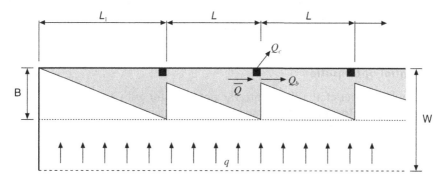

Figure 8.19 Spacing of initial and intermediate gullies.

8.5.4.1 Intermediate gullies

The maximum flooded width B and flow rate $\bar{Q}$ occurs just upstream of a road gully and consists of the sum of the runoff $Q_r = qL$ and the bypass flow Q_b from the previous gully:

$$\bar{Q} = Q_b + Q_r$$

And, from the Rational Method equation (described more fully in Chapter 10),

$$Q_r = iWL$$

where i is the rainfall intensity for a storm duration equal to the time of entry and assuming complete imperviousness (runoff coefficient, $C = 1$), and W is the road width contributing flow to the gully. The flow arriving at the gully is either captured or bypasses it, so

$$\bar{Q} = Q_b + Q_c$$

$$Q_c = Q_r$$

Thus, the captured flow is equal to the runoff generated between gullies. Hence, substituting Equation (8.10) gives

$$E\bar{Q} = iWL$$

$$L = \frac{E\bar{Q}}{iW} \tag{8.16}$$

8.5.4.2 Initial gullies

The most upstream gully in the system is a special case as it does not have to handle carryover from the previous gully; thus, $Q_b = 0$ and, so:

$$\bar{Q} = Q_r$$

$$L_1 = \frac{\bar{Q}}{iW} \tag{8.17}$$

Example 8.6 shows how gully spacing can be calculated.

A second special case is the terminal gully that can have no carryover. These act as weirs under normal conditions and as orifices under large water depths. Methods to design such gullies are outlined in Highways England (2020).

8.5.4.3 Potential optimisation

A UK study by Doncaster et al. (2012) into the potential for gully optimisation suggests that it may be appropriate for gullies to be spaced more widely than traditionally specified (at least from a hydraulic point of view). In a particular study catchment, they found there was a good case for a 50% reduction in the number of gullies, provided gully gratings could be made to perform more effectively.

EXAMPLE 8.6

Determine the spacing of the initial and subsequent gullies on a road in the London area. The road is 5 m wide with a crossfall of 1:20 and a longitudinal gradient of 1%. The road surface texture suggests a Manning's n of 0.010 should be used. A design rainfall intensity of 55 mm/h is to be used at which the flood width should be limited to 0.75 m.

Solution

Allowable flow depth = 0.75/20 = 0.0375 m.

Channel criterion (Equation (8.14)),

$$C = \frac{20 \times 0.01^{1/2}}{0.010} = 200$$

Maximum flow rate (Equation (8.13)),

$$\bar{Q} = 0.31 \times 200 \times 0.0375^{8/3} = 0.010 \text{ m}^3 / \text{s}$$

Thus, the spacing of the initial gully should be (Equation (8.17))

$$L_I = \frac{0.010 \times 3600 \times 10^3}{55 \times 5} = 131 \text{m}$$

Read from Table 8.2, $E = 0.99$
$L = EL_I = 130$ m
Allow a 20% reduction of capacity for potential blockage.
Maximum gully spacing is approximately 100 m.

8.6 CULVERTS

8.6.1 Culverts in urban drainage

Where urban drainage systems include open channels, culverts may be needed to carry the flow under a road or railway. Culverts are also common on natural watercourses, though the practice of culverting long lengths is now recognised as having a negative impact on amenity and biodiversity.

Comprehensive practical advice on culvert design and management is provided by Benn et al. (2019). This is accompanied by spreadsheets for modelling hydraulic performance. Screens, for debris or security, are also covered, though the clear recommendation is that screens should be avoided where possible, and guidance for this is given.

Hydraulically, if a culvert is not flowing full, it simply behaves as an open channel. The approach is usually to adopt the principles of open-channel flow even when culverts are flowing full. This is in contrast with the usual approach to part-full pipes (Section 7.4), which is based on the principles of pipe flow even though there is a free surface.

8.6.2 Flow cases

Different longitudinal water surface profiles occur within a culvert depending on conditions (Figure 8.20). It is assumed that if the depth upstream of the culvert is less than 1.2 times the culvert height, the culvert behaves as an open channel. Under these conditions, the shape of the longitudinal water surface profile is influenced by two other factors: whether the slope of the culvert is hydraulically mild or steep (Section 7.5.4), and whether conditions downstream exert an influence on the water depth within the culvert. If they do, this is termed a *downstream surcharge*.

If the depth upstream of the culvert is greater than 1.2 times the culvert height, the flow rate in the culvert may be limited either by the properties of the inlet acting as an orifice (Section 8.1.1) or by the friction and local losses in the culvert. These two conditions can be termed *inlet controlled* or *losses controlled*. For the losses-controlled case, there may or may not be a downstream surcharge.

These seven possibilities are listed in Table 8.3 together with the basis for the standard calculation procedure. There is more detail in Examples 8.7 and 8.8.

The resulting surface profiles are presented in Figure 8.20.

Unfortunately, there is no standard universally applied classification of flow cases in culverts. For example, Benn et al. (2019) and Hamill (2011) use different systems that are both different from the one presented here. Also, the term "inlet control" can be used in different ways. Here, it is used to distinguish from "losses controlled" in determining capacity.

Table 8.3 Conditions for different water surface profiles

Flow condition	See Figure 8.20	Downstream surcharge?	Basis for calculations
Open channel – mild slope	(a)	No	d_n determined from Manning's equation
	(b)	Yes	S_f determined from Manning's equation
Open channel – steep slope	(c)	No	d_n determined from Manning's equation
	(d)	Yes	d_n determined from Manning's equation with a hydraulic jump in the culvert
Inlet controlled	(e)		Orifice Equation ((8.1))
Losses controlled	(f)	No	S_f determined from Manning's equation (full) and local losses
	(g)	Yes	S_f determined from Manning's equation (full) and local losses

Note: d_n normal depth, d_c critical depth, d_u upstream depth.

EXAMPLE 8.7

A rectangular cross-section culvert has a width of 1.2 m, height (H_c) 0.6 m, slope 1:100, and Manning's n 0.013. For a significant length within the culvert, flow is at the normal depth, 0.15 m. At the inlet, local losses can be calculated using $k_L = 0.5$. It can be assumed that upstream velocity is negligible. There is no surcharge downstream.

Determine the depth immediately upstream of the culvert (if the bed of the channel and the invert of the culvert are the same at the inlet).

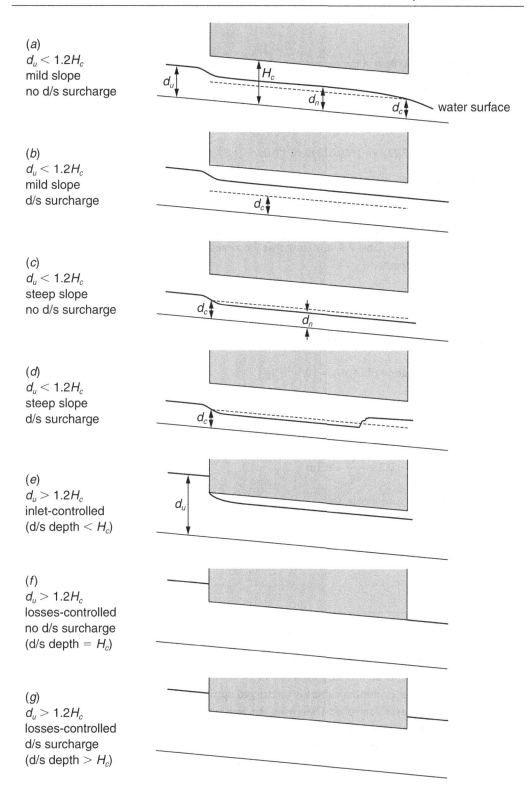

Figure 8.20 Flow cases for culverts.

Solution

There is open channel flow in the culvert since 0.15 m < 0.6 m:

$$A = 1.2 \times 0.15 = 0.18 \, m^2$$

$$R = 0.18 \div 1.5 = 0.12 \, m$$

Manning's Equation (7.23): $v = (1/0.013)0.12^{2/3}0.01^{1/2} = 1.87 \, m/s$

$$Q = v \times A \left(from \, Equation \, (7.3) \right), so \, Q = 1.87 \times 0.18 = 0.337 \, m^3/s$$

Is the slope of the culvert hydraulically steep or mild?

To determine d_c critical depth, use Equation (7.27) $v_c = \sqrt{gd_m}$.

From the definition in Table 7.4, it is clear that for a rectangular channel, d_m equals the actual depth. So at critical depth,

$$v_c = \sqrt{gd_c} \, or \left(using \, Equation \, (7.3) \right) \frac{0.337}{1.2 \times d_c} = \sqrt{g \times d_c}$$

this gives $d_c = 0.2$ m, which is greater than the normal depth, 0.15 m, so the slope is **steep** (Section 7.5.4).

Loss at entry (Equation (7.14)) $h_L = \left(0.5v_c^2 / 2g \right)$.

Using Equation (7.3), $v_c = 0.337 / \left(1.2 \times 0.2 \right) = 1.4 \, m/s$, so entry loss = 0.05 m

In the culvert, specific energy (Equation (7.25))

$$E = y + \frac{v^2}{2g} = 0.2 + \frac{1.4^2}{2g} = 0.3 \, m$$

Upstream velocity is negligible, so

$$d_u = upstream \, E = 0.3 + 0.05 = 0.35 \, m$$

EXAMPLE 8.8

A rectangular cross-section culvert has a width of 1.2 m, height (H_c) 0.6 m, Manning's n 0.013, and slope 1 in 100. The depth immediately upstream (d_u) is 0.9 m. The length of the culvert is 30 m.

Determine whether or not the culvert is surcharged upstream. Determine whether the flow is inlet controlled or losses controlled. Determine the flow rate.

Assume that upstream velocity is negligible, and that there is no surcharge downstream (i.e., depth downstream does not affect depth in the culvert).

For entry loss, use $K = 0.5$. If the entry is acting as an orifice, $C_d = 0.62$.

Solution

$d_u > H_c$, so the culvert is either inlet controlled or losses controlled.

1. Assume inlet controlled:

   Equation ((8.1)), $Q = C_d A \sqrt{2gH} = 0.62 \times 1.2 \times 0.6 \sqrt{2 \text{ g } 0.6} = 1.53 \text{ m}^3 / s$

2. Assume losses controlled:

   Subscripts 1 and 2 denote conditions at upstream and downstream ends of the culvert, respectively.

   E_1 (specific energy) $= d_u = 0.9$ m, since upstream velocity is negligible.

   $v_1 = v_2$; velocity all along the culvert is written as v below.

For this case (with no free surface), the Manning equation should be written

$$v = \frac{1}{n} R^{2/3} S_f^{1/2} \text{ or } S_f = \left[\frac{vn}{R^{2/3}} \right]^2$$

$$R = \frac{1.2 \times 0.6}{1.2 + 0.6 + 1.2 + 0.6} = 0.2 \text{ m}$$

since the wetted perimeter includes the top of the culvert. This gives $S_f = 1.445 \times 10^{-3} \times v^2$.

The height of the upstream water surface above the level of the bed at the outlet can be expressed two ways, so

$$0.9 + S_o L = 0.5 v^2 /2g + S_f L + v^2 /2g + 0.6$$

$$0.9 + 0.01 \times 30 = 1.5 v^2 / 2g + 1.445 \times 10^{-3} \times v^2 \times 30 + 0.6$$

this gives $v = 2.24$ m/s so $Q = 1.61$ m³/s.

Losses limit the capacity of the culvert to 1.61 m³/s, but the inlet acting as an orifice limits the flow rate to 1.53 m³/s.

So, the flow rate cannot exceed 1.53 m³/s, and the culvert is inlet controlled.

PROBLEMS

1. An orifice plate is being designed for flow control at the outlet of a detention tank. The outlet pipe has a diameter of 450 mm, a slope of 0.0018, and roughness k_s of 0.6 mm. The water level in the tank at the design condition varies between 1.5 and 1.7 m above the outlet invert, and the desired outflow is 100 L/s. Select an appropriate orifice diameter. (Assume orifice $C_d = 0.6$; check that the orifice will not be drowned.)   [200 mm, not drowned]

2. A throttle pipe to control outflow (to treatment) from a CSO is being designed. The pipe will have a length of 25 m and a diameter of 200 mm. A check is being carried out to see how well the pipe will control the flow (roughness of the pipe, $k_s = 0.6$ mm). When the difference between the water level at the upstream and downstream end

is 2.5 m, what would be the flow rate in the pipe considering friction losses only? In this condition, is the pipe inlet controlled (assume $C_d = 0.6$)? Use Figure 7.5 for pipe flow.   [120 L/s, no]

3. A rectangular transverse weir in a CSO has a width equal to the width of the chamber itself: 2.2 m. The weir crest is 1.05 m above the floor of the chamber. When the water level is 0.15 m above the crest, determine the flow rate over the weir.   [0.235 m³/s]

4. A tangential vortex intake has $Q_{max} = 15$ m³/s, $B_1 = 2.75$ m, $L = 9$ m, $z = 4$ m, $B_2 = 0.56$ m, $D = 2.25$ m (using the symbols defined in Section 8.3). Comment on the appropriateness of this design.   [$Q_c = 18$ m³/s and $Q_f = 16.5$ m³/s, so while Q_c and Q_f exceed Q_{max}, $Q_c > Q_f$, so the design is not OK.]

5. Estimate the flow rate in the channel of a road with a longitudinal gradient of 0.5% and a crossfall of 1:40 if the width of the flow is 2.5 m. Assume $n = 0.013$.   [42 L/s]

6. A rectangular cross-section culvert has a width of 1.2 m, height of 0.6 m, slope of 1:1000, and Manning's n 0.013. For a significant length within the culvert, flow is at the normal depth, 0.2 m. At the inlet, local losses can be calculated using $K = 0.5$. It can be assumed that upstream velocity is negligible. Determine the depth immediately upstream of the culvert (if the bed of the channel and the invert of the culvert are the same at the inlet). Sketch the shape of the water surface from just upstream of the culvert to just downstream.   [0.236 m]

7. A rectangular cross-section culvert has a width of 1.2 m, height of 0.6 m, Manning's n 0.013, and slope 1 in 1000. The depth immediately upstream is 1 m. The length of the culvert is 30 m. Determine whether or not the culvert is surcharged upstream. Determine whether flow in the culvert is inlet controlled or losses controlled. Determine the flow rate. Assume that upstream velocity is negligible and that there is no surcharge downstream (i.e., depth downstream does not affect depth in the culvert). For entry loss, use $K = 0.5$. If the entry is acting as an orifice, $C_d = 0.62$.   [losses controlled; 1.37 m³/s]

REFERENCES

Ackers, P. and Crump, E.S. 1960. The vortex drop. *Proceedings of the Institution of Civil Engineers*, 16, August, 433–442.

Andoh, R., Osei, K., Fink, J., and Faram, M. 2008. Novel drop shaft system for conveying and controlling flows from high level sewers into deep tunnels. *World Environmental and Water Resources Congress, ASCE*, Honolulu, Hawaii, May, on CD.

Balmforth, D. 2009. *Modelling ancillaries: weir coefficients*. WaPUG User Note No 27.

Balmforth, D.J., Saul, A.J., and Clifforde, I.T. 1994. *Guide to the Design of Combined Sewer Overflow Structures*, Report FR 0488, Foundation for Water Research.

Benn, J., Kitchen, A., Kirby, A., Fosbeary, C., Faulkner, D., Latham, D. and Hemsworth M. 2019. *Culvert, Screen and Outfall Manual*, C786, RP1075, CIRIA.

Burrows, R. and Emmonds, J. 1988. Energy head implications of the installation of circular flap gates on drainage outfalls. *Journal of Hydraulic Research*, 26(2), 131–142.

Butler, D. and Parsian, H. 1993. The performance of a vortex flow regulator under blockage conditions. *Proceedings of the 6th International Conference on Urban Storm Drainage*, Niagara Falls, Canada, 1793–1798.

Cárdenas-Quintero, M. and Carvajal-Serna, F. 2021. Review of the hydraulic capacity of urban grate inlet: a global and Latin American perspective. *Water Science and Technology*, 83(11), 2575–2596.

Chanson, H. 2004a. Hydraulics of rectangular dropshafts. *Journal of Irrigation and Drainage Engineering*, 130(6), 523–529.

Chanson, H. 2004b. *The Hydraulics of Open Channel Flow: An Introduction*, 2nd edn. Chapter 20, Elsevier.

Chow, V.T. 2009. *Open-Channel Hydraulics*, The Blackburn Press.

Cosco, C., Gómez, M., Russo, B., Tellez-Alvarez, J., Macchione, F., Costabile, P. and Costanzo, C. 2020. Discharge coefficients for specific grated inlets. Influence of the Froude number. *Urban Water Journal*, 17(7), 656–668.

Crispino, G., Pfister, M. and Gisonni, C. 2019. Hydraulic design aspects for supercritical flow in vortex drop shafts. *Urban Water Journal*, 16(3), 225–234.

Davis, A., Jacob, R.P., and Ellett, B. 1996. A review of road-gully spacing methods. *Journal of Chartered Institution of Water and Environmental Management*, 10, April, 118–122.

Del Guidice, G. and Gisonni, C. 2010. Vortex dropshafts: History and current applications to the sewer system of Naples (Italy). *Proceedings of the First IAHR European Congress*, Edinburgh, May.

Delo, E.A. and Saul, A.J. 1989. Charts for the hydraulic design of high side-weirs in storm sewage overflows. *Proceedings of the Institution of Civil Engineers*, Part 2, 87, June, 175–193.

Doncaster, S., Blanksby, J., Shepherd, W., and Sailor, G. 2012. *Gulley Optimisation – An Investigation into the Potential for Gulley Optimisation to Reduce Maintenance Requirement and to Reduce Surface Water Flood Risk*. Research Report, SKINT (North Sea Skills Integration and New Technologies).

Echavez, G. and Ruiz, G. 2008. High head drop shaft structure for small and large discharges. *Proceedings of the 11th International Conference on Urban Drainage*, Edinburgh, September, on CD.

Gómez, M., Russo, B. and Tellez-Alvarez, J. 2019. Experimental investigation to estimate the discharge coefficient of a grate inlet under surcharge conditions. *Urban Water Journal*, 16(2), 85–91.

Granata, F. 2016. Dropshaft cascades in urban drainage systems. *Water Science and Technology*, 73.9, 2052–2059.

Granata, F., de Marinis, G., Gargano, R., and Hager, W.H. 2011. Hydraulics of circular drop manholes. *Journal of Irrigation and Drainage Engineering*, 137(2), 102–111.

Hamill, L. 2011. *Understanding Hydraulics*, 3rd edn, Palgrave Macmillan.

Highways England. 2020. CD 526 *Spacing of road gullies*. Highways England.

Lopes, P., Leandro, J., Carvalho, R.F., Páscoa, P. and Martins, R. 2015. Numerical and experimental investigation of a gully under surcharge conditions. *Urban Water Journal*, 12(6), 468–476.

May, R.W.P. 1994. Alternative hydraulic design methods for surface drainage. *Road Drainage Seminar*, H.R. Wallingford, Wallingford, November.

May, R.W.P., Bromwich, B.C., Gasowski, Y., and Rickard, C.E. 2003. *Hydraulic Design of Side Weirs*, Thomas Telford.

Riisnaes, S., Poole, B., Cooper, M., Thornton, C., Digman, C., and Marples, N. 2014. Building vortex drop shafts in dense urban areas – A different approach. *Chartered Institution of Water and Environmental Management Urban Drainage Group Autumn Conference*.

Yu, D. and Lee, J.H.W. 2009. Hydraulics of tangential vortex intake for urban drainage. *Journal of Hydraulic Engineering*, 135(3), 164–174.

Zhao, C.-H., Zhu, D.Z., Sun, S.-K., and Liu, Z.-P. 2006. Experimental study of flow in a vortex drop shaft. *Journal of Hydraulic Engineering*, 132(1), 61–68.

Chapter 9

Foul sewers

9.1 INTRODUCTION

Separate foul sewers (also known as sanitary sewers or wastewater sewers) form an important component of many urban drainage systems. The emphasis in this chapter is on the design of such systems (Section 9.2). In particular, the distinction is made between large (Section 9.3) and small (Section 9.4) foul sewers and their different design procedures. Solids transport is introduced in Section 9.5. Analysis of existing systems using computer-based methods is covered in Chapters 20 and 21. The design of non-pipe-based (sanitation) systems is discussed in Chapter 23.

9.1.1 Flow regime

All foul sewer networks physically connect wastewater sources with treatment and disposal facilities by a series of continuous, unbroken pipes. Flow into the sewer results from random usage of a range of different appliances, each with its own characteristics. Generally, these are intermittent, of relatively short duration (seconds to minutes), and hydraulically unsteady. At the outfall, however, the observed flow in the sewer will normally be continuous and will vary only slowly (and with a reasonably repeatable pattern) throughout the day. Figure 9.1 gives an idealised illustration of these conditions.

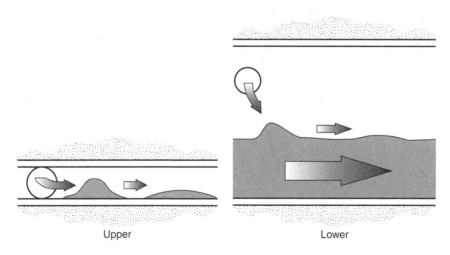

Upper　　　　　Lower

Figure 9.1 Hydraulic conditions in foul sewers in dry weather (schematic).

DOI: 10.1201/9781003408635-9

The sewer network will have zones with continuously flowing wastewater, as well as areas that are mostly empty but are subject to flushes of flow from time to time. It is unlikely, even under maximum continuous flow conditions, that the full capacity of the pipe will be utilised. Intermittent pulses feed the continuous flow further downstream, and this implies that somewhere in the system, there is an interface between the two types of flow. As the usage of appliances varies throughout the day, the interface will not remain at a single fixed location.

9.2 DESIGN

This chapter shows how foul sewers can be designed to cope with conditions as previously described. A general approach to foul (and storm) sewer design is illustrated in Figure 9.2. This should be read in conjunction with Figure 6.8.

Design is accomplished by first choosing a suitable *design period* and *criterion of satisfactory service*, appropriate to the foul *contributing area* under consideration. The type and number of buildings and their population (the maximum within the design period) are then estimated, together with estimates of the unit water consumption. This information is used to calculate *dry weather flow* (DWF) in the main part of the system. Flows in building drains and small sewers are assessed in a probability-orientated discharge unit method, based on usage of domestic appliances. *Hydraulic design* of the pipework is based on safe transportation of the flows generated using the principles presented in Chapter 7. Broader issues of sewer layout including horizontal and vertical alignment are covered in Chapter 6.

9.2.1 Choice of design period

Urban drainage systems have an extended life span and are typically designed for conditions 25–50 years into the future. They may well be in use for very much longer. The choice of design period will be based on factors such as

- Useful life of civil, mechanical, and electrical components
- Feasibility of future extensions of the system
- Anticipated changes in residential, commercial, or industrial development
- Financial considerations

It is necessary to make estimates of conditions throughout the design period that are as accurate as possible.

9.2.2 Criterion of satisfactory service

The degree of protection against wastewater "backing-up" or flooding is determined by consideration of the specified criterion of satisfactory service. This protection should be consistent with the cost of any damage or disruption that might be caused by flooding. In practice, cost–benefit studies are rarely conducted for ordinary urban drainage projects; a decision on a suitable criterion is made simply on the basis of judgement and precedent. Indeed, this decision may not even be made explicitly, but nevertheless, it is built into the design method chosen.

The design choice of the peak-to-average flow ratio implicitly fixes the level of satisfactory service in large foul sewers. For small sewers, the criterion can be used explicitly to determine flows, though standard (and therefore fixed) values are routinely used.

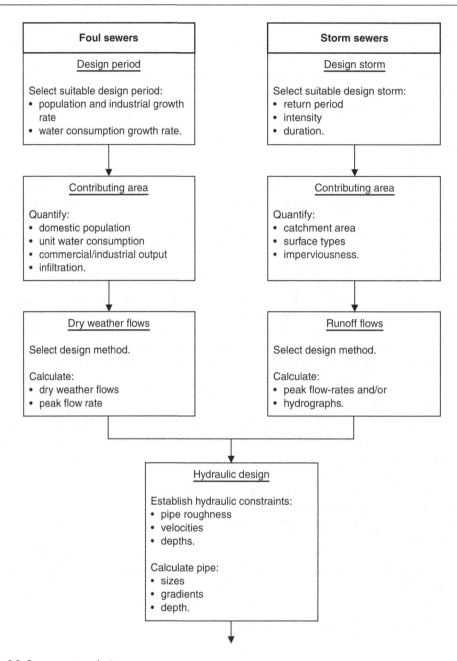

Figure 9.2 Sewer system design.

9.3 LARGE SEWERS

In this text, a distinction is drawn between large and small foul sewers. This is only for convenience, as there is no precise definition to demarcate between the two types. The same pipe may act as both large and small at different times of the day (measured in hours) or at different times in its design period (measured in years).

Flow in large foul sewers is mostly open channel (although in exceptional circumstances this may not be the case), continuous, and quasi-steady. Changes in flow that do occur will be at a relatively slow rate and in a reasonably consistent diurnal pattern. In large sewers, we can say that the inflows from single appliances are not a significant fraction of the capacity of the pipe and that there is substantial baseflow (see Figure 9.1).

9.3.1 Flow patterns

The pattern of flow follows a basic diurnal pattern, although each catchment will have its own detailed characteristics. Generally, low flows occur at night with peak flows during the morning and evening. This is related to the pattern of water use of the community but also has to do with the location at which the observation is made. Figure 9.3 illustrates the impact of three important effects. The inflow hydrograph (a) represents the variation in wastewater generation that will, in effect, be similar all around the catchment (see Chapter 3). If the wastewater was collected at one point and then transported from one end of a long pipe to the other, flow *attenuation* due to in-pipe storage would cause a reduction in peak flow, a lag in time to peak, and a distortion of the basic flow pattern (b). Normal sewer catchments are not like this and consist of many-branched networks with inputs both at the most distant point on the catchment and adjacent to the outfall. Thus, the time for wastewater to travel from the point of input to the point under consideration is variable, and this *diversification* effect causes a further reduction in peak and distortion in flow pattern (c). Additional factors that can influence the flow pattern are the degree of infiltration and the number and operation of pumping stations. These effects can be predicted in existing sewer systems using computational hydraulic models (as described in Chapter 20), but they also need to be predicted in the design of new systems.

The flow is usually defined in terms of an average flow (Q_{av}) – that is DWF – and peak flow. The magnitude of the peak flow can then be related to the average flow (see Figure 9.4). A minimum value can also be defined.

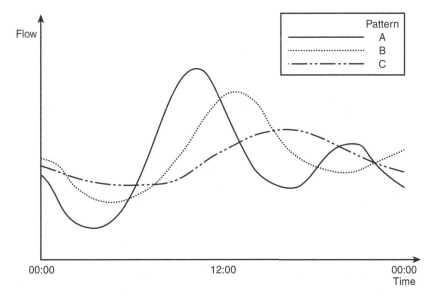

Figure 9.3 Diurnal wastewater flow pattern modified by attenuation and diversification effects.

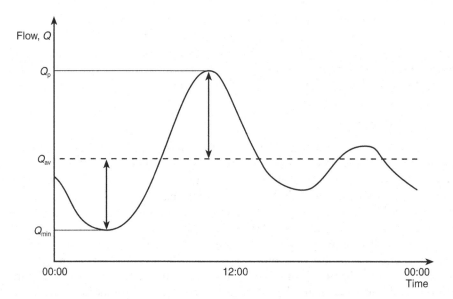

Figure 9.4 Definition of diurnal wastewater flow pattern.

Large sewer design therefore entails estimating the average DWF in the sewer by assuming a daily amount of wastewater generated per person (or per dwelling, or per hectare of development) contributing to the flow, multiplied by the population to be served at the design horizon. Commercial and industrial flows must also be estimated at the design horizon. Allowance should be made for infiltration. The peak flow can be found by using a suitable multiple or peak factor.

9.3.2 Dry weather flow

When the wastewater is mainly domestic in character, measured DWF is defined as

> The average daily flow ... during seven consecutive days without rain (excluding a period which includes public or local holidays) following seven days during which the rainfall did not exceed 0.25 mm on any one day.

(IWEM, 1993)

If the flow contains significant industrial flows, DWF should be measured during the main production days. Ideally, flows during summer and winter periods should be averaged to obtain a representative DWF.

DWF is, therefore, the average rate of flow of wastewater not immediately influenced by rainfall; it includes domestic, commercial, and industrial wastes, and infiltration, but excludes direct stormwater inflow. The quantity is relevant both to foul and combined sewers. DWF can be expressed simply in the following manner (Ministry of Housing, 1970):

$$DWF = PG + I + E \qquad (9.1)$$

where DWF is expressed in litres per day (L/d), P is the population served, G is the average *per capita* domestic water consumption (L/hd.d), I is (dry weather) infiltration (L/d), and E is the average industrial effluent discharged in 24 hours (L/d).

The current definition of DWF does have its weaknesses, particularly the difficulty of finding suitable dry periods, and the lack of direct linkage with Equation (9.1). Alternatively, if a time series of measured total daily wastewater volume data is available (perhaps at a WTP), the nonparametric 80% exceeded flow (Q80) provides a good estimate of DWF. If there are 365 measured volume values in a year ranked from the lowest to the highest, the Q80 is the 73rd value (EA, 2018).

Lund et al. (2021) have demonstrated the potential of using smart water meter data to estimate in-sewer DWF and its spatial and temporal dynamics.

9.3.3 Domestic flow (PG)

The domestic component of DWF is the product of the population and the average *per capita* water consumption.

9.3.3.1 Population (P)

A useful first step in predicting the contributing population that will occur at the end of the design period is to obtain as much local, current, and historical information as possible. Official census information is often available and can be of much value. Additional data can almost certainly be obtained at the local planning authority, and officers should be able to advise on future population trends, and also on the location and type of new industries. Housing density is a useful indicator of current or proposed population levels.

9.3.3.2 Per capita water consumption (G)

In Chapter 3, we discussed in detail the relationship between water use and wastewater production. We also considered typical *per capita* values and discussed that there will be changes in per capita water consumption (PCC) that are independent of population growth.

Where typical discharge figures for developments similar to those under consideration are available, these should be used. In the absence of such data, a conservative figure of 220 L (200 L + 10% infiltration) has been widely used in the UK. However, UK Building Regulations indicate PCC in new buildings should be designed to be no greater than 125 L/hd.d or even 110 L/hd.d where required by planning permission (Approved Document G, 2015). Even lower PCC values are possible (e.g., 80 L/hd.d) with careful design of household systems.

Specific design allowance can be made for buildings such as schools and hospitals as given in Table 9.1. See also Example 9.1.

Table 9.1 Daily volume and pollutant load of wastewater produced from various sources

Category	Volume (L/d)	BOD$_5$ load (g/d)	Per
Day schools	50–100	20–30	Pupil
Boarding schools	150–200	30–60	Pupil
Hospitals	500–750	110–150	Bed
Nursing homes	300–400	60–80	Bed
Sports centre	10–30	10–20	Visitor

EXAMPLE 9.1

Estimate the average daily wastewater flow (L/s) and BOD_5 concentration (mg/L) for an urban area consisting of residential housing (100,000 population), a secondary school (1000 students), a hospital (1000 beds) and a central shopping centre (50,000 m²).

Solution

Area	Magnitude	Unit flow (m³/unit.d)	Flow rate (m³/d)	Unit BOD_5 load (kg/unit.d)	BOD_5 load (kg/d)
Residential	100,000 population	0.20	20,000	0.06	6000
School	1000 students	0.10	100	0.03	30
Hospital	1000 beds	0.75	750	0.15	150
Shopping	50,000 m²	0.004	200	0.0015	75
Total			21,050		6255

Average daily wastewater flow = (21,050 × 1000)/(24 × 3600) = 244 L/s.
Average BOD_5 concentration = (6255 × 1000)/21,050 = 297 mg/L.

9.3.4 Infiltration (I)

The importance of groundwater infiltration and the problems it can cause are discussed in Section 3.4. As mentioned above, the conventional approach in design is to specify infiltration as a fraction of DWF – namely, 10%. Thus, for a design figure of 200 L/hd.d, 20 L/hd.d would be specified. However, evidence (Ainger et al., 1997) suggests this may be too low. In the absence of other evidence, for new systems in high groundwater areas, infiltration figures as high as 120 L/hd.d could be used. In the next section, we will cover approaches to measurement of infiltration in existing systems.

There is a difficulty, however, in making such a large design allowance for infiltration. If an allowance is used, this will increase the design flow rate and may in turn increase the required pipe diameter. A bigger sewer will have a larger circumference and joints, potentially allowing more infiltration to enter the system. Thus, the allowance may well have actually caused more infiltration.

Is there a solution to this dilemma? It is suggested that rather than building-in large design allowances that may cause larger pipes to be chosen, it would be a better investment to ensure high standards of pipe and joint manufacture, installation, and testing. Rehabilitation of existing pipes is considered in Chapter 16.

9.3.4.1 Measurement

The infiltration component of DWF can be estimated in a number of different ways. The simplest way is to assume that night-time flows (Q_{min} in Figure 9.4) represent infiltration. However, with an unknown number of appliances running overnight (e.g., washing machines, dishwashers, dripping taps), this assumption is not recommended, other than as a first approximation. There is also the difficulty of accounting for attenuation of flows at different points in the network.

Other approaches include using artificial tracers or inferring infiltration based on measuring commonly sampled parameters for wastewater quality, such as temperature, conductivity, or nutrients. De Benedittis and Bertrand-Krajewski (2005) found in a study of French sewers that the computed value of the infiltration fraction varied in a range of up to 20% of DWF depending on which technique was used. However, this variability was considered acceptable and still allowed accurate identification of infiltration at the sub-catchment scale.

Further information on infiltration measurement through field surveys and other data sources such as from borehole logs, river monitoring, CCTV surveys and water supply leakage is given in UKWIR (2022).

9.3.4.2 Prevention

The conventional way of addressing infiltration problems in existing systems is by sewer rehabilitation (see Chapter 16) or even full replacement of pipes/joints. However, this can be an uneconomic approach especially when the source or sources are unknown or occur infrequently (see Chapter 3). Alternatives include short-term solutions such as tankering of flows and over-pumping into a nearby watercourse. A longer-term approach is to more clearly identify sources throughout the catchment and address these on a systematic basis.

9.3.5 Non-domestic flows (E)

Background information on non-domestic wastewater flows can be found in Section 4.3. In design, probably the most reliable approach is to make allowance for flows on the basis of experience of similar commerce or industry elsewhere. If these data are not available, or for checking what is known, the following information can be used. Table 9.2 shows examples of

Table 9.2 Daily volume and pollutant load of wastewater produced from various commercial sources

Category	Volume (L/d)	BOD_5 load (g/d)	Per
Hotels, boarding houses	150–300	50–80	Bed
Restaurants	30–40	20–30	Customer
Pubs, clubs	10–20	10–20	Customer
Cinema, theatre	10	10	Seat
Offices	750	250	100 m²
Shopping centre	400	150	100 m²
Commercial premises	300	100	100 m²

Table 9.3 Design allowances for industrial wastewater generation

Category	Volume (L/s.ha)	
	Conventional	Water saving [a]
Light	2	0.5
Medium	4	1.5
Heavy	8	2

[a] Recycling and reusing water where possible.

daily wastewater volume produced by a variety of commercial sources. Table 9.3 provides an areal allowance for broad industrial categories. Water UK (2023) recommends design figures of 0.5 L/s.ha for "normal" industry and 1.0 L/s.ha for "wet" industry.

Most commercial and industrial premises have a domestic component of their wastewater, and ideally, the estimation of this should be based on a detailed survey of facilities and their use. Mann (1979) suggests that a figure of 40–80 L/hd.(8-hour shift) may be appropriate. Water UK (2023) recommends 0.6 L/s.ha of developable land.

9.3.6 Peak flow

Two approaches to estimating peak flows are used. In the first, typically used in British practice, a fixed DWF multiple is used. In the second, a variable peak factor is specified. Both methods aim to take account of diurnal peaks and the daily and seasonal fluctuations in water consumption together with an allowance for extraneous flows such as infiltration.

BS EN 16933-2:2017 recommends that a multiple up to 6 be used. This figure is most appropriate for use in sub-catchments subject to relatively little attenuation and diversification effects. For larger sewers, a value of 4 is more realistic. A still lower figure (2.5) is relevant for predicting DWFs in combined sewers because this flow will determine velocity, not capacity.

Water UK (2023) suggests that a design (peak) flow of 4000 L/dwelling.day (0.05 L/s per dwelling) should be used for foul sewers serving residential developments. This approximates to three persons/property discharging 200 L/hd.d with a peak flow multiple of 6 and 10% infiltration.

Opinions and practice differ on whether the DWF to be multiplied should include or exclude infiltration. If DWF is determined from Equation (9.1), the most satisfactory form of applying a multiple of 4, for example, is $4(DWF-I) + I$.

Peak flows may also be determined by the application of variable peak factors. Figure 9.3 shows that attenuation and diversification effects tend to reduce peak flows, and so the ratio of peak to average flow generally decreases from the "top" to the "bottom" of the network. Thus, the peak factor varies depending on position in the network (see Figure 9.5). Location is usually described in terms of the population served or the average flow rate at a particular point.

The relationship between peak factor (P_F) and population can be described algebraically with equations of the following form:

$$P_F = \frac{a}{P^b} \tag{9.2}$$

where P is the population drained (in thousands) and a, b are constants.

However, there are a number of other such equations, and some of the most well-known are listed in Table 9.4.

Example 9.2 illustrates that the numerical values produced by different equations can vary significantly. Thus, any of the formulae available should be used with caution.

One of the reasons for the disparity in the peak factor predictions is the general variability in diurnal flow patterns. The degree of uncertainty is also illustrated by the confidence limits (dashed lines) in Figure 9.5. Zhang et al. (2005) have used Buchberger and Wu's (1995) Poisson rectangular pulse model for instantaneous residential water demands to provide a theoretical framework to predict the shape and form of Figure 9.5 and a means to estimate the confidence limits. While strictly speaking this work is applicable to water distribution networks (diversification effects are captured but attenuation effects are not), agreement is still reached with several of the equations listed in Table 9.4.

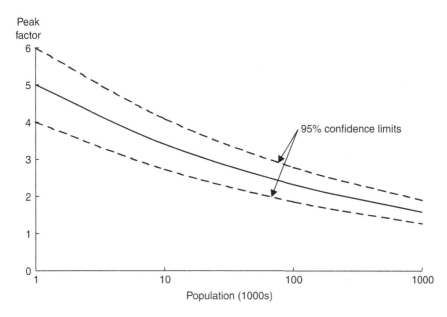

Figure 9.5 Ratio of peak flows to average daily flow (with 95 percentile confidence limits).

Table 9.4 Peak factors

Reference	Method	Notes	Equation
Harman (1918)	$1 + \dfrac{14}{4 + \sqrt{P}}$	1	(9.3)
Gifft (1945)	$\dfrac{5}{P^{1/6}}$	1	(9.4)
Babbitt (1953)	$\dfrac{5}{P^{1/5}}$	1	(9.5)
Fair and Geyer (1954)	$1 + \dfrac{18 + \sqrt{P}}{4 + \sqrt{P}}$	1	(9.6)
—	$4Q^{-0.154}$	2	(9.7)
Gaines (1989)	$2.18Q^{-0.064}$	3	(9.8)
Gaines (1989)	$5.16Q^{-0.060}$	3	(9.9)

[1] Population P in 1000 s.
[2] Flow Q in 1000 m³/d.
[3] Flow Q in litres per second (L/s).

EXAMPLE 9.2

A separate foul sewer network drains a domestic population of 250,000. Estimate the peak flow rate of wastewater at the outfall (excluding infiltration) using both Babbitt's and Gaines's formulas. The daily *per capita* flow is 145 L.

Solution

Average daily flow = (250,000 × 145)/(3600 × 24) = 420 L/s.

    Babbitt (Equation 9.5):

$$P_F = \frac{5}{p^{1/5}} = \frac{5}{250^{1/5}} = 1.66$$

Peak flow = 1.66 × 420 = 697 L/s.

    Gaines (Equation 9.8):

$$P_F = 2.18Q^{-0.064} = 2.18 × 420^{-0.064} = 1.48.$$

    Peak flow = 1.48 × 420 = 622 L/s

9.3.7 Design criteria

9.3.7.1 Capacity

Foul sewers should be designed (in terms of size and gradient) to convey the predicted peak flows. It is common practice to restrict the depth of flow (typically to $d/D = 0.75$) to ensure proper ventilation.

9.3.7.2 Self-cleansing

Once the pipe size has been chosen based on capacity, the pipe gradient is selected to ensure a minimum "self-cleansing" velocity is achieved. The self-cleansing velocity is that which avoids long-term deposition of solids and should be reached at least once per day. BS EN 16933-2: 2017 recommends a minimum of 0.7 m/s for sewers up to DN300 at a flow rate of 2 DWF. Higher velocities may be needed in larger pipes (see Chapter 17). Water UK (2023) requires a velocity of 0.75 m/s to be achieved at one-third design flow (i.e., 2 DWF). Some engineers prefer to specify a higher self-cleansing velocity to be achieved at full-bore flow. Figure 7.8 shows how this allows for the reduction in velocity that occurs in pipes that are flowing less than half full. In practice, the pipe size and gradient are calculated together to obtain the best design.

9.3.7.3 Roughness

For design purposes, it is conservatively assumed that the pipe roughness is independent of pipe material. This is because, in foul and combined sewers, all materials will become slimed during use (see Chapter 7). BS EN 16933-2: 2017 recommends a k_s value of 0.6 mm (for use in the Colebrook–White equation) where the peak DWF exceeds 1 m/s, and 1.5 mm where it is between 0.76 and 1 m/s.

9.3.7.4 Minimum pipe sizes

The minimum pipe size is generally set at DN75 or DN100 for house drains and DN100 to DN150 for the upper reaches of public networks, and this choice is based on experience.

9.3.8 Design method

The following procedure should be followed for foul sewer design:

1. Assume pipe roughness (k_s).
2. Prepare a preliminary layout of sewers, including tentative inflow locations.
3. Mark pipe numbers on the plan according to the convention described in Chapter 6.
4. Define contributing area DWF to each pipe.
5. Find cumulative contributing area DWF.
6. Estimate peak flow (Q_p) based on average DWF and peak factor/multiple factor analysis.
7. Make a first attempt at setting the gradients and diameters of each pipe.
8. Check $d/D < 0.75$ and $v_{max} > v > v_{min}$.
9. Adjust pipe diameter and gradient as necessary (given hydraulic and physical constraints), and return to step 5.

Example 9.3 illustrates the design of a simple foul sewer network.

EXAMPLE 9.3

A preliminary foul sewer network is shown in Figure 9.6. Design the network using fixed DWF multiples (6 for domestic flows, and 3 for industrial) based on the availability of an average grade of 1:100. The inflow, Q_a is 30 L/s at peak. For the sake of simplicity, infiltration can be neglected.

Data from the network are contained in columns (1), (2), and (6) of the table below. The maximum proportional depth is 0.75 and the minimum velocity at peak flow is 0.75 m/s. Pipes roughness is $k_s = 1.5$ mm.

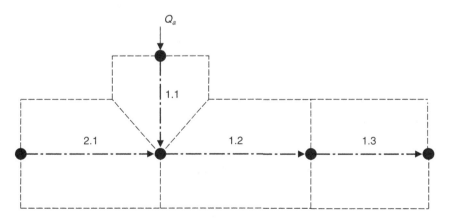

Figure 9.6 System layout (Example 9.3).

Solution

Using the raw data on land use, peak inflow rates are calculated. It is assumed that the commercial and industrial rates specified are peak rates.

(1)	(2)	(3)	(4)	(5)	(6)	(7)	(8)
Pipe number	Number of houses	Peak flow rate $(Q_3)^a$ (L/s)	Commercial area (ha)	Peak flow rate $(Q_5)^b$ (L/s)	Industrial area and type (ha)	Peak flow rate $(Q_7)^c$ (L/s)	Total peak flow rate $(Q_3 + Q_5 + Q_7)$ (L/s)
1.1	200	8.4	–	0	1.65M	19.8	28.2
2.1	250	10.5	–	0	1.70L	10.2	20.7
1.2	140	5.9	1.10	1.1	0.60L	3.6	10.6
1.3	500	21.0	2.80	2.8	–	0	23.8

[a] Based on three persons per house, 200 L/hd.d and DWF multiple of 6 (Q_3 = 0.042 L/s.house).
[b] Based on 300 L/d.100 m² and DWF multiple of 3 (Q_5 = 1 L/s.ha).
[c] Based on 2 and 4 L/s.ha for Light and Medium industry, respectively, and DWF multiples of 3 (Q_7 = 6 or 12 L/s.ha).

Pipe velocities and depths are calculated using the Colebrook–White equation or can be read from Butler–Pinkerton charts (e.g., Figure 7.9). The pipe/gradient combination chosen is shown in bold.

(1)	(2)	(3)	(6)	(7)	(8)	(9)	
Pipe number	Peak flow [L/s]	Cumulative peak flow [L/s]	Assumed pipe size (mm)	Minimum gradient (1:x)	Proportional depth of flow	Velocity (m/s)	Comments
1.1	28.2	58.2	250	90	0.75	1.45	Depth limited
			300	**240**	**0.75**	**1.04**	Depth limited
			375	600	0.67	0.75	Velocity limited
2.1	20.7	20.7	150	47	0.75	1.45	
			225	**270**	**0.64**	**0.75**	
1.2	10.6	89.5	300	95	0.75	1.60	
			375	**320**	**0.75**	**1.02**	
1.3	23.8	113.3	**375**	**200**	**0.75**	**1.27**	
			450	500	0.75	0.90	

9.4 SMALL SEWERS

As described earlier, small sewers are subject to random inflow from appliances as intermittent pulses of flow, such that peak flow in the pipe is a significant fraction of the pipe capacity, and there is little or no baseflow.

As an appliance empties to waste, a relatively short, highly turbulent pulse of wastewater is discharged into the small sewer. As the pulse travels down the pipe, it is subject to attenuation resulting in a reduction in its flow rate and depth, and an increase in duration and length (see Figure 9.1).

9.4.1 Discharge unit method

Building drainage and small sewerage schemes are often designed using the Discharge Unit Method as an alternative to the methods previously described. Using the principles of probability theory, discharge units are assigned to individual appliances to reflect their relative load-producing effect. Peak flow rates from groups of mixed appliances are estimated by addition of the relevant discharge units. The small sewer can then be designed to convey the peak flow. This approach is now explained in more detail.

9.4.1.1 Probabilistic framework

Consider a single type of appliance discharging identical outputs that have an initial duration of t' and a mean interval between use of T'. Hence, the probability p that the appliance will be discharging at any instant is given by

$$p = \frac{\text{duration of discharge}}{\text{meantime between discharges}} = \frac{t'}{T'} \tag{9.10}$$

In most systems, however, there will be more than one appliance. How can we answer a question such as, "What is the probability that r from a total of N appliances will discharge *simultaneously*?" Application of the binomial distribution states that if p is the probability that an event will happen in any single trial (i.e., the probability of success) and $(1-p)$ is the probability that it will fail to happen (i.e., the probability of failure), then the probability that the event will occur exactly r times in N trials ($P[r,N]$) is

$$P(r,N) = {}^N C_r p^r (1-p)^{N-r} \tag{9.11}$$

or

$$P(r,N) = \frac{N!}{r!(N-r)!} p^r (1-p)^{N-r} \tag{9.12}$$

EXAMPLE 9.4

Calculate the probability of discharge of a single WC that discharges for 10 seconds every 20 minutes at peak times. What percentage of time will the WC be loading the system?

Solution

From Equation 9.10,

$$p_{WC} = 10/1200 = 0.0083$$

The WC will be loading the system 0.8% of the time (at peak) and hence will *not* be discharging 99.2% of the time.

Thus, to use the binomial probability distribution in this application, we must assume that

- Each trial has only two possible outcomes: success or failure – that is, an appliance is either discharging or it is not.
- The probability of success (p) must be the same on each trial (i.e., independent events), implying that t' and T' are always the same.

Neither of these assumptions is fully correct for discharging appliances, but they are close enough for design purposes. Example 9.5 illustrates the basic use of Equation 9.11.

9.4.1.2 Design criterion

While this level of detail is of interest, it is not of direct use. In design, we are concerned with establishing the probable number of appliances discharging simultaneously against some agreed standard. Practical design is carried out using a confidence level approach or "criterion of satisfactory service" (J) as introduced in Section 9.2.2. For small sewers, this is defined as the percentage of time that up to c appliances out of N will be discharging. So,

$$\sum_{r=0}^{c} P(r,N) \geq J \tag{9.13}$$

In design terms, we are trying to establish the value of c for a given J. A typical value for J would be 99%, implying actual loadings will only exceed the design load for less than 1% of the time (see Example 9.6).

At a given criterion of satisfactory service, each individual appliance will therefore have a unique relationship between

- The number of connected appliances and the number discharging simultaneously
- The number of connected appliances and flow rate (because the discharge capacity of each appliance is known, and assumed constant)

Figure 9.7 illustrates the relationship between the number of connected appliances and simultaneous discharge for three common devices, prepared using the binomial distribution

---

EXAMPLE 9.5

What is the probability that 20 from a total of 100 WCs ($p = 0.01$) will discharge simultaneously?

Solution

N = number of trials = total number of connected appliances = 100
p = probability of success = probability of discharge = 0.01
Using the binomial expression with the above data gives (Equation 9.12)

$$P(20,100) = \frac{100!}{20!80!} 0.01^{20} 0.99^{80} = 2.4 \times 10^{-20}$$

In other words, this eventuality is extremely unlikely.

EXAMPLE 9.6

For a criterion of satisfactory service of 99%, determine the number of water widgets discharging simultaneously from a group of five, if their probability of discharge is 20% (unusually high, but used for illustrative purposes). If each widget discharges $q = 0.5$ L/s, find the design flow.

Solution

Now, $N = 5$, $p = 0.2$, and $J = 0.99$. Using Equation 9.12 for increasing values of r, we get the following:

r	P(r,N)	Σ P(r,N)	Σ q (L/s)
0	0.327	0.327	0
1	0.410	0.737	0.5
2	0.204	0.941	1.0
3	0.051	0.992	1.5

So, since at $r = 3$, $\Sigma P(r,N) > 0.99$, up to three water widgets will be found discharging 99% of the time, and more than three will discharge just 1% of the time (i.e., during one peak period every hundred days). Design for $c = 3$ simultaneous discharges, $q = 1.5$ L/s.

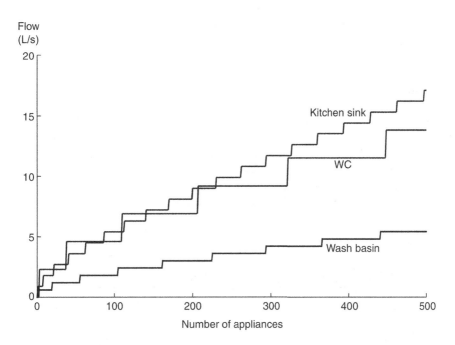

Figure 9.7 Simultaneous discharge of WC, sink, and basin at 99% criterion of satisfactory service.

Table 9.5 Typical UK appliance flow and domestic usage data

Appliance	Flow rate q (L/s)	Duration t′ (s)	Recurrence use interval T′ (s)	Probability of discharge p
WC (9 L)*	2.3	5	1200	0.004
Wash basin	0.6	10	1200	0.008
Kitchen sink	0.9	25	1200	0.021
Bath	1.1	75	4500	0.017
Washing machine	0.7	300	15,000	0.020

* Also see Table 9.6 and discussion in the text of the changing WC flush volumes.
Source: Wise, A.F.E. and Swaffield, J.A. 2002. *Water, Sanitary and Waste Services for Buildings*, 5th edn, Taylor and Francis.

and data from Table 9.5. The stepped appearance of the plots does not reflect the resolution of the calculations used to produce them but is inherent in the calculations.

9.4.1.3 Mixed appliances

In a practical design situation, there will be a mix of appliance types rather than the single types previously discussed. The basic binomial distribution does not take into account the interactions in a mixed system between appliances of different frequency of use, discharge duration, and flow rate.

To overcome this problem, the *discharge unit (DU) method* has been developed based on the premise that the same flow rate may be generated by a different number of appliances depending on their type. DUs are, therefore, attributed uniquely to each appliance type, and the value will depend on

- The rate and duration of discharge
- The criterion of satisfactory service

Recommended values are given in Table 9.6. Note, in particular, that the discharge volume of the WC is 6 L (as compared with 9 L in Table 9.5) to reflect the maximum allowable in the UK under the current Water Supply (Water Fittings) Regulations (Defra, 1999). It is possible to express all appliances in terms of DUs using a family of design curves, based only on intensity of use. BS EN 12056-2: 2000 recommends a power law be used to approximate the relationship between design flow rate Q and the cumulative number of discharge units DU, so

$$Q = k_{DU} \sqrt{\Sigma n_{DU}} \qquad\qquad (9.14)$$

Table 9.6 Discharge unit ratings for domestic appliances

Appliance	Discharge units, DU
WC (6 L)	1.2–1.8
Wash basin	0.3–0.5
Sink	0.5–1.3
Bath	0.5–1.3
Washing machine (up to 6 kg)	0.5–0.8

Source: BS EN 12056-2: 2000. *Gravity Drainage Systems Inside Buildings. Part 2: Sanitary Pipework, Layout and Calculation.*

where Q is peak flow (L/s), k_{DU} is the dimensionless frequency factor, and n_{DU} is the number of discharge units. The value of k_{DU} depends on the intensity of usage of the appliance(s) and is given in Table 9.7. Design curves are given in Figure 9.8 and are used in Example 9.7.

BS EN 16933-2:2017 indicates that if there is only a small number of dwellings and the number and type of appliances are not known, a value of 5.0 DUs/dwelling should be used, subject to a minimum calculated flow rate of 1.6 L/s.

9.4.2 Design criteria

In small sewers and drains, design criteria relate principally to the capacity of the pipe and the requirements of self-cleansing. Sewers are normally designed so that the design flow (at the relevant confidence level) can be conveyed with a proportional depth $d/D < 0.75$. This is done assuming steady, uniform flow conditions as described in Chapter 7.

In small sewers, where solids are transported by being pushed along the pipe invert, self-cleansing is difficult to assess on a theoretical basis (as considered further in Section 9.5). Even if the flow is assumed to be steady and uniform (which is not), such low flows may require relatively steep gradients to achieve self-cleansing velocities. At the heads of runs, the pipe gradient is usually based on "accepted practice" and can be "relaxed" somewhat (as

Table 9.7 Frequency of use factors

Frequency of use	k_{DU}
Intermittent: dwellings, guest houses, offices	0.5
Frequent: hospitals, schools, restaurants	0.7
Congested: public facilities	1.0

Source: BS EN 12056-2: 2000. *Gravity Drainage Systems Inside Buildings. Part 2: Sanitary Pipework, Layout and Calculation.*

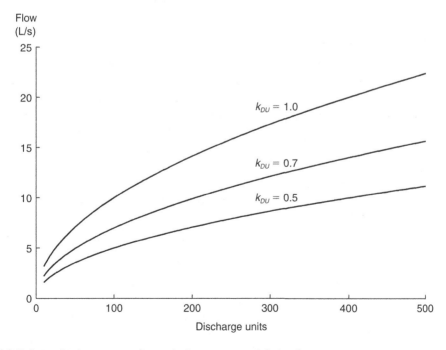

Figure 9.8 Relationship between appliance discharge units and design flow rate.

EXAMPLE 9.7

A residential block is made up of 20 flats, each fitted with a WC, wash basin, sink, bath, and washing machine. It is estimated that in any one flat, between 08:00 and 09:00, all of the appliances are likely to be in use on a Monday. Calculate the design flow rate using the DU method.

Solution

Taking the most conservative values, the discharge units for all appliances = 1.8 + 0.5 + 1.3 + 1.3 + 0.8 = 5.7. Hence, for 20 flats the total of discharge units is 114. Assuming k_{DU} = 0.5, from Equation 9.14 or Figure 9.8,

$$Q = 0.5\sqrt{114} = 5.3 \, \text{L} / \text{s}$$

---

shown in Table 9.8) to a minimum gradient and number of connected WC, depending on the required pipe size. This is in recognition of the flush wave produced by the WC in transporting solids.

The implication of Table 9.8 is that for a public sewer with a diameter of 150 mm or greater, the maximum gradient that needs to be used is 1:150, provided there are at least five connected dwellings.

The major factors influencing minimum pipe diameter are its ability to carry gross solids and its ease of maintenance. Large solids frequently find their way into sewers, either accidentally or deliberately, particularly via the WC and property access points. The minimum pipe size is as set out in Section 9.3.7.4.

An application of the small sewer design method is given in Example 9.8.

9.4.3 Choice of methods

As mentioned earlier in the chapter, the two different design methods (for large and small sewers) represent the different flow regimes in foul sewers. If a large network is to be designed in detail, there comes a point where a change must be made from one method to another. The point at which the change takes place depends on local circumstances, but its location is important as it has an impact on pipe sizes and gradients, and hence cost.

Table 9.8 BS EN 16933-2: 2017 deemed to satisfy self-cleansing rules for small sewers

Design flow (L/s)	DN (mm)	Gradient	Connected WCs
<1	≤100	≥1:40	–
>1	75	≥1:80	–
	100	≥1:80	1
	150	≥1:150	5

EXAMPLE 9.8

Design the foul sewer diameter and gradients for the small housing estate shown in Figure 9.9. Data on the network are shown in columns (1) and (2) of the table below. Use the following design data:

Minimum diameter (mm):	150
Minimum velocity (m/s):	0.75
Minimum gradient:	1:150 (provided number of WCs ≥ 5)
Maximum proportional depth of flow:	0.75
Pipe roughness (mm):	0.6

Solution

For each sewer length, use the minimum pipe diameter and calculate the minimum gradient required to achieve the necessary capacity + self-cleansing. Use Tables 9.6–9.8, Equation 9.14, and the Butler–Pinkerton charts.

Assume each dwelling has (WC + basin + sink):

$$DUs = 1.8 + 0.4 + 1.3 = 3.5$$

For individual pipe lengths draining at least five dwellings, reduce the gradient to 1:150. Take $k_{DU} = 0.5$.

(1)	(2)	(3)	(4)	(5)	(6)	(7)	(8)	(9)	
Pipe number	Number of houses	Number of discharge units	Cumulative number of discharge units	Design flow rate (L/s)	Assumed pipe size (mm)	Minimum gradient (1:x)	Proportional depth of flow	Velocity (m/s)	Comments
1.1	4	14	14	1.9	150	55	0.21	0.75	
1.2	9	31.5	45.5	3.4	150	85	0.32	0.75	*
						150	0.37	0.62	
2.1	10	35	35	3.0	150	75	0.28	0.75	*
						150	0.34	0.59	
1.3	1	3.5	84	4.6	150	100	0.37	0.75	*
						150	0.43	0.65	
3.1	6	21	21	2.3	150	70	0.23	0.75	*
						150	0.29	0.54	
1.4	5	17.5	122.5	5.5	150	120	0.43	0.75	*
						150	0.46	0.69	
1.5	2	7	129.5	5.7	150	125	0.45	0.75	*
						150	0.47	0.70	

* Gradient relaxed to 1:150 as Q > 1 L/s, number of WCs ≥ 5.

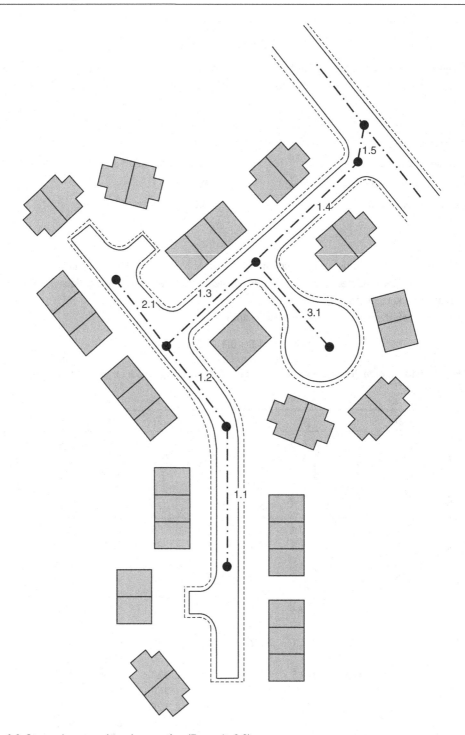

Figure 9.9 System layout and catchment plan (Example 9.8).

EXAMPLE 9.9

Using the data in Table 9.5, calculate the discharge probability of an equivalent single appliance for one household. Use this to estimate the minimum number of connected households that will generate continuous wastewater flow at peak times. What will be the expected flow rate?

Solution

Appliance	Flow rate q (L/s)	Probability of discharge p	pq
WC	2.3	0.004	0.0092
Wash basin	0.6	0.008	0.0048
Kitchen sink	0.9	0.021	0.1890
Bath	1.1	0.017	0.0187
Washing machine	0.7	0.020	0.0140
Total	5.6		0.0656

Assuming a household has one of each appliance, a single equivalent appliance will have a probability of discharge, $p = 0.0656/5.6 = 0.0117$.

If continuous flow occurs when $Np = 1$, $N = 1/0.0117 = 85$ households.

Expected peak flow rate = 5.6 L/s (from table) = Npq.

A suggested approach is to interpret the probability of appliance discharge as a measure of flow intermittency. So, if $Np > 1$, flow is continuous, and the large sewer approach can reasonably be used, otherwise the small sewer method is appropriate. Based on Example 9.9, this should occur during the morning peak in a sewer where there are more than 85 connected dwellings. Continuous flow throughout the day is expected where there are more than 560 connected dwellings (see Problem 12).

Research studies have attempted to combine both approaches. Butler and Graham (1995) used household appliance usage survey data to estimate probabilistically derived input flows into sewers and hence the DWF, and Bailey et al. (2019) used household water meter data to calibrate a stochastic wastewater discharge model also to generate in-sewer DWF patterns.

9.5 SOLIDS TRANSPORT

It is surprising that the transport of gross solids is not routinely and explicitly considered in the design of large or small sewers. However, research has filled the gaps in our understanding of the movement of solids in the different hydraulic regimes encountered, and is giving some important feedback to practical design and operation.

The main characteristics of gross solids transport in sewers are as follows:

- There is a wide variety of solids, and the physical condition of some types varies widely, influencing the way the solids are transported.
- Some solids change their condition as they move through the system, as a result of physical degradation and contact with other substances in the sewer.
- In some hydraulic conditions, solids are carried with the flow, yet at lower flow rates they may be deposited.

- During movement, solids do not necessarily move at the mean water velocity.
- Some solids affect the flow conditions within the sewer.

9.5.1 Large sewers

When solids are advected (moved while suspended in the flow) in large sewers, forces acting on the solids position them at different flow depths depending on their specific gravity and the hydraulic conditions. Figure 9.10 indicates how some solids can be carried along at levels where the local velocity is greater than the mean velocity (v). This means that solids may "overtake" the flow and arrive at combined sewer overflows and wastewater treatment plants before the peak water flow.

Figure 9.11 shows laboratory results for a solid plastic cylinder (artificial faecal solid) plotted as longitudinal solid velocity against mean water velocity, for two contrasting gradients. A linear relationship fits all the data well ($R^2 = 0.98$), and this was found to be the case for all the artificial solids studied and for various "real" gross solids (Butler et al., 2003). This linear relationship can be expressed as

$$v_{GS} = av + \beta \tag{9.15}$$

where v_{GS} is the velocity of a particular gross solid (m/s); v is the mean water velocity (m/s); and α, β are coefficients. Laboratory results indicate β to typically be small enough to neglect, but α varies from 0.98 to 1.27 depending on solid type, with lower specific-gravity solids generally having the higher values. Based on field experiments where the movements of individual artificial solids were tracked, Penn et al. (2018) established $\alpha = 1.01$ and $\beta = 0.11$ ($R^2 = 0.84$). It has also been recommended (Davies et al., 1996) that for the modelling of solids movement in unsteady flow, the relationship between the mean water velocity and the average velocity of any solid type can be assumed to be the same in unsteady (gradually varied) flow as it is in steady (uniform) flow.

Generally, solid size has not been found to be an important variable, except at low flow depths. In this case, larger solids tend to be retarded more than smaller ones by contact with the pipe wall.

Under certain hydraulic conditions (typically low flows, such as overnight), solids may be deposited. Davies et al. (1996) found that a solid's propensity to deposit is based on critical hydraulic parameters of flow depth and mean velocity. They argued that (at least for modelling purposes) deposition of solids takes place when the value of *either* depth or mean velocity goes below the critical value, and re-suspension takes place when that level is subsequently exceeded. Figure 9.12 shows a graph of mean velocity plotted against depth, with points representing the conditions for the deposition of a sanitary towel observed in a laboratory study. The dotted lines indicate suitable values for the critical depth (vertical) and velocity (horizontal). Above and to the right of the dotted lines are conditions in which these types of solids are carried by the flow (both depth and velocity exceeding the critical value). Below or to the left of the dotted lines are the conditions in which they would be deposited. Table 9.9 gives depth/velocity values results for various gross solid types.

With the increasing use of water-saving devices and the consequent reduction in water volumes entering the sewer, what are the implications for solid transport in large foul sewers? Blanksby (2006) indicates that the main impacts are the possibility of increased gross solids deposition and the prospect of increased sedimentation in flatter sewers. Some evidence of this has been found for combined sewers (see Box 24.3). However, model studies have shown that although in some scenarios sewer flows, velocities, and proportional depths may be reduced, sewer *blockage* rates are not expected to increase significantly (Penn et al., 2013).

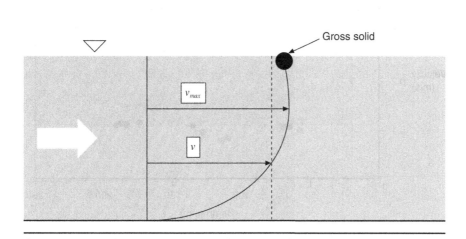

Figure 9.10 Movement of gross solids in large sewers.

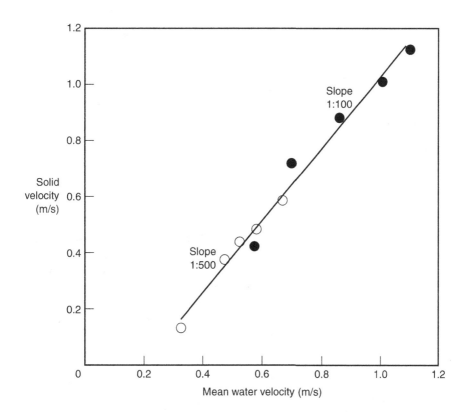

Figure 9.11 Artificial faecal solid velocity versus mean velocity, with linear fit. (After Butler et al. 2003. *Proceedings of Institution of Civil Engineers, Water, Maritime and Energy*, 156[WM2], 165–174.)

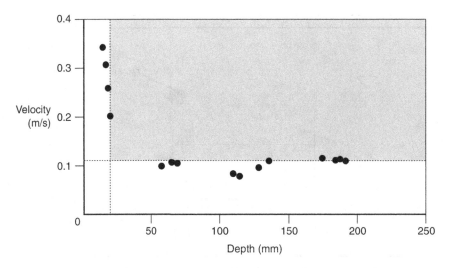

Figure 9.12 Hydraulic conditions for deposition of solids (sanitary towels). (After Butler, D., et al. 2003. *Proceedings of Institution of Civil Engineers, Water, Maritime and Energy*, 156[WM2], 165–174.)

Table 9.9 Critical depth/velocity for various solid types

Solid type	Critical depth (mm)	Critical velocity (m/s)
Solid plastic cylinders:		
Length 80 mm, diameter 37 mm	30	0.20
Length 44 mm, diameter 20 mm	22	0.13
Length 22 mm, diameter 10 mm	20	0.10
Cotton wool wipe	10	0.08
Sanitary towel	20	0.11

9.5.2 Small sewers

The movement of solids in small sewers is somewhat different than that in large sewers. Laboratory experiments demonstrate that there are two main modes of solid movement: *floating* and *sliding dam*. The floating mechanism occurs when the solid is small relative to the pipe diameter and flush wave input. The solid moves with a proportion of the wave velocity and has little effect on the wave itself. Solids that are large compared with the flush wave and pipe diameter move with a sliding dam mechanism (Littlewood and Butler, 2003). In this case, the flush wave builds up behind the solid, which acts as a dam in the base of the pipe. When the flow's hydrostatic head and momentum overcome the friction between the solid and the pipe wall, the solid begins to move along the pipe invert. The amount of movement that occurs depends on how "efficient" the solid is as a dam: the higher the efficiency, the further the solid will move for the same flush wave. The two modes of movement are illustrated in Figure 9.13. Photograph (a) shows toilet tissue alone in the flow, and photograph (b) shows toilet tissue and an artificial faecal solid in combination. Note the pool of water forming behind the solid and propelling it along. The role of the toilet tissue in forming the "dam" is also noteworthy. Solids tend to move farthest in the sliding dam mode.

(a)   (b)

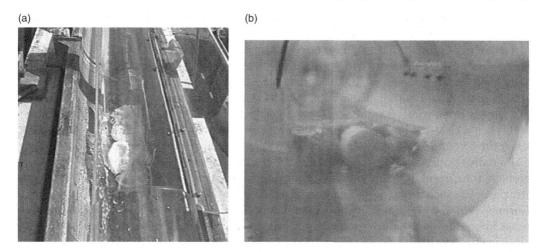

Figure 9.13 Floating (a) and sliding dam (b) mechanisms of solid movement. (Courtesy of Dr. Richard Barnes.)

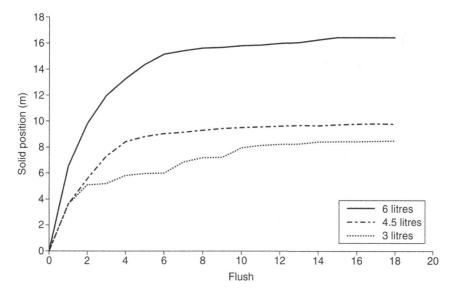

Figure 9.14 Limiting solid transport distance for a gross solid in a 100 mm diameter pipe at a gradient of 1:100 for various toilet flush volumes. (After Memon, F., et al. 2007. *Water Science and Technology*, 55[4], 85–91.)

Eventually, whichever mode of movement prevails, the solid will deposit on the pipe invert, some distance away from its entry point. It will remain there until another wave enters the pipe, travels along to meet the stranded solid, and resuspends it. The solid will move further downstream but for a distance less than the initial movement. The distance moved under the influence of each subsequent flush decreases until the solid is no longer moved at all by the attenuated flush wave (Swaffield and Galowin, 1992). Thus, each solid, flush wave, pipe diameter, and gradient combination has a *limiting solid transport distance* (LSTD). Figure 9.14 shows that for a 6 L flush volume WC, the solid is not moved much more than 16 m even after 20 flush waves have been passed down the pipe. In fact, very little further movement is noted beyond 10 flushes. On the basis of extensive laboratory tests, Walski et al. (2011) concluded that gross solids in small sewers are more likely to be transported when there is more flow and flow of longer duration, when the carrying pipe slope is greater, and when the solids have a lower specific gravity.

A similar question to that asked for large sewers can also be posed for small ones: what are the implications for the more widespread adoption of water-saving devices, especially WCs? Tests have shown (Memon et al., 2007) that, for example, when a 6 L WC discharges into a 1:100 gradient, 100 mm diameter pipe, the LSTD is 16.5 m, but for a 3 L WC discharging to a similar drain configuration, LSTD is reduced to 8.5 m (Figure 9.14). The interpretation of this is not clear-cut but suggests that the *propensity* for blockage formation is increased at lower flush volumes. Drinkwater et al. (2008) agree and argue that "available data suggests that a reduction from six- to three-litre flushes in a conventional WC could pose a significant problem for current drainage systems." Lauchlan et al. (2004) suggest that in 150 mm diameter pipes any such problems would only manifest themselves at gradients of 1:150 or lower. Gormley et al. (2013) go further and assert that on the basis of laboratory studies, under low water use conditions, shallow gradients actually promote solid transport and hence minimise blockage propensity.

PROBLEMS

1 Explain how you would go about the preliminary investigation and design of a foul sewer network for a large housing development.

2 What are the main differences in the hydraulic regime between large and small foul sewers? What implications do these have on the design procedures adopted?

3 Explain the main factors affecting the shape of the DWF diurnal profile.

4 Explain what is meant by *dry weather flow*. Define how it is measured, and discuss the limitations of the current approach.

5 An urban catchment is drained by a separate foul sewer network and has an area of 500 ha and a population density of 75/hd.d. At the outfall of this catchment, calculate the following:
   a. The average DWF (in L/s) assuming water consumption is 160 L/hd.d, trade effluent is 10 m^3/ha.d over 10% of the catchment, and infiltration is 20 L/hd.d   [84 L/s]
   b. The peak DWF using Babbitt's formula.   [201 L/s]

6 If the outfall sewer in Problem 9.5 is 500 mm in diameter with a gradient of 1:200, calculate the following:
   a. The depth of peak flow, assuming k_s = 1.5 mm   [325 mm]
   b. The additional population that could be served, assuming that proportional depth does not exceed 0.75.   [2922]

7 Redesign the foul sewer network specified in Example 9.3 on a steep site with an inflow of Q_a = 45 L/s.

8 Explain how the binomial probability distribution forms the basis of the DU small sewer design method.

9 It has been estimated that in an office block, each WC is used at peak times every 5 minutes and discharges for 10 seconds. In a group of five WCs, calculate the maximum number of WCs discharging simultaneously at the 99.9% confidence level.   [2]

10 Redesign the foul sewer network specified in Example 9.3 to serve the residential housing only, using the DU method.

11 Calculate the total number of dwellings that can be drained by a 150 mm diameter pipe (k_s = 1.5 mm) running with a proportional depth of 0.75 at a gradient of 1:300, using both large and small sewer design methods. Assume 3.5 DUs or 0.05 L/s per dwelling.   [174, 73]

12 How many connected households would there need to be on a sewer network to produce continuous flow in the early hours of the morning if at that time the household single equivalent appliance $pq = 0.01$?   [560]

13 Explain the main differences in the way gross solids are transported in large and small sewers. How would you expect more widespread use of low flush toilets to affect solid transport?

KEY SOURCES

Bizier, P. Ed. 2007. *Gravity Sanitary Sewer Design and Construction*, 2nd edition, ASCE Manual and Report No. 60, WEF Manual No. FD-5.

Butler, D. and Graham, N.J.D. 1995. Modeling dry weather wastewater flow in sewer networks. *American Society of Civil Engineers, Journal of Environmental Engineering Division*, 121(2), 161–173.

McDermott, R., Strong, A. and Griffiths, P. 2019. Solid transfer in low flow sewers, the distance travelled so far is not enough. *Journal of Environmental Protection*, 10, 164–207.

Swaffield, J.A. and Galowin, L.S. 1992. *The Engineered Design of Building Drainage Systems*, Ashgate.

REFERENCES

Ainger, C.M., Armstrong, R.A., and Butler, D. 1997. *Dry Weather Flow in Sewers*, CIRIA R177.

Approved Document G. 2015. Sanitation, Hot Water Safety and Water Efficiency, The Building Regulations 2010, HM Government.

Babbitt, H.E. 1953. *Sewerage and Sewage Treatment*, 7th edn, John Wiley and Sons.

Bailey, O., Arnot, T.C., Blokker, E.J.M., Kapelan, Z. Vreeburg, J. and Hofman, J.A.M.H. 2019. Developing a stochastic sewer model to support sewer design under water conservation measures. *Journal of Hydrology*, 573, 908–917.

Blanksby, J. 2006. Water conservation and sewerage systems, Chapter 5, in *Water Demand Management* (eds. D. Butler and F.A. Memon), IWA publishing, Blackpool, UK.

BS EN 12056-2: 2000. *Gravity Drainage Systems Inside Buildings. Part 2: Sanitary Pipework, Layout and Calculation.*

BS EN 16933-2: 2017. *Drain and Sewer Systems Outside buildings – Design. Part 2: Hydraulic Design.*

Buchberger, S.G. and Wu, L. 1995. Model for instantaneous residential water demands. *American Society of Civil Engineers, Journal of Hydraulic Engineering*, 121(3), 232–246.

Butler, D., Davies, J.W., Jefferies, C., and Schütze, M. 2003. Gross solids transport in sewers. *Proceedings of Institution of Civil Engineers, Water, Maritime and Energy*, 156(WM2), 165–174.

Davies, J.W., Butler, D., and Xu, Y.L. 1996. Gross solids movement in sewers: Laboratory studies as a basis for a model. *Journal of the Institution of Water and Environmental Management*, 10(1), 52–58.

De Benedittis, J. and Bertrand-Krajewski, J.-L. 2005. Infiltration in sewer systems: Comparison of measurement methods. *Water Science and Technology*, 52, 219–227.

Defra. 1999. *The Water Supply (Water Fittings) Regulations. Statutory Instrument 1999 No. 1148.* HMSO, London.

Drinkwater, A., Chambers, B., and Carmen, W. 2008. *Less Water to Waste. Impact of Reductions in Water Demand on Wastewater Collection and Treatment Systems.* Environment Agency Science Report SC060066.

Environment Agency (EA). 2018. *Calculating dry weather flow (DWF) at waste water treatment works. Guidance.* www.gov.uk/government/publications/calculating-dry-weather-flow-dwf-at-waste-water-treatment-works/calculating-dry-weather-flow-dwf-at-waste-water-treatment-works

Fair, J.C. and Geyer, J.C. 1954. *Water Supply and Waste-Water Disposal*, John Wiley and Sons.

Gaines, J.B. 1989. Peak sewage flow rate: Prediction and probability. *Journal of Pollution Control Federation*, *61*, 1241.

Gifft, H.M. 1945. Estimating variations in domestic sewage flows. *Waterworks and Sewerage*, *92*, 175.

Gormley, M., Mara, D.D., Jean, N., and McDougall, I. 2013. Pro-poor sewerage: Solids modelling for design optimization. *Municipal Engineer*, *166*(ME1), 24–34.

Harman, W.G. 1918. Forecasting sewage in Toledo under dry-weather conditions. *Engineering News-Record*, *80*, 1233.

Institution of Water and Environmental Management (IWEM). 1993. *Glossary*. Handbooks of UK Wastewater Practice.

Lauchlan, C., Griggs, J., and Escarameia, M. 2004. *Drainage Design for Buildings with Reduced Water Use*, BRE Information paper, IP 1/04.

Littlewood, K. and Butler, D. 2003. Movement mechanisms of gross solids in intermittent flow. *Water Science and Technology*, *47*(4), 45–50.

Lund, N. S. V., Kirstein, J. K., Madsen, H., Mark, O. Mikkelsen, P.S. and Borup, M. 2021. Feasibility of using smart meter water consumption data and in-sewer flow observations for sewer system analysis: a case study. *Journal of Hydroinformatics*, *23*(4), 795–812.

Mann, H.T. 1979. *Septic Tanks and Small Sewage Treatment Works*, WRc Report No. TR107.

Memon, F., Fidar, A., Littlewood, K., Butler, D., Makropoulos, C., and Liu, S. 2007. A performance investigation of small-bore sewers. *Water Science and Technology*, *55*(4), 85–91.

Ministry of Housing and Local Government. 1970. *Technical Committee on Storm Overflows and the Disposal of Storm Sewage*, Final Report, HMSO.

Penn, R., Schütze, M., and Friedler, E. 2013. Modelling the effects of on-site greywater reuse and low flush toilets on municipal sewer systems. *Journal of Environmental*, *114*, 72–83.

Penn, R., Schütze, M., Alex, J. and Friedler, E. 2018. Tracking and simulation of gross solids transport in sewers. *Urban Water Journal*, *15*(6), 584–591.

UK Water Industry Research (UKWIR). 2022. *Ground Water Infiltration Model. Project Report and Guidance Manual. Report no. 22/SW/01/22.* https://ukwir.org/ground-water-infiltration-model-project-report-and-guidance-manual

Walski, T., Falco, J., McAloon, M., and Whitman, B. 2011. Transport of large solids in unsteady flow in sewers, *Urban Water Journal*, *8*(3), 179–187.

Water UK. 2023. *Sewerage Sector Guidance.* https://www.water.org.uk/sewerage-sector-guidance-approved-documents

Wise, A.F.E. and Swaffield, J.A. 2002. *Water, Sanitary and Waste Services for Buildings*, 5th edn, Taylor and Francis.

Zhang, X., Buchberger, S.G., and van Zyl, J.E. 2005. A theoretical explanation for peaking factors. *Proceedings of American Society of Civil Engineers, World Water and Environmental Resources Congress: Impacts of Global Climate, Anchorage*, May, 51.

Chapter 10

Storm sewers

10.1 INTRODUCTION

Separate storm sewers (also called surface water sewers) form a key component of many urban drainage systems. The emphasis in this chapter is on the design of such pipe-based systems (Section 10.2). Section 10.3 covers the concept of contributing area, and this leads to a detailed discussion of the key stormwater runoff design methods: rational (Section 10.4), time–area (Section 10.5), and hydrograph (Section 10.6). The chapter concludes with an (Section 10.7) an introduction to the estimation of runoff from undeveloped sites. Computer model-based analysis of existing systems is covered in Chapters 20 and 21. Design of non-pipe-based systems is covered next in Chapter 11 (Sustainable drainage systems) followed by surface water flooding in Chapter 12.

10.1.1 Flow regime

All storm sewer networks physically connect stormwater inlet points (such as road gullies and roof downpipes) to a discharge point, or outfall, by a series of continuous and unbroken pipes. Flow into the sewer results from the random input over time and space of rainfall-runoff. Generally, these flows are intermittent, of relatively long duration (minutes to hours), and hydraulically unsteady.

Separate storm sewers (more than foul sewers) will stand empty for long periods of time. The extent to which the capacity is taken up during rainfall depends on the magnitude of the event and conditions in the catchment. During low rainfall, flows will be well below the available capacity, but during very high rainfall the flow may exceed the pipe capacity inducing pressure flow and even surface flooding. Unlike in foul sewer design (Chapter 9), no distinction is made between large and small sewers in the design of storm systems.

10.2 DESIGN

The magnitude and frequency of rainfall are unpredictable and cannot be known in advance, so how are drainage systems designed? The general method has been illustrated in Figure 9.2 in Chapter 9 as a flowchart and should be read in conjunction with Figure 6.8.

Design is accomplished by first choosing a suitable *design storm*. The physical properties of the storm-*contributing area* must then be quantified. A number of methods of varying degrees of sophistication have been developed to estimate the *runoff flows* resulting from rainfall. *Hydraulic design* of the pipework, using the principles presented in Chapter 7,

DOI: 10.1201/9781003408635-10

ensures sufficient, sustained capacity. Broader issues of sewer layout including horizontal and vertical alignment are covered in Chapter 6.

10.2.1 Design storm

The concepts of statistically analysed rainfall, the design storm and climate change were introduced in Chapter 4. These give statistically representative rainfall (appropriately corrected for climate change using a climate change allowance (CCA) that can be applied to the contributing area and converted into runoff flows. Once flows are known, suitable pipes can be designed.

The choice of design storm return period therefore determines the degree of protection from stormwater flooding provided by the system. This protection should be related to the cost of any damage or disruption that might be caused by flooding. In practice, cost–benefit studies are rarely conducted for routine urban drainage projects, a decision on the design storm return period is made simply on the basis of judgement and precedent.

Standard practice in the UK (Water UK, 2023) is to use storm return periods of 1 year or 2 years for most schemes (for steeper and flatter sites, respectively), with 5 years being adopted where property in vulnerable areas would be subject to significant flood damage. Where a building or its contents require additional protection, BS EN 752: 2017 / BS ISO 24536: 2019 recommends a storm return period of either 1.5 × design life of the building or exceptionally, 4.5 × design life.

Although we can assess and specify the design *rainfall* return period, our greatest interest is really in the *flooding* return period. It is normally assumed that the frequency of *rainfall* is equivalent to the frequency of *runoff*. However, this is not completely accurate. For example, antecedent soil moisture conditions, areal distribution of the rainfall over the catchment, and movement of rain all influence the generation of stormwater runoff (see Chapters 4 and 5). These conditions are not the same for all rainfall events, so rainfall frequency cannot be identical to runoff frequency. However, comprehensive storm runoff data are less common than rainfall records and the assumption is therefore usually the best reasonable approach available.

It is *not* the case, however, that the frequency of rainfall is equivalent to the frequency of flooding. Sewers are almost invariably laid at least 1 m below the ground surface and can, therefore, accommodate a considerable surcharge before surface flooding occurs (see Chapter 7). Hence, the capacity of the system under these conditions is increased above the design capacity, perhaps even doubled. Inspection of Figure 4.2 in Chapter 4 illustrates that a 10-year storm will give a rainfall intensity approximately twice that of a 1-year storm for most durations. It follows, therefore, that where a sewer has been designed to a 1-year standard, a surcharge may increase that capacity up to an equivalent of a 10-year storm without surface flooding. That said, this holds for a single pipe whereas in practice different pipes will surcharge at different times. Thus, there is likely to be some additional capacity but this would need to be confirmed using a sewer simulation model (see Chapter 20).

Current practice is to design systems (or perhaps more precisely, confirm the design of systems) such that surface flooding is prevented for storms with a return period of 30 years (BS EN 752: 2017). Flooding from combined sewers into housing areas is likely to be more hazardous than storm runoff flooding of open land, so the type of flooding likely to occur will influence the selection of a suitable return period. Whatever return period is chosen, if a more extreme (i.e., higher return period) event occurs, flooding is likely. More detailed information and advice on designing for stormwater flooding (exceedance flows) is given in Chapter 12.

10.2.2 Optimal design

Most design is carried out by trial and error, as described in the rest of this chapter. System properties (e.g., pipe diameter and gradient) are proposed and then tested for compliance with design constraints or rules (e.g., self-cleansing velocity). If the rules are violated, a new design is proposed and retested iteratively until satisfactory performance is demonstrated. No real attempt is made to achieve an optimum design, rather one that is fit for purpose. There is a body of academic literature on optimal design of storm sewer systems (e.g., Alfaisal and Mays, 2021; Argaman et al., 1973; Diogo and Graveto, 2006), but the concepts and techniques developed have not, as yet, found their way into routine practice. For a thorough review of research and an analysis of why formal optimisation is rarely used, see Guo et al. (2008).

10.2.3 Return period and design life probability of exceedance

As mentioned in Chapter 4, the T-year return period of an annual maximum rainfall event is defined as the long-term average of the intervals between its occurrence or exceedance. Of course, the actual interval between specific occurrences will vary considerably around the average value T, with some intervals being less than T and others greater.

The probability that an annual event will be exceeded during the design life of the drainage system is derived as follows. The probability that, in any 1 year, the annual maximum storm event of magnitude X is greater than or equal to the T-year design storm of magnitude x is

$$P(X \geq x) = \frac{1}{T} \tag{10.1}$$

This is the annual exceedance probability (AEP). So, the probability that the event will not occur in any 1 year is

$$P(X < x) = 1 - P(X \geq x) = 1 - \frac{1}{T}$$

and the probability that it will not exceed the design storm in N years must be

$$P^N(X < x) = \left(1 - \frac{1}{T}\right)^N$$

The probability r that the event will equal or exceed the design storm at least once in N years is, therefore,

$$r = 1 - \left(1 - \frac{1}{T}\right)^N \tag{10.2}$$

If the design life of a system is N years, there is a probability r that the design storm event will be exceeded at some time in this period. The magnitude of the probability is given by Equation 10.2. Example 10.1 explains how they may be used.

EXAMPLE 10.1

What is the probability that at least one 10-year storm will occur during the first 10-year operating period of a drainage system? What is the probability over the 40-year lifetime of the system?

Solution

First 10 years: $T = 10, N = 10$.

The answer is not

$$r = 1 / T = 0.1$$

or

$$r = 10 \times 1 / T = 1.0$$

It is (Equation 10.2)

$$r = 1 - (1 - 0.1)^{10} = 0.651$$

Thus, there is a 65% probability that at least one 10-year design storm will occur within 10 years. In fact, it can be shown that for large T, there is s 63% risk that a T-year event will occur within a T-year period.

Lifetime: $T = 10, n = 40$.

$$r = 1 - \left(1 - \frac{1}{10}\right)^{40} = 0.985$$

In general, a very high return period is required if the probability of exceedance is to be minimised over the lifetime of the system.

10.3 CONTRIBUTING AREA

The following characteristics of a contributing area are significant for storm sewers: physical area, shape, slope, soil type and cover, land use, roughness, wetness, and storage. Of these, the catchment area and land use are the most important for good prediction of stormwater runoff.

10.3.1 Catchment area measurement

The boundaries of the complete catchment to be drained can be defined with reasonable precision either by field survey or the use of contour maps. They should be positioned such that any rain that falls within them will be directed (normally under gravity) to a point of discharge or outfall.

After the preliminary sewer layout has been produced, the catchment can be divided into sub-catchment areas draining towards each pipe or group of pipes in the system. The sub-areas can then be measured using a GIS-based package, normally integrated into sewer simulation software. Aerial photographs or satellite images may also be used. For simplicity, it is assumed that all flow to a sewer length is introduced at its head (that is, at the upstream manhole).

Table 10.1 Approximate percentage imperviousness of land-use types in London

Land-use category	PIMP
Dense commercial	100
Open commercial	65
Dense housing	55
Flats	50
Medium housing	45
Open housing	35
Grassland	<10
Woodland	<10

10.3.2 Land use

Once the total catchment area has been defined, estimates must be made of the extent and type of surfaces that will drain into the system. The *percentage imperviousness* (PIMP) of each area is measured by defining impervious surfaces as roads, roofs, and other paved surfaces (Equation 5.5). Measurement can be carried out using maps, aerial photographs, or satellite images (Castelluccio et al., 2015; Finch et al., 1989; Scott, 1994). Table 10.1 and Figure 10.1 illustrate a land-use classification in London.

Alternatively, the PIMP can be related approximately to the density of housing development using the following relationship:

$$PIMP = 6.4\sqrt{J} \quad 10 < J < 40 \tag{10.3}$$

where J is the housing density (dwellings/ha).

10.3.3 Urban creep

Urban creep is the term used to denote the gradual increase in imperviousness (*PIMP*) of *existing* urban areas. This is caused by the paving over of residential front gardens (typically for parking), and the construction of patios, extensions, conservatories, and so on. Perry and Nawaz (2008) found that between 1971 and 2004, the imperviousness of an existing suburban area of Leeds, UK, increased by 13%, with 75% of this due to the paving of front gardens. The average rate of urban creep in five UK areas between 1999 and 2006 was estimated as 0.38–1.09 m²/dwellings/year (Allitt et al., 2010). It has also been shown to be related to the existing housing density (see Figure 10.2).

The impact of this increased imperviousness will clearly be felt in the magnitude of runoff from any given storm or series of storms (Kelly, 2016). In the Leeds study, the creep was estimated to result in an average annual runoff increase of 12% (Perry and Nawaz 2008). Wright et al. (2011) point out that urban creep can also be expected to have an impact on water quality, due to higher levels of deposited pollutants being washed off during rainfall events.

In the absence of solid local evidence or regulatory requirements, Table 10.2 specifies the percentage increase to be applied to the catchment impervious area (A_i), where permitted development could occur in the future, to account for future urban creep.

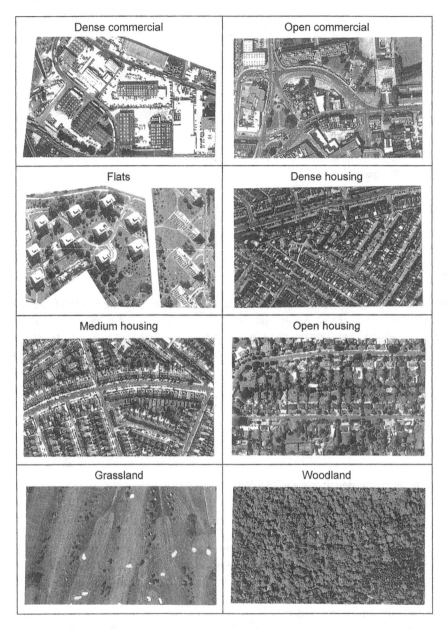

Figure 10.1 Various land use categories in London. Developed by Thames Water Utilities Ltd. on their Beckton and Crossness sewerage modelling projects in association with consulting engineers BGP Reid Crowther Ltd. and Montgomery Watson Ltd. and reproduced with permission.

10.3.4 Runoff coefficient

The dimensionless *runoff coefficient* C is defined in Chapter 5 as the proportion of rainfall that contributes to runoff from the surface. Early workers such as Lloyd-Davies (1906) assumed that 100% runoff came from impervious surfaces and 0% from pervious surfaces, so $C = PIMP/100$, and this assumption is still commonly adopted.

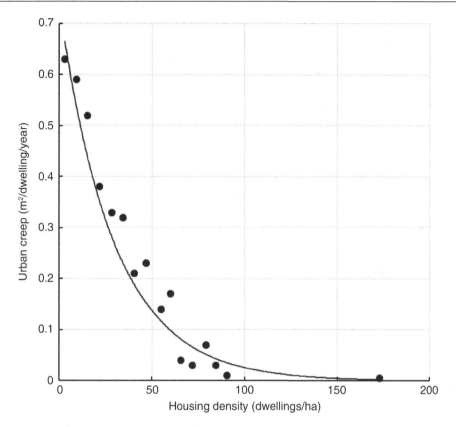

Figure 10.2 The relationship between urban creep and housing density. (Based on HRWallingford. 2012. *Development of Spatial Indicators to Monitor Changes in Exposure and Vulnerability to Flooding and the Uptake of Adaptation Actions to Manage Flood Risk in England*, Report TN-MCS0743-01 R2, www. theccc.org.uk)

Table 10.2 Urban creep design allowance

Residential development density (dwellings/ha)	Urban creep design allowance (percentage [%] of impervious area)
≤ 25	10
30	8
35	6
45	4
≥ 50	2
Flats and apartments	0

Source: Based on LASOO. 2016. Non-statutory Technical Standards for Sustainable Drainage: Practice Guidance, Local Authority SuDS Officer Organisation in association with the House Builders Association and Home Builders Federation.

However, the coefficient actually accounts for the initial runoff losses (e.g., depression storage) and continuing losses (e.g., surface infiltration), and implicitly accounts for the hydrodynamic effects encountered as the water flows over the catchment surface. Therefore, C must be related to *PIMP* but does not necessarily have to equal it – some runoff will come from pervious surfaces, for example. Equation 5.4 in Chapter 5 shows clearly that $C = PR/100$ is related to *PIMP* plus soil type and antecedent conditions. So, considerable knowledge of the catchment is required for accurate determination.

For design purposes, standard values of C such as those in Table 10.3 are often used. Weighted average coefficients are needed for areas of mixed land use.

10.3.5 Time of concentration

An important term used in storm sewer design is *time of concentration* (t_c). It is defined as the time required for surface runoff to flow from the most remote part of the catchment area to the point under consideration. Each point in the catchment has its own time of concentration. It has two components: the overland flow time, known as the *time of entry* (t_e), and the channel or sewer flow time, the *time of flow t_f*. Thus,

$$t_c = t_e + t_f \tag{10.4}$$

10.3.5.1 Time of entry

The time of entry varies with catchment characteristics such as surface roughness, slope and length of flow path, together with rainfall characteristics. Table 10.4 shows ranges of values dependent on the storm return period; rarer, heavier storms produce more water on the catchment surface and, hence, faster overland flow.

10.3.5.2 Time of flow

Velocity of flow in the sewers can be calculated from the hydraulic properties of the pipe, using one of the methods described in Chapter 7. Pipe-full velocity is normally used as a good approximation over a range of proportional depths. If sewer length is known or assumed, the time of flow can be calculated.

Table 10.3 Typical values of runoff coefficient in urban areas

Area description	Runoff coefficient	Surface type	Runoff coefficient
City centre	0.70–0.95	Asphalt and concrete paving	0.70–0.95
Suburban business	0.50–0.70	Roofs	0.75–0.95
Industrial	0.50–0.90	Lawns	0.05–0.35
Residential	0.30–0.70		
Parks and gardens	0.05–0.30		

Source: Adapted from Urban Water Resources Council. 1992. *Design and Construction of Urban Stormwater Management Systems,* ASCE Manual No. 77, WEF Manual FD-20.

Table 10.4 Time of entry

Return period (year)	Time of entry (minute)
1	4–8
2	4–7
5	3–6

Source: After Department of Energy/National Water Council (DoE/NWC), 1981.

10.4 RATIONAL METHOD

The Rational Method has a long history dating back to the middle of the nineteenth century. The Irish engineer Mulvaney (1850) was probably the first to publish the principles on which the method is based, although Americans tend to credit Kuichling (1889) and the British credit Lloyd-Davies (1906) for the method itself. The method and its further development are described below.

10.4.1 Steady-state runoff

Consider a simple, flat, fully impervious rectangular catchment with area A. A depth of rain, I, falls in a time, t. If there were also an impervious wall along the edges of the catchment and no sewers, this rain would simply build up over the area to a depth, I. The volume of water would be $I \times A$.

Now imagine that the runoff is flowing into a sewer inlet at point X with steady-state conditions: water is landing on the area, and flowing away, at the same rate. The sewer will carry the volume of rain ($I \times A$) at a steady, constant rate over the time (t) of the rainfall. So for flow rate (Q),

$$Q = \frac{A}{t}$$

and since the intensity of rain, $i = I/t$,

$$Q = iA$$

Now, since catchments are not totally impervious, and there will be initial and continuing losses, the runoff coefficient C can be introduced, to give

$$Q = CiA \qquad (10.5)$$

Adjusting for commonly used units gives

$$Q = 2.78CiA \qquad (10.6)$$

where Q is the maximum flow rate (L/s), i is the rainfall intensity (mm/h), and A is the catchment area (ha).

10.4.2 Critical rainfall intensity

For this method to be used for design purposes, the rainfall intensity that causes the catchment to operate at a steady state needs to be known. This should give the maximum flow from the catchment. The Rational Method states that a catchment just reaches steady state when the duration of the storm (and hence intensity i) is equal to the time of concentration of the area.

But why is this? Figure 10.3a shows a hydrograph resulting from uniform rainfall with a duration less than the time of concentration. Figure 10.3b gives the hydrograph for the same catchment, resulting from the same uniform rainfall intensity, but this time with infinite duration. The peak flow in Figure 10.3a is lower because the entire catchment is not contributing together (at steady state): contributions from remote parts of the catchment are still arriving after contributions from near parts have ceased. The maximum flow is reached when all the catchment contributes together – that is, when time is equal to or greater than the time of concentration t_c, as in Figure 10.3b.

The basis of the Rational Method is, therefore, an engineering "worst case." The duration of the storm must be at least the time of concentration; otherwise, the maximum flow would not be reached. However, it should not be longer, because storms with longer durations have statistically lower intensities (see Figure 4.2). Therefore, the worst case is when the duration is equal to the time of concentration (see Example 10.2).

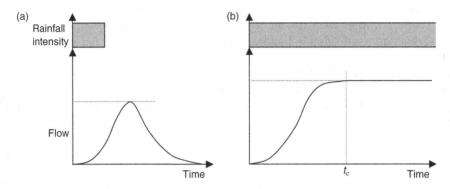

Figure 10.3 Hydrograph response to different duration rainfall of the same intensity.

EXAMPLE 10.2

A new housing estate is to be drained by a separate storm sewer network. The estate is rectangular in plan, 1200 × 900 m, and will consist of approximately 30% paved and roofed surfaces. Determine the maximum capacity required of the sewer carrying stormwater from the whole area. The longest branch leading to this point is 1350 m.

Assume that the average velocity is 1.5 m/s and the time of entry is 4 minutes. Use Equation 4.2 for rainfall calculation where a = 750, b = 10, α = 1, and β = 1.

Solution

$$t_c = 4 + \frac{1350}{1.5 \times 60} = 19 \, min \quad i = \frac{750}{19 + 10} = 26 \, mm / h$$

PIMP = 30%.
Assume 100% runoff from impervious areas and 0% from pervious areas; hence,

$$C = PIMP / 100 = 0.3$$

$$Q = CiA$$

$$Q = 0.3 \frac{26 \times 10^{-3}}{60 \times 60} 1200 \times 900 = 2.34 \, m^3 / s$$

10.4.2.1 Small areas

BS EN 16933-2: 2017 recommends the use of a fixed rainfall intensity of 0.014 L/s.m^2 (equivalent to 50 mm/h) multiplied by the CCA (see Chapter 4) for small paved areas (<= 4000 m^2, main sewer length <200 m). This avoids using unnecessarily high intensities calculated using very low concentration times. It also simplifies design using the Rational Method, since it is not necessary to know the time of concentration for each sub-catchment in order to identify the critical rainfall intensity (as described above). The standard also gives historically derived rainfall intensities (expressed in L/s.m^2) of 5 minutes duration and various return periods mapped for the UK, which can also be used with an appropriate CCA.

10.4.3 Modified Rational Method

Increased understanding of the rainfall-runoff process has led to further development of the Rational Method. The Modified Rational Method is recommended in the *Wallingford Procedure* (Department of Energy/National Water Council [DoE/NWC], 1981) and is shown to be accurate for UK catchment sizes up to 150 ha.

In this approach, the runoff of rainfall is disaggregated from other routing effects. Thus, the runoff coefficient C is considered to consist of two components:

$$C = C_v C_R \qquad (10.7)$$

where C_v is the volumetric runoff coefficient (–) and C_R is the dimensionless routing coefficient (–).

10.4.3.1 Volumetric runoff coefficient (C_v)

This is the proportion of rainfall falling on the catchment that appears as surface runoff in the drainage system. The value of C_v depends on whether the whole catchment is being considered, or just the impervious areas alone. Assuming the latter,

$$C_v = \frac{PR}{PIMP} \qquad (10.8)$$

where PR is given by Equation 5.4 in Chapter 5, and PIMP is given by Equation 5.5. Under summer rainfall conditions, C_v ranges from 0.6 to 0.9, with the lower values pertaining to rapidly draining soils and higher values to heavy clay soils. Note that in this method C_v is calculated, not assumed.

10.4.3.2 Dimensionless routing coefficient (C_R)

The dimensionless routing coefficient C_R varies between 1 and 2, and accounts for the effect of rainfall characteristics (e.g., peakedness) and catchment shape on the magnitude of peak runoff. A fixed value of 1.30 is recommended for design. So, for peak flow Q_p,

$$Q_p = 2.78 \times 1.30 C_v i A_i$$

where A_i is the impervious area (ha):

$$Q_p = 3.61 C_v i A_i \qquad (10.9)$$

10.4.4 Design criteria

It is good practice to follow a number of basic criteria during the design process.

10.4.4.1 Capacity

Storm sewers should be designed (size and gradient) to convey the predicted peak flows. It is conventional to design the pipes to just run full (e.g., to $d/D = 1.0$) but not surcharged. The small extra capacity that can be achieved at just below full flow (see Chapter 7) is neglected.

10.4.4.2 Self-cleansing

In addition to capacity, the pipe should also be designed to achieve self-cleansing. This is achieved by ensuring a specific velocity is reached at the design flow (1 m/s is typically used). In practice, the pipe size and gradient are manipulated together to obtain the best design.

For sewers designed for capacity with long return period storms (e.g., 10+ years), a more conservative approach is to design for self-cleansing with more frequent events (e.g., 1 year).

10.4.4.3 Roughness

As with foul sewers, it is conservatively assumed for design purposes that the pipe roughness will be independent of pipe material, although sliming is not a major issue. BS EN 752-4 recommends a k_s value of 0.6 mm for storm sewers.

10.4.4.4 Minimum pipe sizes

The minimum pipe size is generally set at similar levels to those for foul sewers (see Chapter 9).

10.4.5 Design method

The following procedure should be followed for the Modified Rational Method:

1. Assume design rainfall return period (T), pipe roughness (k_s), time of entry (t_e), and volumetric runoff coefficient (C_v).
2. Prepare a preliminary layout of sewers, including tentative inlet locations.
3. Mark pipe numbers on the plan according to the convention described in Chapter 6.
4. Estimate impervious areas for each pipe.
5. Make a first attempt at setting the gradients and diameters of each pipe.
6. Calculate pipe-full velocity (v_f) and flow rate ($Q_f = \pi r D^2 v_f /4$).
7. Calculate the time of concentration from Equation 10.4. For downstream pipes, compare alternative feeder branches and select the branch resulting in the maximum t_c.
8. Read rainfall intensity from intensity–duration–frequency (IDF) curves (see Chapter 4) for $t = t_c$ (for design T).
9. Estimate the cumulative contributing impervious area.
10. Calculate Q_p from Equation 10.9.
11. Check $Q_p < Q_f$ and $v_{max} > v_f > v_{min}$.
12. Adjust pipe diameter and gradient as necessary (given hydraulic and physical constraints) and return to step 5.
13. Check the design for the impact of climate change by multiplying the rainfall intensity for each pipe by an appropriate CCA.

This calculation procedure can be carried out manually, although in practice, software packages are available to automate the repetitive calculations.

Example 10.3 illustrates how the method may be used to design a simple storm sewer network.

EXAMPLE 10.3

A simple storm sewer network is shown in Figure 10.4. Appropriate rainfall data are given in Table 10.5 and network data in columns (1)–(4) of Table 10.6. Assume pipe gradients are fixed.

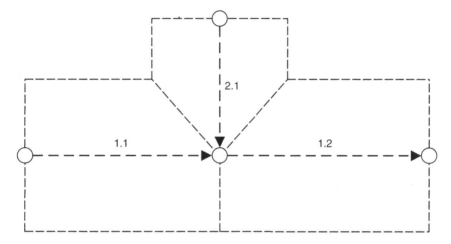

Figure 10.4 System layout (Example 10.3).

Table 10.5 Rainfall intensities (mm/h) at network site (Example 10.3)

| Duration (minutes) | Return period (years) | | |
	1	2	5
6.0	50.0	61.7	81.6
6.2	49.3	60.7	80.4
6.4	48.5	59.8	79.2
6.6	47.8	58.9	78.1
6.8	47.1	58.0	77.0
7.0	46.4	57.2	75.9
7.2	45.8	56.4	74.9
7.4	45.2	55.6	73.9
7.6	44.5	54.8	72.9
7.8	44.0	54.1	71.9
8.0	43.4	53.4	71.0
8.2	42.8	52.7	70.1
8.4	42.3	52.0	69.3
8.6	41.8	51.4	68.4
8.8	41.2	50.7	67.6
9.0	40.8	50.1	66.8

Table 10.6 Design table for Modified Rational Method (Example 10.3)

(1) Pipe number	(2) Pipe length (m)	(3) Pipe gradient (1:x)	(4) Impervious area (ha)	(5) Sum impervious area (ha)	(6) Assumed diameter (mm)	(7) Pipe-full velocity (m/s)	(8) Pipe capacity (l/s)	(9) Time of flow (minutes)	(10) Time of concentration (minutes)	(11) Rainfall intensity (mm/h)	(12) Calculated discharge (L/s)	(13) Comments
1.1	120	200	0.4	0.4	225	0.9	35	2.2	7.2	56.4	73.3	$Q_p > Q_f$
					300	1.1	80	1.8	6.8	58.0	75.4	OK
2.1	100	200	0.6	0.6	300	1.1	80	1.5	6.5	59.4	115.8	$Q_p > Q_f$
					375	1.3	140	1.3	6.3	60.3	117.5	OK[a]
1.2	150	400	0.8	1.8	600	1.2	350	2.1	8.9	50.4	295	OK[a]

[a] More cost-effective solutions could be arrived at by varying pipe gradients.
Return period = 2 years, pipe roughness = 0.6 mm, time of entry t_e = 5 minutes, coefficient C_v = 0.9

Design the network using the Modified Rational Method for a 2-year return period storm using a volumetric runoff coefficient of 0.9 and a time of entry of 5 minutes. Pipe roughness is 0.6 mm. Check the design with a CCA of 1.2.

Solution

Pipe-full velocities and capacities are calculated using the Colebrook–White equation. The design is completed in Table 10.6 (Columns 5–13). With the CCA check, calculated intensities increased by 1.2 in column 11 of Table 10.7.

Table 10.7 Design table for Modified Rational Method with CCA check (Example 10.3)

(1) Pipe number	(2) Pipe length (m)	(3) Pipe gradient (1:x)	(4) Impervious area (ha)	(5) Sum impervious area (ha)	(6) Assumed diameter (mm)	(7) Pipe-full velocity (m/s)	(8) Pipe capacity (l/s)	(9) Time of flow (minutes)	(10) Time of concentration (minutes)	(11) Rainfall intensity (mm/h)	(12) Calculated discharge (L/s)	(13) Comments
1.1	120	200	0.4	0.4	300	1.1	80	1.8	6.8	69.6	90.5	$Q_p > Q_f$
					375	**1.3**	**140**	**1.5**	**6.5**	**71.3**	**92.7**	**OK**
2.1	100	200	0.6	0.6	375	1.3	140	1.3	6.3	72.4	141.1	$Q_p \approx Q_f$, OK
1.2	150	400	0.8	1.8	600	1.2	350	2.1	8.9	60.5	353.4	$Q_p \approx Q_f$, OK

Source: Return period = 2 years, pipe roughness = 0.6 mm, time of entry t_e = 5 minutes, coefficient C_v = 0.9, CCA = 1.2.

In this particular case, inclusion of the CCA only affects the design choice for one leg of the network (1.1), which requires a larger diameter, shown in bold in Table 10.7.

Two points about the methodology are stressed. First, it is important that calculations are carried out for each pipe in turn, and that area and time of concentration refer to the whole upstream contributing area *not* just to the local sub-catchment area. The second point is that each pipe will be designed for a *different* (critical) design storm, with shorter-duration, higher-intensity storms used for upstream pipes (because they have a shorter time of concentration) and longer-duration, lower-intensity storms used for downstream sections.

10.4.6 Limitations

The Rational Method is based on the following assumptions.

1. The rate of rainfall is constant throughout the storm and uniform over the whole catchment.
2. Catchment imperviousness is constant throughout the storm.
3. Contributing impervious area is uniform over the whole catchment.
4. Sewers flow at constant (pipe-full) velocity throughout the time of concentration.

Assumption 1 can underestimate, as can assumption 3 (this is explored further in the next section). But assumption 2 tends to overestimate, as does assumption 4 – sewers do not always run full, and storage effects reduce peak flow. Fortunately, in many cases, these inaccuracies cancel each other out, producing a reasonably accurate result. Thus, the Rational Method and its modified version are simple, widely used approaches suitable for first approximations in most situations and are appropriate for full design in small catchments (<150 ha). For all but the smallest sites, BS EN 16933-2: 2017 recommends that the initial design is checked using an appropriate flow simulation method (Chapter 20), which will also facilitate further analysis to minimise surface flooding (Chapter 12).

10.5 TIME–AREA METHOD

10.5.1 The need

The contributing area in the Rational Method is treated as constant, whereas in fact it is not. For example, during the beginning of rainfall, the area builds up with time, closest surfaces contributing first, and more distant ones later. In many cases, the Rational Method is appropriate even though it does not take this type of effect into account; in others, it is not. This can be demonstrated by considering the three simple cases that follow.

Case (1)
The main storm sewers for a proposed industrial estate are shown in plan in Figure 10.5a. The capacity required at X is calculated using the Rational Method. Assume a catchment of area $A = 100,000$ m², $C = 0.6$, $t_e = 4$ minutes, $v_f = 1.5$ m/s, and the length of the longest sewer

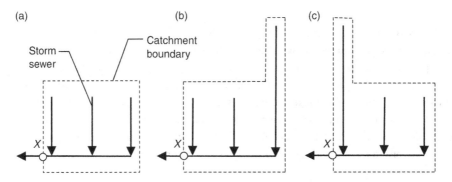

Figure 10.5 Layout of storm sewers.

is 450 m. Now, by using Equation 4.2 ($a = 750$, $b = 10$, $\alpha = 1$, and $\beta = 1$), for rainfall intensity (i) related to duration (D),

$$i = \frac{750}{D+10}$$

We have

$$t_c = 4 + \left[450 / (1.5 \times 60)\right] = 9 \text{ minutes}$$

$$i = \frac{750}{9+10} = 39 \text{ mm / h}$$

$$Q = \frac{39}{1000 \cdot 60 \cdot 60} \times 100{,}000 \times 0.6 = 0.65 \text{ m}^3 / \text{s}$$

Case (2)
An alternative layout (Figure 10.5b) is now considered in which the area is increased and one of the sewers is extended. How much additional flow will there be at X?

The same rainfall formula is used, C is still 0.6, and the assumed velocity and time of entry are unchanged. However, now A has been increased to 112,000 m² and the length of the longest sewer increased to 720 m. This gives

$$t_c = 4 + 720 / (1.5 \times 60) = 12 \text{ minutes}$$

$$i = \frac{750}{12+10} = 34 \text{mm / h}$$

$$Q = \frac{34}{1000 \cdot 60 \cdot 60} \times 112{,}000 \times 0.6 = 0.63 \text{m}^3 / \text{s}$$

It appears that the flow at X is now less, even though the area has increased. So, it is the increase in the time of concentration, and consequent reduction in i, rather than the increase in area, which has had the greatest effect on Q.

In this case, the Rational Method is inappropriate because it does not consider the "worst case." If rainfall lasting 9 minutes fell on this catchment, the whole of the original 100,000

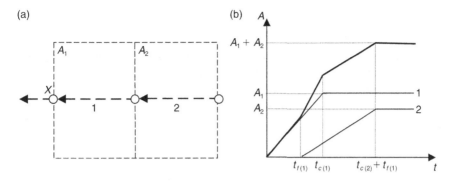

Figure 10.6 Example construction of time–area diagram: (a) simple storm sewer network; (b) time–area diagram for point X.

m² would contribute together, and a flow of 0.65 m³/s, as determined for case (1) would be produced. A flow of 0.63 m³/s would underestimate the capacity required.

Case (3)
Now consider another alternative. In Figure 10.5c, *A* has been increased to 112,000 m² (again), but because of the shape of the extended catchment, the length of the longest sewer is no greater than it was for case (1). Therefore, *i* remains at 39 mm/h, so

$$Q = \frac{39}{1000 \cdot 60 \cdot 60} \times 112,000 \times 0.6 = 0.73 \, \text{m}^3 / \text{s}$$

This is higher than 0.65 m³/s (case (1)), which makes more sense. The difference is that in case (3), the extra area contributes rapidly (because it is close to *X*) and does not increase the time of concentration. So, the appropriateness of the Rational Method for a particular catchment is affected by *the way the contributing area builds up with time.*

10.5.2 Diagram construction

A time–area diagram attempts to overcome one of the main limitations of the basic Rational Method by representing the rate of contribution of area. The diagram is used for storm sewer design by assuming that the time–area plot for each individual pipe sub-catchment is linear. However, the design of each pipe is not only concerned just with the local sub-catchment (in a similar way to the Rational Method), but also with the "concentrating" flows from upstream pipes. The combined time–area diagram for each pipe can be produced using the principle of linear superposition. This is illustrated by the case of a simple two-pipe system (and assuming $C = 1$) in Figure 10.6a.

In this network, the time of concentration of pipe 1 is $t_{c(1)}$, and this is plotted directly onto the time–area plot (Figure 10.6b). Pipe 2 has a sub-catchment time of concentration of $t_{c(2)}$, relative to its own outfall. However, this diagram is not directly overlaid on the previous one but is lagged by the time of flow in pipe 1, $t_{f(1)}$. The ordinates of the two separate diagrams are added to produce the complete diagram.

The resulting time–area diagram is made up of linear segments, but as more individual pipes are added, the shape tends to become non-linear. Example 10.4 illustrates a more complex network.

EXAMPLE 10.4

A storm sewer network has been designed initially with the layout shown in Figure 10.7, and data are given (in columns (1)–(5)) in Table 10.8. Construct the time–area diagram at the network outfall assuming a time of entry of 2 minutes and a Manning's *n* of 0.010.

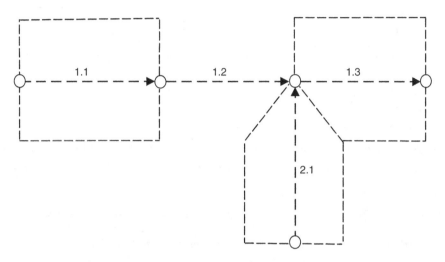

Figure 10.7 System layout (Example 10.4 and 10.5).

Table 10.8 Data for Example 10.4

(1)	(2)	(3)	(4)	(5)	(6)	(7)
Pipe number	Diameter (mm)	Pipe length (m)	Pipe gradient (1:x)	Impervious area (ha)	Pipe-full velocity (m/s)	Time of flow (minutes)
1.1	375	1000	300	2.5	1.2	13.9
1.2	450	500	400	0	1.2	7.1
2.1	300	620	250	2.5	1.1	9.2
1.3	600	300	500	1.0	1.3	4.0

Solution

For simplicity, assume the pipes run full to calculate flow velocities and, hence, time of flow (columns (6) and (7) in Table 10.8).

Time of flow data and time of entry information are used to derive the time–area diagram (Figure 10.8):

Pipe	Time of concentration (mins)	Flow contribution starts (mins)
1.3	$t_{c(1.3)} = 2 + 4.0 = 6$	$t = 0$
2.1	$t_{c(2.1)} = 2 + 9.2 = 11.2$	$t_{f(1.3)} = 4.0$
1.1	$t_{c(1.1)} = 2 + 13.9 = 15.9$	$t_{f(1.2)} + t_{f(1.3)} = 7.1 + 4.0 = 11.1$

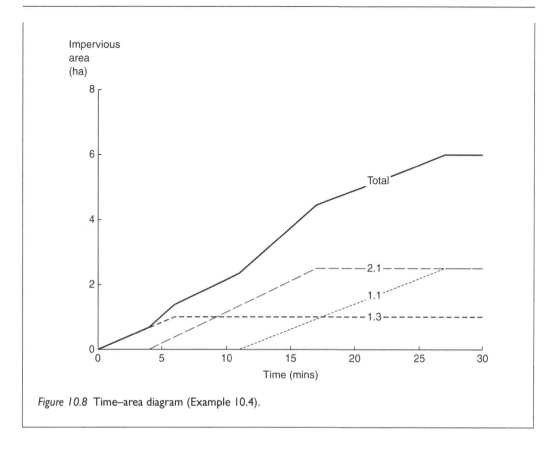

Figure 10.8 Time–area diagram (Example 10.4).

10.6 HYDROGRAPH METHODS

One of the major weaknesses of the Rational Method is that it only produces worst-case design flow and not a hydrograph of flow against time. Hydrograph methods have been developed to overcome this limitation.

10.6.1 Time–Area Method

The Time–Area Method uses the time–area diagram to produce not only a peak design flow, but also a flow hydrograph. The method also allows straightforward use of time-varying rainfall – the design storm (see Chapter 4).

Equation 5.8 from Chapter 5 is repeated below, and this gives the basic equation for finding flow $Q(t)$ when a continuous time–area diagram is combined with rainfall depth increments, $I_1, I_2, \ldots I_N$.

$$Q(t) = \sum_{\omega=1}^{N} \frac{dA(j)}{dt} I_\omega \tag{10.10}$$

where $Q(t)$ is the runoff hydrograph ordinate at time t (m³/s), $dA(j)/dt$ is the slope of the time–area diagram at time j (m²/s), I_ω is the rainfall depth in the ωth of N blocks of duration Δt (m), and j is $t - (\omega - 1) \Delta t(s)$.

Assuming linear incremental change in the time–area diagram, ΔA_1, ΔA_2 ... ΔA_j ... over rainfall time blocks, Δt_1, Δt_2 ... Δt_j ..., runoff is given by

$$Q(t) = \sum_{\omega=1}^{N} \Delta A_j i_{e\omega} \qquad (10.11)$$

where i_e is $I/\Delta t$.

This can be expanded to give

$$Q(1) = A_1 \, i_1$$

$$Q(2) = A_2 \, i_1 + A_1 \, i_2$$

$$Q(3) = A_3 \, i_1 + A_2 \, i_2 + A_1 \, i_3$$

...

The method is summarised as follows:

- Select a suitable integer time interval Δt, typically $t_c /10$.
- Prepare a suitable rainfall hyetograph using Δt as the time interval.
- Produce a time–area diagram for each design point under consideration.

Calculate outflow by reading off the relevant rainfall intensity from the hyetograph (i_1) and contributing area (A_j) from the time–area diagram for each time increment (Δ_t). These need to be accumulated for each time step according to Equation 10.11.

Example 10.5 illustrates how the method may be used to design a simple storm sewer network.

10.6.1.1 Limitations

This method has the advantage over the Rational Method in that it takes some account of the shape of the catchment, allowing an output hydrograph to be produced and to include the effects caused by time-varying rainfall. However, the method still only allows linear translation of the flood wave through the catchment and makes no allowance for storage effects.

10.6.2 Level pool routing method

In the Rational and basic Time–Area Methods, pipe-full velocity is used as the flow routing velocity. However, in reality, this will not be the case, and many parts of the network will operate part-full. In this method, storage in the network is taken into account in a relatively simple way using a *level pool* or *reservoir* routing technique (explored further in 14.4). The method assumes that the retained water acts as a level reservoir of liquid with outflow being uniquely related to water level or storage. This is accomplished in a pipe network by assuming that proportional depth of flow is identical at all points in the system. This assumption is valid if

- The system is designed with a reasonable degree of taper.
- All of the pipes in the system are geometrically similar.

Both of these requirements should be satisfied in new systems.

EXAMPLE 10.5

Using the time–area diagram developed in Example 10.4 and shown in Figure 10.8, evaluate the outfall flow hydrograph for the sewer network in Figure 10.7 under the following design storm.

Time (minutes)	Effective rainfall depth (mm)
0–5	3
5–10	6
10–15	3

Solution

Using a 5-minute time increment (Δt = 5 minutes), read off from Figure 10.7 the cumulative contributing area at each time step and then the incremental area. Convert the rainfall depths to intensities.

Δt	Time (minutes)	ΣA (ha)	ΔA (ha)	i_e (mm/h)
1	5	1.4	1.4	36
2	10	2.6	1.2	72
3	15	4.2	1.6	36
4	20	4.9	0.7	
5	25	5.7	0.8	
6	30	6.0	0.3	
			$\Sigma 6.0$	

From Equation 10.11,

$Q(1) = 1.4 \times 36 = 50.4$
$Q(2) = 1.2 \times 36 + 1.4 \times 72 = 144.0$
$Q(3) = 1.6 \times 36 + 1.2 \times 72 + 1.4 \times 36 = 194.4$
$Q(4) = 0.7 \times 36 + 1.6 \times 72 + 1.2 \times 36 = 183.6$
$Q(5) = 0.8 \times 36 + 0.7 \times 72 + 1.6 \times 36 = 136.8$
$Q(6) = 0.3 \times 36 + 0.8 \times 72 + 0.7 \times 36 = 93.6$
$Q(6) = 0.3 \times 72 + 0.8 \times 36 = 50.4$
$Q(6) = 0.3 \times 36 = 10.8$

The ordinates should be multiplied by 2.78 to obtain the hydrograph flow in litres per second (L/s), from which the peak flow is

$Q_p \approx 550 \, l/s \, @ \, 17 \, \text{minutes}$

Note: this exceeds the capacity of the pipe.

Consider an individual circular pipe of length L carrying flow with proportional depth d/D. Thus, $A = f_1(d/D)$ and $V = f_2(d/D)$, where A is the cross-sectional area of flow in the pipe, and V is the volume of water stored. Now, if d/D is constant everywhere, $S = f_3(d/D)$, where S is the whole system retention (storage volume). For a given slope and pipe roughness, outflow at the design point, $O = f_4(d/D)$. Therefore,

$$O = f_5(S) \text{ or } S = f_6(O) \tag{10.12}$$

Thus, there is a unique relationship between the total volume of water stored (system retention) and the outflow rate at the design point.

So, knowing the inflow I, flow routing can be performed using the basic storage equation to estimate O. Example 10.6 illustrates the application of the method.

EXAMPLE 10.6

Calculate the outflow hydrograph and peak for the sewer network in Figure 10.8 accounting for in-pipe storage.

Solution

Initially, derive a relationship between S and O for circular pipes. Table 7.5 gives

$$A = \frac{D^2}{8}(\theta - \sin\theta) = \frac{\pi D^2}{4}\left(\frac{\theta - \sin\theta}{2\pi}\right)$$

where θ is defined in Equation 7.17.

Now $\pi D^2/4$ is the cross-sectional area of each pipe. So if L is the length of each pipe of diameter D,

$$S = \left(\Sigma L \frac{\pi D^2}{4}\right)\left(\frac{\theta - \sin\theta}{2\pi}\right)$$

Hence, storage can be derived from the system data:

Pipe number	Diameter (mm)	Pipe length (m)	LD^2
1.1	375	1000	140.6
1.2	450	500	101.3
2.1	300	620	55.8
1.3	600	300	108.0
			$\Sigma 405.7$

$$S = 50.7(\theta - \sin\theta)$$

Manning's Equation 7.23 gives

$$O = \frac{1}{n}AR^{2/3}S_o^{1/2}$$

Therefore,

$$O = \frac{1}{n}\frac{D^2}{8}(\theta - \sin\theta)\left(\frac{D}{4}\left[\frac{\theta - \sin\theta}{\theta}\right]\right)^{2/3} S_o^{1/2}$$

$$O = \frac{1}{n}\frac{D^{8/3}}{20.16}\frac{(\theta - \sin\theta)^{5/3}}{\theta^{2/3}} S_o^{1/2}$$

(10.13)

For the outfall pipe (1.3), $D = 0.6$ m, $S_o = 0.002$, and $n = 0.010$ can be substituted into Equation 10.13. So, by varying θ from $0 \rightarrow 2\pi$ (i.e., d/D: $0 \rightarrow 1$), S and O can be plotted together as in Figure 10.9a. Using a time step of 5 minutes, the relationship between $(S/\Delta t) + (O/2)$ and O can be constructed as shown Figure 10.9b.

The calculation now follows the procedure in Example 14.2, with the result shown in Figure 10.10. The routed peak flow is seen to be

$$Q_p \approx 370 \text{L/s} @ 25 \text{ minutes}$$

which is a considerable reduction from the previous estimate and is now within the capacity of the pipe.

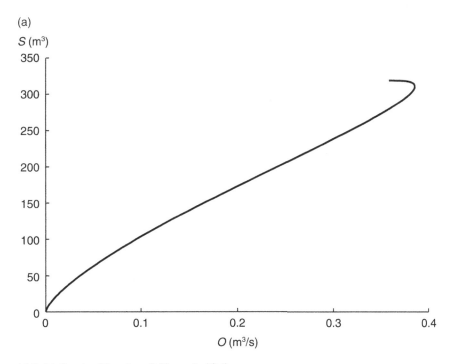

Figure 10.9 (a) Graph of S against O (Example 10.6).

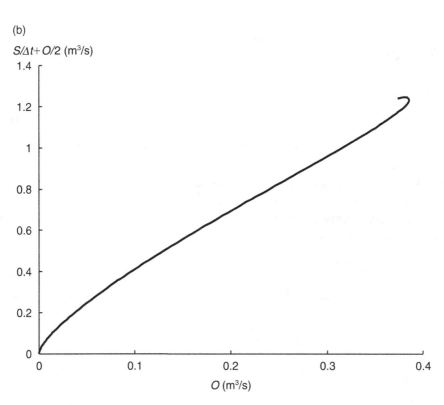

Figure 10.9 (Continued) (b) Graph of $S/\Delta t + O/2$ against O (Example 10.6).

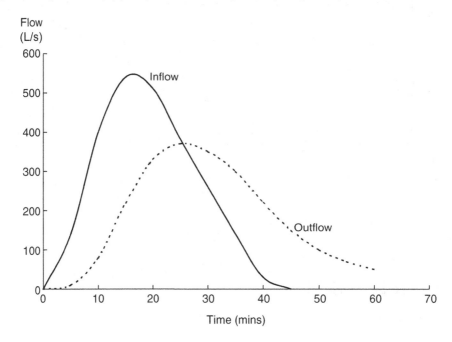

Figure 10.10 Routed hydrograph (Example 10.6).

10.6.3 Limitations

The level pool routing method is an advance over the basic Time–Area Method in that it takes pipe storage effects directly into account. However, the routing assumption is relatively crude, and indeed no account is taken of surface storage effects, and surcharge effects cannot be handled.

Coverage of these aspects effectively requires flow simulation models that have greater power than any of the design-orientated tools described in this chapter. These are described in Chapter 20.

10.7 UNDEVELOPED SITE RUNOFF

A common requirement when developing a particular site is an estimate of its undeveloped (greenfield) runoff response. This can be challenging because different approaches and methods give differing results, however, current good practice is still to use the FSR-based IoH 124 method (Marshall and Bayliss, 1994). This is based on data from 15 small (1.5–22.9 km²) catchments in central southern England and produces the annual average flood based on three readily available variables (Equation 10.14):

$$QBAR_{rural} = 0.00108 \, A^{0.89} \, SAAR^{1.17} \, SOIL^{2.17} \tag{10.14}$$

where $QBAR_{rural}$ is the mean annual peak flow (m³/s), A is the catchment area (km²), $SAAR$ is the standard annual average rainfall (mm) and $SOIL$ is the FSR soil index (–).

$QBAR_{rural}$ can be converted to other return period and duration events using the regional growth curves in the FSR (NERC, 1975). It is recommended that the method is only applied to catchments larger than 50 ha (0.5 km²) drained by a well-defined watercourse. The result from the IoH 124 method can be checked to ensure comparability with flow rate calculated using the Rational Method. Example 10.7 shows how this can be done in practice. For sites with areas less than 50 ha, Equation 10.14 can be solved with $A = 0.5$ km² with the resulting value of $QBAR_{rural}$ reduced in linear proportion to the actual site area.

EXAMPLE 10.7

Calculate the 1-year return period flow from a 200 ha greenfield site. The standard annual average rainfall for the site (1991–2020) is 540 mm, the SOIL index is 0.25 and the runoff coefficient may be taken as 0.1. The length of the main channel is 3 km with a typical flow velocity of 0.15 m/s, and rainfall data on the catchment can be derived from Figure 4.2.

Solution

The flow can be calculated in a straightforward way using equation (11.13).

$$QBAR_{rural} = 0.00108 \times 2^{0.89} \times 540^{1.17} \times 0.25^{2.17} = 0.156 \, \text{m}^3/\text{s} = 56 \, \text{L/s}$$

A rough check can be made by comparing this result with the Rational Method. Assuming the channel flow velocity, $v = 0.15$ m/s during the 1-year event:

$$t_c = 3000 / 0.15 = 20,000 \, \text{s} \approx 5.5 \, \text{h} \text{ giving } i = 4 \, \text{mm} / \text{h from Figure 4.2}$$

From equation 11.5b: $Q = 2.78CiA = 2.78 \times 0.1 \times 4 \times 200 = 222$ L/s

This is certainly in the "same ball park" as the previous total. However, it does indicate the difficulty of using the Rational Method in this way due to its sensitivity to C and v, both of which are assumed variables.

The *Flood Estimation Handbook* has been introduced in Chapter 4 in the context of rainfall estimation. Kjeldsen, et al. (2008) proposed a revision to the statistical method contained within the handbook which has been recommended for use as a relatively straightforward approach to greenfield site runoff estimation (Kellagher, 2013). A free-to-use online tool for both the IoH and FEH methods is available at: www.uksuds.com/tools/greenfield-runoff-rate-estimation

In an update to the original handbook, and as a replacement for the original FSR/FEH rainfall-runoff method, Kjeldsen et al. (2008) also developed the Revitalised Flood Hydrograph (ReFH) rainfall-runoff method. ReFH2 is the most up-to-date version of this method and includes a rural catchment model component which can be used for greenfield site runoff estimation. The model allows data to be used from either the FEH99 or FEH13 rainfall model (see Section 4.3.6) and is being updated for FEH22 data. Further details of the method, its development, and its application can be accessed here: refhdocs.hydrosolutions.co.uk. ReFH2 has been integrated with commonly used urban drainage software packages and is commercially available from WHS.

PROBLEMS

1 Describe and justify the main stages in storm sewer design.
2 "The frequency of rainfall is neither equivalent to the frequency of runoff nor of flooding." Discuss this statement with reference to recommended design storm return periods.
3 If a storm sewer surcharges during heavy rainfall, was it under-designed? Explain your answer.
4 A storm sewer network has been designed based on a 2-year return period storm for a 50-year design life. What is the probability that the network will
   a. Surcharge at least once in 2 years (the probability of failure)?
   b. Surcharge at least once during its design life?
   c. Flood in any 1 year, assuming flooding is caused by the 10-year return period storm?
   d. Flood at least once in 10 years?   [0.75, 1.00, 0.10, 0.65]
5 What is urban creep, and why is it an important issue for urban drainage?
6 What is percentage imperviousness, and how is it related to the runoff coefficient?
7 Explain what you understand by time of entry, time of flow, and time of concentration. Why is the duration of the design storm in the Rational Method taken as the time of concentration?
8 Demonstrate that a rainfall intensity of 0.014 L/s.m² is equivalent to 50 mm/h.
9 Explain the concept of the Rational Method. What are its main limitations?
10 A small separate storm sewer network has the following characteristics (Figure 10.11):

Sewer	Length (m)	Contributing impervious area (m²)
1.1	180	2000
2.1	90	6000
3.1	90	9000
1.2	90	4000

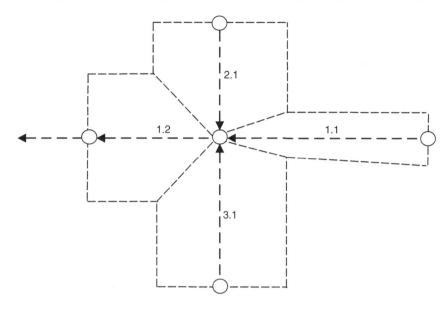

Figure 10.11 System layout (Problem 10.10).

Use the Rational Method to determine the capacity required for each pipe in the network. Assume a time of entry of 4 minutes, that the pipe-full velocity in each pipe is 1.5 m/s, and that design rainfall intensities can be determined from Equation 4.2, where $a = 750$, $b = 10$, $\alpha = 1$, and $\beta = 1$. Further, assume 100% runoff from impervious areas. If a CCA of 1.3 is specified, what impact would this have on the required capacities?

[26 L/s, 83 L/s, 125 L/s, 257 L/s, × 1.3]

11 What is a time–area diagram? Explain how it is used in the Time–Area Method.

12 Construct the time–area diagram (at point X) for each of the equally sized and graded catchments (1–3) shown in Figure 10.12.

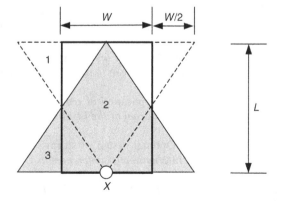

Catchment 1: dashed triangle
Catchment 2: rectangle
Catchment 3: shaded triangle

Figure 10.12 Three catchments (Problem 10.12).

13 Estimate the hydrograph at the outlet to pipe 1.2 (Problem 10.10) for a short storm with the following profile:

Time (min)	0–1	1–2	2–3
Intensity (mm/h)	20	28	64

Use the Time–Area Hydrograph Method.   [Peak 127 L/s]

14 Describe the basis of the level pool routing method, and discuss its strengths and weaknesses.

15 Discuss the challenges in estimating runoff from undeveloped sites.

KEY SOURCES

Department of Energy/National Water Council (DoE/NWC). 1981. *Design and Analysis of Urban Storm Drainage. The Wallingford Procedure*, 4 vol., Standing Technical Committee Report No. 28.

Shaw, E.M., Beven, K.J., Chappell, N.A., and Lamb, R. 2010. *Hydrology in Practice*, 4th edn, CRC Press.

Woods-Ballard, B., Wilson, S., Udale-Clark, H., Illman, S., Scott, T., Ashley, A., and Kellagher, R. 2016. *The SuDS Manual*, 5th edn, CIRIA C753.

REFERENCES

Alfaisal, F.M. and Mays, L.W. 2021. Optimization Models for Layout and Pipe Design for Storm Sewer Systems. *Water Resources Management*, 35, 4841–4854

Allitt, R., Allitt, M., and Allitt, K. 2010. *Impact of Urban Creep on Sewerage Systems*. Report 10/WM/07/14, UK Water Industry Research, London.

Argaman, Y., Shamir, U., and Spivak, E. 1973. Design of optimal sewerage systems. *American Society of Civil Engineers Journal of Environmental Engineering*, 99(5), 703–716.

BS EN 752: 2017. *Drain and Sewer Systems Outside Buildings. Drain and sewer - Sewer system management.*

BS EN 16933-2: 2017. *Drain and Sewer Systems Outside buildings – Design. Part 2: Hydraulic Design.*

BS ISO 24536: 2019. *Service activities relating to drinking water supply, wastewater and stormwater systems — Stormwater management — Guidelines for stormwater management in urban areas.*

Castelluccio, M., Poggi, G., Sansone, C., and Verdoliva, L. 2015. Land Use Classification in Remote Sensing Images by Convolutional Neural Networks. *arXiv.org*, arXiv: 1508.00092 [cs.CV].

Diogo, A.F. and Graveto, V.M. 2006. Optimal layout of sewer systems: A deterministic versus a stochastic model. *American Society of Civil Engineers Journal of Hydraulic Engineering*, 132(9), 927–943.

Finch, J., Reid, A., and Roberts, G. 1989. The application of remote sensing to estimate land cover for urban drainage catchment modelling. *Journal of the Institution of Water and Environmental Management*, 3, 558–563.

Guo, Y., Walters, G., and Savic, D. 2008. Optimal design of storm sewer networks: Past, present and future. *Proceedings of the 11th International Conference on Urban Drainage, Edinburgh, September*, on CD-ROM.

HRWallingford. 2012. *Development of Spatial Indicators to Monitor Changes in Exposure and Vulnerability to Flooding and the Uptake of Adaptation Actions to Manage Flood Risk in England*. Report TN-MCS0743-01 R2, www.theccc.org.uk

Kellagher, R.. 2013. *Rainfall Runoff Management for Developments*. Environment Agency Report SC030219.

Kelly, D.A. 2016. Impact of paved front gardens on current and future urban flooding, *Journal of Flood Risk Management*. https://doi.org/10.1111/jfr3.12231

Kjeldsen, T.R., Jones, D.A. and Bayliss, A.C. 2008. *Improving the FEH Statistical Procedures for Flood Frequency Estimation*, Environment Agency, Science Report: SC050050.

Kuichling, E. 1889. The relation between the rainfall and the discharge of sewers in populous district. *Transactions of American Society of Civil Engineers*, 20, 1–56.

LASOO. 2016. *Non-statutory Technical Standards for Sustainable Drainage: Practice Guidance*, Local Authority SuDS Officer Organisation in association with the House Builders Association and Home Builders Federation.

Lloyd-Davies, D.E. 1906. The elimination of storm water from sewerage systems. *Proceedings of Institution of Civil Engineers*, 164(2), 41–67.

Marshall, D.C.W. and Bayliss, A.C. (1994) *Flood Estimation for Small Catchments*. IH Report 124, Centre for Ecology and Hydrology, Wallingford.

Mulvaney, T.J. 1850. On the use of self-registering rain and flood gauges in making observations on the relation of rainfall and flood discharges in a given catchment. *Transactions of Institution of Civil Engineers Ireland*, 4(2), 18.

Natural Environment Research Council (NERC). 1975. *Flood Studies Report*, 5 volumes, Institute of Hydrology.

Perry, T. and Nawaz, R. 2008. An investigation into the extent and impacts of hard surfacing of domestic gardens in an area of Leeds, United Kingdom. *Landscape and Urban Planning*, 86(1), 1–13.

Scott, A. 1994. Low-cost remote-sensing techniques applied to drainage area studies. *Journal of Institution of Water and Environmental Management*, 8, 498–501.

Urban Water Resources Council. 1992. *Design and Construction of Urban Stormwater Management Systems*, ASCE Manual No. 77, WEF Manual FD-20.

Water UK. 2023. *Sewerage Sector Guidance*. https://www.water.org.uk/sewerage-sector-guidance-approved-documents

Wright, G.B., Arthur, S., Bowles, G., Bastien, N., and Unwin, D. 2011. Urban creep in Scotland: Stakeholder perceptions, quantification and cost implications of permeable solutions. *Water and Environment Journal*, 25, 513–521.

Chapter 11

Sustainable Drainage Systems (SuDS)

11.1 INTRODUCTION

Stormwater management or *surface water management* (using non-pipes systems) has already been referred to a number of times in this text where we look to manage runoff from developed areas in a more natural way at its source, using the infiltration, evapo-transpiration and storage capacities and processes of engineered components to "slow the flow," and by preserving, restoring, and recreating natural landscape features. In short, we attempt to mimic the natural functioning of the site prior to its development by minimising and/or offsetting increases in surface imperviousness and by treating stormwater as a resource to be reused, retained, and enjoyed.

This approach has been given different names in different countries. In the US, the techniques are called *Low Impact Development* (LID), *Stormwater Control Measures* (SCMs), or *Green Stormwater Infrastructure* (GSI). In Australia, the general expression *water-sensitive urban design* (WSUD) communicates a philosophy for water engineering in which water use, reuse, and drainage, and their impacts on the natural and urban environments, are considered holistically. China has developed the *Sponge City* approach (see Box 1.4). In the UK, since the mid-1990s, the label has been *SuDS* (sustainable drainage systems) although the term *blue–green infrastructure* (BGI) is also increasingly used, sometimes interchangeably with SuDS. Further information on sustainability, including definitions, is given in Chapter 24. Several other terms are in use around the world conveying similar meanings (see Fletcher et al., 2015, for more detail and historical context).

Preceding chapters, especially Chapter 10, have dealt in detail with piped systems for stormwater, so in this chapter, we consider the alternatives to pipes, namely, SuDS, including their wider benefits above and beyond water quantity and quality control. Section 11.2 describes the common types of SuDS components in turn, Section 11.3 gives guidance on aspects of their design including multiple benefits, and Section 11.4 considers how they can be linked together. Section 11.5 summarises the importance of addressing public concerns and engaging affected communities. A range of important issues still surround SuDS and their implementation, and these are considered in Section 11.6. Finally, Section 11.7 covers site runoff control – how to achieve it and what the relevant standards are.

11.2 COMPONENTS

Details of many individual SuDS components and approaches to their selection to maximise benefits based on land use, site and catchment characteristics, and required performance are set out in the comprehensive Construction Industry Research and Information Association (CIRIA) document, *The SuDS Manual* (Woods-Ballard et al., 2016). Another resource on

DOI: 10.1201/9781003408635-11

SuDS is freely available online at: www.susdrain.org. This section summarises the most common types of individual SuDS components:

- Inlet controls
- Infiltration components
- Vegetated surfaces
- Pervious pavements
- Filter drains
- Infiltration basins
- Detention basins
- Ponds
- Constructed wetlands

Sometimes, these components are further differentiated into hard SuDS such as pervious pavements and soft SuDS such as swales.

11.2.1 Inlet controls

Runoff can be managed at or near its source by a variety of means including blue roofs, green roofs, rainwater harvesting (RWH) and water butts, and paved area ponding.

11.2.1.1 Blue roofs

Stormwater can be retained on flat roofs, thus exploiting their reservoir storage potential by using flow restrictors on the roof drains. This will induce an additional live load to be taken into account in the structural design depending on the maximum depth of the retained water. A 75 mm depth system will induce a loading of 0.75 kN/m² (equivalent to a typical snow load). Particular care is needed in specifying and maintaining the roof waterproof membranes. In practice, flow restrictors can become blocked, leading either to overtopping or prolonged ponding, pointing to the need for regular maintenance and exceedance routes. Blue roofs can also be specified in combination with green roofs.

11.2.1.2 Green roofs

A green roof is a planted area that has rainfall storage potential, encourages evapotranspiration, and provides the added benefit of water quality improvement as stormwater travels through the soil layer (Figure 11.1). There are two broad types: extensive and intensive.

Extensive green roofs take the form of a "carpet" of plants, supported by a lightweight growing media and overlying a drainage layer. They vary from thin, low-diversity sedum mats to "biodiverse green roofs" with a variety of sedums and wildflowers and thicker substrates. Intensive green roofs (roof gardens) typically incorporate more deeply rooted vegetation such as trees, shrubs, and cultivated flowers. They will also often incorporate seating areas. Extensive roofs are the most common due to their lower relative maintenance and roof load requirements but offer little plant diversity and are less resilient to droughts. Intensive roofs have a wider vegetative diversity and more co-benefits but require a greater substrate depth to adequately support such growth. They also require greater maintenance, including watering of the vegetation (The GRO Green Roof Code, 2021).

During a storm event, rainfall is intercepted by the plant layer of the green roof, infiltrated and stored in the substrate, and then transported downstream in the drainage layer. Extensive

Figure 11.1 Extensive green roof, Portland, Oregon.

green roofs have been shown to retain between 50% and 100% of precipitation over the long term, but this depends on the intensity and duration of the rainfall events and is influenced by the roof slope, substrate depth, substrate moisture content, evapo-transpiration rates, and plant morphology (Whittinghill et al., 2015). In a UK context, Stovin et al. (2012) found that over a 29-month period, cumulative retention of a trial green roof was 50%. However, this varied widely between storm events, ranging from 0% to 100%, and the roof provided only 13% retention for the largest event (16-year return period). For storms with rainfall depth > 2 mm, mean peak flow attenuation was 60%. In their UK study, Nawaz et al. (2015) found the overall retention to be 66%. On the basis of a comprehensive review of studies in 21 countries, retention was found to vary widely with an average of 62% (Zheng et al., 2021). Also in a UK context, typical evapo-transpiration rates have been found to vary from 0.6 to 1.25 mm/day depending on roof design and the environmental conditions (Poë et al. 2015).

Green roofs have numerous wider benefits, including reduced air and noise pollution, increased habitat and biodiversity, improved aesthetics, increased roof lifespan, energy savings, and mitigation of the urban heat island effect (Berardi et al., 2014; Early et al., 2007).

11.2.1.3 Water butts and rainwater harvesting

Water butts (or *water barrels, cisterns*) are simple storage containers provided at the foot of a building rainwater downpipe, either above or below ground (Figure 11.2a). These typically have a capacity of 200–350 L and in large numbers can provide significant runoff storage. However, traditional butts are designed as a means of harvesting rainwater for gardens and other outside use and if not regularly emptied, they will have limited stormwater management functionality. A simple solution to this is to fit "*leaky butts*" with low-level outlets which slowly drain to the drainage system so there is some capacity for the next rainfall event. In the system shown in Figure 11.2b, a proportion of the roof runoff is diverted into a small internal tank, which can be used to tend the vegetation in the planter and other outside uses. When this is full, the excess flow spills into a larger tank inside the planter, which slowly discharges the surplus runoff into the sewer network via a vortex flow controller.

An even simpler alternative is to discharge runoff from the downpipe away from the building and over stable pervious areas (such as lawns, swales, and porous pavements) rather than directly to the underground pipe system. In this way, the surface runoff is delayed, infiltration is increased, and pollutants are removed to a certain extent. This can be carried out in existing urban areas as well as new-build.

Figure 11.2 (a) Water butt (Courtesy of Keith Herbert)

Figure 11.2 (Continued) (b) Thames Water Planter. (Courtesy of David Walters and Dejan Vernon).

RWH consists of a system to collect, filter, and store rainfall-runoff from roofs with the intention of using the stored water for non-potable applications around the house and garden. Particularly in the drier months, tanks may be fully or partly empty for some of the time and, therefore, some storage volume will be available to retain flows and provide attenuation providing water demand is greater than runoff yield. It is difficult to accurately predict the performance of individual existing systems due to the wide variety of factors in play, but evidence suggests this can be significant (Campisano et al., 2014). The design of standard RWH systems is well established (BS EN 16941-1:2018) but design to include a stormwater management function is less so. However, Kellagher (2012) has developed a method for sizing the volume of single tanks based on water use behaviour, rainfall statistics, and roof size. The applicability of the method was demonstrated in a UK case study by using the number of bedrooms and average property occupancy as a proxy for water use.

An alternative approach partitions the rainwater storage tank as shown in Figure 11.3. The retention and throttle concept includes dedicated storage for retaining runoff and limiting outflow while protecting the volume required for non-potable water supply. The throttle can

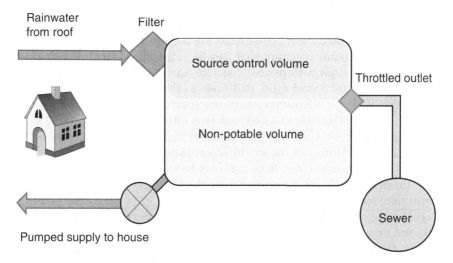

Figure 11.3 Dual-purpose rainwater harvesting system.

be either passive (e.g., an orifice or vortex regulator) or active (e.g., powered and controlled). Melville-Shreeve et al. (2016a) compared a traditional RWH system with a dual-purpose RWH system when both were combined with that of a downstream storage tank for runoff control from a new development. While the total storage volume for the latter system was found to be greater, its overall cost was lower due to cheaper RWH tanks. In a small-scale UK pilot study of a dual-use system, results clearly showed the interaction between water use within buildings and the stormwater management function, and specifically the importance of the former to achieve the latter (Quinn et al., 2020). Ricardo (2020) argues that RWH systems suitable for UK houses could be developed that provide 95% of the user's non-potable water demand whilst also maintaining sufficient attenuation capacity to control stormwater runoff during the 1 in 100 year design storm.

There is a number of innovations in this area and many different RWH system configurations are under test or coming onto the market (Melville-Shreeve et al., 2016b). In particular, enhancements to system performance can potentially be made by adding sensors and automatic control valves to the tank system equipment (so-called "*smart butts*"). Indeed, there is potential for widespread deployment under central control with the goal of reducing flooding and overflow frequencies across the urban catchment (Oberascher et al., 2021; Wei et al., 2021).

11.2.1.4 Paved area ponding

In principle, similar benefits to rooftop ponding (i.e., blue roofs) can be gained by exploiting the storage available on paved areas. Potential sites include car parks, paved storage yards, and other large impervious surfaces (also see Chapter 12 on surface water flooding). The advantage over roof storage is that much larger surfaces are available and ponding depth can be greater. Disadvantages relate particularly to the nuisance value of ponded water and, in extreme cases, possible damage and safety issues. In addition, unless the system is properly maintained, it will not function. Methods to mobilise ground-level storage usually involve restricting flow into the sewer system via gullies with orifices or vortex regulators (see Chapter 8).

11.2.2 Infiltration components

The two most common infiltration components are soakaways and infiltration trenches. A *soakaway* is an underground structure that can be stone filled, formed with plastic mesh boxes, dry wall lined, or built with precast concrete ring units (see Figure 11.4). It is recommended that any filling has a void ratio, e (defined as the ratio of interstitial volume to soil volume), of at least 30%. An *infiltration trench* is a linear excavation lined with a filter fabric, backfilled with stone, and possibly covered with grass. Runoff is diverted to the soakaway or trench and either infiltrates the soil or evaporates (see Figure 11.5). The component provides storage and enhances the ability of the soil to accept water by creating a surface area of contact. Soakaways and infiltration trenches should not be located within 5 m of the foundations of buildings, or under roads. As a result of their shape characteristics, trenches are usually more efficient than soakaways at controlling runoff.

Soakaways and trenches can be used in any area that has pervious subsoils such as gravel, sand, chalk, and fissured rock. Soil types across the UK have been mapped in detail and can be accessed via the LandIS database (www.landis.co.uk/soilscapes). From this, it is possible to identify areas with permeable and slowly permeable soils suitable for infiltration SuDS. These systems are only suitable in areas where the water table is low enough to allow a free

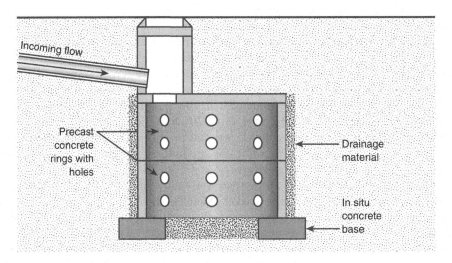

Figure 11.4 Soakaway.

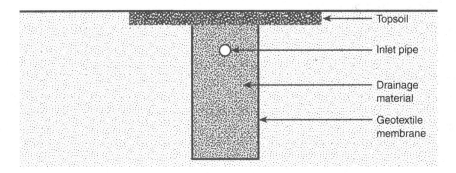

Figure 11.5 Infiltration trench.

flow of stormwater into the subsoil at all times of the year. The base of the soakaway or trench should therefore be at least 1 m above the groundwater level. Trenches installed on land gradients greater than about 4% need "flow checks" at regular intervals.

In addition to disposing of the stormwater, soakaways and trenches can reduce the concentration of some of the pollutants it contains, by processes of physical filtration, absorption, and biochemical activity. Mean annual removal efficiencies for suspended solids, metals, polyaromatic hydrocarbon (PAH), oil, and chemical oxygen demand (COD) of 60–85% have been recorded for infiltration trenches draining highway runoff (Colwill et al., 1984).

A potential problem is clogging by sediments carried with the stormwater. Siriwardene et al. (2007) have studied the processes involved and have shown that even very fine sediments can cause this effect.

There are some restrictions on using infiltration SuDS where there is a risk of pollution to groundwater resources (see Section 11.6.5), in source protection zones (SPZs).

11.2.3 Vegetated surfaces

The main types of vegetated surfaces used in stormwater management are filter strips, grassed swales and bioretention systems. *Swales* (also known as *bioswales*) are grass-lined channels used for the conveyance, storage, infiltration, and treatment of stormwater (see Figure 11.6). Runoff enters directly from adjoining buildings or other impermeable surfaces. The runoff is stored either until infiltration takes, or until the filtered runoff is conveyed elsewhere, to the sewer system, for example. Vegetated bioswales include taller growing plants and shrubs. *Filter strips*, also known as *vegetative buffer strips*, are gently sloping areas of ground designed to promote sheet flow of stormwater runoff.

Figure 11.6 Swale. (Courtesy of Prof. Chris Jefferies.)

To function well, swales require shallow slopes (< 5%) or the introduction of check dams and soils that drain well. Typically, they have side slopes of no greater than one in three, allowing them to be easily maintained by grass-cutting machinery. The bottom width is usually between 0.5 m and 2 m, they are 0.25–2 m deep, and can be readily incorporated into landscape features. Filter strips should allow a minimum flow distance of about 6 m. Swales and strips delay stormwater runoff peaks and provide a reduction in runoff volume due to infiltration and evapo-transpiration. Typical velocities should be below 0.3 m/s to encourage settlement.

They are often used as a pre-treatment in combination with other control measures. Pollutants are removed by sedimentation, filtration through grass and adsorption onto it, and infiltration into the soil. Runoff quality can be considerably improved, and Ellis (1992) found that a swale of length 30–60 m could retain 60–70% of solids and 30–40% of metals, hydrocarbons, and bacteria. Controlled experiments on grass swales in Australia (Fletcher et al., 2002) have demonstrated reductions in total suspended solids *concentrations* of between 73% and 94% and in total suspended solids *loads* of between 57% and 88%, together with significant reductions in total nitrogen and total phosphorus. Revitt et al. (2017) found that infiltration from swales had negligible impact on groundwater quality if properly maintained.

Bioretention systems (and *rain gardens*) are shallow landscaped depressions that typically consist of a filter medium (usually sandy), underlain by a gravel drainage layer, and may be lined. Their role is to intercept runoff from frequent storm events and treat stormwater by mechanical filtration, sedimentation, adsorption, and plant and microbial uptake (Hatt et al., 2009). De-Ville et al. (2021) have shown how the filter media can be replaced with 100% recycled waste components which have a very different particle size distribution without any loss in hydrological performance.

Implementation of vegetative SuDS components may give rise to local leaf fall with the potential to block stormwater inlets and/or gullies. This, in turn, may lead to exceedance flows (see Section 11.3.5). Maintenance issues are discussed later in the chapter (Section 11.6.2).

11.2.4 Pervious pavements

Pervious pavements or *permeable pavements* are used mostly for car parks (Figure 11.7) but can also be used for other surfaces where there is no traffic or very light traffic. A typical arrangement for the pavement structure is illustrated as a vertical section in Figure 11.8. There are a number of alternatives for the surface layer. It could consist of one of a variety of types of blocks or could be a layer of porous asphalt. Blocks may be *porous*, allowing water to seep through them via pores in the material, or *permeable*, where the material is not porous but the blocks are laid in such a way that water can pass *between* them. Permeable blocks may fit tightly, with water passing through narrow slots between blocks (Figure 11.9), or may be laid with a pattern of larger voids, which are filled with soil and grass, or gravel.

Below the surface layer of blocks is a bedding layer of sand or small-size gravel, separated from the sub-base below by a layer of geotextile. (The bedding layer is not necessary with porous asphalt.) The sub-base consists of crushed rock or other sufficiently hard material, or a plastic mesh structure.

At the lower surface of the sub-base is a pervious geotextile, if infiltration to the ground is intended, or an impervious geomembrane if it is not. If there is no infiltration, the pavement structure is a *tanked system*, providing considerable attenuation to the storm flow but still requiring arrangements for outflow.

The storage potential of a pervious car park structure can also be exploited by receiving further runoff from roof surfaces or other impermeable surfaces. There is also the potential for runoff stored in the sub-base to be used for applications such as toilet flushing and

Figure 11.7 Car park with pervious pavement. (Courtesy of Formpave Limited.)

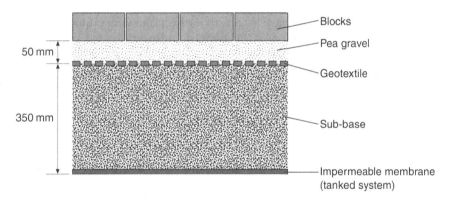

Figure 11.8 Typical pervious pavement structure, vertical section.

garden/landscape watering. Antunes et al. (2016) confirm the great potential for reuse but provide caution on the quality of the water produced.

Infiltration rates through permeable pavement surfaces, especially new ones, are generally high. For new surfaces, rates are commonly in excess of 1000 mm/h, and sometimes considerably higher, depending on the type of surface. Studies from a number of countries reported by Pratt et al. (2002) indicate that the infiltration rate of a pavement surface may reduce over its life to 10% of its original (new) value. To limit blockage at the surface, it is recommended (Woods-Ballard et al., 2016) that the surface should be cleaned three times a year. However, Rommel et al. (2001) showed that other parts of the system, especially the geotextile, were more likely than the surface to limit infiltration through the structure as a whole.

Monitored sites have consistently demonstrated substantial reduction and delay in storm peaks and reduction in overall volumes. The sites have included tanked systems (Abbott et al., 2003; Schlüter and Jefferies, 2001) and systems with infiltration to the ground (Macdonald and Jefferies, 2001).

Figure 11.9 Permeable surface blocks. (Courtesy of Formpave Limited.)

Pervious pavements may also have a positive effect on water quality by providing mechanisms that encourage filtration, sedimentation, adsorption, and chemical/biological treatment, as well as storage (Pratt et al., 2002). Laboratory and field tests have demonstrated excellent performance in the retention and degradation of oils (Pratt et al., 1999). Effluent pollutant levels have been shown to be significantly correlated with the type of sub-base used (Sañudo-Fontaneda et al., 2014).

11.2.5 Filter drains

Filter drains are linear components consisting of a perforated or porous pipe in a trench of filter material. They have traditionally been constructed beside roads to intercept and convey runoff, but they can be used simply as conveyance components. They may or may not allow infiltration to the ground, in the same way as pervious pavements.

11.2.6 Infiltration basins

Infiltration basins are depressions with vegetative cover that store runoff for infiltration into the ground. They are used where there is sufficient capacity for infiltration and where infiltration is appropriate. The bottom of the basin is flat to provide uniform infiltration. The side slopes should be no steeper than one in four. They are generally used for relatively small catchments.

11.2.7 Detention basins

Detention basins provide storage for stormwater with controlled outflow to the next stage of stormwater management or to a watercourse. They are effectively storage facilities formed out of the landscape. They are not intended to encourage infiltration to the ground and may be lined if infiltration is to be prevented completely. After controlled outflow has taken place the basin is commonly left dry until the next rainfall. In many cases, detention basins are hardly noticed by the public as a consequence of thoughtful landscaping and choice of vegetation. They are often multifunctional, also operating as recreation areas and only filling during storms. Detention basins may be online or offline. They have relatively low pollutant removal efficiencies because of the re-erosion of previously deposited solids during filling.

11.2.8 Ponds

Ponds provide storage and treatment for stormwater within a permanent volume of water (Figure 11.10). They have aesthetic, recreational value (e.g., sailing, fishing) and environmental benefits such as returning wildlife habitats into urban areas, in addition to their flood control function. They can also play a significant role in pollution control since sedimentation and biological processes may enhance the water quality of the outflow. This is because many pollutants are attached to suspended solids, which are themselves captured by sedimentation. The depth of the pool is usually limited to 1.5–3 m to avoid thermal stratification. Shallow side slopes and well-chosen marginal vegetation help ensure safety. McLean et al. (2005) give practical guidance on ponds and detention basins.

11.2.9 Constructed wetlands

Constructed or *artificial wetlands* (including *reed beds*, *reed marshes*, and *vegetative systems*) are shallow areas of excavated land filled with earth, rock, or gravel, saturated with water or covered by shallow flowing water at some time during the growing season, and planted with selected aquatic plants. The key roles of the plants are to transmit oxygen from the atmosphere to the root system (thus the soil) and to encourage microbial growth. Wetlands require relatively large areas of flat to gently sloping (less than about 5%) land.

It is generally agreed that wetlands are simple and inexpensive to build, but there is some disagreement on their ease of operation. In order to remain effective, wetlands need long-term programmes of maintenance, involving the planting and extraction of the aquatic plants. It has been estimated that a wetland has a life of about 15–20 years.

In addition to substantially attenuating and reducing runoff flows, wetlands can significantly improve water quality by removing large quantities of particulate and soluble contaminants (including suspended solids, metals, excess nutrients, and bacteria) through biological action and sedimentation. Wetlands also trap silt and promote recovery of DO in the effluent. Results show that wetlands can, on an annual basis, remove almost all bacteria and suspended solids, and just under half the total phosphorus and nitrogen load (Ellis, 1992). They also have valuable secondary uses, in providing wildlife habitats and recreational/educational areas. However, wetlands are sensitive and must be carefully managed to avoid vegetation die-off.

There should be a mixture of water depths. Typically, depths over 1 m should cover no more than 20% of the area, and depths of under 0.5 m should cover more than 50%. There

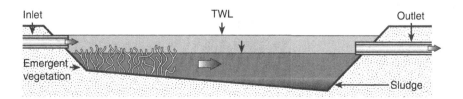

Figure 11.10 Simplified wet detention pond.

should also be continuous baseflow, but velocities should be low (less than 0.1 m/s). A sediment removal forebay is normally included. Further details are given in Scholz (2006).

11.2.10 Oil separators

Oil separators are underground chambers with compartments designed to separate oil and water, and retain the oil fraction. They are usually provided for small catchments, particularly where heavy oil or petrol spills are expected (e.g., petrol stations). Outflow is routed to the drainage system.

In most types of interceptors, the flow rate of the effluent is reduced, and the lighter oil fraction is separated by gravitational means. Interceptors suitable for treating stormwater should also contain a storage section for silt and suspended matter. The captured oil and sediment must be regularly removed as, otherwise, a heavy storm event may cause oil accumulated from previous storms to be discharged to the receiving water. Current designs include more efficient tilting plate separators and coalescing filters.

Information on applications of oil separators, on types and classes of separator, and on design, sizing, and operation, are given in GPP 3, one of the Guidance for Pollution Prevention documents by NIEA/SEPA (2022).

11.3 ELEMENTS OF DESIGN

The ultimate aim in design is to find the right tools for the job, used in the best combination. The result may be a system of drainage based entirely on SuDS and involving no conventional piped drainage. Alternatively, it may contain elements of source control in combination with oversized sewers or underground storage tanks, discharging heavily attenuated storm flows to a conventional drainage system. In a densely developed area, specifying and installing SuDS may well require far greater effort and imagination than relying solely on a standard piped solution.

We consider here elements in the design of SuDS components. The detailed design of a whole SuDS system is beyond the scope of this book, but sources of further guidance are given throughout this chapter.

11.3.1 Rainfall

Clearly, all SuDS design work involves some prediction of rainfall. Appropriate methods have been presented in Section 4.3. Predicting and allowing for the impact of climate change (Section 4.6) is as relevant to SuDS as it is to other methods of drainage.

11.3.2 Runoff

Methods for predicting runoff have been presented in Sections 5.2 and 5.3. Apart from requiring predictions of runoff from developed areas, SuDS design often calls for values of greenfield runoff (before development). Further guidance is given in Section 10.7.

11.3.3 Conveyance

Clearly, any pipes that form part of a SuDS system can be designed using the principles set out in Sections 7.3 and 7.4. Flow in swales and over filter strips can be treated as open channel flow, and calculations can be based on the Manning equation (Section 7.5). It is recommended that Manning's n is taken as 0.25 when the depth of flow is below the top of the grass and 0.1 when it is above. When estimating the roughness value to be used, vegetative growth should be considered, as at times, it may be short and hence increased conveyance rates and at other times considerably longer, and therefore reduced conveyance rates (and the potential to improve water quality).

11.3.4 Inlets and outlets

Inlet and outlet arrangements are important details in SuDS design. Elements of outlet design that relate to the control of outflow are covered in Sections 8.1 and 8.2. Inlet arrangements may include sediment detention, and also arrangements for flow distribution, for example, kerb details to guide runoff from a road onto a filter strip, or overflow arrangements for different levels of runoff. Details are given in *The SuDS Manual* (Woods-Ballard et al., 2016).

11.3.5 Exceedance flows

It was noted in Chapter 10 that rainfall events producing runoff flows greater than the design capacity of the pipe system (typically high intensity, short duration storms) may well cause flooding. This is equally true of SuDS systems although they typically have longer duration critical events and tend to be able to buffer flows more effectively than pipes, even when their design capacities are notionally exceeded. The topic of exceedance is covered in much more detail in Chapter 12.

11.3.6 Storage volume related to inflow and outflow

A simple method for determining storage volume by considering storms of different durations is given in Section 14.3. More detailed methods of considering the interaction between inflow, outflow, and storage are demonstrated in Sections 14.4 and 14.5.

11.3.7 Infiltration from a pervious pavement sub-base

CIRIA Report 156 (Bettess, 1996) gives the following method for *plane infiltration systems*. The formula below gives the maximum depth of water in the sub-base:

$$h_{max} = \frac{D}{n}(Ri - f) \tag{11.1}$$

h_{max} is the maximum depth (m), D is the duration of rainfall (h), n is the porosity of sub-base (volume of voids ÷ total volume), R is the ratio of drained area to infiltration area, i is the rainfall intensity (m/h), and f is the infiltration rate (m/h).

Accurate estimation of soil infiltration rate is difficult because rates depend on numerous factors, such as

- Soil particle size and grading
- The presence of organic material
- Plant, animal, and construction activity
- Soil history

In particular, considerable differences in value may occur at different times of the year and with different antecedent conditions. The best, but still not ideal, information can be obtained from undertaking on-site trials. General soil data, for use in preliminary calculations, is given in Table 11.1. Infiltration systems are not suitable in soils with infiltration rates less than 0.001 mm/h. Porosity ranges from 0.2 for graded sand to as high as 0.9 for some proprietary geocellular products.

The recommendation in CIRIA 156 is that the infiltration rate is determined from on-site tests and that a factor of safety is then applied in calculations to account for progressive siltation (Table 11.2).

Table 11.1 Typical soil infiltration rates

Soil type	Rate (mm/h)
Gravel	10–1000
Sand	0.1–100
Loam	0.01–1
Chalk	0.001–100
Clay	<0.0001

Source: After Bettess, R. 1996. *Infiltration Drainage—Manual of Good Practice*, CIRIA 156.

Table 11.2 Factor of safety applied to the measured infiltration rate

Area drained (m²)	Factor of safety, related to consequence of failure		
	No damage or inconvenience	Minor inconvenience	Major consequences
<100	1.5	2	10
100–1000	1.5	3	10
>1000	1.5	5	10

EXAMPLE 11.1

Determine the maximum depth of water in the sub-base of a pervious car park for the following storm event: rain intensity 50 mm/h, duration 1 hour. The soil infiltration rate has been measured as 15 mm/h. The area drained is 500 m² and the area of pavement is 250 m². Surface flooding would cause minor inconvenience. The porosity of the sub-base material is 0.4.

Repeat for a tanked system.

Solution

Assume a safety factor of 3. Therefore, $f = 15/3 = 5$ mm/h. From Equation (11.1),

$$h_{max} = \frac{1}{0.4}\left(2 \times 50 \times 10^{-3} - 5 \times 10^{-3}\right) = 0.238\,m$$

For a tanked system,

$$h_{max} = \frac{1 \times 2 \times 50 \times 10^{-3}}{0.4}0.250\,m$$

If there is no infiltration from the bottom of the sub-base (a tanked system), and outflow by other means can be ignored, the expression is even simpler:

$$h_{max} = \frac{DRi}{n}$$

11.3.8 Infiltration from a soakaway or infiltration trench

The main methods available for the design of non-plane infiltration systems are BRE Digest 365 (BRE, 1991) and CIRIA Report 156 (Bettess, 1996). These have some differences, which include procedures for the use of test pits to measure infiltration rates. Also, as described above, CIRIA 156 recommends factors of safety to be applied to the infiltration rate in calculations, whereas BRE Digest 365 assumes that infiltration takes place through the sides but not through the base (which gives an implicit factor of safety by assuming that the base will become clogged with fine particles). BRE Digest 365 assumes that infiltration from the system into the soil is constant and corresponds to that when the system is half full of water, whereas CIRIA 156 includes infiltration through the base and uses a more complex representation of infiltration overall. The method in BRE Digest 365 (which leads to more straightforward calculations than CIRIA 156) is presented in most detail here.

11.3.8.1 BRE Digest 365 method

Storage for runoff from the critical storm, at a given return period (e.g., 10 years), is given by

Storage = Runoff volume – Infiltration during storm   (11.2)

So, if the critical storm has duration, D (h), storage, S (m³), is given by

$S = iA_iD - fa_{50}D$ (11.3)

I is the rainfall intensity (m/h), A_i is the impervious area (m²), f is the soil infiltration rate (m/h), and a_{50} is the effective surface area for infiltration (m²).

This can be applied to an infiltration trench as follows. A trench has a width of B_d, an effective depth (i.e., depth below the invert of the incoming pipe) y, and a length L. Assuming discharge *from* the trench *to* the soil is through the sides (length and ends) only, and that the average water level is half the effective trench depth, the effective infiltration surface area is

$a_{50} = y(B_d + L)$ (11.4)

Active storage volume S in the trench is

$S = yB_dLn$ (11.5)

n is the porosity of the trench material (free volume ÷ total volume).

After the rainfall event, it is recommended that the infiltration rate should be sufficient to empty at least half the stored volume within 24 hours. The suggested design procedure is

1. Determine the soil infiltration rate
2. Adopt a component cross section
3. Determine the required storage volume by considering a range of durations of 10-year design storms
4. Review the suitability of the design and check that the component will half empty within 24 hours

A calculation using this method is presented as Example 11.2.

EXAMPLE 11.2

Design an infiltration trench to serve an individual property draining an impermeable area of 100 m² if it is filled with an aggregate, which gives a free volume that is 0.3 of the total volume. The soil infiltration rate is estimated from an on-site trial pit at 10 mm/h. Rainfall statistics for the 10-year storm are as derived in Example 4.2.

Solution

1. $f = 10$ mm/h (no factor of safety is used, but it is assumed that there is no infiltration through the base).

2. Say: Width (B_d) = 1 m
   Effective depth = 1 m
   ... determine L

3. Equation 11.3:

$$S = iA_iD - fa_{50}D$$

Substitute using Equations 11.4 and 11.5:

$$yB_dLn = iA_iD - fy(B_d + L)D$$

Sample calculation for $D = 1$ hour: $i = 24.8$ mm/h

$$1 \times 1 \times L \times 0.3 = 24.8 \times 10^{-3} \times 100 \times 1 - 10 \times 10^{-3} \times 1 \times (1 + L) \times 1$$

giving $L = 7.97$ m.

Considering the range of durations of 10-year design storms from Example 4.2:

(h)	(min)	Storm duration, D	
		Intensity, i (mm/h)	Length, L (m)
	5	112.8	3.13
	10	80.4	4.44
	15	62	5.12
	30	38.2	6.25

	Storm duration, D		
(h)	(min)	Intensity, i (mm/h)	Length, L (m)
1		24.8	7.97
2		14.9	9.23
4		8.6	9.96
6		6.1	9.93
10		4	9.64
24		2	8.27

The critical case is at 4 hours: length 9.96 m – say 10 m.

4. Time for emptying is evaluated assuming half the stored volume discharges through the effective surface area for infiltration:

$$t = \frac{0.5B_d L_n}{y(B_d + L)f} = \frac{0.5 \times 1 \times 10 \times 0.3}{1 \times (1+10) \times 10 \times 10^{-3}} = 13.6 \text{ hours}$$

which is acceptable (< 24 hours).

11.3.8.2 CIRIA 156 method

For a particular rainfall event discharging to an infiltration component, the maximum depth of water in the component is given by

$$h_{\max} = a\big(\exp(-bD) - 1\big) \tag{11.6}$$

in which

$$a = \frac{A_b}{P} - \frac{iA_D}{Pf}$$

and

$$b = \frac{Pf}{nA_b}$$

D is the storm duration (h), A_b is the area of the base (m²), P is the perimeter of infiltration component (m), i is the rainfall intensity (m/h), A_D is the impermeable area from which run-off is received (m²), and f is the infiltration rate (m/h) = measured rate ÷ factor of safety.

Time for half emptying (t in hours) is given by

$$t = \frac{nA_b}{fP} \ln \left[\frac{h_{\max} + \dfrac{A_b}{P}}{\dfrac{h_{\max}}{2} + \dfrac{A_b}{P}} \right] \tag{11.7}$$

11.3.9 Water quality

11.3.9.1 Mass balance

The basis of water quality design for SuDS techniques is essentially a matter of mass balance of pollutants, in many ways similar to hydraulic design with an additional term:

$$\text{Accumulated mass} = \text{Mass inflow} - \text{Mass outflow} \pm \text{Mass change} \qquad (11.8)$$

The mass inflow for a given time period is the product of the flow volume and the mean pollutant concentration. The mass change is a function of the chemical, biological, or physical changes that take place within the component. Change can be both positive and negative, and it is possible that, for a given storm event, there is a net *export* of pollutants due to the release of pollutants in some way bound within the component (e.g., in sediments or biomass).

The difficulties in the design become apparent when the complexities of urban rainfall and runoff are compounded with the great number of pollutant types and concentrations, and with the many reactions that can take place. In addition, the pollutant removal rates of components vary considerably.

For practical purposes, it can be assumed that the percentage of pollutant mass captured is proportional to the percentage of runoff flow volume captured (i.e., diverted away from the piped system). Capture of about 10–15 mm of effective rainfall (i.e., runoff) should capture in the range of 80–90% of the annual effective rainfall and thus that percentage of the pollution. The reason is that the majority of the storms in any given year only produce a small amount of runoff. Therefore, unlike stormwater quantity management, where infrequent major storms are of most concern, in water quality control it is total *volumes* that are of most significance.

11.3.9.2 Treatment volume

A feature of storage is that it encourages the settlement of some of the suspended sediment and associated pollutants contained in the incoming stormwater. In practice, removal efficiencies vary considerably, but well-designed ponds have the capacity to make reductions in many pollutants.

For significant treatment to take place, the captured runoff must be retained for at least 24 hours. CIRIA (2000) suggest the following expression to estimate the required treatment volume, based on data obtained from the Wallingford Procedure:

$$V_t = 9(M5 - 60\,\text{min})(SOIL\,/\,2 + PIMP(1 - SOIL\,/\,2)\,/\,100) \qquad (11.9)$$

V_t is the basic treatment volume (m³/ha), $SOIL$ is the soil index for the UK based on the Winter rain acceptance potential (WRAP) classification, $M5 - 60$ min is a 5-year return period, 60-minute duration (standard) rainfall depth, and $PIMP$ is the percentage imperviousness of the catchment served.

Dry detention ponds are designed to fully contain the treatment volume V_t, and to drain down within 24 hours. Wet retention ponds require a volume of $4\ V_t$ to allow time for biological treatment in addition to physical processes.

Other important factors for optimum pollutant reduction (Ellis, 1992) are that

- The ratio of pond volume to mean storm runoff volume is between 4 and 6.
- The storage volume exceeds 100 m³ per hectare of effective contributing area.
- The ratio of pond surface area to effective contributing area is 3–5%.

Reliance on a single pond (or tank) is not regarded as best practice in providing water quality protection for receiving waters. Rather, the drainage of a site should be designed using the management train concept described in Section 11.4.1, delivered using multiple storage and evaporative (e.g., vegetated) components.

11.3.9.3 Nutrient neutrality

Nutrients such as nitrogen and phosphorus are often present in runoff and can result in the eutrophication of water environments (see Sections 2.3.5 and 2.3.6). Wastewater discharges from new housing developments discharged at sites in *"unfavourable condition"* are currently controlled in England (Natural England, 2023). This takes the form of provision of suitable onsite or offsite mitigation measures to avoid any additional damage. Development plans can be considered *"nutrient neutral"* where they can demonstrate that they will cause no overall increase in nutrient pollution affecting specified habitat sites. One approach to achieving this is through the appropriate deployment of SuDS treatment trains (see Section 11.4.1). CIRIA have produced good practice guidance on nutrient minimisation for nitrogen (Bradley, 2023) and phosphorus (Bradley et al., 2022).

11.3.10 Amenity

Amenity is defined in *The SuDS Manual* (Woods-Ballard et al., 2016) as "a useful or pleasant facility or service." Many SuDS facilities can, with careful design, detailing, and ongoing maintenance contribute towards providing amenity benefits in many ways, including environmental quality, health and well-being, education opportunities, recreation, visual appeal, and more. Each of these can, in principle, be expressed as design criteria. That said, because of its rather broad definition and appeal to personal preference, amenity is probably the most difficult benefit to crystallise in practice. More detailed advice and examples are given in *The Manual*.

11.3.11 Biodiversity

Biodiversity, while linked to amenity, is a separate concept and refers to "the number and variety of different organisms found within a particular region." This is expressed locally as the character of animals and plants that share the same living space as humans. Design objectives include supporting and enhancing local habitats, contributing to habitat connectivity, and providing diversity of local ecosystems (Woods-Ballard et al., 2016). Biodiversity benefits are closely linked to the enhancement of *ecosystem services* and *natural capital*, and so contribute to improved sustainability; although, a consistent, positive link to amenity benefit is less well established. New developments in England have a mandatory *biodiversity net gain* obligation of 10%, which should result in more or better quality natural habitat than there was prior to the development (Schedule 7A of the Town and Country Planning Act). SuDS is one way of achieving or contributing to this uplift. Biodiversity can also provide a useful pathway towards community engagement activities and awareness-raising (see Section 11.5).

11.3.12 Multiple benefits

In addition to the core benefits of flood risk management, water quality management, biodiversity, and amenity, SuDS can also deliver a myriad of additional benefits to a greater or lesser extent (Susdrain, 2023):

- Air quality improvement
- Building resilience
- Health and well-being
- Pumping reduction

- Carbon reduction and sequestration
- Crime reduction
- Economic growth
- Educational enhancement
- Enabling development
- Groundwater recharge

- Recreation
- Temperature moderation
- Tourism
- Traffic calming
- Water reuse

Figure 11.11 shows a detailed mapping of the benefits accruing from a range of BGI/SuDS components and their impact pathway to beneficiaries and capitals improved.

CIRIA have developed a tool (ciriabest) and comprehensive guidance to evaluate, quantify, and compare the multiple benefits that SuDS can create (Horton et al., 2019). The online, spatial, cost-evaluation tool is commercially available and can be accessed here: www.ciriabest.com.

11.3.13 Modelling

The main applications for models of drainage systems are in *design* and *simulation* and these are described and explained in Chapter 20 (Modelling in Practice). That chapter includes conventional sewered catchments, sewered systems that include SuDS and drainage systems based on SuDS only.

11.4 SUDS APPLICATIONS

11.4.1 Management train

The use of SuDS components in combination is important to maximise their overall performance, termed a *management train* or also referred to as a *treatment train*. The recommended sequence of possibilities, with components appropriate for each stage, is given in Figure 11.12. The basic principle is that different elements can be combined to reduce runoff at source, channel runoff between SuDS elements, allow infiltration and storage, and progressively improve water quality to the point where it can be discharged into a watercourse. It is preferable to use drainage solutions as close to the top of this diagram as possible, but if all drainage needs cannot be achieved at a particular stage, the designer must move further down the list.

Maqbool and Wood (2022) suggest a key advantage is that successive SuDS components can progressively improve water quality and Huang et al. (2020) argue that a more diverse range of co-benefits can be accrued. Bastien et al. (2010) concluded that the use of a treatment train approach provides a more flexible solution than individual SuDs or end-of-pipe solutions taking into account cost and performance objectives. Flexibility and adaptability are key elements of resilience (see Chapter 24).

Jefferies et al. (2008) describe a tool and scoring system for providing guidance on the appropriate level and combination of SuDS depending on the type and scale of a development and the nature of the receiving water. Intended to provide greater consistency in meeting regulatory requirements for SuDS, the tool balances the risks of pollution to the receiving water with the treatment provided in the management train.

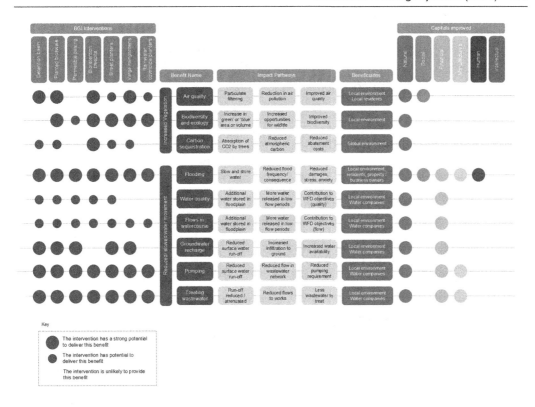

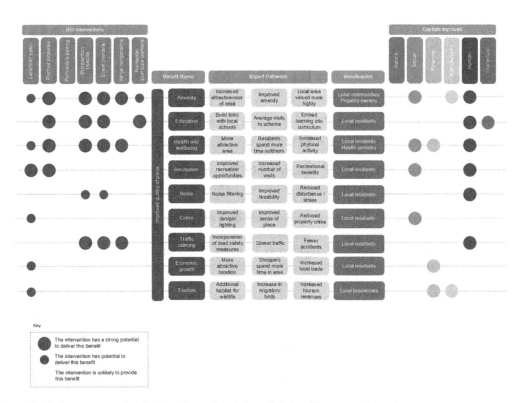

Figure 11.11 Assessment of wider benefits and their beneficiaries. (Courtesy of Arup)

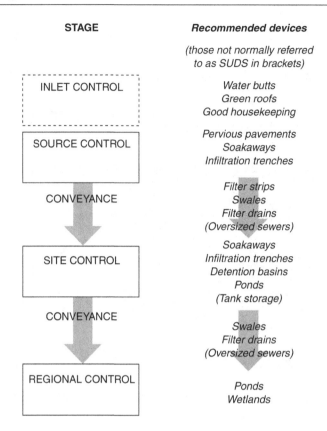

STAGE

INLET CONTROL

SOURCE CONTROL

CONVEYANCE

SITE CONTROL

CONVEYANCE

REGIONAL CONTROL

Recommended devices

*(those not normally referred
to as SUDS in brackets)*

*Water butts
Green roofs
Good housekeeping*

*Pervious pavements
Soakaways
Infiltration trenches*

*Filter strips
Swales
Filter drains
(Oversized sewers)*

*Soakaways
Infiltration trenches
Detention basins
Ponds
(Tank storage)*

*Swales
Filter drains
(Oversized sewers)*

*Ponds
Wetlands*

Figure 11.12 SuDS management train.

11.4.2 Retrofit

In existing sewered catchments, there are also benefits in *retrofitting* SuDS components. Retrofitting can be "needs-driven" in response to a particular driver (e.g., flooding) and delivered through a defined scheme. It can also be opportunistic as part of a regeneration or urban improvement scheme. Retrofit SuDS can also be cheaper than traditional solutions (11.6.4), and nearly always provide more benefits (11.3.12). There can also be specific challenges concerning space requirements, ownership, and ongoing maintenance (see Section 11.6: "Issues"). Dolowitz et al. (2018) posed the question "Retrofitting urban drainage infrastructure: grey or green?" and found different post-industrial cities had answered it differently depending on local context. Influencing factors included access to capital investment, institutional flexibility, local leadership, regulatory frameworks, and social context. Public ownership of the drainage infrastructure appeared to enable green retrofit (i.e., SuDS) more readily.

Stovin et al. (2013) presented a specific example of the viability of SuDS retrofit in the context of addressing CSO discharges in London (which was not implemented). However, things moved on and since then progress has been made with a map of implemented retrofit SuDS in London, now available at: apps.london.gov.uk/suds.

Susdrain (www.susdrain.org) contains numerous examples of retrofit solutions and schemes, and CIRIA's guidance (Digman et al., 2012) *Retrofitting to Manage Surface Water* provides practical steps to support the retrofitting of SuDS in urban areas.

11.5 PUBLIC ATTITUDES AND COMMUNITY ENGAGEMENT

SuDS are far more obviously part of the urban landscape than underground pipe systems, and therefore, to be successful, they must be welcomed by the public. Amenity benefits are only genuine if they are perceived as such by the public. A survey in Scotland of public attitudes to SuDS (Apostolaki et al., 2001) found that public opinion varied according to the type of component, with more positive attitudes towards ponds than swales, for example. The only real concerns over ponds were over safety aspects. A poor understanding of the purpose of swales was thought to contribute to the negative attitudes. In a study of six English housing developments, residents appreciated the wildlife and green space but had concerns over pests and litter (Williams et al., 2019).

Public education and awareness raising have the potential to increase the popularity and acceptance of SuDS. This can be facilitated pre- and post-installation through a variety of means including clear and visually attractive information about local SuDS systems in the form of leaflets, information boards, and signs. Schools and their teachers have an important role to play here (Ward et al., 2023). This can help local people understand the purpose and function of SuDS, appreciate the services they can provide, and so engage in good behaviour around installations. However, for even greater benefit, consideration should be given to involving the community in the co-design of installations so they can offer informed opinion and local experience, and options can be tailored, where possible, towards local preferences (Everett and Lamond, 2016).

11.6 ISSUES

We consider here some of the wider issues that are involved when SuDS are considered as drainage options. Some of these issues are described as "barriers," and the growth in the use of SuDS has been associated with their progressive removal. Most commentators point to the importance of partnerships between developers, local planning authorities, and environmental agencies, potentially also including consulting engineers, landscape architects, water service providers, highway authorities, third-sector organisations, residents, and the wider community.

11.6.1 Long-term performance

Concerns have been expressed about the performance of SuDS components over the long term and reported experiences have been mixed. Jefferies (2004), for example, collated the results of monitoring of a wide range of SuDS sites in Scotland including ponds, pervious pavements, detention basins, filter drains, and swales. All systems were found to operate effectively in terms of flow attenuation, with systems closest to the source of runoff having the most impact. No evidence was found to suggest that the systems studied would not continue to operate as designed, provided they continued to be maintained properly. However, Bergman et al. (2011) examined the performance of two stormwater infiltration trenches installed in the late 1990s in Copenhagen and found that the life span of the infiltration trenches was much shorter than expected due to siltation and clogging. Similar concerns were expressed by Achleitner et al. (2011) on the hydraulic permeability of an infiltration and swale system.

This evidence points to the conclusion that SuDS are not "fit and forget" systems, and there is an ongoing need for appropriate maintenance as discussed next.

11.6.2 Maintenance

Accepting responsibility to maintain SuDS components in perpetuity is seen by some as a risk because of a lack of data on long-term performance. Possible deterioration in performance, perhaps due to blockage by silt, makes others question the appropriateness of the word *sustainable* for such components. However, all infrastructures require maintenance and the obvious fact that SuDS also need to be maintained does not necessarily put them at a disadvantage. The key is understanding what maintenance is required, how to carry it out and at what frequency. The maintenance needs of both piped systems and SuDS are described in Chapter 18.

11.6.3 Adoption

As described in Section 6.4.2, if a developer constructs a piped drainage system, it is normal for the system to be "adopted" by the sewerage undertaker who will then include the system as a revenue-earning asset and accept responsibility for operation and maintenance.

It is not clear that the same procedure is suitable for all SuDS components because many could be considered as public space landscape features (rather than drainage features) and therefore more appropriately the responsibility of the local authority. The Flood and Water Management Act 2010 proposed a framework for the mandatory approval and adoption of drainage systems, an approving body, and national standards on the design, construction, operation, and maintenance of SuDS (so-called "Schedule 3"). This was subsequently implemented in Wales. In England, the government addressed the issue through planning policy from 2015. However, a review (Defra, 2023) concluded that this approach had proven to be ineffective and that Schedule 3 should instead be implemented with approval and adoption being the responsibility of the unitary authority or county council on the basis of statutory standards (see Sections 11.7.2 and 11.7.3).

11.6.4 Costs

It is inevitable that cost comparisons are made between piped drainage solutions and SuDS solutions. Cumulative evidence suggests that for many new developments, it is cost-beneficial to install a SuDS scheme in preference to a traditional drainage system. In fact, SuDS can be *significantly* cheaper to construct depending on the site conditions (30% according to Defra, 2011) and maintenance costs are also typically lower. Conditions that favour the installation of SuDS include (Royal Haskoning DHV, 2012; Environmental Policy Consulting, 2017):

- Larger sites
- Flatter sites
- Pervious ground conditions
- Lower density development

Retrofit costs are often higher than new-build solutions, but this is predominantly because they will often seek to provide a wider range of benefits. It is becoming more common for these benefits along with the costs to be considered in making investment decisions. The ciria-best tool (see Section 11.3.12) helps to quantify and monetise these benefits taking account of scale, location, and timing. *The SuDS Manual* recommends and presents a whole life costing approach, and provides some typical unit costs. The issue of costs is returned to in more detail in Chapter 19 (Planning).

11.6.5 Groundwater pollution

The risk of groundwater pollution from infiltrated runoff is rightly a potential concern. The Environment Agency (2018) gives detailed guidance in terms of the required aquifer protection. There is generally no objection to infiltration of roof drainage, but, depending on groundwater vulnerability, an oil interceptor may be required for other impermeable areas, or infiltration simply may not be allowed. However, Ellis (2006) cautions that there is contradictory evidence as to whether infiltration systems pose a long-term threat to groundwater quality.

11.6.6 Failures

SuDS do not always function as designed and failures can occur. Figure 11.13 gives a summary of some common examples of this. Vollaers et al. (2021) found three common issues contributing to these failures. SuDS

- make use of different technologies compared with conventional pipe-based solutions and hence require different knowledge and skillsets
- require crossing of conventional system boundaries and there is only limited knowledge of the interfaces between previously unconnected systems
- require the involvement of more decision-makers and stakeholders adding to relational complexity.

The authors make the convincing point that the key to preventing such failures in the future is to anticipate what malfunctions might occur and how to avoid them. This can only be achieved through experience built on thorough investigation and documentation of any failures that do occur in practice

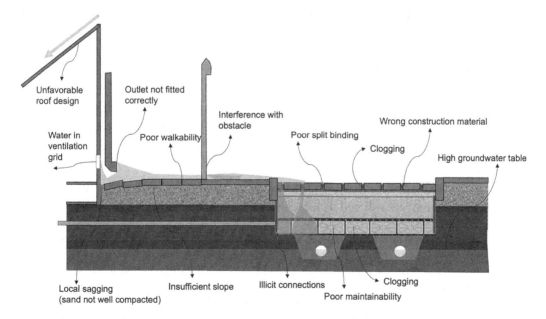

Figure 11.13 Common failures in SuDS. (Reproduced from Vollaers et al. 2021. Root causes of failures in sustainable urban drainage systems (SUDS): an exploratory study in Section 11 municipalities in the Netherlands. *Blue-Green Systems*, 3(1), 31, under a CC BY 4.0 DEED. Attribution 4.0 International Licence.)

11.7 SITE RUNOFF CONTROL

11.7.1 Design principles

The fundamental principle applied to control runoff from any development site is that the hydrological response of the developed site should, as far as possible, replicate the undeveloped state. To achieve this, or at least come close to it, Kellagher (2013) proposed adherence to three key criteria:

- Interception – no runoff from the site for up to 5mm of rainfall
- 1-year return period *flow* control – aimed at morphological and ecological protection of receiving waters
- 100-year return period *flow* and *volume* control – aimed at flood protection of those living downstream

In addition to design for peak flows, as commonly used in pipe sizing (Chapter 10), these principles contain volumetric requirements aimed at better replicating runoff from greenfield sites.

11.7.1.1 Interception storage (small rainfall events)

Kellagher (2013) states that approximately 50% of all rainfall events are less than 5mm in depth and cause no measurable runoff from greenfield areas. In contrast, runoff from a development takes place during virtually every rainfall event resulting in receiving waters receiving frequent discharges with polluted washoff from urban surfaces. Replication of the greenfield runoff from small events will result in far fewer polluted discharges, thereby limiting the impact on the receiving environment. Certain SuDS components or combinations of components (e.g., RWH, swales, porous pavements) are able to partially or fully achieve this limit. Further details of component performance in this respect are given in *The SuDS Manual* (Woods-Ballard et al., 2016).

11.7.1.2 Long-term storage (extreme rainfall events)

The total volume of runoff from a developed site subject to extreme rainfall events is typically between 2 and 10 times the runoff volume from the same site in its greenfield state (Kellagher, 2013). It is important to control this additional volume from the developed site because

- a large proportion of runoff tends to be released much more quickly than during greenfield runoff
- even if this volume was released at the peak rate of the greenfield runoff, flood depths and extents in rivers will still be increased

In theory, the 100-year flood runoff from a site should therefore be controlled to both greenfield volume and rates of runoff to ensure the same conditions occur downstream after development. In practice, this is very difficult to achieve for all rainfall events but should still remain the objective.

A simple approach to approximating this is to use the 100-year, 6-hour event to estimate runoff volumes before and after development, with the difference captured using *long-term storage* and, ideally, preventing it being discharged. In principle, this volume could be infiltrated (or reused) within the site boundary but in practice this is also difficult to achieve. If this is the case, the long-term storage volume needs to be specifically designed for and discharged at a maximum rate of 2 L/s.ha.

11.7.1.3 Attenuation storage (flow rate limitation)

Development runoff through traditional pipe networks will discharge into receiving waters at much greater flow rates and velocities than the undeveloped site. For significant rainfall events, this could cause scour and erosion seriously affecting the morphology and ecology of the receiving water channel, and potentially exacerbate downstream flooding. Both of these aspects are addressed by applying a discharge limit from the developed site, thus requiring attenuation storage (discussed in further detail in Chapter 14).

The design principle is to match the peak runoff flow rate for events of similar frequency of occurrence from the developed site to that of the undeveloped site. Achieving an exact replication would be difficult, time-consuming, and probably unjustified, given the uncertainties involved in greenfield site runoff estimation. Kellagher (2013) therefore recommended an equivalence rule is applied (as a matter of expediency) for both the 1-year event and the 100-year event using the critical duration storm to find the maximum attenuation storage volume.

In practice, there are two conditions where the greenfield flow rate would not be applied to define the limiting discharge rates:

- If discharge limits based on calculated greenfield runoff $QBAR_{rural}$ (see Section 10.7) are lower than 1 L/s.ha, use 1 L/s.ha instead. This has the effect of limiting excessive storage volumes.
- If storage outflow rates are calculated to be less than 5 L/s, use 5 L/s instead. This has the effect of minimising blockage problems of discharge control devices (see Section 8.1)

11.7.1.4 Treatment volume

The concept of treatment volume has been discussed in Section 11.3.9.2. Kellagher (2013) recommended the amount of storage provided should equal or exceed the equivalent volume of runoff of 15 mm of rainfall. This does not have to be provided as a single tank or pond at the site outfall but could be distributed throughout the site.

11.7.2 Non-Statutory Technical Standards (NSTS)

The site runoff control standards for SuDS applicable in England are summarised in Table 11.3 (Defra, 2015). These follow the principles described above in so far as they make a distinction between allowable peak runoff rates (Q_p) and volumes (V) with calculations required for the 1-year and 100-year return period rainfall events for the former and the 1-year return period, 6-hour rainfall event and 100-year return period, 6-hour event for the latter. In addition, different criteria apply for greenfield (undeveloped) and brownfield (previously developed) sites. In the brownfield case, peak flows and volumes should be less than or equal to the equivalent greenfield values where reasonably practicable. Otherwise, they should be less than the current discharge rates from the site; so-called "betterment." *Missing* is any specification for site interception or treatment volumes.

Table 11.3 Summary of SuDS Non-statutory technical standards – England

Control standard	Clause	Greenfield (Q_g, V_g)	Clause	Brownfield (Q_b, V_b)
Peak runoff rate (Q_p) • 1:1 yr & 1:100 yr rainfall events	S2	$Q_p \leq Q_g$	S3	$Q_p \leq Q_g^*$ $Q_p \leq Q_b^{**}$
Runoff volume (V) • For 1:1 yr, 6 hr & 1:100 yr, 6 hr rainfall events	S4	$V \leq V_g$	S5	$V \leq V_g^*$ $V \leq V_b^{**}$
	S6	If not reasonably practicable to constrain V (S4 & S5), it must be discharged at a rate Q that does not adversely affect flood risk.		

* Where reasonably practicable.
** Otherwise (betterment)

Source: Adapted from Defra. 2015. *Sustainable Drainage Systems. Non-statutory Technical Standards for Sustainable Drainage Systems*, Department for Environment, Food and Rural Affairs.

A number of issues have arisen since these standards were published, and they include the following:

- Their non-binding status
- Lack of definition of what "reasonably practicable" means in practice
- Lack of definition of what degree of betterment should be achieved
- Lack of full uptake in practice
- Lack of standards for water quality and other benefits.

The content and uptake of these and other relevant standards have been reviewed with recommendations made for improvement and rationalisation. These will form the basis for revised statutory standards (Defra, 2021, UKWIR, 2021) as discussed in Section 11.6.3.

11.7.3 Statutory standards

In contrast to the situation in England, the Welsh Government (2018) has published a full set of statutory standards for SuDS which covers design construction, operation and maintenance (Table 11.4). The standard is notable for, and different to the NSTS, in including an initial hierarchy standard (S1) which addresses the use of surface water by and within the development and where it should be discharged. Equally notable are the standards that address water quality (S3), amenity (S4), and biodiversity (S5). Further detail is given in Welsh Government (2018), which also makes extensive reference to *The SuDS Manual* (Woods-Ballard et al., 2016).

Table 11.4 Summary of SuDS statutory standards – Wales

Standard		Aim
S1	Surface water runoff destination	To ensure that runoff is treated as a resource and managed in a way that minimises the negative impact of the development on flood risk, the morphology and water quality of receiving waters, and the associated ecology.
S2	Surface water runoff hydraulic control	To manage the surface water runoff from and on a site to protect people on the site from flooding from the drainage system for events up to a suitable return period, to mitigate any increased flood risk to people and property downstream of the site as a result of the development, and to protect the receiving waterbody from morphological damage.

(Continued)

Table 11.4 (Continued)

Standard		Aim
S3	Surface water quality management	To minimise the potential pollution risk posed by the surface water runoff to the receiving waterbody.
S4	Amenity	To ensure that, where possible, the design of SuDS components enhances the provision of high-quality, attractive public space which can help provide health and wellbeing benefits, improve liveability for local communities and contribute to improving the climate resilience of new developments.
S5	Biodiversity	To ensure, where possible, the design of SuDS components creates ecologically rich green and blue corridors in developments and enriches biodiversity value by linking networks of habitats and ecosystems together. Biodiversity should be considered at the early design stage of a development to ensure the potential benefits are maximised.
S6	Design of drainage for Construction, O & M	To design robust surface water drainage systems so they can be easily and safely constructed, maintained, and operated, taking account of the need to minimise negative impacts on the environment and natural resources.

Source: Adapted from Welsh Government. 2018. Statutory Standards for Sustainable Drainage Systems – Designing, Constructing, Operating and Maintaining Surface Water Drainage Systems, WG36005.

PROBLEMS

1 What are SuDS, and why is their use so important?
2 Outline the range of SuDS components and describe how the main types of components operate.
3 Explain the function and operation of constructed wetlands in stormwater management.
4 How and to what extent are SuDS components able to improve the water quality of runoff?
5 What is meant by the term *SuDS management train*?
6 A 0.9 m wide, 1 m effective depth infiltration trench is required to drain an impermeable area of 120 m². The trench fill material is known to have a porosity of 0.33, but the soil infiltration rate can only be estimated as 25–50 mm/h. Calculate the required length of the trench. Rainfall statistics are the same as those used in Example 11.1. [10.1 m]
7 Describe and discuss the multiple benefits provided by SuDS.
8 How might the local community be more actively engaged in SuDS design and operation?
9 Explain why SuDS schemes are more cost-effective for large, low-density developments. What other factors will influence their capital and operating costs?
10 Summarise the main standards needed for SuDS design and operation.

KEY SOURCES

Ashley, R., Walker, L., D'Arcy, B., Wilson, S., Illman, S., Shaffer, P. 2015. UK sustainable drainage systems: Past, present and future. *Proceedings of the Institution of Civil Engineers, Civil Engineering,* *168(CE3)*, 125–130.
BS 8582: 2013. *Code of Practice for Surface Water Management for Development Sites.*
BS EN 16941-1: 2018. *On-site non-potable water systems. Part 1: Systems for the use of rainwater.*

Campisano, A., Butler, D., Ward, S., Burns, M.J. Friedler, E., DeBusk, K., Fisher-Jeffes, L.N., Ghisi, E., Rahman, A., Furumai, H., Han, M. 2017. Urban rainwater harvesting systems: Research, implementation and future perspectives. *Water Research, 115*, 195–209.

Charlesworth, S.M., and Booth, C.A. Eds. 2016. *Sustainable Surface Water Management. A Handbook for SUDS*, Wiley Blackwell.

Krivtsov, K., Arthur, S., Allen, D. and O'Donnell, E. Eds. 2019. *Blue green infrastructure – perspectives on planning, evaluation and collaboration*. Report C780a, CIRIA, London, UK.

National Research Council. 2009. *Urban Stormwater Management in the United States*, National Academies Press, Washington, DC.

Woods-Ballard, B., Wilson, S., Udale-Clark, H., Illman, S., Scott, T., Ashley, R., and Kellagher, R. 2016. *The SUDS Manual*, 5th edn, Report C753, CIRIA, London, UK.

REFERENCES

Abbott, C.L., Weisgerber, A., and Woods-Ballard, B. 2003. Observed hydraulic benefits of two UK permeable pavement systems. *Proceedings of the Second National Conference on Sustainable Drainage*, Coventry University, June, 101–111.

Achleitner, S., Engelhard, C., Stegner, U., and Rauch, W. 2011. Local infiltration devices at parking sites—Experimental assessment of temporal changes in hydraulic and contaminant removal capacity. *Water Science and Technology, 55*, 193–200.

Antunes, L.N., Thives, L.P., and Ghisi, E. 2016. Potential for potable water savings in buildings by using stormwater harvested from porous pavements. *Water, 8*, 11.

Apostolaki, S., Jefferies, C., and Souter, N. 2001. Assessing the public perception of SUDS in two locations in Eastern Scotland. *Proceedings of the First National Conference on Sustainable Drainage*, Coventry University, June, 28–37.

Bastien, N., Arthur, S., Wallis, S., and Scholz, M. 2010. The best management of SuDS treatment trains: A holistic approach. *Water Science and Technology—WST, 61*(1), 263–272.

Berardi, U., GhaffarianHoseini, A., and GhaffarianHoseini, A. 2014. State-of-the-art analysis of the environmental benefits of green roofs. *Applied Energy, 115*, 411–428.

Bergman, M., Hedegaard, M.R., Petersen, M.F., Binning, P., Mark, O., and Mikkelsen, P.S. 2011. Evaluation of two stormwater infiltration trenches in central Copenhagen after 15 years of operation. *Water Science and Technology, 63*, 2279–2286.

Bettess, R. 1996. *Infiltration Drainage—Manual of Good Practice*, Report 156, CIRIA, London, UK.

Bradley, J. 2023. *Using SuDs to Reduce Nitrogen in Surface Water Runoff*. Report C815, CIRIA, London, UK.

Bradley, J., Haygarth, P., Stachyra, K. and Williams, P. 2022. *Using SuDS to Reduce Phosphorus in Surface Water Runoff*. Report C808, CIRIA, London, UK.

Building Research Establishment. 1991. *Soakaway Design*, Digest 365.

Campisano, A., Di Libertoa, D., Modicaa, C., and Reitanob, S. 2014. Potential for peak flow reduction by rainwater harvesting tanks. *Procedia Engineering, 89*, 1507–1514.

Construction Industry Research and Information Association (CIRIA). 2000. *Sustainable Urban Drainage Systems—Design Manual for Scotland and Northern Ireland*, Report C521 and *Sustainable Urban Drainage Systems—Design Manual for England and Wales*, Report C522, CIRIA, London, UK.

Colwill, D.M., Peters, C.J., and Perry, R. 1984. *Water Quality in Motorway Run-off*, TRRL Supplementary Report 823, Transport and Road Research Laboratory.

Defra. 2011. *Commencement of the Flood and Water Management Act 2010, Schedule 3 for Sustainable Drainage – Impact Assessment*. Report for Department for Environment, Food and Rural Affairs.

Defra. 2015. *Sustainable Drainage Systems. Non-statutory Technical Standards for Sustainable Drainage Systems*, Department for Environment, Food and Rural Affairs.

Defra. 2021. *Recommendations to Update Non-Statutory Technical Standards for Sustainable Drainage Systems (SuDS)*, Final Report, Department for Environment, Food and Rural Affairs.

Defra. 2023. *The Review for Implementation of Schedule 3 to the Flood and Water Management Act 2010*. Department for Environment, Food & Rural Affairs.

De-Ville, S., Green, D., Edmondson, J., Stirling, R., Dawson, R., Stovin, V. 2021. Evaluating the potential hydrological performance of a bioretention media with 100% recycled waste components. *Water*, *13*, 2014.

Digman, C.J., Ashley, R.M., Balmforth, D.J., Balmforth, D., Stovin, V., and Glerum, J. 2012. *Retrofitting to Manage Surface Water*, Report C713, CIRIA, London, UK.

Dolowitz, D.P., Bell, S. and Keeley, M. 2018. Retrofiting urban drainage infrastructure: green or grey? *Urban Water Journal*, *15*(1), 83–91.

Early, P., Gedge, D., Newton, J., and Wilson, S. 2007. *Building Greener—Guidance on the Use of Green Roofs, Green Walls and Complementary Features on Buildings*, Report C644, CIRIA London, UK.

Ellis, J.B. 1992. Quality issues of source control. *Proceedings of CONFLO 92: Integrated Catchment Planning and Source Control*, Oxford.

Ellis, J.B. 2006. A UK and European perspective of Sustainable Urban Drainage (SUDS) with particular reference to infiltration systems and groundwater pollution. It Doesn't Just Go Away, You Know…, SUDS and Groundwater Monitoring, *Proceedings of the IAH, 26th Annual Groundwater Conference*, Co. Offaly.

Environment Agency. 2018. *The Environment Agency's Approach to Groundwater Protection, Version 1.2.*

Environmental Policy Consulting. 2017. *Sustainable Drainage Systems on New Developments. Analysis of Evidence including Costs and Benefits of SuDS Construction and Adoption*, Final Report for the Welsh Government.

Everett, G. and Lamond, J. 2016. SuDS and human behaviour: Co-developing solutions to encourage sustainable behaviour. *FLOODrisk 2016, Third European Conference on Flood Risk Management*, E3S Web of Conferences, doi:10.1051/e3sconf/2016

Fletcher, T.D., Peljo, L., Fielding, J., Wong, H.F., and Weber, T. 2002. The performance of vegetated swales for urban stormwater pollution control. *Global Solutions for Urban Drainage: Proceedings of the Ninth International Conference on Urban Drainage*, Portland, Oregon, September, on CD-ROM.

Fletcher, T.D., Shuster, W., Hunt, W.F., Ashley, R.A., Butler, D., Arthur, S. 2015. SUDS, LID, BMPs, WSUD and more—The evolution and application of terminology surrounding urban drainage. *Urban Water Journal*, *12*(7), 525–542.

Hatt, B.E., Fletcher, T.D., & Deletic, A. 2009. Hydrologic and pollutant removal performance of stormwater biofiltration systems at the field scale. *Journal of Hydrology*, *365*(3), 310–321.

Horton, B., Digman, C.J., Ashley, R.M. and McMullan, J. 2019. *B£ST Guidance – Guidance to assess the benefits of blue and green infrastructure using B£ST*, Release version 5, Report W047b, CIRIA, London, UK.

Huang, Y., Tian, Z., Ke, Q., Liu, J., Irannezhad, M., Fan, D. Hou, M. and Sun, L. 2020. Nature-based solutions for urban pluvial flood risk management. *WIREs Water*, 7, e1421.

Jefferies, C. 2004. *SUDS in Scotland—The Monitoring Programme of the Scottish Universities SUDS Monitoring Group*. Scotland and Northern Ireland Forum for Environmental Research (SNIFFER) Report SR(02)51.

Jefferies, C., Duffy, A., Berwick, N., McLean, N., and Hemmingway, A. 2008. SUDS treatment train assessment tool. *Proceeding of the 11th International Conference on Urban Drainage*, Edinburgh, September, on CD-ROM.

Kellagher, R. 2012. *Stormwater Management using Rainwater Harvesting. Testing the Kellagher / Gerolin Methodology on a Pilot Study*. Report SR736, Release 2.0, HR Wallingford.

Kellagher, R. 2013. *Rainfall Runoff Management for Developments*. Report SC030219, Environment Agency.

Macdonald, K. and Jefferies, C. 2001. Performance comparison of porous paved and traditional car parks. *Proceedings of the First National Conference on Sustainable Drainage*, Coventry University, June, 170–181.

Maqbool, R. and Wood, H. 2022. Containing a sustainable urbanized environment through SuDS devices in management trains. *Science of the Total Environment*, *807*, 150812.

McLean, N., Campbell, N., Bray, R., and D'Arcy, B.J. 2005. New approaches to detention basins and ponds. *Proceedings of the Third National Conference on Sustainable Drainage*, Coventry University, June, 211–221.

Melville-Shreeve, P., Ward, S., and Butler, D. 2016a. Dual-purpose rainwater harvesting system design, in *Sustainable Surface Water Management. A Handbook for SUDS* (eds. S.M. Charlesworth and C.A. Booth), Wiley Blackwell, 255–268.

Melville-Shreeve, P., Ward, S., and Butler, D. 2016b. Rainwater harvesting typologies for UK houses: A multi criteria analysis of system configurations. *Water, 8*, 129.

Natural England. 2023. *Nutrient Neutrality and Nutrient Mitigation*. Catalogue Code NE776.

Nawaz, R., McDonald, A., and Postoyko, S. 2015. Hydrological performance of a full-scale extensive green roof located in a temperate climate, *Ecological Engineering, 82*, 66–80.

NIEA/SEPA. 2022. *Use and Design of Oil Separators in Surface Water Drainage Systems, GPP 3. Northern Ireland Environment Agency* / Scottish Environment Protection Agency.

Poë, S., Stovin, V. and Berretta, C. 2015. Parameters influencing the regeneration of a green roof's retention capacity via evapotranspiration. *Journal of Hydrology, 523*, 356–367.

Pratt, C.J., Newman, A.P., and Bond, P.C. 1999. Mineral oil bio-degradation within a permeable pavement: Long term observations. *Water Science and Technology, 39*(2), 103–109.

Pratt, C.J., Wilson, S., and Cooper, P. 2002. *Source Control Using Constructed Pervious Surfaces—Hydraulic, Structural and Water Quality Performance Issues*. Report C582, CIRIA, London, UK.

Quinn, R., Melville-Shreeve, P., Butler, D. and Stovin, V. (2020). A critical evaluation of the water supply and stormwater management performance of retrofittable domestic rainwater harvesting systems. *Water, 12*, 1184, 1–14.

Oberascher, M. Rauch, W. and Sitzenfrei, R. 2021. Efficient integration of IoT-based micro storages to improve urban drainage performance through advanced control strategies. *Water Science and Technology, 83*(11), 2678–2690.

Revitt, D.M., Ellis, J.B., and Lundy, L. 2017. Assessing the impact of swales on receiving water quality. *Urban Water Journal, 14*(8), 839–845.

Ricardo. 2020. *Independent review of the costs and benefits of rainwater harvesting and grey water recycling options in the UK*. ED 13617100, Final Report, Waterwise.

Rommel, M., Rus, M., Argue, J., Johnston, L., and Pezzaniti, D. 2001. Carpark with '1 to 1' (impervious/permeable) paving: Performance of 'Formpave' blocks. Novatech 2001 *Proceedings of the Fourth International Conference on Innovative Technologies in Urban Drainage*, Lyon, France, June, 807–814.

Royal Haskoning DHV. 2012. *Costs and Benefits of Sustainable Drainage Systems*, Committee on Climate Change, Final Report, 9X1055.

Sañudo-Fontaneda, L.A., Charlesworth, S.M., Castro-Fresno, D., Andrés-Valeri, V.C.A., and Rodriguez-Hernandez, J. 2014. Water quality and quantity assessment of pervious pavements performance in experimental car park areas. *Water Science and Technology, 69*(7), 1526–1533.

Schlüter, W. and Jefferies, C. 2001. Monitoring the outflow from a porous car park. *Proceedings of the First National Conference on Sustainable Drainage*, Coventry University, June, 182–191.

Scholz, M. 2006. *Wetland Systems to Control Urban Runoff*, Elsevier.

Siriwardene, N.R., Deletic, A., and Fletcher, T.D. 2007. Clogging of stormwater gravel infiltration systems and filters: Insights from a laboratory study. *Water Research, 41*, 1433–1440.

Stovin V., Vesuviano G., and Kasmin H. 2012. The hydrological performance of a green roof test bed under UK climatic conditions. *Journal of Hydrology, 414–415*, 148–161.

Stovin, V.R., Moore, S.L., Wall, M., and Ashley, R.M. 2013. The potential to retrofit sustainable drainage systems to address combined sewer overflow discharges in the Thames Tideway catchment. *Water and Environment Journal, 27*, 216–228.

Susdrain. 2023. *Benefits of SuDS*. https://www.susdrain.org/delivering-suds/using-suds/benefits-of-suds/SuDS-benefits.html

The GRO Green Roof Code. 2021. Green Roof Organisation. www.greenrooforganisation.org/2021/03/05/the-gro-code-of-best-practice-2021

UK Water Industry Research (UKWIR). 2021. *Surface Water Drainage from New Developments. Report no. 21/SW/01/21*. https://ukwir.org/surface-water-drainage-from-new-developments

Vollaers, V., Nieuwenhuis, E., van de Ven, F. and Langeveld, J. 2021. Root causes of failures in sustainable urban drainage systems (SUDS): an exploratory study in 11 municipalities in the Netherlands. *Blue-Green Systems*, *3*(1), 31.

Ward S., Paling N. and Rogers A. 2023. Mobilising sustainable, water-resilient communities in the UK: evidence and engagement across scales. *Proceedings of the Institution of Civil Engineers – Engineering Sustainability*, *176*(4), 171–179

Wei, X., Burns, M.J., Cherqui, F. and Fletcher, T.D. 2021. Enhancing stormwater control measures using real-time control technology: a review. *Urban Water Journal*, *18*(2), 101–114.

Welsh Government (2018) *Statutory Standards for Sustainable Drainage Systems – Designing, Constructing, Operating and Maintaining Surface Water Drainage Systems*, WG36005.

Whittinghill, L.J., Rowe, D.B., Andresen, J.A., and Cregg, B.M. 2015. Comparison of stormwater runoff from sedum, native prairie, and vegetable producing green roofs. *Urban Ecosystems*, *18*, 13–29.

Williams, J.B., Jose, R., Moobela, C., Hutchinson, D.J., Wise, R. and Gaterell, M. 2019. Residents' perceptions of sustainable drainage systems as highly functional blue green infrastructure. *Landscape and Urban Planning*, *190*, 103610.

Zheng, X., Zou, Y. Lounsbury, A.W., Wang, C. Wang, R. 2021. Green roofs for stormwater runoff retention: A global quantitative synthesis of the performance. *Resources, Conservation & Recycling*, *170*, 105577.

Chapter 12

Surface water flooding

12.1 INTRODUCTION

In the previous two chapters, we looked at methods to design storm sewers and SuDS to operate effectively under defined hydrological conditions. The concept of the design storm was introduced, the choice of which broadly determines the degree of protection afforded against flooding. What should be the engineer's approach to the situation when design flows are exceeded to the extent that flow either emerges onto the urban surface from the underground pipe system, spills from overground SuDS components or can no longer enter the already full pipes and components? Until relatively recently, little account was taken of this phenomenon because the system was considered to "have failed." It is instructive to revisit the limited advice on this topic given in the first edition of this book (2000): "Generally, when piped systems are being designed, care should be taken to 'define' overland flow flood routes to minimise damage to properties."

Over the last 25 years, however, much more attention has been given to this topic as a consequence of the following:

- Significant and frequently occurring surface water flooding events (see Box 12.1)
- An increased understanding of the importance of surface water flooding. For example, 325,000 properties in England are in areas at high risk of surface water flooding (more likely than 1 in 30 annual exceedance probability [AEP]), even more than those at risk from rivers and the sea (NIC, 2022)
- A greater understanding of the implications of climate change (as discussed in Chapter 4)
- Population growth and increasing urbanisation resulting in more runoff and properties at risk
- Increasing urban creep (as discussed in Chapter 10)
- Significant costs (the expected annual damage costs for surface water flooding in England are £1.26 billion; Sayers et al. (2022))

Box 12.1 summarises a range of surface water flooding incidents in England.

Surface water flooding is defined as inundation caused by rainfall or other precipitation such that the runoff caused exceeds the conveyance capacity of the urban drainage system, and *exceedance flow* is generated on the urban surface. The terms *stormwater flooding* or *pluvial flooding* are sometimes used synonymously with surface water flooding, or sometimes used to denote urban flooding where there is no sewer or drainage network. The term *urban flash flooding* is also used, which refers to events with a rapid or sudden onset. *Sewer flooding* refers to the specific case when wastewater floods into properties or gardens from foul, combined, or even storm sewers and may or may not be caused by rainfall-runoff.

DOI: 10.1201/9781003408635-12

BOX 12.1   Surface water flooding in England

Below is a selection of surface water flooding incidents over the last few years (NIC, 2022).

- London 2021: Widespread flooding affected 24 of London's 32 boroughs, with over 1500 properties flooded. This affected homes, businesses, health infrastructure, and transport networks.
- Rochdale and Greater Manchester, 2020: Flooding caused by Storms Ciara and Dennis affected numerous properties and local businesses.
- England, 2019: Heavy rain caused flash flooding and travel problems, disrupting road and rail travel.
- The Midlands, 2018: Storms hit parts of the West Midlands, Worcestershire, and Milton Keynes. Collectively, this resulted in over 750 properties being flooded.
- Kent and Cornwall, 2017: Over 50 properties were flooded in at least four areas of Kent, and a further 50 properties were flooded and roads damaged in Cornwall.
- Woking, 2016: 45 properties were flooded, and three schools and one road were closed.
- Canvey Island, 2014: Surface water flooding impacted over 200 properties.
- Newcastle, 2012: Flooding resulted in over 500 properties being flooded, including shops and schools. Transport networks including roads and trams were also impacted.

This chapter covers the concept of exceedance flow and what it means (Section 12.2); it reviews appropriate standards (12.3) and defines the notion of flood risk in urban areas (12.4). Appropriate management strategies are discussed together with recommended strategic approaches (12.5) and the chapter concludes with an explanation of flood resilience and how it can be enhanced (12.6).

12.2 EXCEEDANCE

Before exceedance flow can be formally defined, it is important to establish that the drainage system is more than just pipes in the ground. This is emphasised by formally considering the system as having minor and major components. The *minor system* consists of drainage hardware such as gully inlets, manholes, pipes, and SuDS components used to control more frequent storm flows and is generally under the control of the system designer and subsequent owner. The *major system* consists of *surface flood pathways* such as roads and paths and includes temporary storage areas such as car parks and playing fields. These may not have been specifically designed as such and are known as *default pathways*. *Design pathways* are constructed or designated at the planning stage and will additionally consist of floodways, retention basins, flood-relief channels, and open and culverted watercourses.

For any particular rainfall event, if the conveyance capacity of the piped or open channel minor system is exceeded (either due to lack of capacity or blockage), the excess flow that appears on the surface in the major system is termed *exceedance flow*. Figure 12.1 shows exceedance in progress as the surcharged minor combined sewer system spills its flow onto the default major system road surface.

Figure 12.1 Sewer system surcharge and exceedance flow generation. (Courtesy of Professor Dušan Prodanović.)

12.2.1 Systems interface

A key aspect of exceedance is the interface between the minor and major systems. The points of contact between the two systems and their mutual interactions are as follows.

- Gully inlets: normally ingress points to the minor system for surface water. If the conveyance capacity is reached, the flow will not be able to enter at these points and will be retained on the surface. If conveyance capacity is exceeded, the flow may exit at these points into the major system (see Figure 12.1). Similar effects are caused if the inlet or inlet connection becomes blocked.
- Manholes: normally personnel access points into the minor system. If the conveyance capacity is exceeded and the system is surcharged, the flow may exit at these points, even to the point of lifting heavy manhole covers.
- Open SuDS: include swales and detention basins, which may overtop, spilling flow onto surrounding ground (see Chapter 11 for further details).
- WCs and other household appliances: in extreme cases or if WCs are sited in building basements, surcharged flow in combined sewers may reverse up the building drains and be forced backwards through the toilet and into properties.
- Receiving water outfalls: normally used as egress points for the minor system. However, if water levels rise in the receiving watercourse, a backwater can be formed reducing the effective capacity of the minor system and potentially inducing exceedance. In some cases, where no flap valve exists, water can also enter the minor system.

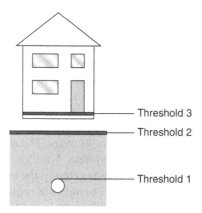

Figure 12.2 Urban drainage flooding threshold level standards.

The cause or type of flooding is difficult to diagnose after the event. In many cases, flooding is due to multiple effects at multiple points of contact. However, *Lead Local Flood Authorities* (designated in England and Wales under the 2010 Flood and Water Management Act) have a duty to investigate the cause of flooding and sewerage undertakers routinely investigate flooding incidents, maintaining detailed records.

12.2.2 Exceedance flow

Calculating the degree and extent of the interaction between the minor and major systems is complex and resource-consuming. Developments in modelling software have made this easier to represent, but significant effort is still required to validate such models. Defra (2010) recommends three levels of assessment and detail (strategic, intermediate, and detailed) depending on the needs of the particular study or engineering project. While the focus is on existing problems, level 3 could also be applied to new developments. The three levels are as follows.

1. Strategic: indicates where exceedance may accumulate over a large geographical scale (unlikely to include the minor system)
2. Intermediate: identifies not only where exceedance occurs but also more accurately establishes where it comes from (will include the minor system)
3. Detailed: represents the topography and increasing levels of detail applied typically to a township or localized problem.

Above-ground modelling approaches are covered in more detail in Chapter 20.

12.3 STANDARDS

Three threshold standards can be defined (Figure 12.2) when designing or analysing drainage systems, set to *avoid* the following:

1. Drainage system overload (minor system)
2. Surface flooding (minor system)
3. Property flooding (major system)

Threshold 1 standards are those conventionally used in design practice, for the minor system, such as given in BS EN 752: 2017 / BS ISO 24536: 2019 and discussed in Section 10.2.1 (Design storm). This denotes the normal hydraulic capacity of the drainage system (e.g., pipe-full gravity flow) based on a particular rainfall event return period. Once this threshold is crossed, the system will become progressively overloaded. In the case of pipes, this includes becoming surcharged which can generate additional flow capacity. *Threshold 2* standards refer to the maximum capacity of the system to convey the surface water without exceedance flow being generated. BS EN 752: 2017 specifies that surface flooding should be prevented for storms with a return period of up to 30 years. The *threshold 3* standard effectively denotes the maximum capacity of the major system. National Planning Policy Guidance for England and Wales requires that consideration is given to the effect of rainfall in excess of the 30-year return period storm, with threshold 3 protection required for events up to the 100-year return period (Department for Communities and Local Government [DCLG], 2012).

12.4 FLOOD RISK

Chapter 10 considers the concept of the design storm and how it can be expressed either in terms of return period or AEP. Design is then based on probabilistic thresholds as given in the previous section. However, the management of coastal, river, and (increasingly) surface water flooding follows a risk-based approach, where both the likelihood *and* outcome of flooding are taken into account. Put simply, the risk of flooding R is given by

$$R = P \cdot C \tag{12.1}$$

where R is the total risk under management over n years (£), P is the probability of a particular occurrence, and C is an undesirable consequence (or damage) of the occurrence (£).

The basic idea is to move away from considering the probability of a storm event in isolation, and instead to relate it formally and explicitly to the detailed impact or consequence it will have. For example, flooding may occur regularly in one location (hence with a high probability) but have very little impact, so the risk is low. The same probability of flooding may occur at another location, this time causing high impact due to the particular local circumstances, giving high risk. A probability-based approach would not directly distinguish between these two situations, whereas a risk-based approach would. Thus, investment can be targeted towards the maintenance and improvement of those assets that contribute most towards risk reduction.

High-probability events with high consequences will be classified as high risk; similarly, low-probability, low-consequence events will be low risk. As Figure 12.3 shows, however,

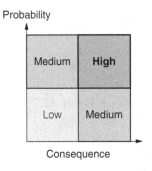

Figure 12.3 The relationship between probability, consequence, and risk.

low probability–high consequence situations are classified as medium risk, as are those with high probability and low consequence. These two cases are likely to be viewed very differently by the public (Merz et al., 2009) and may well need to be managed differently even though they pose similar (medium) risks.

However, to use such a risk assessment approach a more comprehensive analysis is required, which combines hydrological knowledge about the frequency of different storm events with hydraulic modelling information regarding inundation frequency of flood water, and flood damage evaluation (see Section 12.4.2).

12.4.1 Flood damage

Flood-induced damage can be classified as tangible or intangible, distinguished by the ease or otherwise of assigning a monetary value. Tangible damage can also be subdivided into direct or indirect (Hammond et al., 2015). Damage to property (e.g., floors, walls, ceilings, and all internal items) is direct, whereas interruption and disruption of economic and social activities are indirect, including loss of production and profit, disruption to road traffic, and additional costs for emergency and cleaning work. Intangible damage also includes health and psychological impacts. Whittle et al. (2012) highlight a range of experiences suffered by families including their post-flooding vulnerability and emotional recovery. In practice, only tangible effects are considered in any formal analysis, although with consideration of multiple benefits and impact assessment becoming more commonplace, this is likely to change in the future.

12.4.1.1 Flood depth

Although not the only factor related to property damage, flood depth or stage is the key indicator. If the floodwater depth is below ground-floor level, the damage is likely to be limited for most properties, although it may be significant if the water enters basements, cellars, or voids under floors. Once the floodwater level rises to 0.5 m above the ground-floor level, properties can be significantly affected, including damage to internal surfaces, electrical sockets and equipment, kitchen cupboards, carpets, furniture, and personal belongings. Much greater damage still can occur when flood depth is greater than 0.5 m above the ground-floor level, and, in extreme cases, the structural integrity of the building can be compromised.

The *Multi-Coloured Handbook* (Priest et al., 2023) contains relationships between flood damage and flood depth, duration, property types, and social categories for both urban and rural areas in the UK, based on a regularly updated and extensive dataset. Examples of several depth–damage curves are given in Figure 12.4.

There is a relationship at every location between flood stage and flood probability, and between flood stage and flood damage. These can be combined as shown in Figure 12.5 to give a direct flood probability–flood damage relationship. The *annual average risk* of flooding is the integral of the product of probability and damage (Equation (12.1)), which is the (shaded) area under the curve in Figure 12.5c.

A further important consideration, over and above direct flood damage, is the consequence of damage or disruption to important facilities. For example, buildings providing essential services (e.g., hospitals, emergency services) or critical infrastructure (e.g., water, electricity, transport) should have a higher consequence attached to being flooded. This could be implemented by setting different standards of protection such as a 1 in 200 AEP event for national infrastructure with 1 in 1000 year protection for the most critical facilities (Cabinet Office, 2010).

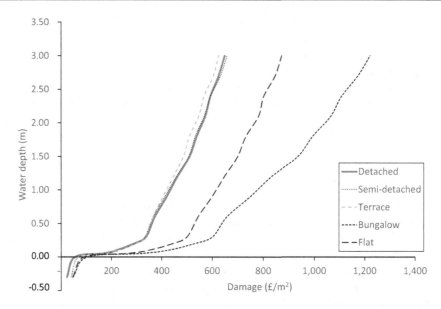

Figure 12.4 Typical relationship between flood depth and direct damage in residential properties. (Based on Data from the Multi-Coloured Handbook (Priest et al., 2023) and the English Housing Survey 2018 (MHCLG, 2020))

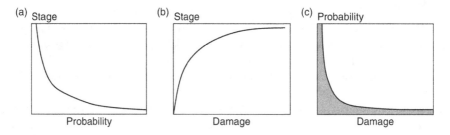

Figure 12.5 Interrelation of flood probability, stage, and damage.

12.4.2 Risk assessment

Defra (2010) proposes a three-level risk assessment approach taking into account contemporary software capabilities. Key steps to applying a risk assessment include the following:

- Identify the level of assessment and understand the appropriateness of the available models and software to determine the risk.
- Select the *storm* return periods to be used as the threshold 2 standard, and determine the critical duration for the network under consideration. Note that for larger systems, this may require more than one duration.
- Begin the assessment with the 30-year return period (for example) critical duration storm, and determine locations that surcharge or flood. Focus modelling efforts in known flooding locations. If no flooding occurs, increase the storm return period.

- Estimate the consequences of the flooding at each location, with the degree of detail linked to the level of study. This may range from enumerating the number of properties impacted, to establishing total damage costs using the depth-damage relationships. A number of different tools are now available to help support this process (e.g., Chen et al., 2016).
- Present the results across an area appropriate for the type of assessment either qualitatively using a risk matrix (Figure 12.3) or quantitatively by monetising the expected annual damage.
- Test solutions or designs repeating the process to check if risk is increased or reduced (noting that the critical storm duration for the solutions may be different to those that cause the worst flooding).

Ryu and Butler (2008) argue that continuous simulation is more appropriate for urban drainage flood risk assessment than considering single storm events. This allows a more complete and accurate evaluation of flood risk (i.e., the area under the curve in Figure 12.5c). However, this thoroughness comes at the expense of computational and physical time. Dawson et al. (2008) have developed an approach based on urban flood risk assessment but with the additional goal of attributing the risk between key stakeholders.

UK Water Industry Research (UKWIR, 2011) has developed a risk-based framework based on defining the probability and consequence of (frequent) surface water flooding for actual and predicted flooding. This includes a method to assess actual incidents, with the probability defined by recorded flood frequency rather than using a hydraulic model. The method can be used to help prioritise investments.

12.4.3 Surface flood risk maps

Surface water flood risk has been mapped across Great Britain. This has been modelled at 2 m resolution or finer for over 90% of urban areas and 5 m elsewhere, taking into account the more detailed effect of roads, buildings, bridges, and embankments wherever possible. Flooding is mapped for scenarios as a result of rainfall with the following AEPs: 3.3%, 1.0% and 0.1%. Information provided for each scenario is extent, depth, velocity (including flow direction at maximum velocity), and hazard (as a function of depth and velocity). It also includes information about the source of the data (i.e., whether it was from the nationally produced modelling or locally produced modelling) and the confidence in the data outputs (EA, 2019). Figure 12.6 shows an example of the output. Access to the data for England is freely available at: www.gov.uk/check-long-term-flood-risk.

The impact of urban (surface water) flash flooding in England has also been compiled based on historical data since 2010 and mapped in detail here: www.climatenode.org/maps/about_UFF_maps.html.

12.5 MANAGEMENT

12.5.1 Options

A key aspect of surface water flooding is the need to devise effective and sustainable approaches to managing the risk it creates. In terms of new systems, the basic approach is the careful design of both the minor and major systems against appropriate threshold standards and equally careful planning of the interaction between them. For existing systems, a range of possible solutions is available, and these are summarised in the following list.

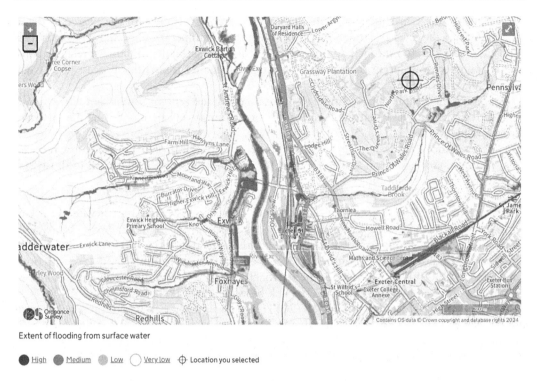

Extent of flooding from surface water

● High ● Medium ● Low ○ Very low ⊕ Location you selected

Figure 12.6 Example of surface water flood depth mapping. (Reproduced from https://www.gov.uk/check-long-term-flood-risk, under the Open Government Licence v3.0)

- Reduce or limit inflow (*minor system*): Chapter 11 explains the options available for this, key among them is the use of SuDS measures to infiltrate flow locally (whilst being mindful of where this flow will go to) or to store and reuse it. An alternative approach is to divert or disconnect flows to alternative systems or outfalls.
- Increase system capacity (*minor system*): Improved and/or targeted cleaning of pipes and gullies can increase system capacity if it is limited by blockage or sediment (Chapter 18). Alternatively, rehabilitation may be considered (Chapter 16), possibly including pipe upsizing.
- Store more flow (*minor or major system*): The key aim is to attenuate flow and therefore reduce peak flow rates. Storage can be provided at source (Chapter 11), within the minor system (Chapter 14), or within the major system (see Section 12.5.2). Storage performance can be enhanced by active management using real-time control (see Section 22.4).
- Better deploy surface flow features (*major system*): This key area is discussed in Section 12.5.2.
- Improve building flood resilience: This non-structural response is covered in Section 12.5.3.

The NIC (2022) proposes a solutions hierarchy to maximise the range of benefits and reduce costs. It prioritises system maintenance and technological enhancement, followed by above-ground interventions, with below-ground interventions (additional pipes and sewers) considered last.

12.5.2 Surface flow features

The surface conveyance system for exceedance flow can be made up of many elements, including gully inlets, pavements, roads, drives, and other less formal flow pathways, and these are discussed in the next few sections. CIRIA reports by Balmforth et al. (2006) and Digman et al. (2014) provide a variety of examples and case studies.

12.5.2.1 Gully inlets

Road gullies were considered in Chapter 6 and further in Chapter 8 as important parts of the minor urban drainage system. As discussed, they are also a key interface between the major and minor systems. The design of road gully sizing/spacing recognises and indeed utilises the fact that a proportion of the flow passes by, and over, the gully grating. Under extreme rainfall events, higher exceedance flows generate higher depths of flow and greater flow widths, and could also have a negative impact on the hydraulic efficiency of the gully itself. So, in design, attention needs to be given to selecting the appropriate capacity and spacing of inlets. In analysing the effectiveness of existing gullies, effort should focus on establishing whether sufficient inlet capacity is available given limiting kerb heights and/or flow widths.

Only in exceptional circumstances will the capacity of the gully grating restrict inflow into the gully pot (unless it is deliberately designed to do so or becomes blocked with debris). However, the gully outlet or lateral pipework may be the limiting component. Escarameia and May (1996) have determined this capacity experimentally at 11 L/s and 19 L/s for 100-mm and 150-mm gully outlets, respectively, with results corroborated by Shepherd et al. (2012). Example 12.1 demonstrates the point numerically.

12.5.2.2 Surface pathways

On undeveloped sites, the underlying topography defines natural flow pathways. When designing new sites, the most effective approach is to follow natural pathways as far as possible, providing linked, free passage for flowing excess water. This includes the identification of points of entry/exit to the drainage system plus the location of effective flow barriers such as kerbs, walls, buildings, and other relevant urban features. These pathways will only rarely convey significant flows and will be used for other purposes day to day. Legal safeguards

---

EXAMPLE 12.1

Based on the conditions described in Example 8.5, calculate the captured flow by each gully and compare this with the capacity of a freely discharging 100 mm diameter gully lateral. Comment on the performance of the gully under exceedance flows.

Solution

Captured flow is equal to the runoff generated between gullies. So from Equation 8.10,

$$Q_c = E\bar{Q} = 0.99 \times 0.010 = 9.9 \, L/s$$

This can be compared with the lateral capacity of 11 L/s, indicating a well-matched gully and lateral. However, very little (lateral) capacity is available to cope with extreme flows.

(e.g., recording structures or features on registers held by the Lead Local Flood Authority) and appropriate maintenance are needed to ensure their continued availability.

The channel forming a surface pathway can be designed by standard application of open channel hydraulics as set out in Chapter 7, typically using Manning's equation. An appropriate case is given later in the book, in Example 23.1. Complications arise when conditions deviate significantly from uniform, steady flow. Details on the design of flow transitions and junctions are given in Balmforth et al. (2006). The concept of using the road as a drain is further explored in Digman et al. (2014) and in Chapter 23.

12.5.2.3 Surface storage

Surface storage capacity can be developed to replace or reduce the need for extensive surface pathways. The key points about major system storage areas are that they are above-ground, multi-functional facilities. Their goal is to retain the flood volume temporarily and release it slowly at a lower flow rate. A wide range of alternatives is available as listed in Table 12.1. The capacity design of such facilities is comparable to minor system storage design, discussed in detail in Chapter 14.

An example of this type of "sacrificial" storage area is a car park, where water can be allowed to accumulate up to kerb height over a defined area. Although vehicle owners may be temporarily inconvenienced, there is unlikely to be any significant structural damage to the site after flooding, safety risks are minimal due to the low depths, and as long as the area can be adequately drained after the event normal use can be quickly resumed. Figure 12.7 illustrates a good example of this from the Netherlands.

Table 12.1 Types of major system storage

Storage type	Description	Maximum water depth (m)
Car parks	Used to temporarily store exceedance flows. Depth restricted due to potential hazards to vehicles, pedestrians, and adjacent property.	0.2
Recreational areas	Hard surfaces used as basketball, five-a-side football and hockey pitches, or tennis courts.	Typically 0.5, up to 2.0 in extreme cases
Minor roads	Roads with speed limits up to 30 mph where the depth of water can be controlled by design.	0.1
Playing fields	Used for sports such as football and rugby. Set below the ground level of the surrounding area and may cover a wide area, offering significant flood volume.	Typically 0.5, up to 2.0 in extreme cases
Parkland	Has a wide amenity use. Often may contain a watercourse. Care is needed to keep floodwater separate and released in a controlled fashion to prevent downstream flooding.	Typically 0.5, up to 2.0 in extreme cases
School playgrounds	Playgrounds can provide significant flood storage. Extra care should be taken to ensure the safety of the children.	0.3
Industrial areas	Low-value storage areas. Care should be taken in the selection as some areas could create significant surface water pollution.	0.6

Source: Adapted from Balmforth, D. et al. 2006. *Designing for Exceedance in Urban Drainage—Good Practice*, CIRIA C635.

Figure 12.7 Multi-functional flood storage scheme in Benthemplein Water Square, Rotterdam. (Courtesy of C40 Cities.)

12.5.2.4 Safety

Allowing surface water flooding does pose the risk of contact with the public and hence safety issues should be addressed. For flow across surfaces that could be used by pedestrians, hazard regimes as a function of depth and velocity should be carefully considered in accordance with the recommendations in Table 12.2. In addition, the depth of flow in any channel should be limited to 0.3 m or 0.2 m if trafficked. See also Example 12.2.

In flood storage areas, depth should generally be limited to a maximum of 0.6 m. It may be appropriate to increase this depth up to 1.5 m or even 2 m in extreme circumstances (Woods-Ballard et al., 2016). In all cases, an appropriate design risk assessment should be completed, considering key aspects such as location, signposting, safe access, and egress.

Table 12.2 Flood safety criteria

Hazard criteria	Max depth (m)	Depth-velocity (m²/s)
Low hazard for children	0.5	<0.4
Moderate hazard for adults	1.2	<0.8
Significant hazard for adults	1.2	<1.2

Source: Adapted from Russo, B. et al. 2013. *Natural Hazards*, 69(1), 251–265.

EXAMPLE 12.2

A footpath will be used as a flow conveyance route during flooding. The path is of rectangular cross section, 100 mm deep, 2.5 m wide, and with a gradient of 1%. Check the path will be safe for pedestrians. $n = 0.018$.

Solution

Using Manning's Equation 7.23:

$$v = \frac{1}{n}R^{2/3}S_\circ^{1/2} = \frac{1}{0.018}\left(\frac{2.5 \times 0.1}{2.5 + 2 \times 0.1}\right)^{2/3} 0.01^{1/2} = 1.14\,\text{m}/\text{s}$$

We check with Table 12.1:

$$d \cdot v = 0.11 < 0.4\,\text{m}^2/\text{s}$$

This result indicates the path would still be safe for all pedestrians to use while flooded.

12.5.3 Flood protection of buildings

Buildings are not normally within the province of urban drainage engineers. However, two areas related to urban drainage flooding are particularly pertinent and should be considered: layout and fabric.

12.5.3.1 Layout

In many cases, the effects of flooding can be minimised by the careful positioning of buildings in relation to the topography and defined flood pathways, and by the sympathetic design of landscaping features. Figure 12.8 shows a small housing development where the above-ground flow pathways have been defined taking into account site topography. A temporary, landscaped flood storage area is identified at the low point of the site.

A second key aspect is correctly setting the level of housing relative to the major system, including threshold levels. Valuable flood protection can be achieved by careful detailing of the property and surrounds including entrance details, driveway slopes, and drop kerbs.

12.5.3.2 Fabric

An important aspect of mitigating flood risk is improving the flood protection of buildings both at the design stage and retrofit, so-called *property flood resilience* (PFR). Detailed advice on this has been published (Bowker et al., 2007, Kelly et al., 2020). The basic strategy proposed is that the best approach at low flood depths (<0.3 m outside the building) is to avoid flooding by excluding the building from the flood risk zone (such as by raising or moving the property). Failing this, building flood protection needs to be improved. At low depths (<0.6 m) an attempt should be made to exclude the water, but at high depths, water is best allowed to enter and then exit relatively unimpeded. Table 12.3 summarises this.

For existing buildings that have been subjected to flooding, appropriate repair techniques for damaged floors, walls, services, fittings, and building services can be found in Garvin et al. (2005).

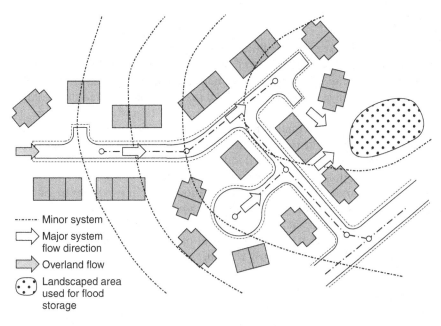

------ Minor system
⇨ Major system flow direction
⬜ Overland flow
⬚ Landscaped area used for flood storage

Figure 12.8 Identification of major system flood flow pathways.

Table 12.3 Building fabric design strategy

Strategy	Design water depth (m)	Approach	Mitigation measures
Avoidance		• Remove building/development from flood hazard	• Land raising, landscaping, raised thresholds
Resistance and resilience	< 0.3	• Attempt to keep water out – "water exclusion strategy"	• Materials and constructions with low permeability
	0.3–0.6	• Attempt to keep water out, in full or in part, depending on structural assessment • If structural concerns exist, follow the approach below (i.e., assume depth > 0.6m)	• Materials with low permeability to at least 0.3 m • Flood-resilient materials and designs • Access to all spaces to permit drying and cleaning
	> 0.6	• Allow water through property to avoid the risk of structural damage • Attempt to keep water out for low depths of flooding – "water entry strategy"	• Materials with low permeability up to 0.3 m • Accept water passage through building at higher water depths • Design to drain water away after flooding • Access to all spaces to permit drying and cleaning

Source: Bowker, P. et al. 2007. *Improving the Flood Performance of New Buildings. Flood Resilient Construction.* Communities and Local Government, RIBA Publishing.

Research has found these measures to be economically worthwhile for properties with an annual probability of flooding of 2% or greater (Thurston et al., 2008). In a UK city, individually targeted PFR measures were found to be more cost-effective than blanket provision in specific high-risk areas (Webber et al., 2021).

12.6 FLOOD RESILIENCE

Resilience is a concept that concerns how systems of all types prepare for, respond to, and recover from shocks of all types (Zhou et al., 2010). Gersonius (2008) argues that flood resilience incorporates four capacities: to avoid damage through the implementation of structural measures, to reduce damage in the case of a flood that exceeds a desired threshold, to recover quickly to the same or an equivalent state, and to adapt to an uncertain future. Djordjević et al. (2011) state resilience equates to resisting, recovering, reflecting, and responding. Zevenbergen et al. (2008) make the point that the development of flood resilience requires an understanding of how cities respond to flooding across varying spatial and temporal scales from surface water flooding to river and coastal flooding.

Butler et al. (2016) argue that resilience can be thought of in two separate but interrelated ways: properties and performance. In the former case, the emphasis is on providing properties or attributes of the system that are assumed to build resilience. In the latter case, the focus is on performance, that is, what standard or level of service should be provided to achieve acceptable resilience? The two are clearly related.

12.6.1 Attributes

Attributes that build or promote resilience include increasing flexibility, diversity, connectivity, and redundancy (Cabinet Office, 2011; Gersonius et al., 2013). Based on these general attributes, it is possible to provide recommendations on how to build flood resilience based on a raft of measures as shown in Table 12.4. Batica and Gourbesville (2014) have developed a Flood Resilience Index that can be applied on a case-by-case basis to indicate the extent and rate of change of resilience measures. While appealing and relatively straightforward to carry out, the effect of specifying particular attributes or measures on the performance of a system is uncertain. Increased connectivity, for example, may provide resilience for some threats but

Table 12.4 Flood resilience measures

Category	Measure	
Capacity building	• Flood maps (inundation and risk) • Informational material (brochures, public presentations, Internet portals) • Communication	• Face-to-face learning • Web-based learning • Training • Collaborative platforms
Land use control	• Spatial planning • Flood-risk-adapted land use • Building regulations	• Building codes • Zoning ordinances
Flood preparedness	• Flood-resistant buildings • Wet-proofing • Dry-proofing	• Flood contingency plans (local scale) • Infrastructure maintenance
Financial preparedness	• Insurance of residual risk	• Reserve funds
Emergency response	• Evacuation and rescue plans • Forecasting and warning services	• Emergency operations control • Emergency response staff
Emergency infrastructure	• Allocation of temporary containment structures • Telecommunications network	• Transportation and evacuation facilities
Recovery	• Disaster recovery plans	• Governmental pecuniary provisions

Source: Adapted from Batica, J. and Gourbesville, P. 2014. Flood resilience index—Methodology and implementation. *Proceedings of the 11th International Conference on Hydroinformatics,* August, New York.

actually decreases resilience to others, such as targeted attacks (Albert et al., 2000). Lamond et al. (2015) make the case for how provision of SuDS measures can develop greater flood resilience but this is still controversial in the academic literature (see Chapters 11 and 24).

12.6.2 Performance

The second approach requires that resilience performance is agreed upon and quantified. Butler et al. (2016) suggest this can be defined as "the degree to which the system minimises level of service failure *magnitude* and *duration* over its design life when subject to exceptional conditions." Mugume et al. (2015) and Mugume and Butler (2017) have used this definition to specify an urban drainage resilience index (*Res*) that links failure in terms of above-ground flood magnitude and duration with system residual functionality using a *stress-testing* procedure called Global Resilience Analysis (GRA):

$$Res = 1 - \left(V_F / V_I \cdot t_F / t\right) \qquad (12.2)$$

where V_F is the total flood volume (m³), V_I is the total inflow volume into the system (m³), t_F is the mean duration of flooding (hours), and t is the elapsed time (hours).

The causes of failure (flooding) in this context can be either functional (e.g., rainfall-induced) or structural (e.g., pipe blockage).

Mugume et al. (2015) have applied the index to a large open channel system in Kampala, Uganda, by modelling the effect of systematically failing (in this context, blocking) multiple channels (links) and computing the flood volume and duration at all nodes. Figure 12.9a illustrates the initial resilience index of the system and how this deteriorates with the degree of failure experienced. Figures 12.9b and 12.9c indicate how two different strategies or measures perform relative to an assumed minimum acceptable resilience level of service threshold of 0.7 (the dashed horizontal line).

The main advantage of the GRA method is that the focus is transferred from the accurate quantification of the probability of occurrence of sewer failures (e.g., Egger et al., 2013) to the evaluation of the effect of different sewer failures, on the ability of an urban drainage system to minimise the resulting flooding impacts (e.g., Kellagher et al., 2009). The main

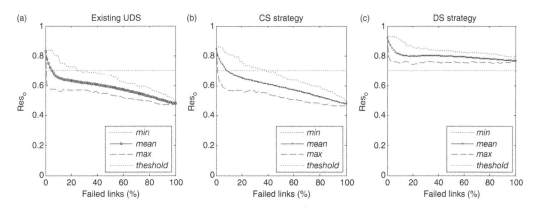

Figure 12.9 Mean, minimum, and maximum values of Res_0 for various degrees of link failure for (a) the existing urban drainage system, (b) a centralised storage strategy, and (c) a distributed storage strategy. The horizontal dashed line is an assumed minimum acceptable resilience level of service threshold of 0.7. (Reproduced from Mugume et al. (2015) *Water Research*, 81, 15–26, under the terms of the Creative Commons Attribution License CC BY).

disadvantage of this general performance-based approach is the computational burden it demands.

12.6.3 Combined approach

Atkins (2017) have proposed a combined approach to resilience assessment. This relies on using an attribute-based approach to establish the vulnerability of a specific sewered catchment (see Table 12.5). Those with higher vulnerability grades (3–5) are then subjected to a detailed modelling performance assessment for potential surface flooding from an extreme rainfall event with a return period of 50 years. This is chosen to effectively stress test systems beyond their current design criteria (i.e., threshold 2 – 30 years).

The whole issue of resilience and its relationship to sustainability is returned to in more detail in Chapter 24.

Table 12.5 Vulnerability characteristics and associated grades

Description	Grade
Catchment topography funnelling all flows into one area	5
Catchments with a rapid response	5
Unknown asset data	5
Only drainage system in catchment/high proportion of combined sewers	5
Sewer flooding risk from historic reported incidents	4
Repeated blockage risk from historic reported incidents	4
Urban density (high population concentration)	4
Proximity to sea/river level	3
Large complex networks with many dependencies	3
Dependence on pumping	3
Proximity to water table	3
Growth potential (unplanned)	3
Consequence of flood risk management by others	2
Growth potential (planned)	2
Catchments with a slow response – flat sewers and septicity	2
Where no key issues identified	1

Source: Adapted from Atkins (2017). *Developing and Trialling Wastewater Resilience Metrics.* Final Report for Water UK. Report 5160627/DG/001.

PROBLEMS

1 Why was the issue of urban drainage flooding neglected for so many years, and why is it now of such interest?

2 Explain the concept of flow exceedance. Is it a good thing, or should it be eliminated?

3 Three surface water flooding threshold standards are used in practice. Explain what they are and comment on the magnitude you think they should be set at.

4 A 100-year return period flood has caused extensive property damage of £25 million in area A. Area B is subject to regular flooding on average once a year. Each minor inundation causes an average of £250,000 in damage. Which area would you prioritise for investment and why? [£250,000]

5 Explain the differences between tangible and intangible damage, and direct and indirect damage. How would you attempt to quantify each category?

6 An extreme storm event causes a residential area to be flooded to an average depth of 0.5 m. For the same storm, by what depth must flooding be reduced to cut the flood risk by half? [0.25 m]

7 Surface water flood maps are available for flood probability, extent, depth, velocity, and hazard. Which of the five parameters are the most important and how would you use them?

8 What options are available to reduce urban drainage flood risk? Discuss when and where each solution would be most applicable.

9 Calculate the width of flooding just upstream of the initial gully in the road specified in Problem 5 in Chapter 8 if the design rainfall intensity is doubled to 100 mm/h. [1.90 m]

10 An area has been flooded. Rather than reconstruct the sewer system it is proposed that building modifications are carried out. What types of modification would you specify? How could this reduce flood risk?

11 What is the difference between resilience attributes and resilience performance?

12 What measures would you put in place to improve flood resilience and why?

KEY SOURCES

Ashley, R., Gersonius, B. and Horton, B. 2020. Managing flooding: from a problem to an opportunity. *Philosophical Transactions of the Royal Society*, A378, 20190214.

Balmforth, D., Digman, C., Kellagher, R., and Butler, D. 2006. *Designing for Exceedance in Urban Drainage—Good Practice*, CIRIA C635, London.

Digman, C., Balmforth, D., Hargreaves, P., and Gill, E. 2014. *Managing Urban Flooding from Heavy Rainfall: Encouraging the Uptake of Designing for Exceedance*, CIRIA C738, London.

Mcbain, W., Wilkes, D. and Retter, M. 2010. *Flood Resilience and Resistance for Critical Infrastructure*, CIRIA C688, London.

WHO. 2017. *Flooding: Managing Health Risks in the WHO European Region*. World Health Organization, Regional Office for Europe.

Zevenbergen, C., Cashman, A., Evelpidou, N., Pasche, E., Garvin, S., and Ashley, R. 2010. *Urban Flood Management*. CRC Press.

REFERENCES

Albert, R., Jeong, H., and Barabasi, A.L. 2000. Error and attack tolerance of complex networks. *Nature*, 406(6794), 378–382.

Atkins (2017). *Developing and Trialling Wastewater Resilience Metrics*. Final Report for Water UK. Report 5160627/DG/001.

Batica, J. and Gourbesville, P. 2014. Flood resilience index—Methodology and implementation. *Proceedings of the 11th International Conference on Hydroinformatics*, August, New York.

Bowker, P., Escarameia, M., and Tagg, A. 2007. *Improving the Flood Performance of New Buildings. Flood Resilient Construction*. Communities and Local Government, RIBA Publishing.

BS EN 752: 2017. *Drain and Sewer Systems Outside Buildings. Drain and sewer – Sewer system management*.

BS ISO 24536: 2019. *Service activities relating to drinking water supply, wastewater and stormwater systems — Stormwater management — Guidelines for stormwater management in urban areas*.

Butler, D., Ward, S., Sweetapple, C., Astaraie-Imani, M., Diao, K., Farmani, R. and Fu, G. 2016. Reliable, resilient and sustainable water management: The Safe and SuRe approach. *Global Challenges*, doi:10.1002/gch2.1010

Cabinet Office. 2010. *Strategic Framework and Policy Statement on Improving the Resilience of Critical Infrastructure to Disruption from Natural Hazards*. London.

Cabinet Office. 2011. *Keeping the County Running: Natural Hazards and Infrastructure. A Guide to Improving the Resilience of Critical Infrastructure and Essential Services*. London.

Chen, A.M., Hammond, M.J., Djordjevic, S., Butler, D., Khan, D.M., and Veerbeek, W. 2016. From hazard to impact: Flood damage assessment tools for mega cities. *Natural Hazards*, 82(2), 857–890.

Dawson, R.J., Speight, L., Hall, J.W., Djordjevic, S., Savic, D., and Leandro, J. 2008. Attribution of flood risk in urban areas. *Journal of Hydroinformatics*, 10(4), 275–288.

DCLG. 2012. *Technical Guidance to the National Planning Policy Framework*, Department for Communities and Local Government.

Defra. 2010. *Surface Water Management Plan Technical Guidance*, Department for Environment, Food and Rural Affairs.

Djordjević, S., Butler, D., Gourbesville, P., Mark, O., and Pasche, E. 2011. New policies to deal with climate change and other drivers impacting on resilience to flooding in urban areas: the CORFU approach. *Environmental Science and Policy*, 14(7), 864–873.

Egger, C., Scheidegger, A., Reichert, P., and Maurer, M. 2013. Sewer deterioration modeling with condition data lacking historical records. *Water Research*, 47, 6762–6779.

EA. 2019. What is the Risk from Surface Water Map? Report version 2.0, Environment Agency.

Escarameia, M. and May, R.W.P. 1996. *Surface Water Channels and Outfalls: Recommendations on Design*, Report SR406, HR Wallingford.

Garvin, S., Reid, J., and Scott, M. 2005. *Standards for the Repair of Buildings Following Flooding*, CIRIA Report C623.

Gersonius, B. 2008. Can resilience support integrated approaches to urban drainage management? *Proceedings of 11th International Conference on Urban Drainage*. Edinburgh.

Gersonius, B., Ashley, R., Pathirana, A., and Zevenbergen, C. 2013. Climate change uncertainty: Building flexibility into water and flood risk infrastructure. *Climatic Change*, 116, 411.

Hammond M.J., Chen A.S., Djordjević S., Butler D., and Mark O. 2015. Urban flood impact assessment: A state-of-the-art review. *Urban Water Journal*, 12(1), 14–29.

Kellagher, R.B.B., Cesses, Y., Di Mauro, M., and Gouldby, B. 2009. An urban drainage flood risk procedure—A comprehensive approach. *WaPUG Annual Conference*, Blackpool.

Kelly, D., Barker, M., Lamond, J., McKeown, S., Blundell, E. and Suttie, E. 2020. *Guidance on the Code of Practice for Property Flood Resilience*. CIRIA C790B, London.

Lamond, J.E., Rose, C.B., and Booth, C.A. 2015. Evidence for improved urban flood resilience by sustainable drainage refit. *Proceedings of the Institution of Civil Engineers, Urban Design and Planning*, 168(DP2), 101–111.

MHCLG. 2020. *English Housing Survey 2018: size of English homes – fact sheet*. Ministry of Housing, Communities & Local Government, https://www.gov.uk/government/statistics/english-housing-survey-2018-size-of-english-homes-fact-sheet

Merz, B., Elmer, F. and Thieken, A.H. 2009. Significance of "high probability/low damage" versus "low probability/high damage" flood events. *Natural Hazards and Earth Systems Science*, 9, 1033–1046.

Mugume, S.N. and Butler, D. 2017. Evaluation of functional resilience in urban drainage and flood management systems using a global analysis approach. *Urban Water Journal*, 14(7), 727–736.

Mugume, S.N., Gomez, D.E., Fu, G., Farmani, R., and Butler, D. 2015. A global analysis approach for investigating structural resilience in urban drainage systems. *Water Research*, 81, 15–26.

NIC. 2022. *Reducing the Risk of Surface Water Flooding*. National Infrastructure Commission.

Priest, S., Viavattene, C., Penning-Rowsell, E., Stenhausen, M., Parker, D., Joyce, J, Morris, J. and Chatterton, J. 2023. *Flood and Coastal Erosion Risk Management: for Economic Appraisal*. Available online: http://www.mcm-online.co.uk/handbook

Russo, B., Gómez, M., and Macchione, F. 2013. Pedestrian hazard criteria for flooded urban areas. *Natural Hazards*, 69(1), 251–265.

Ryu, J. and Butler, D. 2008. Managing sewer flood risk. *Proceedings of the 11th International Conference on Urban Drainage*, Edinburgh, September, on CD-ROM.

Sayers, P.B., Ashley, R., Carr, S., Eccleston P., Horritt, M, Horton, B. and Miller, J. 2022. *Surface Water – Future Risk and Investment Needs*. A report by Sayers and Partners for the National Infrastructure Commission, London.

Shepherd, W., Blanksby, J., Doncaster, S., and Poole, T. 2012. Assessment of road gullies. *Proceedings of the 10th International Conference of Hydroinformatics, HIC 2012*, Hamburg.

Thurston, N., Finlinson, B., Breakspear, R., Williams, N., Shaw, J. & Chatterton, J. 2008. *Developing the Evidence Base for Flood Resistance and Resilience: R&D Summary Report* FD2607/TRI, Environment Agency/Defra.

UKWIR. 2011. *A Risk Based Approach to Flooding*, Report 11/WM/17/2. https://ukwir.org/661d92a4-6197-419d-933b-1618de9ef6b9

Whittle, R., Walker, M., Medd, W., and Mort, M. 2012. Flood of emotions: Emotional work and long-term disaster recovery. *Emotion, Space and Society*, 5, 60–69.

Woods-Ballard, B., Wilson, S., Udale-Clark, H., Illman, S., Scott, T., Ashley, A., and Kellagher, R. 2016. *The SuDS Manual*, 5th edn, CIRIA C753, London.

Webber, J.L., Chen, A.S., Stevens, J., Henderson, R., Djordjevic, S. and Evans, B. 2021. Targeting property flood resilience in flood risk management. *Journal of Flood Risk Management*. 14, e12723.

Zevenbergen, C., Veerbeek, W., Gersonius, B., and Van Herk, S. 2008. Challenges in urban flood management: Travelling across spatial and temporal scales. *Journal of Flood Risk Management*, 1(2), 81–88.

Zhou, H., Wang, J., Wan, J., and Jia, H. 2010. Resilience to natural hazards: A geographic perspective. *Natural Hazards*, 53, 21–41.

Chapter 13

Combined sewers and combined sewer overflows

13.1 INTRODUCTION

Combined sewer systems are discussed in broad terms in Chapter 1. The essential feature of combined sewers is that they carry both wastewater and stormwater in the same pipe. As it is not possible to carry the full combined flow to treatment and there is a need to limit the frequency of flooding, it is necessary for *combined sewer overflows* (CSOs) to discharge a proportion of the flow to a local watercourse. Note that the term *storm overflow* is also used but this typically has a somewhat wider meaning and includes not just CSOs but also emergency overflows at pumping stations and settled storm tank overflows at wastewater treatment plants (WTPs).

As with surface water flooding, discussed in the previous chapter, much more attention has been given over the last decade to storm overflows into the environment as a consequence of the following:

- Realisation of the large number of extant, permitted CSOs
- The high number and duration of CSO spills discovered since routine monitoring has been enforced (see Section 2.5.5)
- The concern that these spills can have a significant environmental, public health and social impact
- A greater understanding of the implications of climate change (as discussed in Chapter 4)
- Population growth and increasing urbanisation resulting in more wastewater and more surface water runoff
- Increasing urban creep resulting in more surface water runoff (as discussed in Chapter 10)
- Changing public attitudes and expectations

This chapter deals with the special characteristics of combined sewers, and, in particular, with CSOs. In Section 13.2, we discuss how flows are accommodated in combined sewers; in Section 13.3, we go on to describe the specific role of CSOs leading to Section 13.4, which discusses how and to what extent pollution is controlled using this system. Various CSO designs are explained in Section 13.5, their design details are described in Section 13.6, and their performance is discussed in Section 13.7. The chapter is concluded with Section 13.8 which considers the management of CSOs including their problems and potential solutions.

DOI: 10.1201/9781003408635-13

13.2 SYSTEM FLOWS

The inflow to a combined sewer system consists of both wastewater (see Chapter 3) and stormwater (see Chapter 5). At the point of inflow, the flow rates can be calculated from the methods given in Chapters 9 and 10.

A typical layout of a small combined sewer system including a CSO is given in Figure 13.1. All connections of stormwater and wastewater are made to the single combined sewer.

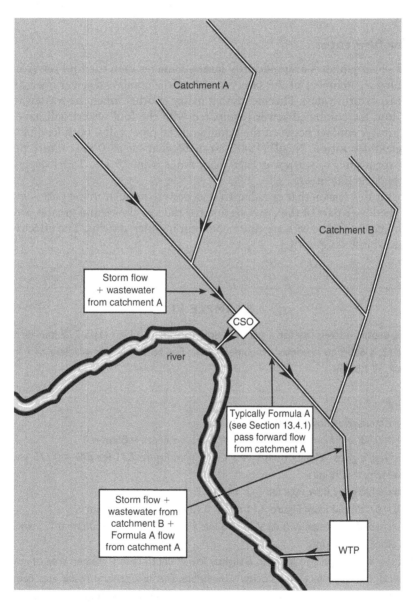

Figure 13.1 Typical layout of a combined sewer system (schematic plan).

Upstream of the CSO, the pipe carries the full combination of stormwater and wastewater from the upstream catchment. If the combined flow does not exceed the *setting* of the CSO (see Section 13.4.1 for further details), all continues to the WTP. If the combined flow exceeds the CSO setting, there will be overflow to the watercourse, and the flow retained in the system downstream will be determined by the CSO setting. Therefore, at different points throughout a combined sewer system in storm conditions, there can be significant differences in both the rate and composition of flow. In addition, the downstream system can substantially influence the ability of the CSO to pass forward flow (PFF). Note, that the WTP itself will typically have storm tank storage and a high-level overflow to the watercourse.

13.2.1 Low flow rates

A combined sewer pipe has a significantly larger diameter than the foul sewer in a separate system draining a catchment of the same size (since the combined sewer must also have the capacity to carry stormwater). This means that in dry weather, when the wastewater flow rate is relatively low, the combined sewer (compared with the foul sewer) will have lower flow depths and greater contact between the liquid and the pipe wall – both leading to a greater risk of sediment deposition. Nicoll (1988) estimates that for a 10-year return period storm, the ratio of stormwater to wastewater flow can range from 25:1 to 150:1 depending on the time of day and catchment size.

It is partly for this reason that egg-shaped sewers were popular in the past – with a smaller diameter in the lower part of the cross section for the low flows, and greater area above for storm flows. Egg-shaped sewers are quite common in older systems. The effect is illustrated and investigated in Example 13.1.

EXAMPLE 13.1

Determine depth and velocity for a part-full flow rate of 15 L/s in (1) a 225 mm diameter foul sewer, and (2) a 600 mm diameter combined sewer. For both, take the gradient as 1 in 100, and roughness k_s as 1.5 mm.

Solution

1. Chart/table gives flow rate full $Q_f = 45$ L/s
   $Q/Q_f = 0.33$, and from Figure 7.11 $d/D = 0.375$, so depth = 85 mm
   Chart/table gives velocity (full) $V_f = 1.1$ m/s from Figure 7.11 for $d/D = 0.375$, $v/v_f = 0.9$, so velocity = 0.99 m/s
2. Chart/table gives flow rate full $Q_f = 610$ L/s
   $Q/Q_f = 0.025$, and from Figure 7.11 $d/D = 0.125$, so depth = 75 mm
   Chart/table gives velocity (full) $V_f = 2.15$ m/s from Figure 7.11 for $d/D = 0.125$, $v/v_f = 0.45$, so velocity = 0.97 m/s.

Note that the larger pipe size leads to a slightly lower depth (and a greater area of contact with the pipe wall), but the effect on velocity is negligible. This is a general result and demonstrates that egg-shaped sewers do not have such a significant effect on low-depth velocities as might be expected.

13.3 THE ROLE OF CSOS

13.3.1 Flow

A CSO has two main functions. The first is hydraulic: to take an inflow and divide it into two outflows, one to the WTP (the continuation flow, "pass forward flow," or "setting") and one to the watercourse (the spill flow). This is illustrated in Figure 13.2 which is an expanded diagram of the CSO shown in Figure 13.1. Here, Q_u is the combined sewer inflow rate (m³/s), Q_d is the continuation flow rate (m³/s), and Q_s is the spill rate (m³/s). Also shown is Q_1 which is the watercourse flow rate upstream of the CSO (m³/s) and Q_2 which is the watercourse flow rate downstream of the CSO (m³/s).

The most common means of achieving the flow split is a weir. If the surface of the flow passing through the CSO is below the crest of the weir, flow continues to the WTP only. As the flow rate increases, so does the level of the water surface. When the water surface is above the weir crest, some flow passes over the weir, while the rest continues to the WTP. The flow rate over a weir is related to the depth of water above the crest, so if the water surface continues to rise, so does the spill flow. The continuation flow is also likely to rise slightly as a result of the increase in head.

The hydraulic design of a CSO requires care (as considered in more detail later). There could be a number of effects of poor hydraulic design. If a spill takes place prematurely, the capacity of the continuation pipe will be under-used, and an unnecessarily large volume of polluted flow could be discharged to the watercourse. But if the weir is set too high, an excessive surcharge of the upstream system might be caused, potentially leading to flooding, and too much flow might be forced down the continuation pipe causing flooding elsewhere in the sewer system including at the WTP. In a good hydraulic design, the spill will take place at the optimum level, and the continuation flow will not increase greatly while the spill flow increases with rising water level.

13.3.2 Pollution

The second function of a CSO is related to pollution. The ideal would be that all pollutants continue to the WTP (i.e., are retained within the sewer system) but this is not achieved as we will see later. Figure 13.2 also shows a non-specific pollutant concentration in the combined sewer flow X (mg/L), where X_1 and X_2 are the concentration of the same non-specific pollutant in the watercourse just upstream and downstream of the CSO, respectively (mg/L). We assume that the CSO chamber has *no impact* on the discharged pollutant concentration

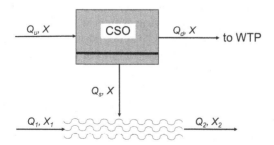

Figure 13.2 Schematic representation of flows and concentrations of wastewater at a CSO

(X), that the upstream watercourse pollutant concentration, $X_1 = 0$ and that the conditions in the watercourse just downstream of the CSO are fully mixed (Aspinall, 1981; Nicoll, 1988).

Let's use multiples of DWF to represent the combined sewer inflow ($\alpha \times DWF$), the continuation flow or setting ($\beta \times DWF$), and the watercourse flow ($\gamma \times DWF$). If DWF = average dry weather flow rate (m³/s) and x = DWF pollutant concentration (mg/L), then

$$Q_u = \alpha.DWF, \quad Q_d = \beta.DWF, \quad Q_s = (\alpha - \beta).DWF$$

$$Q_1 = \gamma.DWF, \quad Q_2 = [\gamma + (\alpha - \beta)].DWF$$

$$X = x / \alpha.$$

Load in upstream sewer = $X.Q_u = (x/\alpha).\alpha.DWF$
Load in downstream sewer = $X.Q_d = (x/\alpha).\beta.DWF$
Load in spill flow = $(x/\alpha).(\alpha - \beta).DWF$

Therefore, the pollutant concentration in the watercourse immediately downstream of the CSO X_2 (mg/L) is given by Equation 13.1.

$$X_2 = \frac{\left(\dfrac{x}{\alpha}\right)(\alpha - \beta).DWF}{[\gamma + (\alpha - \beta)].DWF}$$

$$X_2 = x\frac{(\alpha - \beta)}{\alpha[\gamma + (\alpha - \beta)]} \tag{13.1}$$

Equation 13.1 illustrates the main factors affecting the operation of the CSO and its impact on the receiving watercourse; namely, the flow rate in the sewer, the CSO setting, the flow rate in the receiving water, and the relationship between them. Example 13.2 illustrates its application. It should be recognised, however, that this model is a simplification with the key assumption being that the stormwater pollutant concentration is zero, so the combined sewer pollutant concentration X (and hence in the overflow spill) is reduced by dilution in proportion to the flow multiplier α. This model also neglects any first foul flush effects, discussed in Section 13.3.3.

The performance of CSOs with respect to a range of pollutants is discussed later in the chapter (Section 13.7.1). However, the main message is that although CSOs with screens demonstrate success in retaining larger gross solids, fine suspended and dissolved material, and microorganisms tend to be split between the continuation flow and spill flow in the same proportion as the split in the flow rates (as assumed in Equation 13.1).

The impact of CSO discharges on receiving waters was considered in Section 2.5. These impacts are likely to be most serious when CSOs are poorly designed, operating ineffectively, are overloaded or cannot pass an appropriate amount of flow downstream. Also, sewers that back up as a result of sediment deposition or other blockage problems may cause CSOs to operate prematurely (before the inflow has reached the CSO setting), even in apparently dry weather conditions (so-called "dry spills"). This may cause pollution of receiving waters, particularly as dilution could be limited. Indeed, the evidence shown in Table 13.1 indicates recorded pollution incidents linked to storm overflows are predominantly caused by blockages in the sewer network (Section 18.5) rather than in response to unusually heavy rainfall (i.e., authorised spills).

EXAMPLE 13.2

A CSO is located on a combined sewer that conveys wastewater and stormwater with BOD_5 concentrations of 300 mg/L and 0 mg/L, respectively. During a heavy rainfall event, the combined sewer flow rate increases to 25 × DWF. Calculate the CSO spill BOD_5 concentration and comment on it. Also, determine the BOD_5 concentration in the watercourse immediately downstream of the CSO with

1. a CSO continuation flow (setting) limited to $6DWF$ and assuming the receiving watercourse provides a dilution rate of $8DWF$
2. an $8DWF$ continuation flow assuming the same watercourse dilution
3. a $6DWF$ continuation flow but assuming half the watercourse dilution.

Comment on the results.

Solution

$x = 300$ mgBOD$_5$/L, $\alpha = 25$

$$X = \frac{300}{25} = 12.0 \, mg/L$$

This concentration is lower than compliance limits for secondary treated effluent @ 25 mg/L BOD_5 (EA, 2019). Using Equation 13.1:

1. $\beta = 6, \gamma = 8$

$$X_2 = x \frac{(\alpha - \beta)}{\alpha \left[\gamma + (\alpha - \beta) \right]}$$

$$X_2 = 300 \frac{(25 - 6)}{25 \left[8 + (25 - 6) \right]} = 8.4 \, mg/L$$

2. $\beta = 8, \gamma = 8$

$$X_2 = 300 \frac{(25 - 8)}{25 \left[8 + (25 - 8) \right]} = 8.2 \, mg/L$$

3. $\beta = 6, \gamma = 4$

$$X_2 = 300 \frac{(25 - 6)}{25 \left[4 + (25 - 6) \right]} = 9.9 \, mg/L$$

Most of the reduction in BOD in this example is caused by dilution by the stormwater.

Table 13.1 Causes of sewer overflow pollution events in England 2015–2020 (Adapted from NAO, 2021)

Overflow cause	Fraction (%)
Authorised activity	7
Blockages and failures	79
Unauthorised activity	11
Unknown or other	3

13.3.3 First foul flush

In some systems, a significant feature is the first foul flush in early storm flows, which may contain particularly high pollutant loads. These are likely to have been derived from the following:

1. *Catchment surface washoff and gully pots.* A first flush from this source would be expected as a result of the early rainfall washing off pollutants accumulated on the catchment surface and in gully pots since the last rainfall.
2. *Wastewater flow (including gross solids).* Since the storm wave moves faster than the wastewater *baseflow*, the front of the wave can consist of an ever-increasing volume of overtaken undiluted baseflow. However, this is normally diluted to some extent by the inflow from intermediate branches. Solids entering may float, be neutrally buoyant, or move along the bed (contributing to point 3).
3. *Near-bed solids.* In many sewer systems, high concentrations of organic solids have been observed in a layer moving just above the bed. The added turbulence as the storm flow increases causes these solids to become mixed with the stormwater.
4. *Pipe sediments.* Increasing storm flows provide suitable conditions for the re-erosion of the deposited material.

A first foul flush can be identified on hydrographs and pollutographs recorded in the system. An obvious sign would be a sharp increase in pollutant concentration near the start of a storm. In fact, even if concentration remained constant as the flow rate increased, this would signify an increase in pollutant load rate. A first flush can also be identified by plotting cumulative load against cumulative flow volume (Figure 13.3). The 45° line indicates that pollutants are uniformly distributed throughout the storm. If the line for a particular storm is above the 45° line, a first flush is suggested. Flushes from different conditions or catchments can be compared in this way. In some catchments, the effect is pronounced.

13.4 CONTROL OF POLLUTION

13.4.1 CSO permits and standards

All CSOs have (or should have) a permit to discharge. In England, for example, this is issued by the Environment Agency (EA, 2018). Permits are based on a combination of standards depending on the water environment being protected and these are determined in conjunction with the regulator using the partnership approach presented in the *Urban Pollution Management Manual* (FWR, 2019). Standards specified are of three different types: overflow design, water quality, and spill frequency.

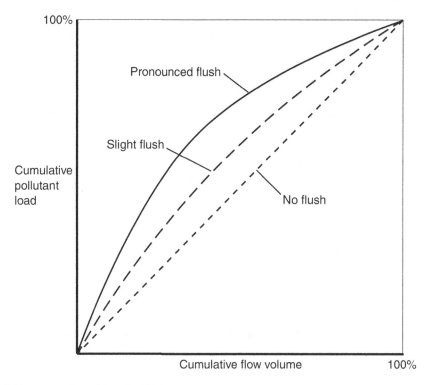

Figure 13.3 Representation of first foul flush.

13.4.1.1 Technical Committee Formula A

Until 1970, the traditional CSO setting (or continuation flow or PFF) had been 6 × DWF (6DWF) in the UK. This implies a 6-fold dilution of the wastewater with stormwater (less during peak water use times and more overnight) prior to discharge. This was based on the principle that the quality of the overflow was as good if not better than the quality of the treated effluent from the WTP assuming the diluting stormwater was not similarly polluted (but see Chapter 5). An extensive study of the effects of CSOs was carried out by the government-appointed "Technical Committee on Storm Overflows and the Disposal of Storm Sewage," whose final report was published in 1970 (Ministry of Housing and Local Government, 1970). Among the conclusions was that it was illogical to base a CSO setting *solely* on a multiple of DWF, and that, because of the harmful effects of CSO pollution, the new standard setting should give a "modest improvement" (i.e., divert less pollution to watercourses).

The setting of 6DWF had allowed for diurnal variations in wastewater flow, plus some stormwater (so that for *low-intensity* rainfall there would be no overflow). If people in one area happened to use more water than they did in another, that was no reason for more stormwater to be retained in the sewer system. It was considered appropriate for a CSO setting to be based on DWF plus some storm allowance (related to population and hence approximately to impervious area, but not to water consumption). The committee also felt there were ambiguities about the inclusion of infiltration and industrial flows in traditional practice. The proposed new standard CSO setting was given by their Formula A:

$$\text{Setting} = \text{DWF} + 1360P + 2E \tag{13.2}$$

for which, from Equation 9.1

$$DWF = PG + I + E$$

where P is the population, G is water consumption per person (L/d), I is the pipe infiltration rate (L/d), and E is average industrial effluent (L/d).

In a catchment where G is 200 L/head.day and there is no infiltration or industrial flow, Formula A gives a setting of 7.8 DWF (see Example 13.3).

The report recommended that the coefficients could be treated with flexibility. In situations where discharge was to a large river (i.e., where dilution was more than 8:1), $1360P$ could be decreased, or it could be increased if to a small stream; industrial effluent of high strength

---

EXAMPLE 13.3

A combined sewer catchment serves a population of 50,000. Determine the overflow setting required upstream of the main outfall sewer discharging to a river using the following approaches:

1. 6DWF
2. Formula A
3. River water quality adjacent to the overflow is limited to a BOD_5 concentration of 5 mg/L for a storm generating a sewer flow rate of 12DWF.

Additional information:
Per capita DWF rate = 200 L/d
DWF BOD_5 concentration = 250 mg/L
River flow rate upstream of CSO = 10DWF
River BOD_5 concentration upstream of CSO = 0 mg/L
Pipe infiltration and industrial flows are negligible.

Solution

1. DWF = 50,000 × 200 = 10.0×10^6 L/d or 116 L/s

   Setting = $6 \times 116 = 696$ L/s

2. Setting = $DWF + 1360P + 2E$

   $$= 10.0 \times 10^6 + 1360 \times 50,000$$

   $$= 903 \text{ L/s} \left(= 7.8DWF \right)$$

3. Solve Equation 13.1 for β

   $X_2 = 5$ mgBOD/L, $x = 250$ mgBOD/L, $\alpha = 12, \gamma = 10$

   $$\beta = \frac{\alpha \left(x - x_2\gamma - x_2\alpha \right)}{x - x_2\alpha}$$

   $$\beta = \frac{12 \left(250 - (5 \times 10) - (5 \times 12) \right)}{250 - (5 \times 12)} = 8.8$$

   Setting = $\beta.DWF = 8.8 \times 116 = 1021$ L/s

might require an increase in the term $2E$. A typical Formula A overflow, which causes "no harm" to the water environment, will spill approximately 40 times a year (Martin Osborne, personal communication).

The report also contained many other detailed recommendations about CSO design.

13.4.1.2 Scottish Development Department (SDD)

The report of the Working Party on Storm Sewage (Scotland), *Storm Sewage: Separation and Disposal* (SDD, 1977), gave details of another significant study of CSOs. Guidelines for CSO setting and storage volume were related to the amount of dilution when combined sewer flow was overflowed to a watercourse (a factor not covered explicitly in Formula A). Dilution was defined as the *minimum flow* in the receiving water (the flow rate exceeded 95% of the time) compared with the sewer DWF. For dilution of more than 7:1, Formula A was considered satisfactory. For dilution of 6:1, it was recommended that either the setting should be enhanced to Formula A + 455P, or that Formula A should be used in its original form and storage should be provided. For any lower dilution, increasing amounts of storage were recommended, in conjunction with a Formula A setting. The key contribution of this work was to emphasise the importance of the link between the CSO setting (PFF) and the receiving water dilution and to propose design approaches to address it (i.e., greater PFF or provision of sewer storage).

13.4.1.3 Freshwater standards

The following standards are relevant to protecting freshwaters from the impact of overflows:

- Formula A (Section 13.4.1.1)
- 99-percentile standards (Section 2.6.3.1)
- Fundamental intermittent standards (FIS) (Section 2.6.3.1)

Guidance on which standard to apply is based on both the significance of the discharge (in terms of the characteristics of the receiving water and the sewer system) and the environmental concerns being addressed (EA, 2018). Table 13.2 indicates the standards and modelling approaches to be adopted depending on receiving water dilution, interactions with other discharges, and the freshwater fisheries present. The basic assumption is that, in most cases, simple standards and models produce more conservative solutions.

Table 13.2 Standards and models to assess the significance of CSOs on freshwater

Significance	Dilution*	Interactions	Population equivalent	Fisheries	Standard	Modelling
Low	> 8:1	None	—	—	Formula A	Verified sewer model
Medium	> 2:1	Limited	< 10,000	Cyprinid	99 percentile	Verified sewer model. Simple stochastic river models
High	< 2:1	Significant	> 10,000	Cyprinid or salmonid	FIS	Detailed, verified sewer hydraulic and quality model. Calibrate driver quality impact model

* 5 percentile river flow: DWF(Adapted from EA, 2018).

13.4.1.4 Estuaries and coastal water standards

Formula A will normally apply as a minimum. If a discharge is made into the freshwater end of an estuary, modelling to assure 99-percentile or FIS standards are met may be required especially for high-significance discharges. For transitional and coastal waters impact modelling to demonstrate compliance with the relevant Water Framework Directive (WFD) standards may be required (EA, 2018).

13.4.1.5 Shellfish waters standards

The following standards are relevant to protecting shellfish waters from the impact of overflows:

- spill frequency emission standards
- environmental quality design standards (Section 2.6.3.2)

Overflows discharging into a designated shellfish water should have *no more than 10 significant spills on average per year* assuming each significant spill impacts for up to 24 hours. Alternatively, the faecal coliform or *E. coli* standard given in Section 2.6.3.2 can be used in conjunction with marine impact modelling (EA, 2018).

13.4.1.6 Bathing waters standards

In a similar way, standards relevant to protecting bathing waters from the impact of overflows are:

- spill frequency emission standards
- environmental quality design standards (Section 2.6.3.3)

For overflows that discharge into designated bathing waters with a target of sufficient or good status, the emission standard calls for *no more than three significant spills per bathing season* (May–September) on average. For those discharging into waters targeting excellent status the standard is *two spills per season*. For marine discharges, this frequency should be even lower depending on the soffit level of the outfall pipe relative to the tide mean water levels. Bathing water EQSs are based on limiting the concentration of intestinal *enterococci* and *E. coli* to the levels given in Table 2.6 in Section 2.6.3.3, as demonstrated by comprehensive marine impact modelling and site investigations (EA, 2018). Some 8% of overflows in the UK are close to designated bathing waters (Defra, 2023).

13.4.1.7 Aesthetic control standards

Aesthetic control standards have been based on the amenity of the receiving water, the expected spill frequency and appropriate screen design. It is now recommended that screens (providing separation from the effluent of a "significant quantity" of solids greater than 6 mm in any two dimensions) are fitted to *all* overflows as a matter of course (EA, 2018).

13.4.2 Storm Overflows Discharge Reduction Plan

Water companies in England have been given a legally binding duty under the Environment Act 2021 to progressively reduce the adverse impacts of discharges from storm overflows. The Storm Overflows Discharge Reduction Plan (Defra, 2023) sets out three headline targets.

13.4.2.1 Protecting the environment

Water companies are only permitted to discharge from a storm overflow where they can demonstrate that there is no "local adverse ecological impact." The target must be achieved for at least 75% of storm overflows discharging in or close to high priority locations (such as Sites of Special Scientific Interest) by 2035, for 100% by 2045, and for all remaining storm overflow sites by 2050.

13.4.2.2 Protecting public health in designated bathing waters

Water companies must significantly reduce harmful pathogens from storm overflows discharging into and near designated bathing waters by either applying disinfection or reducing the frequency of discharges to meet Environment Agency spill standards by 2035.

13.4.2.3 Ensuring storm overflows operate only in unusually heavy rainfall events

Storm overflows will not be permitted to discharge above an *average of 10 rainfall events per year* by 2050. In addition, *all* storm overflows must have screening controls to limit the discharge of gross solids, that is persistent inorganic material, faecal and organic solids. This changes the recommendation mentioned in Section 13.4.1.7 into a requirement.

In this plan, "No local adverse ecological impact" is defined in terms of achieving the UPM FIS or 99-percentile standards for ammonia and dissolved oxygen downstream of the discharge point, and the bathing water standards are those set out by the Environment Agency. Both are discussed in Section 2.6.3.

13.4.3 Monitoring

Event duration monitoring of CSOs is now universally practised in the UK for the following reasons:

- To build up a picture of their performance and reliability
- To demonstrate whether overflows are meeting their permit requirements
- To provide near real-time emergency response data
- To provide near real-time information to the public.

Further details on the monitoring regime and equipment are provided in Section 18.4.2.

13.5 CSO DESIGN

In this section, we will discuss the different CSO designs that are currently available or in use. It should be noted that CSOs have not been routinely built into sewers since combined sewers were superseded by separate sewers and later SuDS. However, many existing overflows have been improved since then and many more will require further upgrades. They are a ubiquitous legacy component.

13.5.1 High side weir

13.5.1.1 Principles

Most modern CSOs are high side weir overflows with screens. We consider first the basic design of a high side weir overflow without screens, used predominantly in the 1980s and 1990s before the development of robust fine screens. An overflow with high weirs, scumboards, and a stilling zone upstream can provide reasonable retention of both floating and sinking solids. Double-side weirs can provide good hydraulic control. High side weirs are often associated with storage. This can be a storage zone downstream of the weir for retention of floating solids, or a large storage volume for retention of the first flush.

13.5.1.2 Dimensions and layout

Recommendations are given by Balmforth et al. (1994) (Figure 13.4) and Wastewater Planning Users Group (WaPUG, 2006). Flow in the chamber must be subcritical. Many poorly performing CSOs (with screens) result from unsatisfactory hydraulic conditions due to their basic design leading to a hydraulic jump in the chamber, potentially resulting in pollution and expensive rehabilitation. Designing high sided weir chambers and their screens are discussed in greater detail in Sections 13.5.2 and 13.6, and Example 13.5.

13.5.2 Screens

13.5.2.1 Principles

Screens are incorporated as a matter of routine within CSO structures. They provide a physical barrier to solids with the aim of preventing them from entering the overflow pipe. The

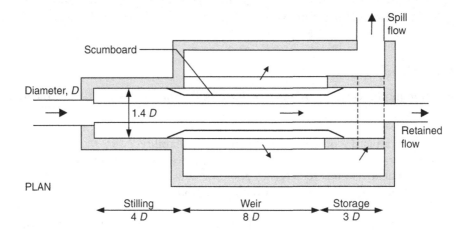

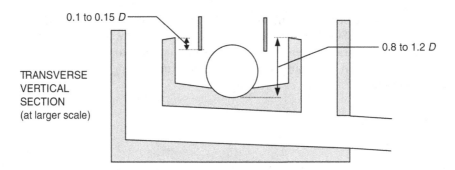

Figure 13.4 High side weir CSO: general arrangement and dimensions.

focus is on gross solids, defined in this context as being larger than 6 mm in two dimensions and of sewage origin. Common screens are meshes or perforated plates (Figure 13.5), which prevent these solids from being discharged by retaining them in the sewer system. "Static" screens rely on flow patterns within the CSO and the design of the screen itself to prevent blocking or blinding but typically have to be cleaned manually as part of planned and reactive maintenance. Powered screens are mechanically cleaned during storm flows to prevent blinding, usually by brushing. However, many different approaches to CSO screen design have been tried, as described in the next section.

13.5.2.2 Development

A field study of screens at CSOs (Meeds and Balmforth, 1995) concluded that mechanically raked bar screens are unlikely to achieve retention efficiencies of greater than 50% (of all gross solids). Screens typically now consist of a mesh rather than parallel bars, partly as a result of solids separation requirements. A laboratory study (Saul et al., 1993) demonstrated that a mesh screen is more effective at retaining solids than a bar screen with the same spacing, though a field study using actual wastewater concluded that 6-mm mesh screens are unlikely to achieve retention efficiencies of greater than 60% (Balmforth et al., 1996).

Figure 13.6 shows a summary of the results from a testing programme at a dedicated CSO research facility. The resulting plots of solids separation efficiency for 6-mm mesh screens are presented as standards against which any alternative or novel approach can be compared using a defined test procedure. *Total efficiency* is the percentage of the total mass of solids entering the CSO that is retained within the sewer system (in the continuation flow or in the

Figure 13.5 Example of a static screen, Hydro-Static® Screen. (Courtesy of Hydro International.)

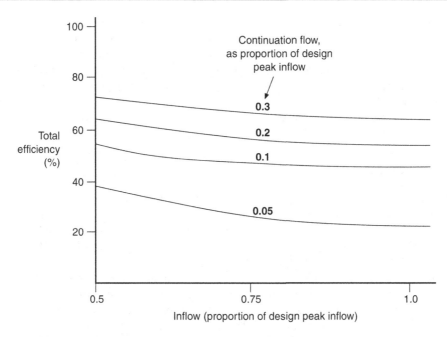

Figure 13.6 Efficiency for 6-mm solids separation. (Based on FWR. 2019. *Urban Pollution Management Manual*, 3rd edn, Foundation for Water Research, http://www.fwr.org/UPM3. With permission of Foundation for Water Research, Marlow.)

chamber). It should be noted that even under the most favourable conditions, total efficiency hardly exceeds 70%. A wide range of proprietary screen arrangements have since been tested at the same facility and compared with these standards in order to help engineers choose suitable devices (Saul, 2000; UKWIR, 2006).

13.5.2.3 Dimensions and layout

It is possible to fit screens to any of the CSO types described in this chapter. However, the standard UK recommendation (WaPUG, 2006) is for CSOs incorporating screens to be of the high sided weir type.

The recommended dimensions and layout in the WaPUG guide to *The Design of CSO Chambers to Incorporate Screens* (WaPUG, 2006) are based on physical tests and on computational fluid dynamics (CFD) modelling. The inlet length of the chamber allows flows to turn onto the screen, with a minimum value of $0.4D_{in}$ but must be at least 0.5 m. Outlet lengths enable solids from the screens to be returned to the outgoing flow and continue through the network. Outlet lengths are based on $1.5D_{in}$. The recommended chamber width is $1.4D$ (where D is the diameter of the incoming pipe), which is the same as the recommendation by Balmforth et al. (1994) for a high sided weir CSO without a screen (Figure 13.4). All these dimensions must be sufficient to enable maintenance of screens; therefore, in smaller CSO chambers, it is common for larger than the minimum size chambers being built.

The weir length should be no more than $6D$ (compared with $8D$ by Balmforth et al. [1994] in Figure 13.4), and the stilling and storage lengths are not needed, though shorter inlet and outlet lengths are still recommended. Either double- or single-side weirs are appropriate. The basic layout of a single high side weir overflow suitable for incorporating screens, using the principles of the WaPUG (2006), is given in Figure 13.7.

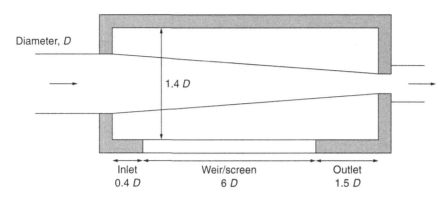

Figure 13.7 Side weir CSO for screens: general arrangement and dimensions with a typical weir/screen length.

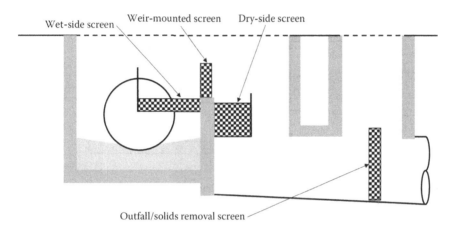

Figure 13.8 Possible positions of CSO screens.

The guide refers to three possible positions for the screen in a high sided weir CSO given in Figure 13.8. A fourth alternative position is on the outfall, similar to a screen used at the inlet works to WTP. The three common positions for screens are as follows:

- Mounted vertically on the weir, so that flow over the weir passes horizontally through it
- Mounted horizontally above the main channel, so that flow over the weir must first pass upward through the screen
- Mounted on the downstream face of the weir, so flow that has passed over the weir then passes down through the screen

13.5.3 Stilling pond

13.5.3.1 Principles

The main principles of a stilling pond CSO are illustrated in Figure 13.9. In dry weather and low-intensity rain, the flow enters via the inlet pipe, passes along a channel through the overflow, and leaves via the throttle pipe. In heavier rainfall, as the inflow increases, the capacity

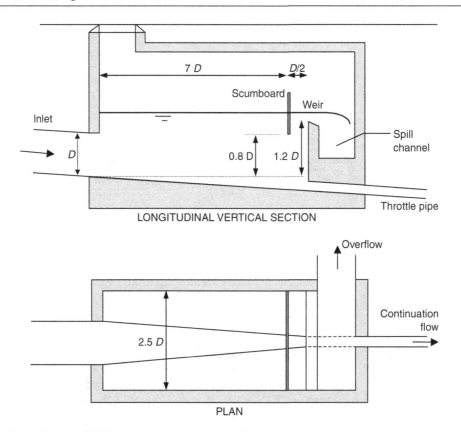

Figure 13.9 Stilling pond CSO: general arrangement and dimensions.

of the throttle pipe is exceeded and flow backs up inside the chamber. The level usually has to rise above the top of the inlet pipe before it reaches the crest of the weir. This causes the inflow to become stilled, which helps to ensure that sinking solids are not carried over the weir. When the water level is above the weir crest, water spills over the weir and out via the spill channel and pipe. The scumboard is positioned to limit the passage of floating solids over the weir. In fact, it does more: it sets up a pattern of circulation in the chamber, which brings many floating solids back to the upstream end of the chamber – making it even less likely that they will flow over the weir.

13.5.3.2 Development

The first major investigation was by Sharpe and Kirkbride (1959). The results are much quoted and had a genuine impact on engineering practice. They concluded that the best conditions were achieved when the inlet velocity was low and the upstream sewer was well flooded in order to create stilling conditions in the chamber. The scumboard created a reverse surface flow that took the floating solids away from the weir. Their recommendation was that the distance from the inlet to the scumboard be at least 4.2 times the diameter of the inlet pipe (*D*). Other recommended dimensions included a chamber width of 2.5*D*, a distance from the scumboard to the weir of 0.5*D*, and a weir level similar to the soffit of the incoming pipe.

Frederick and Markland (1967) carried out laboratory studies of model stilling pond arrangements. Many of their conclusions confirmed those of Sharpe and Kirkbride (1959), "in particular, the incoming sewer needs to be surcharged in order to produce a favourable reverse current near the surface." The main difference in their conclusions was in terms of length, which they recommended should be as great as possible, with an overall length no less than seven times the inlet diameter.

Balmforth (1982) studied the separation of a wide variety of solids in a model stilling pond. A particular aim was to resolve differences in the recommendations for chamber dimensions between Sharpe and Kirkbride and Frederick and Markland. Balmforth confirmed that there were significant advantages to the longer length recommended by Frederick and Markland.

13.5.3.3 Dimensions and layout

Recommendations for chamber dimensions, based on the development described above and best knowledge of CSO operation, are given in the *Guide to the Design of Combined Sewer Overflow Structures* by Balmforth et al. (1994). Their recommended dimensions for a stilling pond are given in Figure 13.9. They are based on the diameter of the incoming pipe. A method of determining this diameter (common to a number of different CSO types) is given in Section 13.6.

A DWF channel runs along the centre of the chamber, contracting in area from the inlet to the throttle pipe. It should have sufficient size and longitudinal slope to carry a flow rate equal to the capacity of the throttle pipe and avoid sediment deposition. The base on either side of this channel slopes towards the centre to drain liquid to the DWF channel. The capacity of the throttle pipe is crucial in determining the setting of the CSO. The upstream head is a function of the crest level and characteristic of the weir.

13.5.4 Hydrodynamic vortex separator

13.5.4.1 Principles

Several types of overflow arrangement exploit the separation of solids that occurs in the circular motion of a liquid. When such a flow is considered in two dimensions, by studying a horizontal section, theory suggests that heavier solids will follow a path towards the outside of the circle. This leads to a design of *vortex overflow* in which the overflow weir is placed on the inside of the chamber and the continuation pipe on the outside (where heavier solids tend to congregate). Floating solids are prevented from flowing over the weir by a baffle.

Studies in three dimensions, with particular chamber shapes, have suggested that heavy solids collect at the bottom of the chamber, in the centre. These have led to designs with the opposite arrangement: the weir on the outside, and the continuation pipe at the centre. Other, more elaborate chamber arrangements have led to flow patterns with even more complex properties.

13.5.4.2 Development

Smisson (1967) carried out pioneering work on models and full-scale vortex overflows, giving detailed descriptions of flow patterns, and design recommendations. The weir was positioned on the inside of the vortex and the continuation pipe on the outside.

A different type of vortex chamber was proposed by Balmforth et al. (1984), called a "vortex overflow with peripheral spill." This design makes use of the known ability of vortex motion to separate settleable solids, but differs from earlier designs in that the foul outlet pipe is set in the centre of the chamber floor, and the overflow occurs over a weir formed in the peripheral (outer) wall.

Modern descendants of the vortex overflow are called *hydrodynamic separators*. A patented design, the Storm King® Overflow, in which separation of solids takes place within a

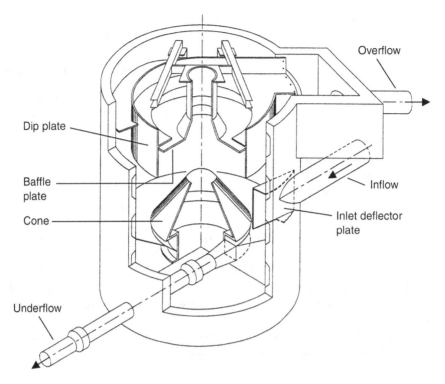

Figure 13.10 Storm King® Overflow. (Courtesy of Hydro International.)

complex flow pattern of upwards and downwards helical flow before passing through a static screen, has been applied in the UK and elsewhere. The arrangement is shown in Figure 13.10. These devices are claimed to have very high gross solids (2D 6 mm) and sediment removal rates with relatively small footprints.

Similar use of these principles has been made in other countries, the US Environmental Protection Agency's (EPA) "swirl regulator" (Field, 1974), and the German HYDROVEX® Fluidsep vortex separator (Brombach, 1987, 1992), for example.

13.5.4.3 Dimensions and layout

Hydrodynamic separators are designed and fabricated by their manufacturers based on performance specifications.

13.5.5 WRcCSO

13.5.5.1 Principles

Tzinaetis and Moy (2019) describe an alternative CSO design (Patent GB2504927, Trademark UK00003922037). It is based on an online, in-series, double (large and small) chamber with a throttle downstream of each chamber and weirs set at different levels in each chamber. Spill from the first chamber (stilling pond) is unscreened and is connected to the downstream sewer system via a bypass pipe. Spill from the second chamber is to the receiving watercourse and is passed through a *static* screen. The second weir is set at a higher level than the weir in the first chamber allowing preferential flow from the first chamber, carrying the floating

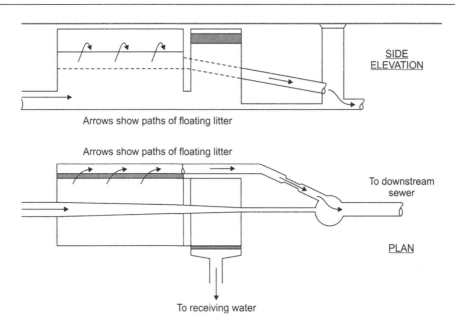

Arrows show paths of floating litter

SIDE ELEVATION

Arrows show paths of floating litter

To downstream sewer

PLAN

To receiving water

Figure 13.11 WRcCSO (Courtesy of WRc).

gross solids, via the bypass pipe to the downstream sewer. The design PFF is achieved by a combination of flow via the bypass pipe from the first chamber and continuation flow from the downstream chamber.

13.5.5.2 Development

The overall benefits of this configuration include:

- No mechanical raking is required due to reduced solids on the screen
- If the screen overtops in extreme events, as there are minimal floating solids in the spill chamber, the aesthetic pollution and cost of litter picking are reduced
- Monitoring of the bypass spill gives an early warning of imminent dry weather discharges before they occur
- Unpublished CFD modelling and field trial results show a gross solids total efficiency > 90% and > 80%, respectively.

13.5.5.3 Dimensions and layout

An elevation and plan of the CSO are shown in Figure 13.11. The detailed dimensions are proprietary.

13.5.6 Storage

13.5.6.1 Principles

The aim of providing storage at a CSO is to retain more of the pollutants in the sewer system rather than allowing the pollutants to be overflowed to a waterbody, even after a weir on the main sewer has come into operation during a storm. When flows in the system have subsided after the storm, the polluted flow retained in the storage can be passed onwards to treatment. Clearly, the larger (and more expensive) the storage, the lower the amount of pollution reaching the waterbody. Optimum sizing of storage ideally needs to take into account the fact that polluting loads during storm flow vary with time. It is common (but not universal) that early flows are particularly polluted as a result of a first foul flush, as considered in Section 13.3.2.

Storage has tended to be provided in conjunction with a high side weir arrangement, but storage can be used to supplement any CSO configuration.

Storage can be provided online or offline. In an *online* arrangement, the flow passes through the tank even in dry weather when the capacity of the tank is not being utilised. When the flow rate increases during a storm, a downstream control will cause the level to rise, to fill up the storage volume, and eventually overflow at the weir (Figure 13.12a). After the storm, the tank empties by gravity into the continuation pipe. In an *offline* arrangement, flow is diverted to the tank via a weir as the level begins to rise (Figure 13.12b). When the tank is full, a higher weir comes into operation and diverts further flows to the watercourse. Screens are typically fitted to either weir but are most commonly found on the diversion weir to reduce solids entering the storage in the first place. After the storm, the tank is emptied into the continuation pipe by gravity or by pumping. The rate at which the storage tank can be emptied is governed by the amount of spare capacity in the pipe and/or WTP.

Storage at a CSO can be provided in a number of forms: rectangular chamber, circular vertical shaft, or oversized pipe or tunnel.

13.5.6.2 Development

In the UPM procedures (FWR, 2019), the size of storage is optimised by the application of sewer quality modelling (either simple or complex). Because of the UPM emphasis on considering the system as a whole, design rules for sizing individual tanks are not proposed. Example 13.4 is an illustration of the way in which model simulations can be used to investigate possible storage proposals. A decision is not made until the full range of design

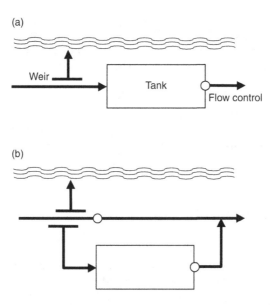

Figure 13.12 Storage tank: (a) online and (b) offline.

EXAMPLE 13.4

Figure 13.13 gives a simulated flow hydrograph and chemical oxygen demand (COD) pollutograph (concentration and load rate) for a catchment, in response to a particular rainfall pattern. Rainfall in this case started at a low intensity, causing a slight increase in flow rate and dilution of COD concentration (from the dry weather level of 470 mg/L). During the early period, the load rate is

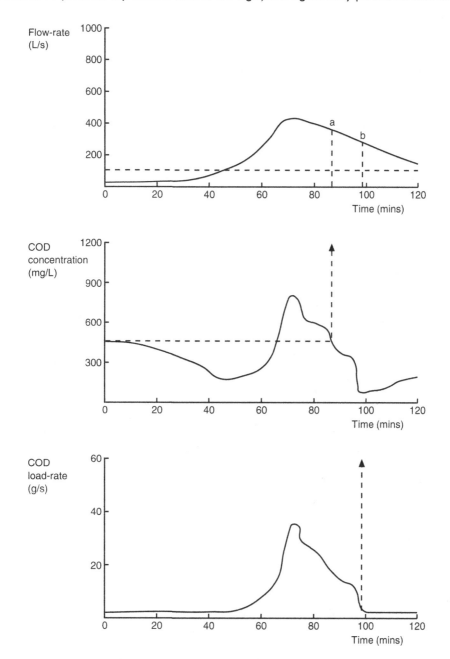

Figure 13.13 Storage: interpreting simulated hydrograph and pollutograph. (Example 13.4)

constant. After 45 minutes, the rain became more intense, causing a significant first flush, apparent from both the concentration and load-rate graphs.

Determine the approximate size of storage that would be needed (1) to retain pollutants until COD concentrations no longer exceed dry weather level and (2) to retain pollutants until COD load rate no longer exceeds dry weather level. Assume that the detention tank will have outflow via a control that limits flow rate to 100 L/s, and via an overflow that operates when the storage is full.

Solution

1. The volume retained would equal the area under the hydrograph (above the 100 L/s line) up to *a*.
   From the graph, the volume ≈ 440 m^3.
2. The volume retained would equal the area under the hydrograph (above the 100 L/s line) up to *b*.
   From the graph, the volume ≈ 580 m^3.

rainfall patterns has been considered, the costs of alternative storage strategies have been determined, and the effects on the rest of the system have been assessed.

Apart from optimising the size of storage, designers need to pay attention to the layout of the chambers to avoid excessive sedimentation. When a tank is full of virtually stationary liquid, conditions are ideal for the deposition of suspended solids (and concentrations are likely to be high in the first flush that the tank will have been designed to retain). In large storage tanks, it is not uncommon to use CFD to analyse and design the storage tanks and maintenance requirements (Hrudka et al., 2021).

13.5.6.3 Dimensions and layout

Saul and Ellis (1992) found that a DWF channel with a steep longitudinal gradient was helpful by creating suitably high velocities. Long narrow tanks with a single DWF channel had better self-cleansing properties than wider tanks with multiple channels. The length-to-width ratio should be as high as possible, and the width should not exceed 4 m. Guidance is also given by the Water Research Centre (1997). CFD modelling has confirmed the importance of the length-to-width ratio (Stovin and Saul, 2000).

13.6 CSO DESIGN DETAILS

13.6.1 Design return period

The design of the inlet pipe and determination of the main chamber dimensions of a CSO with screens are based on the peak flow rate with a 5-year return period being used to satisfy regulatory requirements (EA, 2018). A check should also be made to see how a proposed chamber would respond to more extreme events – including a 30-year return period storm. This is particularly true for the spill channel and outlet pipe, which is the route taken by most of the flow in extreme events.

13.6.2 Diameter of inflow pipe

Dimensions of CSO chambers are based on the diameter of the inflow pipe. The minimum diameter of this pipe is determined from

$$D_{min} = KQ^{0.4} \tag{13.3}$$

where Q is the peak inflow for the design return period (m³/s) and K is the constant taken from Figure 13.14 for a high sided weir.

A relationship (represented in Figure 13.14) between incoming pipe diameter and flow has been developed by physical testing for a range of high sided weir heights, c_w, and weir lengths, L_w (where the weir height is greater than 50% of the incoming diameter). This reduces the likelihood of supercritical flow forming in the chamber (subject to understanding the forces in the incoming pipe/inlet to the chamber).

The general form of Equation 13.3 has its origin in the Darcy–Weisbach Equation 7.8, which can be rearranged to

$$h_f = \frac{\lambda L Q^2}{12.1 D^5} \text{ or } D = \left(\frac{\lambda L}{12.1 h_f} \right)^{0.2} Q^{0.4}$$

where D is the diameter (m), λ is the friction factor (–), L is the length (m), h_f is the friction head loss (m), and Q is the flow rate (m³/s). A design example of the inlet pipe and chamber dimensions for a high sided weir CSO with a screen is given in Example 13.5.

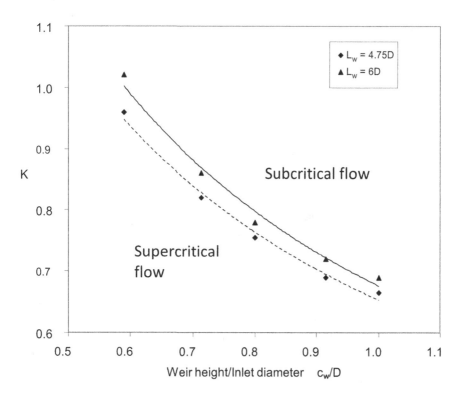

Figure 13.14 Determining values of K to calculate the incoming pipe size from the WaPUG CSO Design Guide. (Courtesy of CIWEM Urban Drainage Group.)

EXAMPLE 13.5

Determine the basic dimensions of a high sided CSO with a screen. Peak flow with a return period of 5 years is estimated as 900 L/s.

The length of the screen required to accommodate the spill flow indicates that the length of the weir (L_w) should be 4.4 m.

Solution

Try inlet pipe diameter (D) of 0.75 m, and weir height (c_w) of 0.6 m:

$$\frac{L_w}{D} = \frac{4.4}{0.75} = 5.9$$

$$\frac{c_w}{D} = \frac{0.6}{0.75} = 0.8$$

From Figure 13.13 (reading from the line for $L_w = 6D$), K should be no less than 0.8. Therefore,

$$D_{min} = KQ^{0.4} = 0.8 \times 0.9^{0.4} = 0.77 \, m$$

We may need to upsize the inlet pipe. An alternative might be to increase the height of the weir. If we upsize the inlet pipe to 825 mm while retaining the weir height of 0.6 m,

$$\frac{L}{D} = \frac{4.4}{0.825} = 5.3$$

$$\frac{c_w}{D} = \frac{0.6}{0.825} = 0.73$$

From Figure 13.13 (now reading between the two lines), K should be no less than 0.83. So, from Equation (13.3),

$$D_{min} = KQ^{0.4} = 0.83 \times 0.9^{0.4} = 0.8 \, m$$

This confirms that an 825 mm diameter inlet pipe, with a weir height of 0.6 m, would be appropriate.

Or, if we retain the 750 mm diameter inlet pipe but increase the weir height to, say, 0.64 m,

$$\frac{L_w}{D} = \frac{4.4}{0.75} = 5.9$$

$$\frac{c_w}{D} = \frac{0.64}{0.75} = 0.85$$

From Figure 13.13 (reading from the line for $L_w = 6D$), K should be no less than 0.76. Therefore,

$$D_{min} = KQ^{0.4} = 0.76 \times 0.9^{0.4} = 0.73 \, m$$

This confirms that the alternative of a 750 mm diameter inlet pipe, with a weir height of 0.64 m, would also be appropriate.

Since the dimensions of the chamber are generally based on the diameter D, let us select this second option. So, with $D = 0.75$ m, the dimensions should be as follows:

Dimension	Calculation	Value (m)	Notes
Chamber width	1.4 D	1.05	May need to be increased to accommodate the screen
Inlet length	0.4 D = 0.3 m	0.5	Minimum value of 0.5 m
Outlet length	1.5 D	1.125	
Weir length	Based on screen	4.4	
Overall length	0.5 + 4.4 + 1.125	6.025	

13.6.3 Creating good inlet flow conditions

A critical element in the design is to ensure that turbulence, particularly because of a hydraulic jump in the chamber, is not created, and this should always be investigated. Calculating the Froude Number (Equation 7.26) indicates if the flow in the incoming pipe is subcritical or supercritical. Where the flow is subcritical throughout the pipe and chambers, a hydraulic jump will not occur.

If supercritical flow meets subcritical flow, then the position of the hydraulic jump should be investigated using the specific force Equation (13.4) for the water in the pipe entering, Fs_1, and the water in the chamber, Fs_2. The design should check that force in the chamber is greater than the force in the pipe $Fs_2 > Fs_1$, so that the jump occurs upstream of the chamber:

$$Fs = \rho Qv + \rho g A \bar{z} \tag{13.4}$$

where $\bar{z}$ is the depth to the centroid of the flow cross section (m).

Flows, velocities, depths, and areas are calculated for the 5-year incoming flow to the pipe or chamber.

13.6.4 Weirs

The hydraulic characteristics of weirs are considered in Section 8.2. The weir length and height are critical in sizing the incoming pipe. There is a balance in selecting an appropriate weir height: if it is too high, it may increase the risk of flooding, but if it is too low it may create unfavourable conditions in the chamber for the screen to perform.

13.6.5 Selecting, sizing and accommodating the screen

Nearly all CSO chambers are designed with a screen unless they only spill in events with a return period greater than 5 years. The 5-year return period design should take into account forecast development and make a suitable allowance for partial screen blinding during spills (e.g., 50%). The chamber must be of sufficient size to accommodate the screen to ensure that

- Unfavourable hydraulic conditions are not created that reduce the effectiveness of the screen, or, in serious cases, damage the screen and lead to unscreened sewage being discharged.
- Maintenance can be completed safely including the ability to replace parts of the screen when all work is undertaken within the closed chamber.

The selection of the screen and its position depend on a number of factors, including:

- Hydraulic headroom – what is happening in the drainage network upstream and downstream
- Design spill flow that will pass through the screen
- Frequency of spill – higher-spilling CSOs require a self-cleaning screen
- Availability of a power supply or ease of providing a power supply
- Planning constraints if a kiosk is required for a powered screen
- Space requirements
- Accessibility for maintenance
- Cost to procure, build, and maintain

13.6.6 Control of outflow

Control of the continuation flow is an important part of the hydraulic design of a CSO. The setting of the overflow is normally defined as the continuation flow when the spill starts – that is, when liquid level reaches the weir crest. As flow rate over the weir increases, so will depth. It is best if the retained outflow does not vary greatly as a consequence. The common methods of control are:

- Fixed orifice
- Adjustable (manual or automated) penstock
- Vortex flow regulator
- Throttle pipe

These are described in Section 8.1. The control is typically designed to "freely" discharge; however, at times, it may become drowned, either because the downstream pipe capacity or the downstream network controls the flows.

13.6.7 Chamber invert

Chambers should be as self-cleansing as possible. Deposition of solids can be limited by suitable longitudinal and lateral slopes; lateral benching should slope between 1:4 and 1:8.

13.6.8 Top water level

The top water level (TWL) in the chamber can be determined from the design maximum inflow and the hydraulic properties of both outflow pipes. The impact of the CSO on the sewer system upstream and downstream may only be fully understood by using a suitable sewer system flow model. The outfall pipe needs to be checked carefully. If it could be drowned at the downstream end or if it discharges to tidal water, this will also need to be considered. Often more detailed hydraulic checks are appropriate beyond a sewer network model for the chamber performance. The TWL is one consideration in deciding the level of the roof of the structure.

13.6.9 Access

Human access is normally via manhole covers at ground level. Where screens are included, there needs to be appropriately sized and positioned access for vertical installation and removal of any machinery. There should be access to clear potential blockages, especially in throttle pipes. Thorough safety precautions are required during maintenance (considered in Section 18.9).

13.7 CSO PERFORMANCE

13.7.1 Pollutant removal

Laboratory tests, full-scale trials, and performance evaluation studies have been carried out on the main CSO types with similar conclusions (Jefferies, 1993; Nicoll, 1988; Saul et al., 1993). The studies demonstrated the importance of the terminal settling velocity distribution of the particulate matter in the flow in determining the efficiency of retaining the pollutants in the sewer system. Results are summarised in Figure 13.15, which demonstrates that pollutants with relatively high terminal velocities such as some gross solids or sewer sediment ("floaters" and "sinkers") are well retained in properly designed CSO chambers. Pollutants in solution, in suspended form or which are neutrally buoyant are not separated efficiently. The most aesthetically sensitive solids, with their close-to-neutral buoyancy can be effectively retained by the use of screens, as previously discussed (Section 13.5.2).

13.7.2 Choice of design

Figure 13.15 also indicates the performance of different CSOs, with each design represented by a different set of "gulls wings." There is no standout "winner" from these results. However, Myerscough and Digman (2008) observe that "the standard WaPUG chamber" – that is, a high side weir overflow fitted with screens using the principles of WaPUG (2006) – is well established as "best practice" but that there is often a need for a "non-standard" design. They point out that physical modelling and CFD play an important role in understanding how CSOs perform, and that such studies can allow these structures to be used with confidence, enable improvements to be made in design, and avoid costly mistakes. Jarman et al. (2008), presenting an overview of CFD modelling in urban drainage, state that "the most extensive application of CFD to urban drainage system analysis has been in relation to CSO chambers." However, the basic design of CSOs has remained largely the same for over 50 years and there is an urgent need for further research and new products.

It should be emphasised again that although most gross solids will be retained by screens and most sediment retained in modern designs, fine suspended particles and dissolved pollutants (including microorganisms) will be discharged roughly in proportion to the flow split at the CSO. Other approaches and options to address the need to minimise pollution impact are summarised in Section 13.8.2.

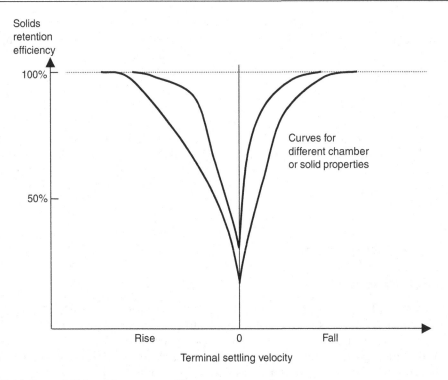

Figure 13.15 Plots of CSO solids retention efficiency (conceptual).

13.8 SYSTEM MANAGEMENT

13.8.1 Problems

A very positive and illuminating benefit of installing event duration monitors on all CSOs in the UK has been in establishing that some operate too frequently causing a potentially negative impact on waterbodies and concern to the public. Some of the underlying causes of this have been mentioned at the beginning of the chapter but a more complete list is as follows:

- Population growth and increasing urbanisation (increasing imperviousness) resulting in more wastewater and more surface runoff, respectively
- Increasing urban creep (further increasing imperviousness) resulting in more runoff
- Changing rainfall patterns as a result of climate change
- Diminished combined sewer capacity to manage surface runoff due to growth in DWF
- Increased groundwater infiltration due to ageing pipework
- Incorrect CSO settings causing overflows to operate more frequently than permitted
- Malfunctioning overflow screens causing aesthetic pollution in waterbodies
- Operational problems such as blockages, sediment deposition, or equipment failure that increase the depth in the PFF causing premature overflow operation.

Any combination of these complex and various causes can and does result in CSOs operating more frequently than expected compared with their design specification. Some of these factors can also result in overflows operating during periods of light or no rainfall when dilution is low in the receiving water (dry spills). It should be clear from this list that a significant reason for these problems is the failure of operators to keep pace with increased flows, increased maintenance needs, customer behaviour, and indeed customer sentiment. However, many aspects are beyond the direct control of operators. Blockages are strongly linked to misuse of the system due to flushing of inappropriate items such as with wipes and FOG, urban creep is typically caused by homeowners building impervious driveways, decking, and plastic grass, and a significant proportion of infiltration enters the public sewer through privately owned lateral connections on customers' properties.

13.8.2 Options

A variety of options can be used to reduce the volume and frequency of CSO discharges and hence reduce their environmental and health pollution impact on waterbodies. The main approaches are listed below:

- Reduce the flow rate and volume of surface runoff entering combined sewers using, for example, SuDS solutions (Chapter 11). This is sometimes called stormwater or surface water separation and is not to be confused with *sewer* separation (see final bullet point below)
- Further encourage water-efficient practices thereby reducing wastewater flows to "free up" system capacity (Chapter 3)
- Further encourage appropriate waste disposal practices to reduce sewer blockages (Chapter 18)
- Increase the hydraulic capacity of the combined sewers to retain more of the flow for full treatment (i.e., increase the CSO setting)
- Increase the hydraulic capacity of the WTPs to treat more flow
- Increase the volumetric capacity of the combined sewers to detain (temporarily store) more of the flow temporarily prior to full treatment (Chapter 14)
- Improve sewer operation and maintenance procedures to reduce infiltration (Chapter 18)
- Improve sewer operation and maintenance procedures to reduce pipe/screen blockages (Chapter 18)
- Use smart systems to actively control flows in the combined sewer system (Chapter 22)
- Treat discharges from CSOs *in situ* to a suitable standard by compact processes or nature-based methods (Section 13.8.2.1).
- Separate the combined sewer system into independent foul and storm systems (Section 13.8.2.2)

In practice, the solution or solutions chosen will depend on the specific case depending on factors such as topography, development, sewer condition, site sensitivity, and so on. Levels of cost, disruption, carbon, and the ability to achieve wider benefits will also inform decisions. Solutions may also depend on whether the focus is on ecological protection, public health concerns, or both. It is probable that a combination of options will provide the most cost-beneficial solution (Quaranta et al., 2022; NEPC, 2024). Drawing on CIWEM (2022) and NIC (2022), to achieve wider climate resilience (including flood risk and water resource management), nature and amenity benefits, as well as water quality improvement, we propose the following solutions hierarchy:

1. system maintenance and technological enhancement (e.g., rehabilitation, smart systems)
2. above-ground interventions (e.g., SuDS, wetlands)
3. below-ground interventions (e.g., additional sewers and storage, sewer separation)

13.8.2.1 Treatment options

A growing range of innovative CSO treatment options are emerging, which can be loosely grouped into nature-based solutions (e.g., constructed wetlands) and technological compact treatment systems (e.g., chemically enhanced sedimentation). Disinfection is also an option, particularly for overflows at WTPs (Pateman et al., 2018). Further details of these systems including their strengths and weaknesses can be found in Botturi et al (2021) and UKWIR (2023).

13.8.2.2 Sewer separation

Sewer separation refers to reconfiguring an existing single-pipe combined sewer system into a two-pipe system of foul and storm sewers. This is typically carried out by constructing an additional foul sewer network and refurbishing the existing combined network as a separate storm sewer network. Clearly, to do this at scale would be highly disruptive and very costly. Although CSOs would be eliminated, stormwater outfalls would not. Thorndahl et al. (2015) showed, in a study of a Danish catchment scheduled for sewer separation, that the process decreases the volumes of stormwater and pollutants diverted to the WTP and discharged to CSOs. However, this happens at the expense of an increase in volumes of stormwater and its pollutant load being diverted to local receiving waters potentially leading to serious impacts. Even if an overall benefit could be established, this would be eroded over time due to illegal misconnections as discussed in Section 5.4.5. Li et al. (2013), for example, found in a field study of 9 separate sewer systems and 3 combined systems that the separate systems had no advantage over the combined systems in terms of pollutant control mainly due to the level of illegal misconnections.

PROBLEMS

1 A combined sewer with a diameter of 750 mm, slope of 0.002, and k_s of 1.5 mm drains a catchment with a DWF of 15 L/s. For a particular rainfall, the flow of stormwater is 750 L/s. Does the sewer have sufficient capacity to carry stormwater and DWF? Would the daily maximum flow in dry weather provide self-cleansing conditions? What are the likely consequences of this? How could the design of this pipe have been improved? [no, no, $v = 0.44$ m/s]

2 What is meant by the "first foul flush"? What may cause it, and what are its implications?

3 What are the main functions of a combined sewer overflow? Explain the common overflow configurations.

4 Define the term *CSO setting*. Describe the importance of a CSO setting and ways in which it can be set.

5 The population of a catchment is 5000 and the average wastewater flow is 180 L/hd.d. Infiltration is 10% of the domestic wastewater flow rate and the average industrial flow is 2 L/s. Determine the DWF and the CSO setting according to Formula A. Express the CSO setting as a multiple of DWF. [13.5 L/s, 96.2 L/s, 7.1]

6 What CSO setting (as a multiple of dry weather flow) is required to limit the BOD_5 in the receiving water just downstream of the overflow to 7 mg/L assuming a dilution of 10 DWF during a spill from an unusually heavy storm? The design storm produces a peak flow in the sewer of 25 DWF and the wastewater dry weather BOD_5 concentration is 250 mg/L. [5 DWF]

7 Explain approaches to CSO design by which discharge to the environment of sewer sediments and gross solids can be reduced.

8 The basic layout of a single high side weir overflow suitable for incorporating screens is being determined using the principles of WaPUG (2006). The diameter of the existing inlet pipe is 525 mm, with an inflow of 480 L/s and a weir height of 0.4 m. Confirm the size of the inlet pipe (adjusting the weir height if appropriate), determine the size of the chamber, and propose a suitable weir length. [diameter 675 mm, weir height 425 mm, width 925 mm, inlet length 500 mm, weir length 3.15 m, and outlet length 1.01 m].

9 What are the main problems associated with CSOs?

10 Compare and contrast the various options available to reduce the volume and frequency of CSO discharges.

11 Compare and contrast the various options available to treat CSO discharges.

12 Some would argue that CSOs are an outdated legacy of our Victorian combined sewer network and should be abolished entirely. How would you go about doing that?

KEY SOURCES

Balmforth, D.J., Saul, A.J., and Clifforde, I.T. 1994. *Guide to the Design of Combined Sewer Overflow Structures*, Report FR 0488, Foundation for Water Research.

Botturi, A., Ozbayram, E.G., Tondera, K., Gilbert, N.I., Rouault, P., Caradot, N., Gutierrez, O., Daneshgar, S, Frison, N., Akyol, C., Foglia, A., Eusebi A.L. and Fatone, F. 2021. Combined sewer overflows: A critical review on best practice and innovative solutions to mitigate impacts on environment and human health. *Critical Reviews in Environmental Science and Technology*, 51(15), 1585–1618.

FWR. 2019. *Urban Pollution Management Manual*, 3rd edn, Foundation for Water Research, www.fwr.org/UPM3

Peters, P.E., and Zitomer, D.H. 2021. Current and future approaches to wet weather flow management: A review. *Water Environment Research*, 93(8), 1179–1193.

Wastewater Planning Users Group (WaPUG). 2006. *The Design of CSO Chambers Incorporate Screens*. WaPUG Guide. http://www.ciwem.org/groups/urban-drainage-group

REFERENCES

Aspinall, D.M.V. 1981. A wet-weather model for evaluating the combined effects of storm-sewage overflows and sewage-works' effluent upon river quality. *Water Pollution Control*, 80, 3.

Balmforth, D.J. 1982. Improving the performance of stilling pond storm sewage overflows. *Proceedings of the First International Seminar on Urban Drainage Systems*, Southampton, September, 5.33–5.46.

Balmforth, D.J., Lea, S.J., and Sarginson, E.J. 1984. Development of a vortex storm sewage overflow with peripheral spill. *Proceedings of the Third International Conference on Urban Storm Drainage*, Gotenborg, June, 107–116.

Balmforth, D.J., Meeds, E., and Thompson, B. 1996. Performance of screens in controlling aesthetic pollutants. *Proceedings of the Seventh International Conference on Urban Storm Drainage, 2*, Hannover, September, 989–994.

Brombach, H. 1987. Liquid–solid separation at vortex storm overflows. *Proceedings of the Fourth International Conference on Urban Storm Drainage, Topics in Urban Storm Water Quality, Planning and Management*, Lausanne, September, 103–108.

Brombach, H. 1992. Solids removal from CSOs with vortex separators. *Novatech 92, International Conference on Innovative Technologies in the Domain of Urban Water Drainage*, Lyon, November, 447–459.

Chartered Institution of Water and Environmental Management (CIWEM). 2022. *River water quality and storm overflows. A systems approach to maximising improvement*, Technical report. https://www.ciwem.org/

Defra. 2023. *Storm Overflows Discharge Reduction Plan*. www.gov.uk/official-documents

EA. 2018. *Water Companies: Environmental Permits for Storm Overflows and Emergency Overflows, Guidance*, Environment Agency, Bristol, UK.

EA. 2019. Waste water treatment works: treatment monitoring and compliance limits *Guidance*, Environment Agency, Bristol, UK.

Field, R. 1974. Design of a combined sewer overflow regulator/concentrator. *Journal of WPCF, 46*(7), 1722–1741.

Frederick, M.R. and Markland, E. 1967. The performance of stilling ponds in handling solids. Paper No. 5, *Symposium on Storm Sewage Overflows*, Institution of Civil Engineers, May, 51–61.

Hrudka, J., Šutúš M., Csóka, M., Raczková, A. and Škultétyová, I. 2021. Combined sewer overflow chamber analysis based on CFD modelling. *IOP Conference Series: Materials Science and Engineering, 1209*, 012019.

Jarman, D.S., Faram, M.G., Butler, D., Tabor, G., Stovin, V.R., Burt, D., and Throp, E. 2008. Computational fluid dynamics as a tool for urban drainage system analysis: A review of applications and best practice. *Proceeding of the 11th International Conference on Urban Drainage*, Edinburgh, September, on CD.

Jefferies, C. 1993. The Performance of Certain Combined Sewer Overflows with Storage. PhD Thesis, *Dundee Institute of Technology*.

Li, T., Zhang, W., Feng, C. and Shen, J. 2013. Performance assessment of separate and combined sewer systems in metropolitan areas in southern China. *Water Science and Technology, 69* (2), 422–429.

Meeds, B. and Balmforth, D.J. 1995. Full-scale testing of mechanically raked bar screens. *Journal of the Chartered Institution of Water and Environmental Management, 9*(6), 614–620.

Ministry of Housing and Local Government. 1970. *Technical Committee on Storm Overflows and the Disposal of Storm Sewage, Final Report*, HMSO, London.

Myerscough, P.E. and Digman, C.J. 2008. Combined sewer overflows—Do they have a future? *Proceedings of the 11th International Conference on Urban Drainage*, Edinburgh, September, on CD.

NAO. 2021. *Understanding Storm Overflows: Exploratory Analysis of Environment Agency Data*, National Audit Office, September.

NIC. 2022. *Reducing the Risk of Surface Water Flooding*. National Infrastructure Commission.

National Engineering Policy Centre (NEPC). 2024. *Testing the waters. Priorities for mitigating health risks from wastewater pollution*, Royal Academy of Engineering. https://nepc.raeng.org.uk/testing-the-waters

Nicoll, E.H. 1988. *Small Water Pollution Control Works: Design and Practice*. Ellis Horwood.

Pateman, D., White, C. and Lincoln, G. 2018. *Plymouth Central STW Stormwater UV Irradiation*, https://waterprojectsonline.com/case-studies/plymouth-central-stw-stormwater-uv-irradiation/

Quaranta, E., Fuchs, S., Jan Liefting, H., Schellart, A. and Pistocchi, A. 2022. Costs and benefits of combined sewer overflow strategies at the European Scale. *Journal of Environmental Management, 318*, 115629.

Saul, A.J. 2000. *Screen Efficiency (Proprietary Designs)*. UKWIR Report 99/WW/08/5.

Saul, A.J. and Ellis, D.R. 1992. Sediment deposition in storage tanks. *Water Science and Technology, 25*(8), 189–198.

Saul, A.J., Ruff, S.J., Walsh, A.M., and Green, M.J. 1993. *Laboratory Studies of CSO Performance*, Report UM 1421, Water Research Centre.

Scottish Development Department. 1977. *Storm sewage: Separation and disposal. Report of the Working Party on Storm Sewage (Scotland)*, HMSO, Edinburgh.

Sharpe, D.E. and Kirkbride, T.W. 1959. Storm-water overflows: The operation and design of a stilling pond. *Proceedings of the Institution of Civil Engineers, 13*, August, 445–466.

Smisson, B. 1967. Design, construction and performance of vortex overflows. Paper No. 8, *Symposium on Storm Sewage Overflows*, Institution of Civil Engineers, May, 99–110.

Stovin, V.R. and Saul, A.J. 2000. Computational fluid dynamics and the design of sewage storage chambers. *Journal of the Chartered Institution of Water and Environmental Management*, 14(2), 103–110.

Thorndahl, S., Schaarup-Jensen, K and Rasmussen, M.R. 2015. On hydraulic and pollution effects of converting combined sewer catchments to separate sewer catchments, *Urban Water Journal*, 12(2), 120–130

Tzinaetis, P. and Moy, F. 2019. WRcCSO: Rethinking the CSO for the 21st Century. *CIWEM Urban Drainage Group Autumn Conference.*

UK Water Industry Research (UKWIR). 2006. *National CSO Test Facility Wigan WWTW: CSO Screen Efficiency 1997–2005.* UKWIR 06/WW/08/14. https://ukwir.org/reports/06-WW-08-14/115213/National-CSO-Test-Facility-Wigan-WWTW-CSO-Screen-Efficiency-1997--2005

UK Water Industry Research (UKWIR). 2023. *Treatment Options for Storm Overflows.* UKWIR 23/WW/22/7. https://ukwir.org/treatment-options-for-storm-overflows

Water Research Centre (WRc). 1997. *Sewerage Detention Tanks—A Design Guide*, Swindon.

Chapter 14

Storage

14.1 FUNCTION OF STORAGE

This chapter is concerned with the hydraulic design of storage: tanks, basins, and ponds, at any point in the urban drainage system. Section 14.1 looks at the various functions of storage, while Section 14.2 discusses overall design and Section 14.3 sizing. Sections 14.4 and 14.5 consider methods of *storage routing* – calculations applied to the variation of inflow with time that predict the effect of storage on outflow. Section 14.4 demonstrates level pool (or reservoir) routing, a traditional method based on a graphical representation. Section 14.5 gives a storage routing method more suited to use on a spreadsheet, based on the variation of water depth with time. Section 14.6 considers optimal location of storage within a drainage system.

Storage can be provided by construction of below-ground detention tanks and chambers, or may exist within the system without being deliberately provided, especially in pipes with spare capacity. Storage for stormwater can also be created above ground, as discussed in Section 8.1 and Chapter 12, and is an integral element in stormwater management (Chapter 11).

In an urban drainage system, storage can have the functions of

- Limiting flooding
- Reducing the amount of polluted storm flow discharged to a watercourse
- Controlling flow to treatment
- Providing a resource for reuse
- Forming part of an active system or real-time control

An example is a new development to be drained by a conventional separate sewer system discharging stormwater to a small stream. To reduce the risk of flooding in the stream, the maximum discharge from the new development must be restricted to a low value. If a detention tank is provided to achieve this, the outflow is likely to be via a flow control (as considered in Section 8.1), often in conjunction with a weir (see Section 8.2) to operate at higher flow rates. The typical relationship between inflow and outflow for a case where outflow is controlled, and does not vary significantly with water level, is shown in Figure 14.1a. The volume of water stored for the case illustrated is given by the shaded area. When the outflow exceeds the inflow, the tank empties.

It is also useful to consider the hydraulic role of storage in more general cases (beyond specific application to detention tanks) where outflow may vary significantly, for example, in reservoirs, or where conceptual "reservoirs" are used to represent more complex systems (as in Sections 5.3.4 and 10.6.2). A general relationship between inflow and outflow is shown in Figure 14.1b. At any value of time, the difference between the inflow and outflow ordinates

DOI: 10.1201/9781003408635-14

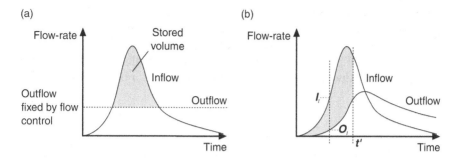

Figure 14.1 Storage: inflow and outflow hydrographs.

Figure 14.2 Installation of a Stormcell storage device. (Courtesy of Hydro International.)

$(I_t - O_t)$ gives the overall rate at which water in the storage is increasing (if inflow exceeds outflow) or decreasing (if outflow exceeds inflow). The total volume of water entering the storage up to any given time, say t' in Figure 14.1b, is given by the shaded area between the curves.

14.2 OVERALL DESIGN

Storage devices come in a number of shapes, sizes, and configurations. Small volumes can be provided in manholes or in oversized pipes. Proprietary concrete, high-density polyethylene or glass-reinforced plastic (GRP) tanks are also available. An alternative to a conventional tank (Andoh et al., 2001) is a system known as *geocellular storage*, based on a three-dimensional plastic matrix with a high void ratio within which the water is stored, removing the need for the structural function of a tank (Figure 14.2). Various types and forms of geocellular storage with the appropriate structural strength can be used within infiltration trenches (Stephenson,

2008) and pervious pavements (Chapter 11). The Construction Industry Research and Information Association (CIRIA) provides guidance on the testing, structural and geotechnical design, installation, and maintenance of such systems (O'Brien et al., 2016).

Larger systems include purpose-built reinforced concrete tanks or multiple-barrelled tank sewers. An important distinction is whether they operate online or offline (see also Chapter 13).

14.2.1 Online

Online detention tanks are constructed in series with the urban drainage network and are controlled by a flow control at their outlet. Flow passes through the tank unimpeded until the inflow exceeds the capacity of the outlet. The excess flow is then stored in the tank, causing the water level to rise. An emergency overflow is provided to cater for high flows (an online storage tank is depicted in Figure 13.12a in Chapter 13). As the inflow subsides at the end of the storm event, the tank begins to drain down, typically by gravity.

The flow control is normally one of those described in Chapter 8: an orifice, weir, vortex regulator, or throttle pipe. An electrically actuated gate linked to a downstream sensor may also be fitted. This will provide more precise control and also enable tank size to be minimised. Details of sewer system control are given in Chapter 22.

A common arrangement for an online tank is an oversized pipe or rectangular culvert. These *tank sewers* are provided with dry weather (in combined systems) or low-flow channels to minimise sediment deposition (Figure 14.3). Benching with a positive gradient is also provided. Another arrangement uses smaller multiple-barrelled sewers operating in parallel. These provide the necessary storage and have better self-cleansing characteristics. As deposition can occur, online tanks still require maintenance.

14.2.2 Offline

Offline tanks are built in parallel with the drainage system as shown in Figure 13.12b. These types of tanks are generally designed to operate at a predetermined flow rate, controlled at the tank inlet. An emergency overflow is provided, as for the online tank. Flow is returned to the system either by gravity or by pumping, depending on the system configuration and levels. A flap valve is normally used for gravity returns.

Offline tanks require less volume than online tanks for equivalent performance and hence less space, but the overflow and throttling devices necessary to divert, regulate, and return flows tend to be more complicated. Maintaining self-cleansing is also more difficult for this type of tank. Regular maintenance or measures to minimise deposition is therefore important.

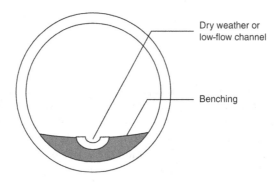

Figure 14.3 Tank sewer.

14.2.3 Flow control

The points at which flow control is required for both online and offline tanks are marked in Figure 13.12. Table 14.1 presents a summary of the flow control requirements. The common devices are described in Chapter 8.

More information on the use of flow control devices in conjunction with storage tanks is given by the Water Research Centre (WRc, 1997) and related to SuDS in Woods-Ballard et al. (2016). Further details on a variety of flow control devices and their application within larger storage facilities are presented by Hall et al. (1993).

Goorden et al. (2021) present work on active control of storage outflow. They model a stormwater storage tank and its catchment as a formalised mathematical decision-making process. Results show that their model can learn optimal strategies for control of outflow to make best use of capacity.

14.3 SIZING

The hydraulic design of a tank or pond serving a new development usually entails limiting the outflow for a specific storm event. So, typical design criteria are

- Rate of outflow – this can be fixed by one of a number of approaches:
  - No greater than estimated values from the undeveloped (or greenfield) site (somewhat problematic as it is difficult to accurately predict runoff flows from small, undeveloped catchments [but see Section 10.7])
  - A value linked to the area of the site and often specified by national or local planning policy (e.g., 4–6 L/s.ha)
  - The capacity of the downstream sewer or watercourse
- Design storm – smaller tanks designed to mitigate flooding are typically designed for 10–30-year return period storms. For large lakes, much higher return periods (typically 100 years) may be specified.

Table 14.1 Flow control for storage devices

Type of storage	Type of control	Purpose	Common devices
Online	Outlet restriction	To match continuation flow to the capacity of the downstream sewer	Orifice Penstock Vortex regulator Throttle pipe
	Relief	To divert excess flow when the capacity of storage (and downstream sewer) has been exceeded	High side weir
Offline	Continuation restriction	To match continuation flow to the capacity of the downstream sewer	Orifice Penstock Vortex regulator Throttle pipe
	Tank inlet	To pass flow into the tank when the downstream capacity has been exceeded	Orifice Penstock High side weir
	Relief	To divert excess flow when the capacity of storage (and downstream sewer) has been exceeded	High side weir
	Tank outlet	To return stored flow to the sewer once the storm has passed	Orifice Penstock Vortex regulator Throttle pipePump

When designing storage to reduce water quality impacts, time series rainfall (Chapter 4) is typically used, combined with water quality modelling (Chapter 20).

The questions in design are as follows: what active storage volume is required to achieve the outflow limitation and which is the critical storm that produces the worst case? It is not simply a case of using the Rational Method (Chapter 10), as the critical storm is usually of longer duration than the one giving maximum instantaneous flow.

A valuable overview of the sizing of storage for new developments and of related concepts is given by Kellagher (2013). The guidance is aimed at developers and local authorities, and the aim is to enable a non-expert "to obtain a quick assessment of the principal drainage requirements needed for a proposed development."

The next section describes a simplified approach to storage sizing. This is followed by an explanation and examples of flow routing procedures, including those used in commercial urban drainage models.

14.3.1 Preliminary storage sizing

A preliminary estimate of storage volume requirements for peak flow attenuation (in online tanks) can be obtained by using

$$S = V_I - V_O \tag{14.1}$$

where S is the storage volume (m^3), V_I is the total inflow volume (m^3), and V_O is the total outflow volume (m^3).

In this case, the outflow is via an outlet restriction using one of the devices in Table 14.1.

Figure 14.4 shows a plot of inflow volume, V_I, versus storm duration, D, for a particular return period. Outflow volume, V_O, has also been plotted, assuming a constant discharge. The difference in the ordinates of the two curves gives the storage, S, required for

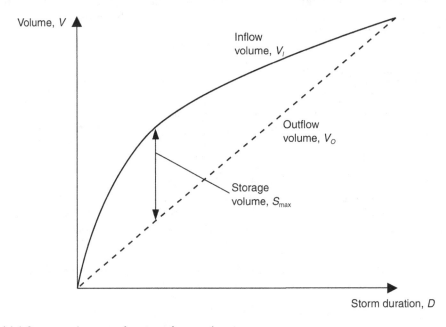

Figure 14.4 Storage volume as a function of storm duration.

EXAMPLE 14.1

A housing development has an impervious area of 25 ha. Determine approximately the volume of storage required to balance the 10-year return period storm event. Downstream capacity is limited, and a maximum outflow of 100 L/s has been specified. Rainfall statistics are the same as those derived in Example 4.2.

Solution

In the table below, column (3) is the inflow volume, which is derived from the product of (1), (2), and the impermeable area (25 ha). Column (4) is the outflow volume; the product of column (1) and the outflow rate (100 L/s). The storage is the difference between (3) and (4).

(1)	(2)	(3)	(4)	(5)
Storm duration, D (h)	Intensity, i (mm/h)	$V_I = iA_ID$ (m³)	$V_O = Q_OD$ (m³)	$S = V_I - V_O$ (m³)
0.083	112.8	2350	30	2320
0.167	80.4	3350	60	3290
0.25	62.0	3875	90	3785
0.5	38.2	4775	180	4595
1	24.8	6200	360	5840
2	14.9	7450	720	6730
4	8.6	8600	1440	**7160**
6	6.1	9150	2160	6990
10	4.0	10,000	3600	6400
24	2.0	12,000	8640	3360

The maximum storage, S_{max}, is 7160 m³, shown in bold in column 5.

any duration storm. The design storage (S_{max}) is the maximum difference between the curves. Example 14.1 shows how S_{max} can be identified using a tabular approach.

14.3.2 Storage routing

A more accurate assessment of the effect of the storage can be obtained by routing an inflow hydrograph through the tank/pond. This can be done using the storage routing methods described in Sections 14.4 and 14.5. In practice, most engineers establish storage volumes using drainage system software (Chapter 20).

14.4 LEVEL POOL (OR RESERVOIR) ROUTING

Calculating the relationship between inflow and outflow as the flow passes through storage (e.g., as shown in Figure 14.1b) is called *routing*. The first of two standard calculation methods with a wide range of applications is now given. The second method is given in Section 14.5.

The difference between inflow and outflow equals the rate at which the volume of water in the storage changes with time, or

$$I - O = \frac{dS}{dt} \qquad (14.2)$$

where I is the inflow rate (m³/s), O is the outflow rate (m³/s), S is the stored volume (m³), and t is time (s).

The simplest application is shown in Figure 14.5. Here, there is one outflow controlled by an arrangement such as a weir, giving a simple relationship between O and H (height of water above the weir crest). S in this case is the *temporary storage*, the volume created when there is outflow. The key to the method is that both O and S are functions of H.

We solve Equation 14.2 for fixed time steps and consider "average" conditions over the period of each time step. Therefore, the average inflow during a time step minus the average outflow equals the change in stored volume during the step:

$$\frac{I_1 + I_2}{2} - \frac{O_1 + O_2}{2} = \frac{S_2 - S_1}{\Delta t} \qquad (14.3)$$

where I_1, O_1, and S_1 are inflow, outflow, and stored volume at the start of the time step, respectively; I_2, O_2, and S_2 are inflow, outflow, and stored volume at the end of the time step, respectively; and Δt is the time step.

A typical application is to calculate outflow for known values of inflow. In each time step, the unknown will be O_2. Since O and S are related via H, we put S_2 with O_2 on the left-hand side of the equation:

$$\frac{S_2}{\Delta t} + \frac{O_2}{2} = \frac{S_1}{\Delta t} - \frac{O_1}{2} + \frac{I_1 + I_2}{2}$$

It is convenient to have the term $\left[(S / \Delta t) + (O / 2) \right]$ on both sides, so we rearrange to

$$\left[\frac{S_2}{\Delta t} + \frac{O_2}{2} \right] = \left[\frac{S_1}{\Delta t} + \frac{O_1}{2} \right] - O_1 + \frac{I_1 + I_2}{2} \qquad (14.4)$$

Now we need to incorporate the way both O and S vary with H. The neatest way of doing this is to create a relationship between $\left[(S / \Delta t) + (O / 2) \right]$ and O (based on the variations of O and S with H). This is demonstrated by Example 14.2.

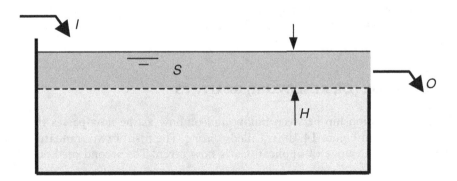

Figure 14.5 Simple application of level pool routing.

EXAMPLE 14.2

Outflow from a detention tank is given by $O = 3.5\,H^{1.5}$. The tank has vertical sides and a plan area of 300 m². Inflow and outflow are initially 0.6 m³/s, and then inflow increases to 1.8 m³/s at a uniform rate over 6 minutes. Inflow then decreases at the same rate (over the next 6 minutes) back to a constant value of 0.6 m³/s. Using a time step of 1 minute, determine the outflow hydrograph.

Solution

We first use the way O and S vary with H to create a relationship between $\left[(S/\Delta t)+(O/2)\right]$ and O, as in Table 14.2.

Table 14.2 Variation with H (Example 14.2)

H (m)	$O = 3.5\,H^{1.5}$ (m³/s)	S = 300 H (m³)	$\dfrac{S}{\Delta t}+\dfrac{O}{2}$ (m³/s)
0	0	0	0
0.2	0.31	60	1.16
0.4	0.89	120	2.44
0.6	1.63	180	3.81
0.8	2.50	240	5.25

We can use the data in Table 14.2 to plot $[(S/\Delta t) + (O/2)]$ against O (Figure 14.6).

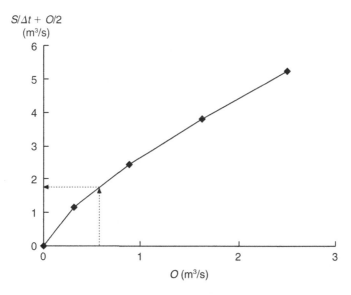

Figure 14.6 Graph of $S/\Delta t + O/2$ against O (Example 14.2).

The calculation now progresses as in Table 14.3. The values of I, and therefore $(I_1 + I_2)/2$, are known. The first value of O is 0.6 m³/s, and from this, the first value of $[(S/\Delta t) + (O/2)]$ can be determined via the graph in Figure 14.6 (giving a value of 1.8 m³/s, as indicated). So for the first time step, $[(S_2/\Delta t) + (O_2/2)]$ is calculated from Equation 14.4 giving 1.8 − 0.6 + 0.7 = 1.9. The corresponding value of O is determined from Figure 14.6, working this time from the y-axis to the x-axis, giving 0.65 m³/s.

Table 14.3 Routing calculation (extract) (Example 14.2)

Time (minutes)	I (m³/s)	O (m³/s)	$\dfrac{S}{\Delta t} + \dfrac{O}{2}$ (m³/s)	$\dfrac{I_1 + I_2}{2}$ (m³/s)
0	0.6	0.6	1.8	0.7
1	0.8	0.65	1.9	0.9
2	1.0	0.76	2.15	

So now we know O after the first time step. For the next time step, $[(S_1/\Delta t) + (O_1/2)]$ is 1.9, and $[(S_2/\Delta t) + (O_2/2)]$ is calculated again from Equation 14.4: 1.9 − 0.65 + 0.9 = 2.15. O_2 is again determined from Figure 14.6, giving 0.76 m³/s – the outflow at 2 minutes. The calculation proceeds in this way until all the values of O have been determined. The resulting outflow hydrograph is given in Figure 14.7.

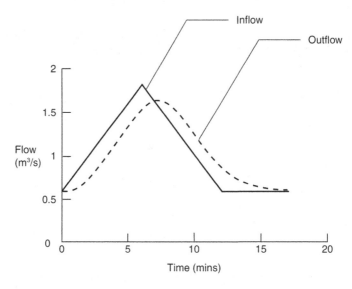

Figure 14.7 Inflow (given) and outflow (calculated) hydrographs (Example 14.2).

14.5 STORAGE ROUTING USING VARIATION OF WATER DEPTH WITH TIME

One disadvantage of the routing method just described is that it is difficult to implement using widely available computational tools such as spreadsheets. However, this can be overcome by transforming the rate of change of storage into the rate of change of head, as follows:

$$\frac{dS}{dt} = A\frac{dH}{dt} \tag{14.5}$$

Typically, there is a relationship between A (plan area of the storage) and H (head over the downstream control device). For storage ponds with vertical sides, A is constant, but for more complex shapes a function could be used, such as

$$A = \alpha H^\beta \tag{14.6}$$

where α and β are constant.
   So, from Equation 14.2,

$$I = O + \frac{dS}{dt}$$

and if the outflow regulator is an orifice outlet (Equation 8.1),

$$O = C_d A_o \sqrt{2gH}$$

then

$$I = C_d A_o \sqrt{2gH} + A\frac{dH}{dt}$$

and, therefore,

$$\frac{dH}{dt} = \frac{I - C_d A_o \sqrt{2gH}}{A} = f(H,t) \tag{14.7}$$

The derivative can be simply represented (using Euler's approximation) as

$$\frac{dH}{dt} \approx \frac{H(t + \Delta t) - H(t)}{\Delta t}$$

So, substituting into Equation 14.7 and solving for $H(t + \Delta t)$ gives

$$H(t + \Delta t) = H(t) + \Delta t . f(H,t) \tag{14.8}$$

which can be solved iteratively as shown in Example 14.3. As Euler's approximation assumes linear change of head over time, it is most accurate when small time increments are used. It is recommended that

$$\Delta t < 0.1 T_p$$

where T_p is the time for the hydrograph to reach peak value.

EXAMPLE 14.3

An online balancing pond is needed to limit the peak storm runoff from the site to 100 L/s. Design a suitable vertically sided storage tank using an orifice plate ($C_d = 0.6$) as the outflow regulator. The maximum head available on the site is 1.5 m. The inflow hydrograph is given below.

Time (h)	0	0.25	0.50	0.75	1.00	1.25	1.50	1.75	2.00
Flow (L/s)	0	75	150	200	300	375	450	525	600
Time (h)	2.25	2.50	2.75	3.00	3.25	3.50	3.75	4.00	
Flow (L/s)	525	450	375	300	225	150	75	0	

Solution

Determine the required orifice diameter D for the maximum outflow (100 L/s) at the available height (1.5 m):

$$A_o = \frac{O}{C_d\sqrt{2gH}} = \frac{0.100}{0.6\sqrt{2g1.5}} = 0.031\,m^2$$

where A_o is the orifice cross-al area.

$$D = \sqrt{\frac{4A_o}{\pi}} = 0.197\,m$$

Thus, use a $D = 200$ mm orifice:

$$O = C_d A_o \sqrt{2gH} = 0.6 \times 0.031\sqrt{2g}H^{0.5} = 0.082H^{0.5}$$

What plan area A is needed? Equation 14.7 gives

$$f(H,t) = \frac{I-O}{A}$$

$\Delta t = 0.25\,h = 900\,s$

In the table below, all of column (2) and the first data point in column (3) refer to data known initially. Column (4) is calculated from the orifice equation. Column (5) is column (2) minus column (4) divided by the plan area of the storage (Equation 14.7). The "new" head in column (6) is the sum of column (3) and column (5) times the time increment (Equation 14.8). Column (3) takes the head from column (6) at the previous time step.

(1)	(2)	(3)	(4)	(5)	(6)
T (h)	I (m³/s)	H(t) (m)	O (m³/s)	f(H,t) (m/s)	H(t + dt) (m)
0.00	0.000	0.00	0.000	0.000000	0.00
0.25	0.075	0.00	0.000	0.000030	0.03
0.50	0.150	0.03	0.013	0.000055	0.08

(1)	(2)	(3)	(4)	(5)	(6)
T (h)	I (m³/s)	H(t) (m)	O (m³/s)	f(H,t) (m/s)	H(t + dt) (m)
0.75	0.225	0.08	0.023	0.000081	0.15
1.00	0.300	0.15	0.032	0.000107	0.25
1.25	0.375	0.25	0.041	0.000134	0.37
1.50	0.450	0.37	0.050	0.000160	0.51
1.75	0.525	0.51	0.059	0.000187	0.68
2.00	0.600	0.68	0.068	0.000213	0.87
2.25	0.525	0.87	0.076	0.000179	1.03
2.50	0.450	1.03	0.083	0.000147	1.16
2.75	0.375	1.16	0.088	0.000115	1.27
3.00	0.300	1.27	0.092	0.000083	1.34
3.25	0.225	1.34	0.095	0.000052	1.39
3.50	0.150	1.39	0.097	0.000021	1.41
3.75	0.075	1.41	**0.097**	−0.000009	1.40
4.00	0.000	1.40	0.097	−0.000039	1.36
4.25	0.000	1.36	0.096	−0.000038	1.33
4.50	0.000	1.33	0.095	−0.000038	1.30

A number of areas were tried iteratively. The above refers to $A = 2500$ m².

At this area (volume) and shown in bold in this table, $H_{max} = 1.41$ m (<1.5 m) and $O_{max} = 97$ L/s (<100 L/s), which is acceptable.

14.6 OPTIMAL LOCATION OF STORAGE

The location of storage within an urban drainage system is traditionally determined by its function. Obvious examples are storage at the location of a CSO, or close to the downstream end of a stormwater connection from a new development.

However, if an urban drainage system is considered as a whole: sewers, CSOs, wastewater treatment plant, and receiving waters, there may be particular distributions of storage that are optimal in terms of minimising discharges and pollution within the whole system. This may, for example, reflect the fact that, in terms of receiving water quality, improvements at a CSO by increasing storage may be more than counterbalanced by negative effects on treatment works effluent quality as a result of increased loading.

A number of studies have explored approaches to identifying optimal locations for storage within an urban drainage system. Fu et al. (2010) investigated the optimal distribution of storage tank volume in combined systems to mitigate the impact of new, additional, developments on receiving water quality. They studied the effect of optimising storage distribution and of optimising flow control and found that a combination of the two was most effective in achieving improvements to receiving water quality.

Concentrating on limiting flooding by stormwater, Li et al. (2015) analysed detention options including tank design (layout and size), using multi-objective optimisation linked to the drainage model SWMM. They presented an application to a case study in China. Cunha et al. (2016), also focusing on stormwater, developed an optimisation model (again linked to SWMM) to investigate the location, dimensions, and flow control properties of storage units. They identified a balance between providing storage and allowing some flooding: beyond a particular total volume of storage, large volume increases resulted in only small flooding reductions.

PROBLEMS

1 A development has an impermeable area of 1.8 ha. Basing rainfall estimation on the formula $i = 750/(t + 10)$, where i is rainfall intensity in millimetres per hour (mm/h) and t is the duration in minutes, determine the volume of storage needed to limit outfall to 70 L/s. (Try storm durations of 8, 12, and 16 minutes.) [72.4 m^3]

2 A detention tank on a sewerage scheme is rectangular in plan: 25 m × 4 m. It is being operated in such a way that the only outflow is over a weir. The flow rate over the weir is given by the standard expression for a rectangular weir, in which $C_D = 0.63$. The width of the weir is 1.5 m. Consider the following case. Initially, the inflow is zero, and the water surface is at the level of the weir crest. Then inflow increases at a uniform rate over 12 minutes to 0.9 m^3/s, and reduces immediately at the same rate back to zero. Determine the peak outflow, using a time step of 2 minutes. [0.8 m^3/s]

3 How much did the tank in Problem 2 attenuate or delay the hydrograph peak? How would the normal operation of the tank differ from that described in Problem 2?

4 Where do you consider the best location will be for storage in an urban drainage system? Should it be centralised or decentralised?

REFERENCES

Andoh, R.Y.G., Faram, M.G., Stephenson, A., and Kane, A. 2001. A novel integrated system for stormwater management. Novatech 2001, *Proceedings of the Fourth International Conference on Innovative Technologies in Urban Drainage*, Lyon, France, 433–440.

Cunha, M.C., Zeferino, J.A., Simoes, N.E. and Saldarriago, J.G. 2016. Optimal location and sizing of storage units in a drainage system. *Environmental modelling and software*, 83, 155–166.

Fu, G., Khu, S. and Butler, D. 2010. Optimal distribution of storage tank volume to mitigate the impact of new developments on receiving water quality. *Journal of Environmental Engineering*, 136(3), 335–342.

Goorden, M.A., Larsen, K.G., Nielsen, J.E., Nielsen, T.D., Rasmussen, M.R. and Srba J. 2021. Learning safe and optimal control strategies for storm water detention ponds. *IFAC-PapersOnLine*, 54(5), 13–18.

Hall, M.J., Hockin, D.L., and Ellis, J.B. 1993. *Design of Flood Storage Reservoirs*, B14, CIRIA.

Kellagher, R. 2013. *Rainfall runoff management for developments*. Report SC030219. Environment Agency.

Li, F., Duan, H.F., Yan, H. and Tao, T. 2015. Multi-objective optimal design of detention tanks in the urban stormwater drainage system: framework development and case study. *Water Resources Management*, 29, 2125–2137.

O'Brien, A.S.O., Lile, C.R., Pye, S.W., and Hsu, Y.S. 2016. *Structural and Geotechnical Design of Modular Geocellular Drainage Systems*, CIRIA C737.

Stephenson, A.G. 2008. A holistic hard and soft SUDS system used in the creation of a sustainable urban village community. *Proceeding of the 11th International Conference on Urban Drainage*, Edinburgh, September, on CD-ROM.

Water Research Centre (WRc). 1997. *Sewerage Detention Tanks—A Design Guide*, Swindon.

Woods-Ballard, B., Wilson, S., Udale-Clark, H., Illman, S., Scott, T., Ashley, R., and Kellagher, R. 2016. *The SUDS Manual*, 5th edn, Report C753, CIRIA, London, UK.

Chapter 15

Pumped systems

15.1 WHY USE A PUMPING SYSTEM?

As indicated in Chapter 1, the need for urban drainage arises from human interaction with the natural water cycle. Sewers usually drain in the same direction that nature does: by gravity. Gravity systems require relatively little maintenance, certainly when compared with systems involving a significant amount of mechanical equipment or the need to maintain fixed pressures. And while neglect is undesirable, so is unnecessary maintenance, and so gravity sewer systems prevail. This can be seen as the result of an implicit decision to accept high capital costs (for deep, large, and expensive sewers) if they result in low operating costs. More explicit consideration of the balance between capital and operational costs to minimise total expenditure (Totex) is now becoming more commonplace (see Chapter 19). However, in some cases, gravity is not enough, usually when it is not cost-effective to provide treatment facilities for each natural sub-catchment. In these circumstances, it is appropriate to pump, and that is the focus of this chapter.

In Section 15.2, we consider the general arrangement of a pumping system; and in Section 15.3, we look at the hydraulics. The pipe carrying the pumped flow, the rising main, is considered in Section 15.4. We cover the pumps themselves in Section 15.5, and the design of pumping stations in Section 15.6. In Section 15.7, we look at alternatives to gravity flow for a whole sewer system; and in Section 15.8, we look at the key issue of energy use and GHG emissions.

15.2 GENERAL ARRANGEMENT OF A PUMPING SYSTEM

Sewer pumping systems have a number of general features:

- In sewer systems based mostly on gravity flow, pumped sections require comparatively high levels of maintenance. Engineers, therefore, prefer to keep the pumped lengths to a minimum to lift the flow as required, and then the system can revert to gravity flow as soon as possible. (Figure 15.1 gives a simple section of a typical arrangement.)
- The liquid being pumped contains solids that vary significantly in size and nature; therefore, pumps must be designed with the risk of clogging, abrasion, and damage in mind. The nature of the liquid also creates risks of septicity, corrosion of equipment, and production of explosive gases (as considered in Chapter 18).

DOI: 10.1201/9781003408635-15

- It is common for centrifugal pumps to deliver flow at a fairly constant rate or, where there are a number of pumps that may work in combination, there may be a number of alternative fixed rates. The exception to this is if variable speed pumps are used. Whatever the flow rate handled by the pumping system, it must generally exceed the rate of flow arriving from the gravity system or be combined with some form of storage, otherwise, there would be a risk of flooding. So pumping systems tend to work on a stop-start basis, with flow arriving at a reception storage (a *wet well* or *sump*), as shown in the simplified pumping station arrangement in Figure 15.2. When the pumps are operating, the wet well empties; when the pumps are not operating the wet well fills. The water level in the wet well is used to trigger the stop and start of the pumps. Where pumps are used specifically as part of a storage scheme, the downstream receiving conditions may dictate when the pumps start and stop.

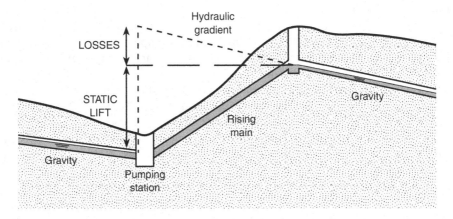

Figure 15.1 Typical sewer pumping arrangement.

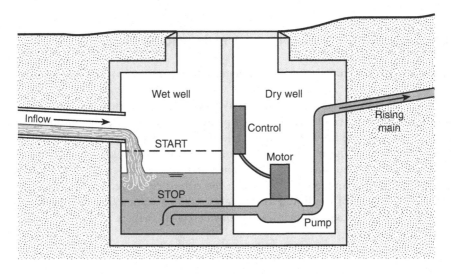

Figure 15.2 Simplified pumping station arrangement.

15.3 HYDRAULIC DESIGN

15.3.1 Pump characteristics

Hydraulically, the function of a pump is to add energy, usually expressed as head (energy per unit weight) to a liquid. The hydraulic performance of a pump can be summed up by the *pump characteristic curve*, a graph of the head added to the liquid, plotted against the flow rate. A typical pump characteristic is given in Figure 15.3a; this shows generally reducing head for increasing flow rate, but it is not a simple relationship – what goes on inside a pump is complex in hydraulic terms. The different types of pumps available are considered in Section 15.5. The characteristics of each type of pump are derived from tests carried out by the manufacturer and are available in the manufacturer's product information.

But at what values of flow rate and head will the pump operate when connected to a particular pipe system? The engineer answers that question at the design stage in the following way.

15.3.2 System characteristics

The pipe system to which the pump will be connected will have a characteristic curve of its own: the *system characteristic*. Water must be given head in order to

- Be lifted physically (the *static lift* – see Figure 15.1).
- Overcome energy losses due to pipe friction and local losses at bends, valves, and so on. As the flow rate increases, velocity increases and energy losses increase in proportion to the square of velocity (as set out in Section 7.3).
- Provide velocity head if the water is to be discharged to atmosphere at a significant velocity.

The system characteristic can be determined from

Head = Static lift + Dynamic losses and velocity head

Since the losses and velocity head are proportional to velocity squared, a typical system characteristic has the shape shown in Figure 15.3b.

Therefore, there are two characteristics: the pump characteristic, which gives the head that a pump is capable of producing while delivering a particular flow rate; and the system

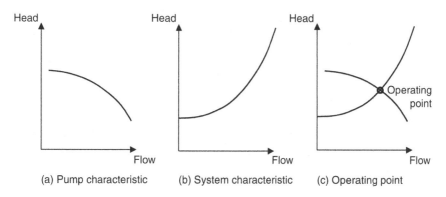

Figure 15.3 Pump and system characteristic curves.

characteristic, which gives the head that would be required for the system to carry a particular flow rate. If a specific fixed-speed pump is going to be connected to a specific system, there is only one set of conditions where what the pump has to offer can satisfy what the system requires: it is the point where the pump characteristic and the system characteristic cross, as shown on Figure 15.3c. This is called the *operating point* or *duty point*.

15.3.3 Power

The power required at the operating point can be derived from the operating flow rate and head in conjunction with the pump's efficiency.

Power (P), *energy per time*, is the product of *weight per time* $\rho g Q$ and head, *energy per weight*. Therefore,

$$P = \rho g Q H \tag{15.1}$$

where ρ is density (taken as 1000 kg/m³ for water); g is gravitational acceleration, 9.81 m/s²; Q is the operating flow rate (m³/s); and H is the operating head (m).

A pump gives power to the water, and it receives power ("power supply"), usually in the form of electrical power. The pump and motor are not 100% efficient at converting the power supply into power given to water. Efficiency (power given to the water divided by power supplied to the pump) varies with flow rate and can be taken from the manufacturer's plot (e.g., Figure 15.4). Therefore,

$$\text{Power supplied} = \frac{\rho g Q H}{\eta} \tag{15.2}$$

where η *is* efficiency (–).

Example 15.1 demonstrates these calculations.

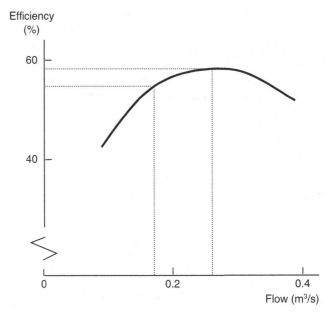

Figure 15.4 Pump efficiency against flow.

EXAMPLE 15.1

A pump in a sewer system is connected to a rising main with an internal diameter of 0.3 m and a length of 105 m. The rising main discharges to a manhole at a level 20 m above the water level in the sump. The roughness k_s of the rising main is 0.3 mm, and local losses total $0.8 \times v^2/2g$. The pump has the following characteristics:

Q (m³/s)	0	0.1	0.2	0.3	0.4
H (m)	33	32	29	24	16
Efficiency (%)		42	56	57	49

Determine the flow rate, the head, and the power supplied at the operating point.

Solution

The system characteristic is given by

Total head required = Static lift + Friction losses + Local losses + Velocity head

This can be expressed as

$$H = 20 + \frac{\lambda L}{D} \frac{v^2}{2g} + 0.8 \frac{v^2}{2g} + \frac{v^2}{2g}$$

We can find λ from the Moody diagram (Figure 7.4), and its value will be constant, provided the flow is rough turbulent. Assuming that it is,

$$\frac{k_s}{D} = \frac{0.3}{300} = 0.001, \text{giving } \lambda = 0.02$$

Therefore,

$$H = 20 + \frac{v^2}{2g} \left[\frac{0.02 \times 105}{0.3} + 0.8 + 1 \right]$$

$$= 20 + 8.8 \frac{v^2}{2g}$$

Of course, $Q = vA$
Therefore,

$$v = \frac{4Q}{\pi 0.3^2}$$

From this, we can determine the relationship between H and Q for the pipe system (the system characteristic).

Alternatively, we can use Wallingford tables (see Section 7.3.4) to determine the system characteristics. Local losses + velocity head = $(0.8 + 1) v^2/2g$, and this can be expressed as an equivalent length using Equation 7.15. Therefore,

$$L_E = D \frac{1.8}{\lambda} = 27 \text{ m}$$

For the system, for any value of Q: $H = 20 + (105 + 27) \times S_f$ (from table).

The system characteristic (from either method) is plotted (as in Figure 15.3c) together with the pump characteristic. The operating point is where the lines cross; at this point, the flow rate is 0.26 m³/s. This gives a velocity of 3.7 m/s, giving R_e of about 10^6 – in the rough turbulent zone – so the assumption about constant λ is valid.

At the operating point, the head is 26 m.

Pump efficiency has been plotted in Figure 15.4. At a flow rate of 0.26 m³/s, efficiency is 57%. Therefore,

$$\text{Power supplied} = \frac{\rho g Q H}{\eta} = \frac{\rho g \times 0.26 \times 26}{0.57} = 116 \, \text{kW}$$

In some systems, the velocity head may be insignificant in relation to the losses and is ignored. In Example 15.1, the velocity head was significant and was rightly included. In another case, instead of discharging to atmosphere at a manhole, the rising main outlet might be "drowned," for example, submerged in a tank into which the liquid is being pumped. In this case, the static lift must be measured up to the liquid surface in the tank. This surface is unlikely to have a velocity, and, therefore, velocity head will not be included. However, there will be exit losses at the point where the rising main discharges to the tank.

It should be noted that the water level in the sump will not be constant because as the pump drains the sump, the level goes down (and therefore static head increases). This may be significant, and if it is, it must be considered in design.

15.3.4 Pumps in parallel

A common arrangement is for two (or more) pumps to be placed in parallel (Figure 15.5). The additional pump(s) can

- Act as a standby to replace others when there is a fault
- Work in a "duty, standby" arrangement where both pumps are used separately but in turn to deliver the required flow
- Operate in an "assist, standby" arrangement where one pump delivers the normal flow but the other comes into operation to reinforce the first pump when high discharges are needed

When two identical pumps are operating in parallel, each delivers flow rate Q and raises the head by H (Figure 15.5), so overall the flow rate is $2Q$, all experiencing an increase in the head of H. The characteristic of two pumps in parallel is given in Figure 15.6. For each value of H, the flow rate is doubled to $2Q$. The new operating point is given by the intersection with the system characteristic (Figure 15.6).

For pumps in parallel, care is needed when determining the efficiency. The operating flow rate in Figure 15.6 is for both pumps together. Half that value gives the flow rate in each pump, and this should be used in determining efficiency from Figure 15.4, as this gives the efficiency for a single pump. Example 15.2 demonstrates this.

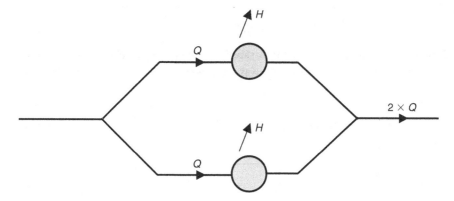

Figure 15.5 Pumps in parallel (schematic).

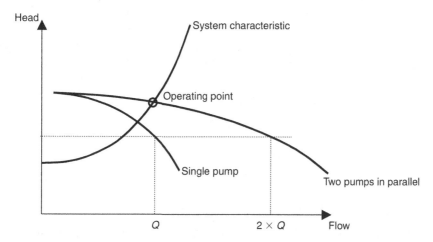

Figure 15.6 Operating point: pumps in parallel.

15.3.5 Suction and delivery pipes

The pipe on the upstream, or inlet side of a pump, is referred to as the *suction pipe*, and the pipe on the downstream, or outlet side, is referred to as the *delivery pipe*. In Examples 15.1 and 15.2, the suction pipe was short and was not considered separately. This arrangement is common in drainage applications. It is also common for the pump volute or body to be below the level of liquid in the sump, as in Figure 15.2, to ensure that the pumps remain "primed" (full of liquid). However, where this is not the case and the suction pipe is long or the pump is at a higher level than the sump level, it is important to ensure that pressure on the suction side of the pump stays well above the vapour pressure of the liquid. This is to avoid cavitation that leads to loss of performance and damage to pumps and ancillaries (Chadwick et al., 2021). Generally, long suction pipes are to be avoided due to the difficulty of maintaining velocity within an appropriate range (see Section 15.4).

EXAMPLE 15.2

For Example 15.1, what would be the flow rate, head, and power supplied at the operating point if an additional pump, identical to the first, was arranged in parallel?

Solution

The pump characteristic for the pumps in parallel is determined by doubling the flow rate for each value of H:

For one pump, Q (m³/s)	0	0.1	0.2	0.3	0.4
Two pumps in parallel, Q (m³/s)	0	0.2	0.4	0.6	0.8
Head, H	33	32	29	24	16

The characteristic for two pumps in parallel can be plotted, together with the system characteristic (and, for the purposes of illustration, the characteristic for one pump) as in Figure 15.6. At the operating point, the flow rate is 0.34 m³/s, and the head is 30 m. As has already been pointed out, care is needed when handling efficiency for pumps in parallel. The flow rate in each pump is 0.17 m³/s; therefore, the efficiency of each pump (Figure 15.4) is 54%. Therefore,

$$\text{Power supplied} = 2 \times \frac{\rho g \times 0.17 \times 30}{0.54} = 185\,kW$$

15.4 RISING MAINS

15.4.1 Differences from gravity sewers

It is useful to consider the ways in which rising mains are different from gravity sewers.

15.4.1.1 Hydraulic gradient

Gravity pipes are designed assuming that the hydraulic gradient is numerically equal to the pipe slope. As shown in Section 7.4, this is because, in part-full pipe flow, the hydraulic gradient coincides with the water surface and, therefore, in uniform flow, is parallel to the invert of the pipe. In a rising main, of course, none of this applies. At the pumps, the flow is given a sudden increase in head, and this is "used" to achieve the static lift and overcome losses along the pipe (Figure 15.1). The slope of the hydraulic gradient is the natural one: downwards in the direction of flow, while the rising main does its job: it rises. The pipe can be laid at a constant depth and follow the profile of the ground.

15.4.1.2 Flow is not continuous

At times there may be no flow in the main, and at other times there may be a number of alternative flow rates, depending on how many pumps are operating. When the pumps are not operating, wastewater stands in the rising main. Therefore, it is important that when pumping resumes, the velocities are sufficient to scour deposited solids.

A standard design range for rising main velocity is between 0.75 and 1.8 m/s; the minimum recommended diameter is 80 mm (Water UK, 2023).

To avoid septicity, wastewater should not be retained in a rising main for more than 12 hours. It is sometimes necessary to arrange for the addition of oxygen or oxidising chemicals to control septicity (considered in Section 18.10.4).

When the range of flows is high, dual rising mains can be used to maintain velocities high enough to prevent deposition. One can also act as standby; but, in this case, both mains must be used regularly to avoid septicity.

One possible consequence of starting and stopping the flow is extremely high or low pressures resulting from surge waves, considered in Section 15.4.3.

15.4.1.3 Power input

We must provide power to create flow in the system. The power is needed year after year for as long as the system operates. This creates trade-offs in the selection of an economic design, including the operating cost of the pumping station and, if also being constructed, the rising main.

A smaller diameter pipe will be cheaper, and the resulting higher velocities will help to ensure the scouring of deposits, but the higher velocities will also create higher head losses (proportion to velocity squared) and, therefore, higher power costs. The economic decision needs to consider design life and time-related comparisons of capital and operating costs.

15.4.1.4 The pipes are under pressure

There is, of course, no open access to rising mains in the way that there is for gravity sewers.

There are some economic advantages of rising mains in comparison with gravity sewers. Because rising mains are under pressure, and always full, the diameter tends to be smaller and the depth of excavation less than a gravity pipe (which is usually not full and must slope downward). The physical gradient of a rising main is not necessarily uniform or even in the same direction (i.e., rising) all the time.

15.4.2 Design features

Common materials for rising mains are ductile iron, steel, and some plastics. Flexible joints are preferable to allow for differential settlement and other causes of underground stress. (There is more detail on pipe material in Chapter 15.)

Valves and other hydraulic features need to be included at key points in a rising main. There should be an isolating valve, normally a "sluice" (or gate) valve, near the start of the rising main, so that the pumping station pipework can be worked on without emptying the main. There must be a non-return (or *reflux*) valve for each pump, which prevents backflow when pumping stops (and prevents circulating flow when there are parallel pumps). Summits (local high points) in the rising main should be avoided if possible, but where unavoidable, should be provided with air release valves. Washout facilities (for emptying the main) should be provided at low points in the main.

Thrust blocks or tied pipework may be needed to withstand the forces created when water is forced to change direction. Their design is beyond the scope of this text but is covered by Thorley and Atkinson (1994) and the American Water Works Association (2009).

15.4.3 Surge

Pressurised pipelines are typically subject to rapid changes in flow velocity, say due to pump operations, and this gives rise to one particular risk – *surge* (also commonly referred to as water hammer).

A change in flow (velocity) in a liquid is always associated with a change in pressure. If flow changes rapidly, these changes in pressure can be significant. The effect is known as surge, and the consequences of ignoring it at the design stage can be catastrophic, with the creation of pressures high or low enough to cause damage to pipework, fittings, or restraints. Not all pumping systems are likely to suffer from serious surge problems, and many devices for overcoming the problems can be simple, but surge must be considered when a pumping system is being designed since a rapid change of flow will always occur with pump failure. There are numerous other events that can lead to elevated surge risks, including valve operations, air valve activity, and sectional priming.

A crucial factor is the rate at which the flow (velocity) changes. If the flow changes gradually, the normal assumption that water (or wastewater) is incompressible can be maintained, and the changes in pressure are not significant. If the flow changes rapidly, the compressibility of water must be considered, with significant pressure changes throughout the system propagating at a velocity dependent on the fluid characteristics and properties of the pipe material. Methods of predicting the changes in pressure are presented in BS EN 16932-2: 2018 and by Hamill (2011). Where surge is identified as a potential risk, modelling should be carried out to determine the possible impact and to evaluate solutions.

Most standard (clean water) surge protection devices (e.g., air vessels or surge tanks) are typically (but not always) deemed inappropriate for wastewater application because of the problems of blockage or stagnation of the stored liquid. Commonly employed methods for managing the surge response of a system include controlled pump operations, controlled airflow management via air valves, pump flywheels, as well as suitable selection of pipe materials and restraint requirements. More detail is given by Chadwick et al. (2021).

15.5 TYPES OF PUMP

As stated, the function of a pump is to add energy to a liquid. There are a number of ways in which this energy can be transferred, but the most common is by a rotating "impeller" driven by a motor (a *rotodynamic pump*).

The most common rotodynamic pump for use with wastewater is a centrifugal pump in which the impeller forces the liquid radially into an outer chamber called a *volute* (Figure 15.7). In effect, the volute converts velocity head into pressure head. The impeller often has a special design to avoid clogging by solids, and this feature means that centrifugal pumps for wastewater tend to have lower specified efficiencies (about 50–85%) than centrifugal pumps for clean water (up to 90%). Centrifugal pumps are suited to relatively high heads and low flow rates: typically heads of up to around 100 m, and flow rates up to around 300 L/s. Centrifugal pumps perform best when primed (filled with water before pumping can begin) and so are normally installed below the lowest level of wastewater to be pumped.

Axial-flow pumps are simpler than centrifugal pumps and have impellers (acting like a propeller) that force the liquid in the direction of the longitudinal axis (Figure 15.8). Axial flow pumps are suited to relatively high flow rates and low heads with efficiencies of 75–90%. Unlike centrifugal pumps, axial-flow pumps suffer a rapid decrease in the head with increased discharge. In *mixed-flow pumps*, the direction in which the water is forced by the impeller is at an intermediate angle, so flow is partially radial and partially axial. Mixed-flow pumps are suitable for medium heads between 4 and 20 m and are capable of pumping flows up to 10,000 L/s. Axial- and mixed-flow pumps are most appropriate for pumping stormwater. *Vortex pumps* are used to handle high solids/grit load where the impeller creates a vortex beyond the volute and into the surrounding water. This helps gather up the solids, including fibrous and stringy material, without coming into contact with the impeller or blocking it.

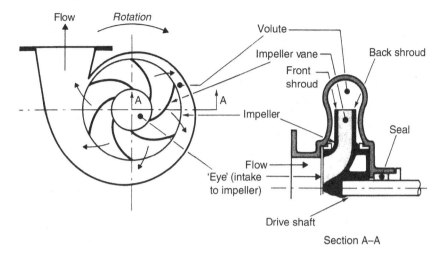

Figure 15.7 Centrifugal pump. (Reproduced from Chadwick, A.et al. 2021. *Hydraulics in Civil and Environmental Engineering*, 6th edn, CRC Press.With permission of CRC Press.)

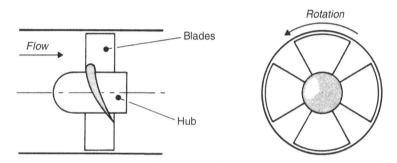

Figure 15.8 Axial flow pump. (Reproduced from Chadwick, A. et al. 2021. *Hydraulics in Civil and Environmental Engineering*, 6th edn, CRC Press.With permission of CRC Press.)

These are relatively low-flow, low-head pumps and are ideal for applications such as providing protection against flooding for individual properties while having the ability to discharge into the downstream system. Figure 15.9 illustrates the shape of pump characteristics for the main types of pumps. In practice, pump selection for a particular application is made by matching requirements to manufacturer's data.

The pump and the motor that drives it are often kept in a "dry well," separate from the wastewater (Figure 15.2). However, an alternative is a *submersible pump* in which the motor is encased in waterproof protection and submerged in the wastewater that is to be pumped (Figure 15.10). This greatly simplifies the design of the pumping station and is common for small- to medium installations.

For very small installations, a rotodynamic pump may not be suitable, because the risk of clogging places a limit on the smallness of a pump. An alternative system is a *pneumatic ejector*, in which the wastewater flows by gravity into a sealed unit and is then pushed out using compressed air. These require little maintenance and are not easily blocked by solids. However, they are of low efficiency and limited capacity (1–10 L/s).

While many pumps operate at a single fixed speed, some types switch between two or more speeds (*multi-speed*), and others can run at continuously variable speeds. The benefit

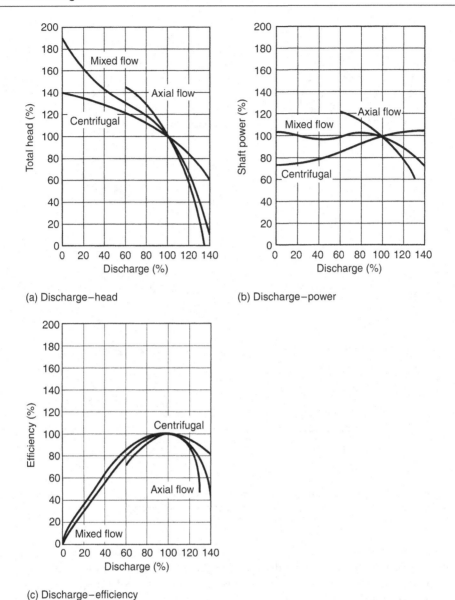

Figure 15.9 Characteristics of various pump types. (Reproduced from Kay, M. 2016. *Practical Hydraulics and Water Resources Engineering*, 3rd edn, CRC Press. With permission of CRC Press and amended.)

of variable speed pumps is that the pumped outflow from the pumping station can be more closely matched to the inflow (from the system); therefore, less storage volume is required. Also, pumps do not have to be stopped and started so frequently; deposition resulting from liquid lying still in the rising main is reduced; and flow rates, velocities, and therefore losses tend to be lower. However, variable speed pumps are more expensive, require more complex control arrangements, and may be very inefficient at some speeds.

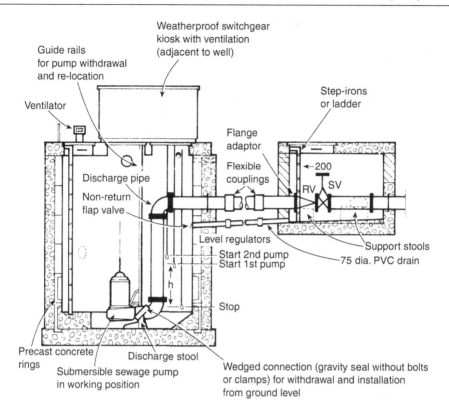

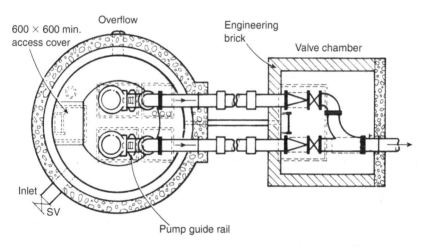

Figure 15.10 Submersible pump. (Reproduced from Woolley, L. 1988. *Drainage Details*, 2nd edn, E & FN Spon. With permission of E & FN Spon.)

15.6 PUMPING STATION DESIGN

15.6.1 Main elements

The design of pumping stations usually involves the integration of a number of branches of engineering, including civil, structural, mechanical, electrical, instrumentation, control, and automation. Also, a pumping station is one of the few elements of an urban drainage system that may be seen above ground, so there may also be significant architectural aspects (Figure 15.11). They require a planning application, in which noise and odour, as well as appearance, may be issues. The extent of all these aspects will, of course, depend on size – pumping stations in urban drainage schemes may be very small, serving just a few people, or they may be large and complex engineering structures serving large populations.

The basic components of a pumping station have already been described in this chapter. Pumps (nearly always more than one) take sewer flow from a reception volume, a sump, and deliver it with increased head into a rising main. The pumps, most commonly centrifugal, are driven by motors, which must be provided with a supply of electrical power. Pumping stations deemed critical to the sewer network performance may need dual or alternative supplies. There must be arrangements for controlling the pumps, usually related to the liquid level in the sump.

15.6.1.1 Wet well–dry well

When the pumps and motors are kept completely separate from the liquid, the sump is referred to as the *wet well*, and the chamber containing pumps, and so on, is referred to as the *dry well*. The motors may be directly beside the pumps or, to provide further remoteness from moisture and for ease of access, may be at a higher level, connected by a long shaft.

Figure 15.11 Pumping station: architectural treatment.

15.6.1.2 Wet well only

The wet well–dry well separation is not needed when submersible pumps are used (as illustrated in Figure 15.10). The wet well in which submersible pumps are placed can be of a simple construction, based on precast concrete segmental rings. For inspection or maintenance, the pumps must be lifted out.

15.6.2 Number of pumps

The appropriate number of pumps is a function of

- The need for standby pumps to be available to cover for faults and planned maintenance
- The flow capacity of the pumps, alone and in parallel, determined from the calculations covered in Section 15.3
- The variation in inflow

The simplest pumping station consists of a duty/standby arrangement. It is common, however, in larger installations to have a number of pumps, arranged in parallel (as previously discussed), which are brought into use successively as inflow increases.

15.6.3 Control

In most systems, while the pumps are running, the level in the sump is falling. At a fixed level, the pumps are turned off and the level starts to rise. Subsequently, the level reaches the point at which pumping is resumed.

All pumping stations require some control. The basic requirement is sensing of upper and lower levels in the sump, and the consequent starting and stopping of the pumps. With more pumps, and more complex starting and stopping procedures, the complexity of the control system increases. Common methods of sensing water level are by ultrasonic detector, float switches, and electrodes. Various level points may be used in order to help prevent "rag rafts" and grit deposits. The safe frequency of operation of the electric motor starter is limited; it is typical to design for between 10 and 15 starts/hour.

Filipe et al. (2019) describe the development of a data-driven optimisation framework for predictive control of pumps based on forecasting wastewater intake rates in order to optimise the pumping cycle. This aims to reduce energy demand by starting to pump when the level in the sump is relatively high (to reduce the lift) while minimising the risk of the sump overflowing.

15.6.4 Sump volume

To determine the required sump volume (V) between "stop" and "start" levels, the time taken to fill the sump while a single pump is idle (t_1) is given by

$$t_1 = \frac{V}{Q_I}$$

where Q_I is the inflow rate. The time taken to pump out the sump (t_2) is

$$t_2 = \frac{V}{(Q_o - Q_I)}$$

where Q_o is the outflow (pump) rate. Thus, the time between successive starts of the pump, the pump cycle (T), is

$$T = \frac{V}{Q_I} + \frac{V}{(Q_o - Q_I)} = \frac{VQ_o}{Q_I(Q_o - Q_I)} \qquad (15.3)$$

Now, to find the minimum sump volume required, V is differentiated with respect to Q_I and equated to zero:

$$\frac{dV}{dQ_I} = \frac{T(Q_o - 2Q_I)}{Q_o} = 0 \qquad (15.4)$$
$$Q_o = 2Q_I$$

Therefore, the pump sump should be sized for a pumped outflow rate that is double the inflow rate. Substituting Equation 15.4 into Equation 15.3 gives

$$V = \frac{TQ_o}{4}$$

If $n = 3600/T$ is the number of motor starts per hour,

$$V = \frac{900Q_o}{n} \qquad (15.5)$$

Thus, the minimum sump working volume is determined from the rate of outflow in conjunction with the allowable frequency of motor starts (see Example 15.3). In practice, there can be other factors that override this, such as the size and arrangement of the pumps. For variable speed drives, this calculation is not strictly required. However, unless the pumping station has an overflow, it is appropriate to specify this volume as a safety measure.

15.6.5 Flow arrangements

Within a pumping station, pipework is usually ductile iron with flanged joints. Flexible joints should be placed outside walls to allow for differential settlement. For each pump, there should be a sluice valve on the suction and delivery sides for isolating the pump.

The base of the sump is usually given quite steep slopes to limit the deposition of solids. For submersible pumping stations, Water UK (2023) states that benching should have a minimum slope of 60° to the horizontal.

In large pumping stations, some form of preliminary treatment to remove large solids, most commonly by means of screens, may be necessary. Alternatively, recirculation pumps may be used to reduce the likelihood of solids settling.

Where possible, pumping stations should have an emergency overflow in case of complete failure of the pumps, with storage for wastewater inflow during emergency repairs. In combined systems, it may be necessary to provide an overflow for storm flows. This would be based on the same principles as other combined sewer overflows (CSOs), described in Chapter 13.

EXAMPLE 15.3

The peak inflow to a sewerage pumping station is 50 L/s. What capacity sump and duty/standby pumps will be required if the number of starts is limited to 10 per hour? How long will the pump operate during each cycle?

Solution

For minimum sump volume, $Q_o = 2Q_I$, so
Capacity of duty pump, $Q_o = 100$ L/s
Capacity of standby pump, $Q_o = 100$ L/s
Pump sump volume (Equation 15.5),

$$V = \frac{900 \times 0.1}{10} = 9m^3$$

Time taken to empty sump,

$$t_2 = \frac{9}{(0.1 - 0.05)} = 180s = 3\,min$$

15.6.6 Maintenance

A pumping station, with mechanical, electrical, instrumentation, control, and automation equipment, is one part of a sewer network that has obvious maintenance needs. And while it is true of any part of a sewer system, it is particularly important that the maintenance needs of pumping systems are taken into account at the design stage. Care and expense in design may reduce the cost of maintenance, and care and expense in maintenance may limit performance deterioration and reduce the cost of replacement.

Priorities in taking maintenance requirements into account in design are to ensure that

- It is possible to isolate and remove the main elements of pipework and equipment. There must be access to allow the pumps to be lifted out vertically; this is especially true for submersible pumps, which it must be possible to lift out with ease.
- Problems caused by solids can be overcome (by suitable wet well design, appropriate impellers, pump selection, pipe sizes, and access to clear blockages)
- Emergencies caused by breakdown, power failure, and so on, can be coped with.

The possibility of power failure needs to be taken into account. Solutions are typically one or more of the following:

- Dual electrical supply
- Standby generator(s)
- Connection points for a temporary generator
- Diesel-powered pumps in larger stations

General requirements for maintenance of pumping systems are given by BS EN 16932-1: 2018.

An important element in maintenance is monitoring performance. Small- to medium-sized pumping stations are usually controlled from the wastewater treatment plant that they serve, by telemetry. The types of information likely to be communicated are

- Failure in the electricity supply
- Pump failure
- Unusually high levels in the wet well
- Flooding of the dry well
- Operation of the overflow

This information is needed for the effective operation of the system, and, in particular, to aid decisions about when to attend to operational issues. Pumping stations are commonly fitted with flow measurement devices to monitor performance and to form part of a planning and management system for the catchment as a whole (covered in Chapters 19 and 20).

More detailed guidance on practical aspects of pumping station design is given in BS EN 16932-1: 2018 and BS EN 16932-2: 2018. Also, for submersible pumping stations, guidance is given in Water UK (2023). Operation and maintenance within the wider urban drainage system are covered in Chapter 18.

15.6.7 Energy demands

One feature of a pumping system is that it has a regular energy requirement, and this contrasts with many other aspects of urban drainage systems. Decisions about the design of sewer systems, and the choices about the use of pumped systems, have always involved a comparison of capital costs against operating costs. When the sustainability of drainage alternatives is also considered, energy consumption must be seen as a significant issue (see Section 15.8).

15.7 NON-GRAVITY SYSTEMS

Where the ground surface is very flat, or where ground conditions make construction of deep pipes difficult, alternative options to gravity drainage are non-gravity systems such as *vacuum sewerage* or *pressure sewerage*. These are systems for entire neighbourhoods, traditionally rural communities, but increasingly small towns or even parts of larger urban areas.

In vacuum sewerage, wastewater is drained from properties by gravity to collection sumps. When the liquid surface in the sump rises to a particular level, an *interface valve* opens and wastewater is drawn into a pipe in which low pressure (in the order of –0.6 bar) has been created by a pump. After the collection sump has been emptied, the interface valve remains open for a short time to allow a volume of air at atmospheric pressure to enter the pipe. The mixture travels at high velocity (5–6 m/s) towards the vacuum source. The vacuum main, in vertical section, typically has lengths that slope downwards in the direction of flow alternating with steeper rising sections. At the end of this pipe, the wastewater is retained in a collection vessel for subsequent pumped removal. Vacuum systems consist of small, shallow pipes with relatively high running costs, compared with the large, deep pipes and low running costs of gravity systems. In the right circumstances, vacuum systems show overall cost advantages (Consterdine, 1995). Other advantages include the fact that the high velocities eliminate sewer blockage, and leaks are easily detected because they cause obvious changes in pressure (Lauwo et al., 2012).

There are many vacuum systems in the UK, for example, in flat areas of East Anglia. There is significant use around the world including in the Middle East, Europe, and parts of Africa, as well as in the US and Australia.

Vacuum systems are covered by BS EN 16932-1: 2018 and BS EN 16932-3: 2018. Further details are given by Read (2004).

In pressure sewerage, wastewater is similarly drained from properties by gravity to local collection sumps. Each sump contains a high head package pump system, with a macerator or grinder pump, that discharges into a foul flow only rising main. Systems can be designed with hundreds of properties connected. Pipe sizes are small in diameter at shallow depths and instrumentation is used to control when the pumps can discharge. Such systems have become more commonplace, for example, in the US (Lauwo et al., 2012) and in New Zealand (Carne et al., 2014). Pressure systems are covered by BS EN 16932-1: 2018.

Vacuum sewerage systems and pressure sewerage systems have a number of characteristics in common. They both tend to involve pipes that have smaller diameters and are shallower pipes than gravity systems. They both tend to reduce infiltration and inflow (see Section 3.4). They both include mechanical systems that require maintenance (although most gravity systems also require maintenance). The main difference, apart from the contrasting methods of generating the movement of wastewater, is that vacuum sewerage involves central infrastructure to create the vacuum, whereas pressure sewerage requires equipment and power at each connection.

15.8 ENERGY USE AND NET ZERO EMISSIONS

The water industry as a whole, including all aspects of water supply and wastewater treatment, as well as drainage, is a large consumer of energy, using up to 3% of the UK's total (Majid et al., 2020). This makes it the fourth most energy-intensive industry nationally. Energy costs are the second largest contributor to overall operating costs (after staff costs) but the percentage at a particular time is dependent on the varying cost of energy.

Reducing energy consumption has always been an industry goal, but there is now added impetus because of the need to minimise carbon emissions. Achieving net zero by 2030 is a published aim of the water industry in England (Water UK, 2019), and by 2040 in the rest of the UK. In an assessment of net zero technology, reviewing the applicability, readiness, and scalability of greenhouse gas (GHG) reduction technologies and interventions applicable to water companies (Jacobs UK Limited, 2022), pump efficiency is identified as an important area. GHG emissions are considered further in Section 18.10.5.

Within urban drainage, the key use of energy is at pumping stations. However, many pumping stations do not operate at optimum energy efficiency (as defined for example in Figure 15.4). Yates and Weybourne (2001) argue that by improving operations, energy reductions of 30–50% can be consistently made. Typical improvements include

- Refurbishing or replacing worn equipment
- Improving the scheduling of multi-pump systems
- Employing high-efficiency motors, efficiency-enhancing coatings, or variable-speed drives
- More effectively matching pump performance to demand
- Using condition-based monitoring to understand the pump performance
- Employing continuous monitoring of energy usage

Jorge et al. (2022) propose a "novel energy balance" for drainage systems which considers the whole system, not just the electrical energy used by equipment. This is to explore the significant potential for water-energy savings that can be found when analysing the whole system. Their approach involves comparing quantified energy in two main blocks: (1) energy inflows, including "intrinsic energy" (hydraulic energy) from inflows (including "undue inflows" such as infiltration), combined with external energy supplied to pumps and treatment processes, etc.; and (2) energy outflows, including those due to hydraulic losses, equipment inefficiency and exceedance flows. The aim is to highlight system inefficiencies and identify specific elements that need to be improved.

PROBLEMS

1 A pumping system has a static lift of 15 m. The pump characteristics are listed below, together with the total losses in the rising main (velocity head can be neglected).

	Q (m³/s)	0	0.05	0.1	0.15	0.2
Pump	H (m)	25	24	20	14	7
	Efficiency (%)		45	55	55	50
Rising main	Losses (m)	0	1	4	9	16

Determine the flow rate, head delivered, and power supplied to the pump at the operating point. If the diameter of the rising main is 250 mm, are conditions suitable for scouring of deposited solids? If the rising main became rougher with age, would the flow rate, and head, increase or decrease?   [0.105 m³/s, 19 m, 36 kW, v = 2.1 m/s OK, Q decrease, H increase]

2 For the same system as in Problem 1, if an additional identical pump is operating in parallel, determine the total flow rate, head delivered, and power supplied to the pumps at the operating point. Which arrangement – one pump or two in parallel – uses power more efficiently?   [0.14 m³/s, 23 m, 64 kW, one pump]

3 There are three main categories of rotodynamic pumps. Describe for each category (1) their basic mode of operation, (2) their advantages and disadvantages, and (3) their application in urban drainage.

4 A pumping station sump is being designed to suit an inflow of 30 L/s. What rate of pumped outflow would give the minimum sump volume? What sump volume would then be required if the pump was to operate at (1) 6 starts/hour and (2) 12 starts/hour? At 12 starts/hour, there is 5 minutes between each start. For how much of that time is the pump operating, and for how long is it idle?   [60 L/s, 9 m³, 4.5 m³, 2.5 minutes]

5 Designing a pumping system presents a different set of challenges from designing a gravity system. Explain why.

6 Compare and contrast vacuum sewerage and pressure sewerage.

7 Considerable quantities of energy are used by the water industry. Identify the main use of energy in urban drainage systems, and discuss ways of reducing it.

REFERENCES

American Water Works Association. 2009. *Ductile-Iron Pipe and Fittings*, 3rd edn, Manual 41, AWWA.

BS EN 16932-1: 2018. *Drain and sewer systems outside buildings. Pumping systems. General requirements.*

BS EN 16932-2: 2018. *Drain and sewer systems outside buildings. Pumping systems. Positive pressure systems.*

BS EN 16932-3: 2018. *Drain and sewer systems outside buildings. Pumping systems. Vacuum systems.*

Carne, S., Salmon, G., and Gamst, C. 2014. Prevention or cure—Designing sewers to achieve low I/I. *Water New Zealand, 183*, March, 59–61.

Chadwick, A., Morfett, J., and Borthwick, M. 2021. *Hydraulics in Civil and Environmental Engineering*, 6th edn, CRC Press.

Consterdine, J.P. 1995. Maintenance and operational costs of vacuum sewerage systems in East Anglia. *Journal of the Chartered Institution of Water and Environmental Management, 9*(6), 591–597.

Filipe, J., Bessa, R.J., Reis, M., Alves, R. and Póvoa, P. 2019. Data-driven predictive energy optimization in a wastewater pumping station. *Applied Energy, 252*, 113423.

Hamill L. 2011. *Understanding Hydraulics*, 3rd edn. Palgrave Macmillan.

Jacobs UK Limited. 2022. *Net Zero Technology Review*. Ofwat.

Jorge, C., Almeida, M.D.C. and Covas, D. 2022. A novel energy balance tailored for wastewater systems. *Urban Water Journal, 19*(5), 441–452.

Kay, M. 2016. *Practical Hydraulics and Water Resources Engineering*, 3rd edn, CRC Press.

Lauwo, S., Sharvelle, S. and Roesner, L. 2012. *A review of advanced sewer system designs and technologies.* Water Environment Research Foundation (WERF).

Majid, A., Cardenes, I., Zorn, C., Russell, T., Colquoun, K., Banares-Alcantara, R. and Hall, J.W. 2020. An analysis of electricity consumption patterns in the water and wastewater sectors in South East England, UK. *Water, 12*, 225.

Read, G.F. 2004. Vacuum sewerage, in *Sewers: Replacement and New Construction* (ed. G.F. Read) (Chapter 16, 327–338), Elsevier Butterworth-Heinemann.

Thorley, A.R.D. and Atkinson, J.H. 1994. *Guide to the Design of Thrust Blocks for Buried Pressure Pipelines*, CIRIA R128.

Water UK. 2019. *Public Interest Commitment update*. Water UK.

Water UK. 2023. *Sewerage Sector Guidance.* https://www.water.org.uk/sewerage-sector-guidance-approved-documents

Woolley, L. 1988. *Drainage Details*, 2nd edn, E & FN Spon.

Yates, M.A. and Weybourne, I. 2001. Improving the energy efficiency of pumping systems. *Journal of Water Supply Research and Technology—Aqua, 50*, 101–111.

Chapter 16

Sewer construction and rehabilitation

This chapter deals with the physical fabric of sewer systems by looking at their construction and rehabilitation.

Section 16.1 gives an overview of construction. In Section 16.2, we describe materials and jointing methods for sewer pipes. Section 16.3 introduces concepts of structural design. An important prerequisite for construction, site investigation, is introduced in Section 16.4. We then look at specific construction methods: open-trench construction (Section 16.5), tunnelling (Section 16.6), and trenchless methods (Section 16.7).

For rehabilitation, we start by considering the context (Section 16.8). We then look at methods of rehabilitation: for structural rehabilitation in Section 16.9 and hydraulic rehabilitation in Section 16.10.

16.1 TYPES OF CONSTRUCTION

Most sewers are constructed underground. This is achieved by one of three general methods:

- Open-trench construction
- Tunnelling
- Trenchless construction

Open-trench construction consists of excavating vertically along the line of the sewer to form a trench, laying pipes in the trench, and backfilling – see Figure 16.1. It is suitable for a wide range of pipe sizes and depths and is the common method for small- to medium-scale sewer construction.

Tunnelling typically involves excavating vertically at a particular location for access and then excavating outwards at an appropriate gradient to form the space for the sewer to be constructed. *Trenchless construction* covers a wide range of technologies that do not include the use of an existing pipeline.

Tunnelling generally involves sizes large enough for human access in which a lining (eventually part of the sewer fabric) is constructed from inside the excavation. Tunnelling tends to be associated with large-scale projects like interceptor sewers.

Underground construction techniques that involve inserting pipes in the ground without a trench are called trenchless or "no-dig" methods. They avoid disruption on the surface and have become increasingly popular as the technology has developed and engineers have become more aware of the costs to business and society of conventional trench construction. Apart from its use in the construction of new sewers, this type of technology is widely used in sewer rehabilitation.

DOI: 10.1201/9781003408635-16

Figure 16.1 Open-trench construction (Courtesy of Thames Water.)

Both tunnelling and trenchless no-dig methods generally require a "Drive" pit or manhole and a "Reception" pit or manhole, set at changes in direction or at suitable or maximum drive distances.

16.2 PIPES

16.2.1 General

The nominal size (DN) of a pipe is the diameter of the pipe in millimetres rounded up or down to a convenient figure for reference. In some materials (including clay and concrete, and some ribbed plastic pipes) the DN refers to the inside diameter, and in some (including smooth plastic), it refers to the outside diameter. The actual diameter may be slightly different from the DN. Of course, the precise diameter must be clear in the manufacturer's product data and must be used in accurate calculations of hydraulic or structural properties.

General requirements for all materials used in gravity sewer systems are given in BS EN 476: 2022 –*General Requirements for Components Used in Drains and Sewers.*

16.2.2 Materials

The main characteristics and applications of the common sewer pipe materials are now considered. Relevant British/European standards are listed at the end of this chapter; these provide more detailed guidance on properties, specifications, and structural behaviour. A detailed survey of pipe materials is also given in the *Materials Selection Manual for Sewers, Pumping Mains and Manholes* (Sewers and Water Mains Committee, 1993).

In general, the most important physical characteristics of a sewer pipe material are

- Durability
- Abrasion resistance
- Corrosion resistance (with special consideration of the effects of H_2S)
- Imperviousness
- Strength
- Ability to be laid to line and level (particularly important for gravity flow pipes)

16.2.2.1 Clay

Vitrified clay is a commonly used material for small- to medium-sized pipes. Its major advantages are its strength and its resistance to corrosion, making it particularly suitable for foul sewers where the generation of H_2S leads to acidic corrosion of many other pipe materials. However, clay is both heavy and brittle and, therefore, susceptible to damage during handling, storage, and installation.

16.2.2.2 Concrete

Plain and reinforced concrete pipes (and culverts) are generally used for medium- to large-sized pipes. It is particularly suited to use in storm sewers because of its size, abrasion resistance, strength, and cost. There is potential for H_2S-generated corrosion (see Chapter 18), which is combatted in the first instance by the use of sulphate-resisting cement, though generally domestic wastewater is initially not harmful to concrete pipes. Non-circular cross-sectional pipes are available, such as egg-shaped and rectangular culverts. A specific type of prestressed concrete pipe is made for pressure applications such as rising mains, and for pipe-jacking (see Section 16.7.1).

16.2.2.3 Ductile iron

Ductile (centrifugally spun) iron pipes are used where significant pressures are expected, such as in pumping mains (Chapter 15) and inverted siphons (Chapter 8), or where high strength is required, such as in onerous underground loading cases and above-ground sewers (BS EN 545:2010). Ductile iron is susceptible to corrosion and needs both internal and external protection, such as zinc coating, cementitious internal lining, and, in some circumstances, extra protection from either epoxy or polyurethane (PU) coatings.

16.2.2.4 Steel

Steel pipes tend to be used in specialist applications where high strength is required. These include sea outfalls, above-ground sewers, and pipe bridges. Steel pipes require protection from corrosion – by internal lining and external coating, often supplemented by cathodic protection.

16.2.2.5 Unplasticised PVC (PVC-U)

PVC-U pipe has found general application in small-size pipes typically <DN300. It is lightweight, making installation straightforward, and is corrosion-resistant. However, as the pipe is flexible, strength relies on support from the bedding, and good construction practice is therefore critical. Smaller sizes are routinely used in building drainage applications but are also used to some extent for public sewers. Structured wall PVC-U pipes utilise a variety of shapes of the pipe wall to increase stiffness for the same volume of material. However, these thinner walls impact the cleaning methods that may be used.

16.2.2.6 Polyethylene (PE)

Polyethylene pipes are used in a wide range of applications for gravity and pressurised systems, being commonly used for stormwater conveyance and storage as well as rising mains. Some PE pipes, especially for gravity systems, have a structured wall. Pipe sizes can typically range between 16 and 3500 mm. Pipes can be welded using different techniques or connected with couplers.

16.2.2.7 Other materials

Other pipe materials in use include structured wall pipe made from polypropylene, glass-reinforced plastic (GRP), pitch fibre, and asbestos cement. Pipes of these materials are known to fail and are increasingly in need of replacement. Significant care is needed with some materials that may contain asbestos fibres, such as pitch fibre and asbestos cement. Many existing sewers are made of brick and remain in good condition, particularly considering their age.

16.2.2.8 Sizes

The range of sizes, and increments in size, depends on the pipe material. For example, clay pipes are available at a number of smaller diameters, with larger diameters at multiples of 100 mm. Traditional sizes for concrete pipes start at 300 mm and increase at 75 mm increments over a wide range. Table 16.1 gives size ranges, together with British/European standards.

16.2.3 Pipe joints

Sewer pipes are usually supplied and laid in the trench in standard straight lengths, and jointed *in situ*. There are several alternative jointing methods, providing either rigid or flexible joints. In most cases, flexible joints are preferred to allow for differential settlement,

Table 16.1 Pipe materials: size ranges, and British/European standards

Material	Normal size range (mm)	Principal British/European standards (full titles are given in References)
Clay	100–300	BS 65, BS EN 295
Concrete	300–2400	BS EN 1916, BS 5911-1, BS 5911-3 to BS 5911-5
Ductile iron	80–1600	BS EN 10224, BS EN 10311
Steel	60–2235	BS EN 10224, BS EN 10311
PVC-U	17–630	BS EN 1401, BS 4660, BS EN 13476
PE	16–2100	BS EN 12201, BS EN 13476

nonuniform support, drying, or other effects, without introducing unacceptable bending moments and stresses in the pipe. The standard jointing methods are as follows.

16.2.3.1 Spigot and socket

This joint is illustrated in Figure 16.2a. It is the normal jointing method for large plastic pipes, concrete pipes, larger clay pipes, and most ductile iron pipes. The spigot is inserted into the socket. A rigid joint can be created using a material such as cement mortar; however, it is much more common for flexibility and watertightness to be provided by an O-ring or shaped seal of rubber, or equivalent material, placed in a groove at the end of the spigot before insertion (Figure 16.2a). This may be pre-fitted in concrete pipes. Insertion causes suitable compression of the O-ring, and seals have a locking capability.

16.2.3.2 Sleeve

An alternative is to use a separate sleeve, as shown in Figure 16.2b. Clay pipes of smaller diameters are commonly jointed using flexible polypropylene sleeves. Plastic pipes use plastic sleeves, including angled flexible strips or O-rings.

16.2.3.3 Bolted flange joints

Simple bolted flanges do not provide flexibility. They are used in rigid installations where exposed pipework (usually ductile iron) needs to be readily dismantled, as in a pumping station (see Section 15.6).

16.2.3.4 Heat fusion jointing

PE piping can be fused to provide a watertight and continuous pipeline. This is a process where electrically heated plates raise the end of the pipes to a temperature that enables the

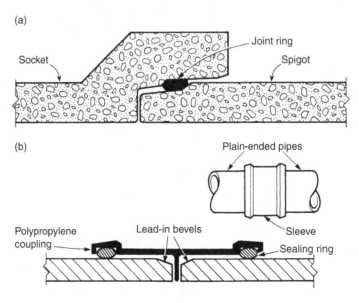

Figure 16.2 (a) Spigot and socket joint (flexible). (b) Sleeve joint. (Reproduced from Woolley (1988) with permission of E & FN Spon.)

pipes to be fused together. WIS 4-32-08 (2002) and BS EN: 12201-1: 2011 provide further information on the specific types of jointing and methods. Generally, the techniques are butt welding or electrofusion.

16.3 STRUCTURAL DESIGN

This section introduces concepts in the structural design of sewers. As an example, we consider the structural design of open-trench sewers. The main components of an open-trench arrangement are shown in Figure 16.3. A pipe laid in a trench has to be strong enough to withstand loads from the soil above it, from traffic and other imposed loads, and from the weight of liquid it carries. With increasing depth of cover, the load from the soil increases, and the load from traffic decreases. The ability to withstand the loads is derived from the strength of the pipe itself, the design of the trench or its absence, and the nature of the *bedding* on which it is laid.

Different calculation procedures are used depending on whether the pipe is considered to be rigid (clay, concrete), semi-rigid (ductile iron), or flexible (plastic), with steel pipe being semi-rigid or flexible depending on its dimensions. There are also differences between trenches considered "narrow" and those considered "wide," and between pipes in trenches and pipes under embankments.

Practising engineers tend to carry out structural design of pipes using standard charts, tables, or software, which relate loads to the properties of the soil and fill material, and the pipe diameter, width of trench, and height of cover. Engineering firms also use their own in-house reference material.

The basis of the procedures is summarised in BS EN 1295-1: 2019 *Structural Design of Buried Pipelines under Various Conditions of Loading*. BS 9295: 2020 *Guide to the Structural*

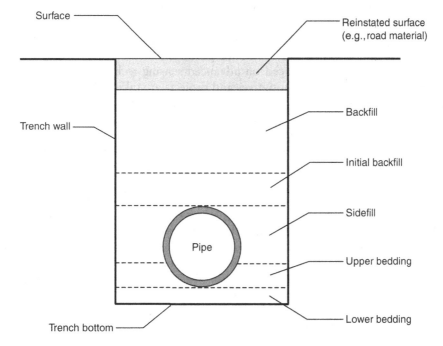

Figure 16.3 Open-trench arrangement.

Design of Buried Pipes provides further information and consolidates previous documents, with a specific focus on UK design methodology. More detailed treatment can be found in Moser and Folkman (2008).

16.4 SITE INVESTIGATION

16.4.1 Site investigation for construction

Site investigation identifies problems with ground conditions and special hazards that may have a significant effect on planning, choice of pipe material, structural design, and construction. The objectives are to provide information useful in the selection of a scheme from a number of alternatives, to inform the detailed design, to estimate costs, and to foresee difficulties. Detailed guidance is given by Whyte (2004).

Site investigation is of great importance for all types of sewer construction but may have particular significance for schemes involving the construction of sewers by tunnelling, augur boring, directional drilling, and for shafts and manholes. In this case, the main areas of interest are geological structure, groundwater, existing services and structures, and special hazards.

Site investigation can typically be divided into three phases:

1. Desk study: analysis of existing data, geological and other maps
2. Site investigation: boreholes, trial excavations, analysis of samples, interpretation
3. During construction: observations and records, probing ahead, further trials and boreholes

16.4.2 Underground assets

All physical work on underground drainage systems, including large-scale construction, requires knowledge of the location of existing underground assets. Clearly, every aspect of the planning, implementation, cost, safety, and effectiveness of this type of work is affected by the other physical assets that share the underground space.

Increasing emphasis is being placed on advanced sensing techniques and on interactive digital mapping, and a number of initiatives and projects are taking this forward.

'Mapping the Underworld' is a research programme based in the UK using advanced sensing technologies to develop a multi-sensor device to locate the position of **all** buried utility assets without excavation. "Through combining a range of technologies – ground penetrating radar, acoustics and both active low-frequency and passive electromagnetic fields – and intelligent data fusion, it is possible to optimise pipe and cable detection and location." (University of Birmingham, 2024)

Once known, the location of underground assets must be recorded and communicated effectively. The National Underground Asset Register (NUAR) is a UK government initiative to establish a standardised interactive digital map of buried assets (Department for Science, Innovation and Technology, 2023). It is due to be fully operational at the end of 2025.

Overall guidance on current procedures for the location of underground assets is provided by PAS 128:2022 *Underground Utility Detection, Verification and Location – Specification*. Four types of utility location survey are defined:

- desktop utility records search (survey type D)
- site reconnaissance (type C) – validating records by visual inspection

- detection (type B) – detection and location by geophysical techniques
- verification (type A) – observation at a manhole or excavation

Guidance on detection methods (survey type B) is given. "In a PAS 128 survey, both electronic locators (EML) and ground penetration radar (GPR) are deployed as in minimum" with other detection methods used as appropriate.

16.5 OPEN-TRENCH CONSTRUCTION

16.5.1 Excavation

In urban areas, all excavation must be carried out with great care so as not to damage existing services (e.g., gas, water, and telecommunications), tree roots, and property. In some areas, services are densely packed. Their location may be indicated on plans held by the responsible authority and, where necessary, diversions may need to be arranged in advance. Guidance on typical locations is given by the National Joint Utilities Group (www.njug.org.uk). However, there are always problems with the precise location of services in the ground and with services that are unknown, omitted, or wrongly located on the plans. Non-intrusive methods (e.g., electromagnetic location or ground-probing radar (Taylor, 2004)) can be used to locate services from the surface, but these may need to be backed up by trial pits – small excavations dug by hand (since their purpose is to prevent machinery from causing damage).

Pipe trenches are generally excavated mechanically, though hand excavation is needed where access is limited and where existing services are a problem.

The minimum trench width specified in BS EN 1610: 2015 is given in Tables 16.2 and 16.3. The width must not exceed any maximum specified in the structural design but needs to be related to the trench support system employed. The normal maximum depth of a trench is restricted by the shoring technique and surrounding environment. Depths in some circumstances can be greater than 6 m.

Table 16.2 Minimum trench width related to pipe diameter, for a supported trench with vertical sides

DN	Minimum trench width (m) (OD = outside diameter)
Below 225	OD + 0.4
225–350	OD + 0.5
350–700	OD + 0.7
700–1200	OD + 0.85
Above 1200	OD + 1

Source: Adapted from BS EN 1610: 2015.

Table 16.3 Minimum trench width related to trench depth

Trench depth (m)	Minimum trench width (m)
Less than 1	No minimum
1–1.75	0.8
1.75–4	0.9
More than 4	1

Source: Adapted from BS EN 1610: 2015.

Where access is needed to the outside of structures like manholes, a working space of at least 0.5 m needs to be provided. Where more than one pipe is laid in the same trench, the working space between the pipes should be 0.35 m for pipes up to 700 mm in diameter, and 0.5 m between larger pipes depending on the jointing technique (although joints can be staggered).

The usual system for temporary support of the sides of the trench consists of vertical steel sheets supported by horizontal timber "walings" and adjustable steel struts (Figure 16.1). Whether the sheets form a continuous wall or are placed at a particular separation depends on the condition of the ground and its need for support, as well as on the possible inflow of groundwater. Alternatives are ready-made frames, boxes, or trench shields, which are moved progressively as excavation, laying, and backfilling proceed. These supports are generally used to prevent wall collapse for the safe access of operatives, and their effect on trench wall dimensions may need to be considered.

Where there is a significant problem of groundwater entering the trench during construction, dewatering may be necessary – either by pumping from the trench bottom or from remote points – to lower the water table.

16.5.2 Pipe laying

Pipes are delivered to site in bulk, and stored by stacking until needed. They must be stacked carefully to avoid damage, not so high as to cause excessive loads on the pipes at the bottom, and far enough from the trench to avoid any threat to the stability of the excavation. Manufacturers normally specify stacking arrangements.

The nature of the bedding will be specified in the design. Where a pipe is being laid directly on the bed, the trench should be carefully excavated to the correct gradient to ensure that the pipe is supported all along its length. Localised sections of poor ground at the base of the trench must be dug out and replaced with suitable material. Small extra excavations are typically needed to accommodate pipe sockets with some clearance, to ensure that the weight of the pipe is not bearing on the socket. Where granular bedding material has been specified, this must be similarly prepared to the correct gradient to give support all along the pipe.

The safe lifting of pipes varies with both size and material, and the manufacturer's advice should always be followed. It is most common to set out sewer pipes in an open trench using a laser, set up either inside the pipe or above the excavation. The pipe invert (defined in Section 6.3) should be used as the reference point, since this is the most important vertical position from a hydraulic point of view, and most pipes are not sufficiently round for setting out to be carried out by reference to the crown of the pipe.

Pipes are generally laid in the direction of the upward gradient so that water in the excavation drains away from the working area. Since the spigot is inserted into the socket, the normal orientation of a pipe is for the socket to be at the upstream end, facing upstream. The specification will indicate tolerances of line and level that must be complied with when the pipe is laid.

Methods of jointing pipes have been described in Section 16.2.3. Where a pipeline passes through a fixed structure, it is normal to include flexible joints and sometimes short pipe lengths (*rocker pipes*) close to the wall of the structure. This pipe is generally short and forms the first two joints outside the structure.

Completed sewer sections are tested for leaks by pressure tests using air or water. Criteria for pressure loss with time are given in BS EN 1610: 2015. Sewers that are subsequently to be adopted by the sewerage undertaker may be subject to closed-circuit television (CCTV) inspection (considered further in Chapter 18).

The trench is backfilled in accordance with the design and specification, typically in carefully compacted layers of specified thickness. During backfilling, the trench support is removed progressively. When backfilling is complete, the surface is reinstated.

Good sources of further information on open-trench construction are by Irvine and Smith (1983) and Read (2004b).

16.6 TUNNELLING

16.6.1 Lining

Common methods of tunnel lining are capable of withstanding loads over a wide range of conditions. Linings of extra strength can be supplied where necessary. The loads experienced during construction may be more critical than those experienced by the completed sewer.

It is common for sewers in tunnel to have a primary lining, most commonly bolted concrete segments, to support construction and permanent loads, and a secondary lining, often *in situ* concrete, to provide suitably smooth hydraulic conditions. An increasingly common alternative is to use a *smoothbore* lining, which is a primary lining that does not need a secondary lining.

16.6.1.1 Primary lining

A precast concrete lining consists of segments that are bolted *in situ* to form a ring, with a narrow key segment at or near the soffit. The excavated area will be slightly larger than the outside of the ring, and the annular space between is filled with grout, injected through holes in the lining.

Standard segments are ribbed and are unsuitable for carrying flow as built. Special concrete blocks can be used to fill the panels between the ribs, but the most common method of achieving a smooth surface is by adding a secondary lining.

16.6.1.2 Secondary lining

In situ concrete can be injected behind circular travelling shutters. Alternatives are ready-made linings of GRPs or fibre-reinforced cement composites, with the annular space between the secondary and primary linings filled with grout.

16.6.2 Ground treatment and control of groundwater

Ground treatment for sewers in tunnels may be needed to control groundwater during construction or to stabilise the ground. The main methods are dewatering, ground freezing, and injection of grouts or chemicals. Groundwater at the tunnel face can be controlled by compressed air.

16.6.2.1 Dewatering

Water is pumped from wells to lower the water table in the area of tunnel construction.

16.6.2.2 Ground freezing

The temperature of the ground is lowered to freeze the groundwater during construction. This is achieved by circulating refrigerated liquids – usually brine or liquid nitrogen – through pipes in the ground.

16.6.2.3 Injection of grouts or chemicals

This may be to reduce permeability or improve the strength of cohesionless soils or broken ground. Injection can be carried out through holes drilled from the tunnel face or from the ground surface. Less drilling may be needed from the tunnel face, but this approach can hold up construction.

16.6.2.4 Compressed air

Groundwater can be held back by balancing the hydrostatic pressure with compressed air inside the tunnel. Pressures are commonly less than 1 atm but can be higher. The part of the tunnel under pressure is sealed off by airlocks, through which personnel and materials pass. Air must be supplied continuously as some of it escapes through the ground. People working in compressed air must have regular health checks, and work is tightly controlled by health and safety requirements.

16.6.3 Excavation

Tunnels are driven from working shafts, usually supporting drives of roughly equal lengths both upstream and downstream.

Most ground can be excavated by a boring machine or by handheld pneumatic tools. Hard rock may need to be drilled and blasted. If the ground cannot be left unsupported during the erection of the primary lining, a tunnel shield, which pushes itself forward from the previously erected primary lining, is used to give continuous support. A tunnel-boring machine may combine the functions of a shield and a mechanical excavator.

Shafts are excavated vertically, mechanically, or by hand, and the ground is usually supported using precast concrete segmental rings similar to those used for tunnels. For sewers in tunnels, working shafts usually become manholes in the completed scheme.

More information on constructing sewers in tunnels is given by Chappell and Parkin (2004a), and detailed information on designing and constructing tunnels is given by Maidl et al. (2014).

16.7 TRENCHLESS METHODS

The choice between open-trench construction and traditional tunnelling can be made on the basis of ease, safety, and cost of construction. Beyond a certain depth and diameter, it is simply cheaper – in purely construction terms – to tunnel, than to excavate and backfill a trench. In contrast, trenchless methods usually become an appropriate alternative to open-trench construction when an additional factor is taken into account: the disruption to business, traffic, and everyday life caused by open-trench construction, or by the need to cross under rail tracks, rivers, and busy roads.

Trenchless methods are also applied commonly to other pipe-laying fields, particularly gas and oil supply, and provision of services such as water and electricity, for which some of the techniques were originally developed.

A brief introduction to some of the principal methods is given here; more information can be found in Thomson (1993), Grimes and Martin (1998), Chappell and Parkin (2004b), Howell (2004a), Howell (2004b), and Syms (2004). The International Society for Trenchless Technology website (www.istt.com) provides a useful source of information on new technologies. Further important applications of trenchless technology – to sewer rehabilitation – are covered later in this chapter.

16.7.1 Pipejacking

Hydraulic jacks are used to push specially made pipes (without protruding sockets, and strong enough to take the jacking forces) through an excavated space in the ground (Figure 16.4a). Ahead of the pipes is a shield at which excavation takes place either mechanically or by hand tool. The jacks push against a thrust wall located in a specially constructed thrust pit. *An Introduction to Pipejacking and Microtunnelling* (PJA, 2017) summarises the key aspects.

16.7.2 Microtunnelling

This is a form of pipejacking for pipe diameters less than 900 mm. Excavation is done by unmanned, remote-controlled equipment. PJA (2017) also provides a good summary of the key aspects of this technology.

16.7.3 Auger boring

Soil is removed from the excavated face by an auger (Figure 16.4b), and pipes are jacked into the excavated space. This is sufficiently accurate to maintain gradients for distances up to 100 m.

16.7.4 Directional drilling (DD)

A pilot bore is made using a guided small-diameter drill pipe. The bore is then enlarged to the required size by further mechanical excavation. This is generally less accurate than auger boring.

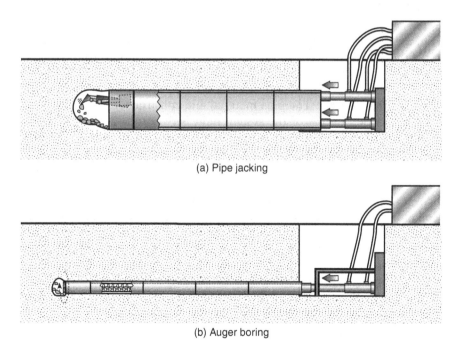

(a) Pipe jacking

(b) Auger boring

Figure 16.4 Pipejacking and auger boring.

16.7.5 Impact moling

An earth displacement mole creates a hole in the ground by pushing the earth outwards. A pipe can then be pushed into the space. This method is used for small diameters and short distances only.

16.7.6 Pipe ramming

A steel casing is driven through the ground by a pneumatic hammer located in a drive pit. The ground material within the casing is removed either during the driving process or, more commonly, when it is complete. Soil removal may be carried out using compressed air or water or by mechanical or manual processes.

16.7.7 Timber headings

Timber sets support an excavation dug by hand, although steel can also be used. This approach is used where there are short distances to tunnel or where trenchless technologies are not appropriate. It can be used when there is an operational issue, and digging down from the surface is not appropriate. Headings must be appropriately designed, and not less than 1.2 m high by 1 m wide. Further information is available in BS 6164: 2019, and the Health and Safety Executive (HSE, 2017) provides guidance on the acceptable size and drive lengths. Pipes are inserted by hand or on a track, and the surrounding is backfilled, typically by concrete or grout.

16.8 REHABILITATION – INTRODUCTION

16.8.1 The need for rehabilitation

How does the need for rehabilitation arise? A typical sewer system in the UK consists of an older part in the centre of the town with newer sections added as the town has expanded. Some parts of the older system may now be undersized, and loadings from traffic, for example, may be far greater than anticipated when the pipe was constructed. The material of the pipe may be old. Overall, about 15% of the UK sewer system is over 100 years old, but in cities, the proportion is higher. The pipe material and construction may be poor quality: a worrying characteristic of 25% of the system built between the World Wars. The system may have been poorly maintained in the past; certainly, it is sometimes possible to "get away with" poor maintenance of sewers because they can still function to some extent in poor structural condition. But this may mean that eventual structural collapse is all the more catastrophic.

Seepage of water into or out of a sewer (infiltration or exfiltration) may increase the risk of sewer collapses and contribute to their seriousness when they eventually occur. Infiltration was discussed in Section 3.4 and is sometimes associated with leakage from water mains, which may also be in a poor structural condition. Any movement of water in the ground may wash soil particles with it, and over a long period of time, this can lead to voids in the ground. A small pipe can cause a large cavity (Figure 16.5). A cracked sewer may be supported by the surrounding ground and not in immediate danger, but if soil is washed in through the cracks by infiltration, then subsidence or collapse becomes far more likely.

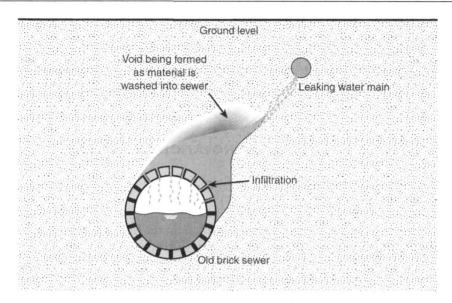

Figure 16.5 Formation of a void around an old sewer.

Symptoms of the need for rehabilitation include

- Collapse
- Flooding of roads and properties (of particular seriousness when wastewater is involved)
- Increased pollution of watercourses
- Blockages
- High levels of infiltration

16.8.2 Rehabilitation approaches

Where sewers deteriorate, replacement with a new sewer is expensive, disruptive and contributes to embedded carbon. The costs of disruption to other services, people, businesses, and traffic in busy towns can be much greater than the costs of construction. This gives rise to a popular saying in sewer rehabilitation: "The greatest asset is the hole in the ground."

The WRc Sewerage Rehabilitation Manual (SRM), first published in 1983, highlighted the value of the hole in the ground and gave methods of renovating and repairing sewers to restore their structural integrity. The current SRM is web-based, and sewer rehabilitation is seen as part of integrated sewer system management and is associated with a risk management approach. Thus, SRM also stands for *Sewerage Risk Management* (WRc, 2023). The web-based material is presented in four sections:

1. About the SRM
2. Sewerage Management Planning
3. Sewer Renovation
4. Surface Water Management

Section 1 is freely available, with the remaining sections for licensed users only. The entire approach should be seen as one element within the overall concept of drainage planning, and this context is explored in Chapter 19.

Methods of renovation and repair are used as part of an integrated rehabilitation programme that might also include, for example, the provision of storage tanks to enhance capacity, and/or the replacement of some sewers, as part of a planned rehabilitation programme.

16.9 METHODS OF STRUCTURAL RENOVATION AND REPAIR

Sewer renovation and repair that causes minimum disruption at the surface is a developing area of technology with highly innovative approaches. Here, we provide an introduction based on some examples of approaches that are currently in use. The intention is to give a flavour of the technology involved. More information on the full range of methods and current developments is available via the SRM website (Water Research Centre, 2023). BS EN 14654-2: 2021 gives an overview of management and control of sewer rehabilitation. BS EN 15885: 2018 gives details on established methods. Read and Vickridge (1997) provide guidance on many aspects of practical implementation.

16.9.1 Renovation

Renovation is defined as "work incorporating all or part of the original drain or sewer by means of which its current performance is improved" (BS EN 14654-2:2021).

Currently, the most commonly used system for the renovation of sewers is *cured-in-place pipe (CIPP) lining*. CIPP linings are positioned in the sewer in a flexible state and then cured in place to form a hard lining, usually in contact with the original sewer wall. One well-established approach involves a felt liner or "sock" impregnated with resin, which is inserted in the sewer by an inversion process (turning inside out) under water pressure. When this lining is in place, it can be cured by circulating hot water (Figure 16.6). An alternative approach involves winching the liner into position and inflating it with air. Alternatives for curing are the use of steam or ultraviolet light. UV curing is a simpler process and has a reduced carbon footprint. For curing with hot water, the liner usually consists of felt impregnated with polyester, vinyl ester or epoxy resin. For curing by UV light the liner is commonly glass matting with added felt and UV-receptive polyester or vinyl ester resin to achieve curing.

A much more long-standing process is *sliplining*. This involves forming a continuous length of plastic lining (such as high-density polyethylene [HDPE]) and pulling it through the existing pipe, resulting in a sewer with a reduced diameter. One approach has been for long lengths of plastic pipe (typically 5 m) to be welded/fused end-to-end on the surface. The resulting continuous length of pipe has some flexibility and can be winched along the sewer via a specially excavated lead-in trench (Figure 16.7a). The winch cable is attached to a nose-cone fixed to the front end of the new lining. As an alternative to welding the lengths on the surface, pipes can be welded in an enlarged trench (Figure 16.7b). The first approach has the disadvantage that it requires space on the surface for assembly of the pipe length; the second requires more excavation (though both approaches require a significant amount). Another alternative is much shorter lengths of pipe (HDPE or polypropylene) with push-fit or screw joints. These are connected within existing manholes (Figure 16.7c). In all cases, the space between the new lining and the old pipe is filled with grout.

As stated, sliplining results in a smaller-diameter pipe. If the existing pipe includes imperfections like distorted cross sections or offset joints, the limitation on the diameter of the new lining may be even greater. It may be appropriate to carry out localised repair of imperfections

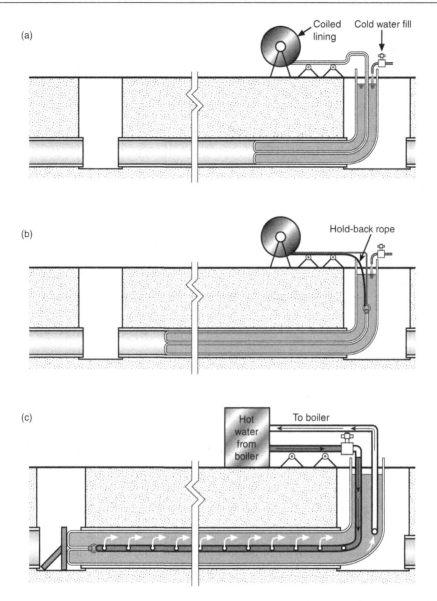

Figure 16.6 Cured in place lining. (a) Commencing insertion; (b) insertion half completed; and (c) insertion completed – curing by hot water circulation

before proceeding with sliplining, in order to increase the practicable diameter of the new lining. The new lining is likely to be smoother than the old, but there may still be a loss of hydraulic capacity.

CIPP lining is becoming much more common for sewers than sliplining for a number of reasons.

- CIPP lining generally requires no excavation whereas sliplining may require a significant amount.
- The reduction in the diameter of the resulting CIPP is less than with sliplining.
- CIPP lining generally reduces friction losses and has no internal jointing.

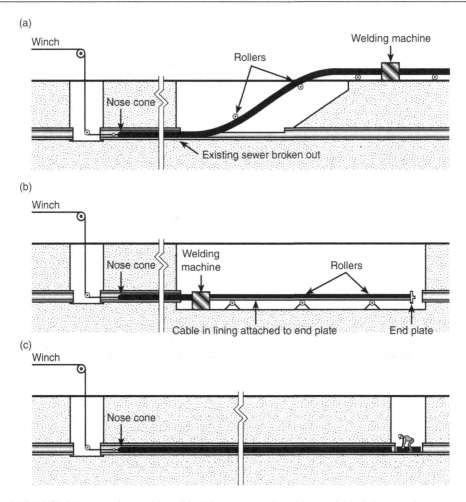

Figure 16.7 (a) Sliplining – surface welding; (b) sliplining – trench welding; and (c) sliplining – short sections

- A CIPP lining may cope with imperfections in the shape of the host pipe more easily than sliplining.
- CIPP lining does not require grouting.
- With modern approaches to construction, CIPP lining is usually a more efficient process.

16.9.1.1 Laterals

Household drain connections, or laterals, pose one of the most awkward problems in sewer renovation for two reasons:

- Badly formed laterals can protrude into the pipe, causing obstruction and potentially preventing the insertion of a lining.
- Once the lining has taken place, existing connections need to be reopened.

Damaged connections can be cut back and repaired prior to lining and reopening. While it is possible to excavate down on poorly connected laterals, this is expensive and time-consuming. Instead, a robotically remote-controlled device typically cuts or grinds the lateral back to the pipe face. This is necessary to limit the liner being stretched/weakened or leaving a gap between the pipe and liner. After lining, the liner is cut at lateral connections, again typically from within the pipe using a robotic remote-controlled cutter.

16.9.1.2 Cleaning

Most rehabilitation operations need to be preceded by sewer cleaning. The available techniques are discussed in Section 18.6. Particularly relevant are the removal of tree roots using high-pressure water cutting, and the use of CCTV to provide a survey of the condition and location of laterals (discussed in detail in Section 18.3)

16.9.1.3 Structural aspects

For new pipes, structural design has been referred to in Section 16.3. The concept of rigid and flexible pipe design, introduced in that section, applies also to renovated sewers. Clearly, structural aspects are a key issue in selecting an appropriate method for sewer renovation; guidance on structural aspects of sewer renovation is available via the SRM website (Water Research Centre, 2023).

16.9.2 Repair

Repair is the localised rectification of damage. Here we give some examples of methods of repairing sewers remotely.

Patch repairs are typically applied at or across a small number of joints/defects and involve remotely placing a patch of fabric containing an appropriate resin material and then curing it in place. In a *resin injection* system, an inflatable packer is used to isolate a pipe defect and to force injected resin into the defect. *Chemical grouting* uses a similar packer device to seal joints and fill any associated voids in the ground. The packer is located at a joint, and collars positioned on either side of the joint inflate to create a seal (Figure 16.8). Applying air or water pressure can first check the effectiveness of the joint. If pressure loss indicates that the joint is unsatisfactory, chemical grout is released under pressure to fill any void outside the pipe and seal the joint.

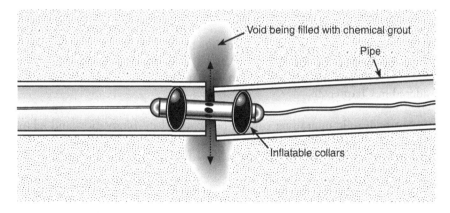

Figure 16.8 Chemical grouting.

An alternative form of chemical grouting treats a whole sewer length at a time. Two chemicals are used, which react together to form a sealing gel. The sewer, laterals, and manholes for a particular length are filled with the first chemical. After a suitable period to allow penetration of the ground around the pipe, the chemical is replaced by a second, which reacts with the residue of the first to form an impermeable mass in the ground around the defects. Any surplus of the second chemical is then pumped out.

Bunting (1997), the *SRM*, and the *UK Society of Trenchless Technology* website provide further information on repair techniques for sewers.

16.10 HYDRAULIC REHABILITATION

Hydraulic rehabilitation of sewer systems involves solving problems caused by hydraulic overloading in order to achieve the performance targets set. A range of approaches can be used. Those that are strongly favoured in current drainage practice involve reducing hydraulic inputs to the piped system, especially stormwater. The most significant method of reducing inputs as a solution to a range of drainage problems is the use of SuDS techniques to divert water away from entering the piped system or to limit its flow.

Retrofitting SuDS, or installing sustainable drainage components in a catchment that is already drained by a conventional pipe system, allows this disconnection of stormwater inputs. It has been shown to be effective in a variety of settings including highly urbanised catchments.

The use of retrofit SuDS in this context has been introduced in Section 11.4.2. In addition, the Susdrain website (www.susdrain.org) provides case studies and describes general approaches that demonstrate the role of retrofit SuDS in the hydraulic rehabilitation of sewer systems. CIRIA's guidance (Digman et al., 2012) *Retrofitting to Manage Surface Water* gives relevant guidance on retrofitting SuDS in urban areas.

The uncertainties associated with climate change and concern about the carbon footprint associated with new construction point clearly towards these approaches to hydraulic rehabilitation. However, other approaches may be considered.

Removal of local constrictions may allow the capacity of significant sections of the system to be increased. Sewer cleaning will also increase capacity, though if the deposition of debris had been caused by the physical nature of the sewer, then proactive maintenance will be required. There is more information on sewer cleaning in Chapter 18.

More efficient use can be made of the existing capacity of the sewer by

- Providing additional storage within the system in the form of detention tanks (described in Chapter 14)
- Installing throttles and controls at key points to make the best use of in-sewer storage (these devices are described in Section 8.1)
- Applying techniques for overall system management, including real-time control, as considered in Chapter 22

When sufficient hydraulic upgrading cannot be achieved by renovation and adaptation, it may be necessary to replace existing sewers with new sewers of larger capacity. This will, of course, bring with it the very significant issues already discussed in Section 16.8.

PROBLEMS

1 Describe the properties of the main sewer pipe materials. What factors affect the selection of a pipe material for a particular case?
2 Describe the main methods of sewer construction. Discuss the factors that influence the selection of the most appropriate method in particular cases.
3 Describe methods of ground treatment for sewer construction and the circumstances in which they might be used.
4 What does *SRM* stand for?
5 Why is sewer rehabilitation more common now than it was 50 years ago?
6 How can flow models be used in the planning of sewer rehabilitation schemes?

KEY SOURCES

Read, G.F. 2004a. *Sewers: Replacement and New Construction*, Butterworth-Heinemann.
Read, G.F. and Vickridge, I.G. (eds.). 1997. *Sewers—Rehabilitation and New Construction; Repair and Renovation*, Arnold.
WRc. 2023. *Sewerage Risk Management*. http://srm.wrcplc.co.uk

REFERENCES

Bunting, N. 1997. Localised repair techniques for non-man-entry sewers, in *Sewers—Rehabilitation and New Construction; Repair and Renovation* (eds G.F. Read and I.G. Vickridge) (Chapter 11, 233–253), Arnold.
Chappell, M., and Parkin, D. 2004a. Tunnel construction, in *Sewers: Replacement and New Construction* (ed. G.F. Read) (Chapter 7, 150–192), Elsevier Butterworth-Heinemann.
Chappell, M., and Parkin, D. 2004b. Pipejacking, in *Sewers: Replacement and New Construction* (ed. G.F. Read) (Chapter 9, 210–223), Elsevier Butterworth-Heinemann.
Department for Science, Innovation and Technology, 2023. *National Underground Asset Register (NUAR)*, www.gov.uk/guidance/national-underground-asset-register-nuar
Digman, C.J., Ashley, R.M., Balmforth, D.J., Balmforth, D., Stovin, V., and Glerum, J. 2012. *Retrofitting to Manage Surface Water*, Report C713, CIRIA, London, UK.
Grimes, J.F., and Martin, P. 1998. *Trenchless and Minimum Excavation Techniques, Planning and Selection*, CIRIA SP147.
Howell, N. 2004a. Impact moling, in *Sewers: Replacement and New Construction* (ed. G.F. Read) (Chapter 11, 236–253), Elsevier Butterworth-Heinemann.
Howell, N. 2004b. The pipe ramming technique, in *Sewers: Replacement and New Construction* (ed. G.F. Read) (Chapter 12, 254–271), Elsevier Butterworth-Heinemann.
Health and Safety Executive (HSE). 2017. *Tunnelling and Pipejacking: Guidance for Designers*, www.hse.gov.uk/construction/pdf/pjaguidance.pdf
Irvine, D.J., and Smith, R.J.H. 1983. *Trenching Practice*, CIRIA R97.
Maidl, B., Thewes, M., and Maidl, U. 2014. *Handbook of Tunnel Engineering, Volumes I and II*, Wiley.
Moser, A.P., and Folkman, S. 2008. *Buried Pipe Design*, 3rd edn, McGraw-Hill Professional.
Pipe Jacking Association (PJA). 2017. *An Introduction to Pipe Jacking and Microtunnelling*. https://www.pipejacking.org/assets/pj/static/PJA_intro.pdf
Read, G.F. 2004b. Open-cut and heading construction, in *Sewers: Replacement and New Construction* (ed. G.F. Read) (Chapter 6, 132–149), Elsevier Butterworth-Heinemann.

Sewers and Water Mains Committee. 1993. *Materials Selection Manual for Sewers, Pumping Mains, and Manholes*, WSA/FWR.

Syms, B. 2004. Horizontal directional drilling, in *Sewers: Replacement and New Construction* (ed. G.F. Read) (Chapter 14, 297–326), Elsevier Butterworth-Heinemann.

Taylor, N. 2004. Site investigation and mapping of buried assets, in *Sewers: Replacement and New Construction* (ed. G.F. Read) (Chapter 3, 95–99), Elsevier Butterworth-Heinemann.

Thomson, J.C. 1993. *Pipejacking and microtunnelling*, Blackie Academic and Professional.

University of Birmingham, 2024. *Mapping the Underworld.* www.birmingham.ac.uk/university/colleges/eps/research/resilience-energy/underworld.aspx

Woolley L. 1988. *Drainage details*, 2nd edn, E & FN Spon.

Whyte, I. 2004. Site investigation, in *Sewers: Replacement and New Construction* (ed. G.F. Read) (Chapter 2, 47–94), Elsevier Butterworth-Heinemann.

STANDARDS AND SPECIFICATIONS

BS 65: 1991. *Specification for Vitrified Clay Pipes, Fittings and Ducts, also Flexible Mechanical Joints for Use Solely with Surface Water Pipes and Fittings.*

BS EN 295: 2013. *Vitrified Clay Pipe Systems for Drains and Sewers.*

BS EN 476: 2022. *General Requirements for Components Used in Drains and Sewers.*

BS EN 545: 2010. *Ductile Iron Pipes, Fittings, Accessories and Their Joints for Water Pipelines. Requirements and Test Methods.*

BS EN 1295-1: 2019 *Structural design of buried pipelines under various conditions of loading.*

BS EN 1401–1: 2019. *Plastics Piping Systems for Non-Pressure Underground Drainage and Sewerage, Unplasticized Polyvinyl Chloride (PVC-U). Specifications for Pipes, Fittings and the System.*

BS EN 1610: 2015. *Construction and Testing of Drains and Sewers.*

BS EN 1916: 2002. *Concrete Pipes and Fittings, Unreinforced, Steel Fibre and Reinforced.*

BS 4660: 2022. *Thermoplastics Ancillary Fittings of Nominal Sizes 110 and 160 for Below Ground Gravity Drainage and Sewerage.*

BS 5911–1: 2021. *Concrete Pipes and Ancillary Concrete Products. Specification for Unreinforced and Reinforced Concrete Pipes (Including Jacking Pipes) and Fittings with Flexible Joints.*

BS 5911–3: 2022. *Concrete Pipes and Ancillary Concrete Products. Specification for Unreinforced and Reinforced Concrete Manholes and Soakaways.*

BS 5911–5: 2004. *Concrete Pipes and Ancillary Concrete Products. Specification for Pre-Stressed Non-Pressure Pipes and Fittings with Flexible Joints.*

BS 6164: 2019. *Code of Practice for Health and Safety in Tunnelling in the Construction Industry.*

BS 9295: 2020. *Guide to the Structural Design of Buried Pipes.*

BS EN 10224: 2002. *Non-Alloy Steel Tubes and Fittings for the Conveyance of Aqueous Liquids Including Water for Human Consumption.*

BS EN 10311: 2005. *Joints for the Connection of Steel Tubes and Fittings for the Conveyance of Water and Other Aqueous Liquids.*

BS EN 12201-1: 2011. *Plastic Piping Systems for Water Supply, and for Drainage and Sewerage Under Pressure. Polyethylene (PE). General.*

BS EN 13476: 2018. *Plastics Piping Systems for Non-Pressure Underground Drainage and Sewerage. Structured-Wall Piping Systems of Unplasticized Polyvinyl Chloride (PVC-U), Polypropylene (PP) and Polyethylene (PE).*

BS EN 14654-2: 2021 *Drain and sewer systems outside buildings – management and control of activities, Part 2 Rehabilitation.*

BS EN 15885: 2018 *Classification and characteristics of techniques for renovation, repair and replacement of drains and sewers.*

PAS 128:2022 *Underground Utility Detection, Verification and Location – Specification.*

WIS 4-32-08: 2002. *Specification for the Fusion Jointing of Polyethylene Pressure Pipeline Systems Using PE80 and PE100 Materials, Water Industry Specification.*

Chapter 17

Sediments

17.1 INTRODUCTION

Sediment is ubiquitously present in urban drainage systems. It is found deposited on catchment surfaces, in gully pots, and in drains and sewers. Drainage engineers have long recognised its presence in stormwater and the problems it may cause. They have sought to exclude larger, heavier sizes from the piped system by the provision of gully pots and designed sewers to limit in-pipe deposition. The theory is that sediment that does enter the system is carried downstream, where it is eventually trapped and removed at the outlet of the system. This may be the case for newly designed systems, but for older (especially combined) networks, sedimentation in sewers is commonplace.

The movement of sediment through a drainage catchment is a complex, multi-stage process. Sediment deposited on roads, for example, is initially freed from the surface and then washed transversely by overland flow to the channel, where it is transported parallel to the kerb-line under open channel flow. The sediment is discharged with the flow into the gully inlet under gravity and is captured in the gully pot (in part) by sedimentation or transferred to the receiving sewer. Once in the sewer, the material is transported under open channel flow as suspended or bed-load. Suspended sediment is carried along in the main body of the flow, while bed-load travels more slowly in contact with the invert of the pipe. Some material may be deposited and/or re-eroded as it progresses downstream.

During transport, sediment may be discharged to a watercourse via a combined sewer overflow (CSO) (if in operation) or settled in the wastewater treatment plant (WTP) grit removal device. At points of deposition (surface, gully, sewer), sediment may be extracted from the system by cleaning. A representation of sediment inputs, outputs, and movement through the system is given in Figure 17.1. Further details on the removal of sediment from systems are given in Chapter 18.

The rate of progress of material through the system depends on factors such as the characteristics of the

- Sediment (physical, chemical)
- Flow (velocity, degree of unsteadiness)
- Drainage network (layout, geometry)

For example, different types of sediment will move through the system in different ways. Particles of very small size or low density may remain in suspension under all normal flow conditions and be transported through the system without being deposited. Sediments with low settling velocities may only form deposits during periods of very low flow, and may

DOI: 10.1201/9781003408635-17

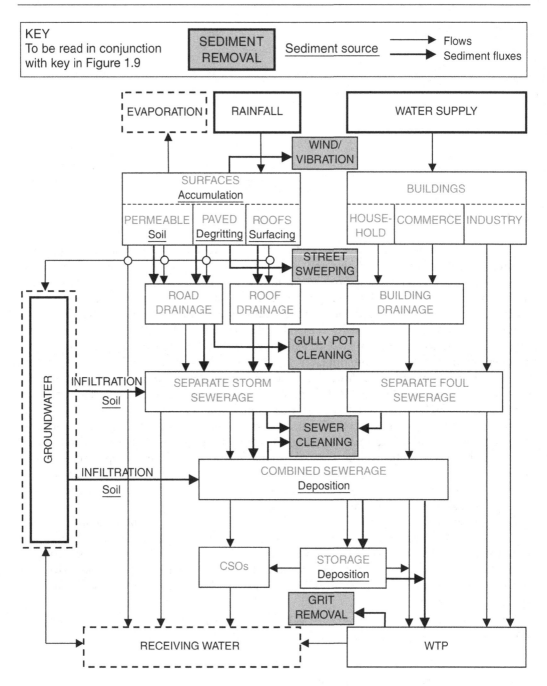

Figure 17.1 Entry and exit points of sediment in drainage systems.

readily be re-entrained when higher velocities occur in the pipes as a result of storms or diurnal variations in flow. By contrast, larger and denser particles may only be transported by peak flows that occur relatively infrequently, and in some cases, they may form permanent stationary deposits near where they enter the sewer system.

Deposition commonly occurs during dry weather flow periods (particularly during low flow at night) and in decelerating flows during storm recessions. Deposits also form at structural and hydraulic discontinuities such as at joints, changes in gradients, and ancillary structures. Only the steepest sewers are immune from deposition.

This chapter reviews the origins (Section 17.2), problems, and effects (Section 17.3) of sediment. Subsequent sections look at how the sediment is transported through the system (Section 17.4) and at its detailed characteristics (Section 17.5). Sewer self-cleansing design methods are presented in Section 17.5.

17.2 ORIGINS

17.2.1 Definition

Sewer sediments are defined as any settleable particulate material that is found in stormwater or wastewater and is able, under appropriate conditions, to form bed deposits in sewers and associated hydraulic structures. Using the basic solids classification presented in Table 2.1, this would include

- Grit
- Suspended solids
- Sanitary
- Stormwater

17.2.2 Sources

Sources of sediment entering sewers are quite diverse. Any particulate-generating material or activity in the urban environment is a potential source. Broadly, three categories can be established: sanitary, surface, and sewer. These are defined in Table 17.1.

Table 17.1 Sources of sewer sediment

Source	Type
Sanitary	• Large faecal and organic matter with specific gravity close to unity • Fine faecal and other organic particles • Paper and miscellaneous materials flushed into sewers (wet wipes/sanitary refuse) • Vegetable matter and soil particles from the domestic processing of food • Materials from industrial and commercial sources
Surface	• Atmospheric fall-out (dry and wet) • Particles from erosion of roofing material • Grit from abrasion of road surfaces or from resurfacing works • Grit from de-icing operations on roads • Particulates from motor vehicles (e.g., vehicle exhausts, rubber from tyres, wear and tear) • Materials from construction works (e.g., building aggregates, concrete slurries, exposed soil) and other illegally dumped materials • Detritus and litter from roads and paved areas (e.g., paper, plastic, cans) • Silts, sands, and gravels washed or blown from unpaved areas • Vegetation (e.g., grass, leaves, wood)
Sewer	• Soil particles infiltrating due to leaks or pipe/manhole/gully failures • Material from infrastructure fabric decay

17.3 EFFECTS

17.3.1 Problems

There are three major effects of sediment deposition leading to a number of serious consequences, and these are summarised in Table 17.2. Table 17.2 also lists parts of the book where further information can be found on the consequences of sediment deposition.

The first effect of deposited sediment is its propensity to initiate blockage.

Larger, gross solids and other matter may build up, leading to partial or total blockage of the pipe bore.

The second effect results from the fact that a deposited bed restricts the flow in the sewer, resulting in a loss of hydraulic capacity. The reasons for this effect are discussed in Section 17.3.2, but the result can be a pipe or manhole surcharge. Another common effect is the premature operation of CSOs.

The third major effect results from the ability of the deposited sediment to act as a pollutant store or generator. The reasons and possible mechanisms for this are discussed later in the chapter. These pollutants are only stored temporarily and can be released under flood flow conditions, probably contributing to the commonly observed first foul flush of heavy pollution (see Section 13.3.2). Biochemical changes in the bed of sediment can result in septic conditions, releasing gas that can be highly corrosive to the sewer fabric.

It should be clear from the previous discussion that excessive sediment deposition should be avoided if at all possible at the design stage. Extensive sediment removal is a difficult, recurrent, and expensive process (Section 18.6).

17.3.2 Hydraulic

The presence of sediment in sewer flows has three hydraulic effects of varying importance: dissipation of energy in keeping solids in suspension, reduction of flow cross-sectional area, and increased frictional losses due to the texture of the bed.

17.3.2.1 Suspension

In the case of a sewer without deposits, the presence of sediment in the flow or moving along the invert causes a small increase in energy loss, and this is observed as a reduction in discharge capacity of about 1% for rough pipes (Ackers et al., 1996a).

Table 17.2 Effects and consequences of sewer sediment deposition

Effect	Consequences	Further information (Section)
Blockage	• Surcharging	7.4.5/18.5
	• Surface flooding	12/18.5
Loss of hydraulic capacity	• Surcharging	17.3.2
	• Surface flooding	
	• Premature operation of CSOs	13.3.1
Pollutant storage	• Washout to receiving waters during CSO operation	2.5.1
	• Shock loading on treatment plants	
	• Gas and corrosive acid production	18.10

For flows in sewers in which there is a deposited bed of sediment, the energy losses associated with keeping the sediment in motion are relatively insignificant compared with the other effects.

17.3.2.2 Geometry

A deposited bed reduces the cross-sectional area available to convey flow and therefore increases the velocity and head loss for a given discharge and depth of flow. The loss of total area is relatively small (<2%) provided that the depth of sediment is less than 5% of the pipe diameter, but becomes important at sediment depths above about 10%.

17.3.2.3 Bed roughness

Usually of greatest significance is the increase in overall resistance caused by the rough texture of the deposited bed. Above the *threshold of movement*, sediment quickly forms into ripples and dunes, and initially, these grow in size as the flow velocity increases. The effective roughness value of dunes kb (k_s in the Colebrook–White equation) can reach 10% or more of the pipe diameter, compared with typical values for the pipe walls that are in the range of 0.15–6 mm (depending on the material and the degree of sliming). The value of k_b can be estimated (May, 1993) from

$$k_b = 5.62R^{0.61}d_{50}^{0.39}$$

(17.1)

where R is hydraulic radius (m) and d_{50} is sediment particle size larger than 50% (by mass) of all particles in the bed (m).

Under these conditions, a 5% depth of deposited sediment with dunes could reduce the pipe-full capacity by 10–20%. However, at higher velocities, the dunes tend to reduce in size until the bed again becomes flat with much lower roughness. The loss of hydraulic capacity due to a deposited bed can, therefore, vary considerably with the flow conditions. The approximate combined effects of shape and roughness are illustrated in Figure 17.2.

17.3.3 Pollution

Solids of sanitary origin are readily mixed, in combined sewers, with sediments entering from surface sources. The organic material tends to adhere to the heavier inorganic sediment and deposits with them. These deposits form a rough surface that encourages further adherence of organic matter. In such an environment, anaerobic conditions are likely to develop, resulting in partial digestion of the sediment. By-product fatty acids are then liberated into the interstitial liquor (as described further in Chapter 18) with the possibility of substantial biochemical oxygen demand/chemical oxygen demand (BOD/COD) loads being generated. Some evidence suggests that degradation processes in sediment can increase pollutant discharges by up to 400% (Binnie and Partners and Hydraulics Research, 1987). Clearly, the presence of sediment deposits encourages the retention of solids and pollutants during low flows. This increases the potential for degradation before such material is flushed away.

The concept and importance of the first foul flush are discussed in Chapter 13. Sediment deposits are commonly considered to be one of its major causes. Field evidence indicates that 70–90% of the pollution load discharged from CSOs can be derived from the erosion of in-sewer deposits (Chen et al., 2022; Crabtree, 1989).

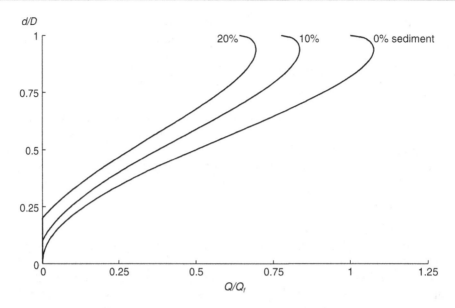

Figure 17.2 Effect of sediment bed on flow capacity of sewers.

17.3.4 Transport

The movement of sediment has three broad phases: entrainment, transport, and deposition.

17.3.5 Entrainment

As wastewater flows over a sediment bed in a sewer, hydrodynamic lift and drag forces are exerted on the bed particles. If these two combined forces do not exceed the restoring forces of sediment submerged weight, interlocking and cohesion (if applicable), then the particles remain stationary. If they exceed the restoring force, then entrainment occurs, resulting in movement of the particles at the flow/sediment boundary. Not all the particles of a given size at this boundary are dislodged and moved at the same time, as the flow is turbulent and contains short-term fluctuations in velocity. The limiting condition, below which sediment movement is negligible (the threshold of movement), is usually defined in terms of either a critical boundary shear stress (τ_o) or critical erosion velocity (v). The two are related as shown in Equation 7.22 reproduced below:

$$\tau_o = \frac{\rho \lambda v^2}{8}$$

where ρ is liquid density (kg/m³) and λ is the Darcy–Weisbach friction factor (–).

In storm sewers, sediments are mainly inorganic and non-cohesive, although some deposits may be cementitious and become permanent if undisturbed for long periods. Sediments in foul sewers generally have cohesive-like properties due to the nature of the particles and the presence of greases and biological slimes. In combined sewers, the sediments tend to be a combination of the first two types.

Cohesion will tend to increase the value of shear stress that the flow needs to exert on the deposited bed in order to initiate movement of particles in the surface layer. In laboratory tests, erosion of synthetic cohesive sediments has been observed to occur at bed shear stresses of 2.5 N/m² (surficial material) and 6–7 N/m² (granular, consolidated deposits). Field studies in large sewers, however, show that lower shear stresses of around 1 N/m² may initiate erosion (Ashley and Verbanck, 1996).

It is possible that cohesion may also alter the way in which the sediment then moves and thus affect the sediment-transporting capacity of the flow. However, experimental research (Nalluri and Alvarez, 1992) using synthetic cohesive sediments suggests that the second factor may not be significant; once the structure of a cohesive bed is disrupted, the particles are stripped away and transported by the flow in a similar way to non-cohesive sediments.

17.3.6 Transport

Once sediment has been entrained into the flow, it travels, as mentioned in the introduction, in suspension or as bed-load. Finer, lighter material tends to travel in suspension and is primarily influenced by turbulent fluctuations in the flow, which in turn are influenced by bed shear. It is advected at mean flow velocity. Heavier material travels by rolling, sliding, or saltating along the pipe invert (or deposited bed) as bed-load. This type of movement is affected by the local velocity distribution, and advection velocities in this mode are considerably lower than the flow mean velocity. In an urban drainage network, with graded materials of differing specific gravity, a combination of these modes exists.

Table 17.3 indicates that the mode of transport depends on the relative magnitude of the lifting effects due to turbulence, as measured by the shear velocity (U^*), and the settling velocity (W_S). Shear velocity is given by

$$U^* = \sqrt{\frac{\tau_o}{\rho}}$$

(17.2)

A large body of knowledge has been built up, including many different predictive equations, for sediment transport, based particularly on loose-boundary channels including rivers (see, e.g., Raudkivi, 1998). Equations are available, normally in terms of the volumetric sediment carrying capacity of the flow, for both suspended and bed-load transport.

Although these equations are useful in highlighting important principles, they should not be used uncritically for sewer design and analysis. Conditions in pipes are different than those in rivers: pipes have rigid boundaries, significantly different and well-defined cross sections, and they transport different materials. However, there have been a number of studies particularly focusing on pipes and sewers, both in the laboratory and in the field, and these have been comprehensively reviewed and compared by Ackers et al. (1996a). Recommended transport equations are given in Section 17.5.

Table 17.3 Mode of sediment transport

W_s/U^*	Mode
>0.6	Suspension
0.6–2	Saltation
2–6	Bed-load

Source: Raudkivi, A.J. 1998. *Loose Boundary Hydraulics*, CRC Press.

EXAMPLE 17.1

Analysis of the sediment in an urban catchment shows it to consist predominantly of grit with a characteristic settling velocity of 750 m/h. Estimate the mode of transport of this sediment in a 0.15% gradient, 1.5 m diameter sewer flowing half-full.

Solution

$$R = \frac{D}{4} = 0.375 \, \text{m}$$

For a pipe flowing half full, the wall shear stress is given by Equation 7.21:

$$\tau_0 = \rho g R S_0 = 1000 \times 9.81 \times 0.375 \times 0.0015 = 5.5 \, \text{N} / \text{m}^2$$

Shear velocity is given by Equation 17.2:

$$U^* = \sqrt{\frac{5.5}{1000}} = 0.074 \, \text{m} / \text{s}$$

So, as

$$\frac{W_s}{U^*} = \frac{750 / 3600}{0.074} = 2.8 > 2$$

transport will be bed-load.

17.3.7 Deposition

If the flow velocity or turbulence level decreases, there will be a net reduction in the amount of sediment held in suspension. The material accumulated at the bed may continue to be transported as a stream of particles without deposition. However, below a certain limit, the sediment will form a deposited bed, with transport occurring only in the top layer (the *limit of deposition* [LOD]). If the flow velocity is further reduced, sediment transport will cease completely (the *threshold of movement*). The flow velocities at which deposition occurs tend to be lower than those required to entrain sediment particles.

17.3.8 Sediment beds and bed-load transport

If an initially clean sewer flowing part-full is subjected to a sediment-laden flow transported under bed-load, but conditions are not sufficient to prevent deposition, a sediment bed will develop. It will increase the bed resistance, causing the depth of flow to increase and the velocity to decrease.

Intuitively, it might be assumed that a reduction in velocity would cause a reduction in the sediment-transporting capacity of the flow, leading to further deposition and possibly blockage. In fact, laboratory evidence has shown (May, 1993) that the presence of the deposited bed actually allows the flow to acquire a *greater* capacity for transporting sediment as bed-load. This is because the mechanism of sediment transport is related to the width of the deposited bed, which can, of course, be much greater than the narrow stream of sediment that is present along the bed of the pipe at the LOD. The effect more than compensates for the reduction in velocity caused by the roughness of the bed. Ultimately, the increased deposited bed depth (and width) and the associated increased sediment transport capacity may balance with the incoming sediment load and prevent further deposition. Thus, in principle, a small amount of deposition may be advantageous in terms of sediment mobility.

17.4 CHARACTERISTICS

17.4.1 Deposited sediment

The characteristics of sewer sediment deposits vary widely according to the sewer type (foul, storm, or combined), the geographical location, the nature of the catchment, local sewer operation practices, history, and customs. Crabtree (1989) proposed that the origin, nature, and location of deposits found within UK sewerage systems could be used to classify sediment under five categories A–E (see Figure 17.3). The characteristics of these deposits are described in the following and are summarised in Table 17.4.

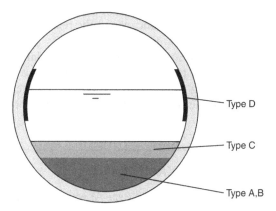

Figure 17.3 Typical sediment deposits in a sewer.

Table 17.4 Physical and chemical characteristics of sewer sediment type classes

	Sediment type				
	A	B	C	D	E
Description	Coarse, loose granular material	As A but concreted with fat, tars, etc.	Fine-grained deposits	Organic slimes and biofilms	Fine-grained deposits
Location	Pipe inverts	As A	Quiescent zones, alone or above A material	Pipe wall around mean flow line	In CSO storage tanks
Saturated bulk density (kg/m³)	1720	N/A	1170	1210	1460
Total solids (%)	73.4	N/A	27.0	25.8	48.0
COD (g/kg)[a]	16.9	N/A	20.5	49.8	23.0
BOD$_5$ (g/kg)[a]	3.1	N/A	5.4	26.6	6.2
NH$_4$$^+$–N (g/kg)[a]	0.1	N/A	0.1	0.1	0.1
Organic content (%)	7.0	N/A	50.0	61.0	22.0
FOG (%)	0.9	N/A	5.0	42.0	1.5

Source: Adapted from Crabtree, R.W. 1989. *Journal of the Institution of Water and Environmental Management,* 3, 569–578.

[a] Grams pollutant per kilogram of wet bulk sediment.

17.4.1.1 Physical characteristics

Type A material is the coarsest material, found typically on sewer inverts. These deposits have a bulk density of up to 1800 kg/m³ and an organic content of 7%, with some 6% of particles typically smaller than 63 μm. The finer material (type C) is typically 50% organic, with a bulk density of approximately 1200 kg/m³ and some 45% of the particles are smaller than 63 μm. Type E is the finest material, although there is no definite boundary between any of the types A, C, or E. This is perhaps to be expected, as the sediment actually deposited depends on the material available for transport and the flow conditions in specific locations.

17.4.1.2 Chemical characteristics

Table 17.4 summarises the mean chemical characteristics of the deposited sediments. A high degree of variability is observed in practice (e.g., coefficient of variation 23–125%). On a mass-for-mass basis, wall slimes are the most polluting in terms of oxygen demand (49.8 gCOD/kg wet sediment). There is a broad decrease in strength among types, in the order D, E, C, and A, with type A material having mean COD levels of just 17.9 g/kg.

However, this does not show the full significance of the relative polluting potential of each type of deposit. This is illustrated by Example 17.2.

Results from Example 17.2 show that although type D material is of higher unit strength, where small quantities are found in practice, it tends to be relatively insignificant. Type A material clearly shows up as having the bulk of the pollution potential (79% in this case) because of its large volume. The actual value varies depending on location. It should be

---

EXAMPLE 17.2

A 1500 mm diameter sewer has a sediment bed (type A) with an average thickness of 300 mm. Above this is deposited a 20-mm type C layer and above that flows 350 mm of wastewater (BOD_5 = 350 mg/L). Along the walls of the sewer at the waterline are two 50 mm × 10 mm thick biofilm deposits (type D). Calculate the relative pollutant load associated with each element in the pipe.

Solution

For each of the sediment types, the cross-sectional area can be calculated from the pipe's geometrical properties (Chapter 7) to give volume per unit length. Combining this information with the bulk density of the sediment and its pollutant strength enables the unit pollution load to be estimated as shown in Table 17.5.

Table 17.5 Data for Example 17.2

Type	Depth (mm)	Volume (m³/m)	Bulk density (kg/m³)	BOD_5 (g/kg)	Unit BOD_5 (g/m length)	Percentage (%) load
A	0–300	0.252	1720	3.1	1344	79
C	300–320	0.024	1170	5.4	152	9
Wastewater	320–670	0.488	1000	0.35	171	10
D	50 × 10 × 2	0.001	1210	26.6	32	2
Total					1699	100

realised, too, that the total polluting load would only be released under extreme storm flow conditions that erode all the sediment deposits. More routine storm events will probably erode only a fraction of the type A deposits. It is also interesting to note that, in this case, the wastewater itself only represents 10% of the potential pollutant load.

17.4.1.3 Significance of deposits

Type A and B deposits are normally associated with loss of sewer capacity, and type A deposits are the most significant source of pollutants. The nature of the sediment appears to vary between areas, with large organic deposits being found nearer the heads of networks and with more granular material (type A) being found in trunk sewers. Larger interceptor sewers typically have more type C material intermixed with type A (Ashley and Crabtree, 1992). Pipe wall slimes/biofilms (type D) are important because they are very common, highly concentrated, easily eroded, and typically increase hydraulic roughness (Fouada et al., 2014).

17.4.2 Mobile sediment

17.4.2.1 Suspension

The predominant particles in suspension during both dry and wet weather flows are approximately 40 μm in size, and primarily attributed to sanitary solids. Most of the suspended materials in combined sewer flow (≈90%) are organic and are biochemically active with the capacity to absorb pollutants. Settling velocities are usually less than 10 mm/s (Crabtree et al., 1991).

17.4.2.2 Near-bed

Under dry weather flow conditions, sediment particles can form a highly concentrated, mobile layer, "dense undercurrent," or "muddy layer" just above the bed (see Figure 17.4). Solids in this region are relatively large (>0.5 mm), organic (>90%) particles and are believed to be trapped in a matrix of suspended flow (Verbanck, 1995). Concentrations of solids of up to 3500 mg/L have been measured, and the corresponding biochemical pollutants are also particularly concentrated (Ashley and Crabtree, 1992). According to Ashley et al. (1994), typically, 12% of total solids are conveyed in the material moving near the bed. The rapid entrainment of near-bed solids is thought to make a significant contribution to first foul flushes (Larrarte et al., 2017). Chapter 13 discusses these flushes in relation to the pollution load that may be discharged from CSOs.

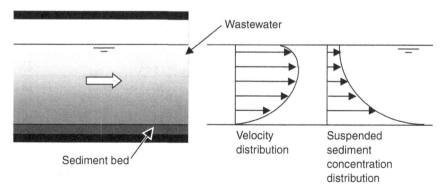

Figure 17.4 Velocity and suspended sediment distributions in dry weather.

17.4.2.3 Granular bed-load

Granular particles (2–10 mm) are transported as "pure" bed-load only in steeper sewers (>2%). In flatter sections (<0.1%), little granular material is observed in motion, presumably because it is deposited.

17.4.2.4 Particle size

In Section 5.4.2, it was shown how smaller particle sizes tend to be associated with a greater proportion of pollutants than might be expected from their mass. The same is true of mobile sewer sediment as shown in Table 17.6.

17.5 SELF-CLEANSING DESIGN

17.5.1 Velocity based

The need for sewers to be designed to carry sediment has been recognised for many years. Conventionally, this has been done by specifying a minimum "self-cleansing" flow velocity that should be achieved at a particular depth of flow or with a particular frequency of occurrence (see Chapters 9 and 10). Although this approach has apparently been successful in many cases, a single value of minimum velocity, unrelated to the characteristics and concentration of the sediment or to other aspects of the hydraulic behaviour of the sewer, does not properly represent the ability of sewer flows to transport sediment. In particular, it is known that a higher flow velocity is needed to transport a given concentration of sediment in large sewers than in small sewers.

It is also important to appreciate that conditions in gravity sewers are extremely variable. Flow rates and the sediment entering a system can vary considerably with time and location, so a sewer designed to be self-cleansing in normal conditions is still likely to suffer sediment deposition during periods of low flow and/or high sediment load.

17.5.2 Tractive force based

US practice (Bizier, 2007) recommends a *tractive force* method for the self-cleansing design of gravity sanitary (foul) sewers. The term is somewhat misleading because what is being referred to is actually a stress rather than a force, with units of Newtons per square metre (N/m^2). To check the impact of specifying a minimum shear stress rather than velocity, the

Table 17.6 Percentage of total pollution load associated with different particle size fractions

Pollutant	Particle size fraction (μm)		
	<50	50–250	>250
BOD	52	20	28
COD	68	4	28
TKN	16	58	26
Hydrocarbons	69	4	27
Lead	53	34	13

Source: Bertrand-Krajewski, J.-L., Briat, P.M., and Scrivener, O. 1993. *Journal of Hydraulic Research*, 31(4), 435–460.

variation of shear stress with depth could be added to Figure 7.8 in Chapter 7, which shows the geometric and hydraulic elements for part-full pipe flow. From Equation 7.21, we can infer that $\tau_o/\tau_{of} \equiv R/R_f$, a plot of which is already shown. As can be seen, the lines for R/R_f and v/v_f are similar, so it does not matter much whether a self-cleansing velocity or a critical shear stress (tractive force) is specified, provided an appropriate value is used. Merritt (2009) suggests using a critical shear stress of 0.87 N/m² (associated with a 1 mm diameter particle) for separate foul sewers. Merritt and Enfinger (2019) argue for its importance. The application of this method is illustrated in Example 17.3.

17.5.3 The CIRIA method

The CIRIA method (Ackers et al., 1996a; Butler et al., 1996a, 1996b) was developed in an attempt to relate minimum velocity to all the factors that affect it most, namely, pipe size and roughness, proportional flow depth, sediment size and specific gravity, degree of cohesion between the particles, sediment load or concentration, and the presence of a deposited bed.

17.5.3.1 Self-cleansing

The method proposes the following definition of self-cleansing:

> An efficient self-cleansing sewer is one having a sediment-transporting capacity that is sufficient to maintain a balance between the amounts of deposition and erosion, with a time-averaged depth of sediment deposit that minimises the combined costs of construction, operation and maintenance.
>
> (Ackers et al., 1996a)

The important aspect of this definition is that it does not necessarily require sewers to be designed to operate completely free of sediment deposits if more economical overall solutions can be achieved by allowing some deposition to occur. This is a viable alternative because, as described in Section 17.4.4, the presence of the deposited bed can significantly increase the sediment transporting capacity of the pipe, despite the adverse effect of the deposits on the geometry and hydraulic roughness.

EXAMPLE 17.3

Using the tractive force method, calculate the gradient of a 150 mm diameter foul sewer required to reach a critical shear stress of 1 N/m² when the pipe runs half full.

Solution

For a pipe flowing half full, $R = D/4 = 0.0375$ m, and wall shear stress is given by Equation 7.21:

$$1.0 = \rho g R S_0 = 1000 \times 9.81 \times 0.0375 \times S_0$$
$$S_0 = 0.0269 \approx 1:370$$

17.5.3.2 Movement criteria

The results of field and laboratory research indicate that, to achieve self-cleansing performance, sewers should be designed to

1. Transport a minimum concentration of fine-grain particles in suspension
2. Transport coarser granular material as bed-load
3. Erode cohesive particles from a deposited bed

A thorough comparison of the application of mainly laboratory-based equations to typical sewer design situations was carried out, based on these three sediment movement criteria.

17.5.3.2.1 Suspended load transport

Macke's (1982) equation was found to provide a reasonable fit to laboratory data for suspended particles moving at the LOD (no deposition) and was recommended for use as the normal design method:

$$C_\upsilon = \frac{\lambda^3 \upsilon_L^5}{30.4\left(S_G - 1\right)W_s^{1.5}A} \tag{17.3}$$

where C_υ is the volumetric sediment concentration (discharge rate of sediment/discharge rate of water), υ_L is the limiting flow velocity without deposition (m/s), S_G is the sediment specific gravity (–), and A is the flow cross-sectional area (m²).

The equation is valid beyond $\tau = 1.07$ N/m².

Where sediment is to be transported over a sediment bed, the Ackers-White (1973) equation, originally developed for alluvial channels, is advocated. This has been applied to sewer design by Ackers (1991) with a reduced effective sediment bed width as follows:

$$C_\upsilon = J\left(W_e \frac{R}{A}\right)^\alpha \left(\frac{d'}{R}\right)^\beta \lambda_c^\gamma \left(\frac{\upsilon}{\sqrt{g\left(S_G - 1\right)R}} - K\lambda_c^\delta \left(\frac{d'}{R}\right)^\varepsilon\right)^m \tag{17.4}$$

where W_e is the effective width of the sediment bed (m), d' is sediment particle size (m), and λ_c is the friction factor for the pipe and sediment bed (–).

The various empirical coefficients ($J, K, M, \alpha, \beta, \gamma, \delta$, and ε) depend on the dimensionless grain size D_{gr} and the mobility parameter at the threshold of movement A_{gr} (defined in more detail by Ackers et al., 1996a).

17.5.3.2.2 Bed-load transport

For bed-load transport at the LOD, no existing equation gave a good fit over a full range of the data, and so a new equation was derived:

$$C_\upsilon = 3.03 \times 10^{-2} \left(\frac{D^2}{A}\right)\left(\frac{d'}{D}\right)^{0.6}\left(1 - \frac{\upsilon_t}{\upsilon_L}\right)^4 \left(\frac{\upsilon_L^2}{g\left(S_G - 1\right)D}\right)^{1.5} \tag{17.5}$$

$$\upsilon_t = \sqrt{0.125g\left(S_G - 1\right)d'}\left(\frac{d}{d'}\right)^{0.47} \tag{17.6}$$

where v_t is the threshold velocity required to initiate movement and d is the depth of flow.

A similar procedure was followed to evaluate equations for bed-load transport in circular pipes with deposited beds. The best fit to the experimental data was provided by a slightly modified version of the method due to May (1993):

$$C_v = \eta \left(\frac{W_b}{D} \right) \left(\frac{D^2}{A} \right) \left(\frac{\theta \lambda_g v^2}{8g(S_G - 1)D} \right)$$

(17.7)

where W_b is the sediment bed width (m), λ_g is the friction factor corresponding to the grain shear factor (–), θ is the transition coefficient for particle Reynolds number (–), and η is the sediment transport parameter (–).

Evaluation of this equation is more complicated because of the need to estimate the flow resistance produced by the deposited bed. Full details of the recommended method are given by Ackers et al. (1996a).

Pipes designed in accordance with these equations should allow the transport of sediment as bed-load at a rate sufficient to avoid deposition or limit it to a specified depth.

17.5.3.2.3 Bed erosion

The effect of cohesion on the shear stress at the threshold of movement is specifically allowed for in criterion 3. Based on field evidence and experimental investigations into erosion shear stresses, various relationships between pipe diameter and minimum full-bore velocity were identified, depending on the chosen bed shear stress and bed roughness. It was recommended that the design flow conditions needed to erode cohesive particles from a deposited bed should have a minimum value of shear stress of 2 N/m^2 on a flatbed with a Colebrook–White roughness value of $k_b = 1.2$ mm (based on 1 mm cohesive sediment particles). The required full-bore velocity (v_f) is given by a modification of Equation 7.21:

$$v_f = \sqrt{\frac{8\tau_b}{\rho \lambda_b}}$$

(17.8)

$$\lambda_b \approx \frac{1}{4 \left[\log_{10} \left(\dfrac{k_b}{3.7D} \right) \right]^2}$$

(17.9)

where τ_b is the critical bed shear stress (N/m^2) and λ_b is the friction factor corresponding to the sediment bed (–).

In many cases, the roughness of the deposited bed will be much higher resulting in bed shear stresses that are higher, making this approach conservative.

17.5.3.3 Design procedure

The CIRIA method proposes a detailed procedure in which design tables are worked up from first principles using the equations and information supplied. As an alternative, a simplified procedure is provided in which selected standard values of sediment characteristics and other parameters have been adopted, allowing the presentation of 10 simplified design tables. These cover foul, storm, and combined sewers; medium and high sediment loads

(Table 17.7); and criteria based on either no deposition (shown as the LOD in Figure 17.5) or an allowable average deposition of up to 2% of the pipe diameter. Figure 17.5 gives the minimum design velocities for foul and storm sewers based on the design tables (Ackers et al., 1996b). Example 17.4 illustrates the method.

Experimental studies (Safari and Shirzad, 2019) have since validated the benefit of allowing a limited deposited bed in larger sewers and recommend this should be fixed at 1% for design purposes.

17.5.3.4 Limitations

Although the CIRIA method is a significant advance over conventional approaches, it should be noted that it does have limitations (Arthur et al., 1999). The key one is that it is based

Table 17.7 Typical sewer sediment characteristics and applicability

Sediment class	Normal transport mode	Types of sewer applicable	Parameter	Sediment load		
				Low (L)	Medium (M)	High (H)
Sanitary solids	Suspension	Foul (F)	X (mg/L)	100	350	500
		Combined (C)	$d50$ (μm)	10	40	60
			S_G	1.01	1.4	1.6
Stormwater solids	Suspension	Foul (F)	X (mg/L)	50	350	1000
		Storm (S)	d_{50} (μm)	20	60	100
		Combined (C)	S_G	1.1	2.0	2.5
Grit	Bed-load	Storm (S)	X (mg/L)	10	50	200
		Combined (C)	d_{50} (μm)	300	750	750
			S_G	2.3	2.6	2.6

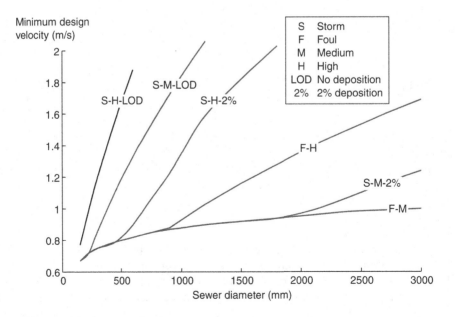

Figure 17.5 Minimum design velocities using the simplified procedure.

EXAMPLE 17.4

A 2100 m long storm sewer is to be designed to carry a discharge of 1500 L/s in a location with a maximum available fall of 7 m. There are no specific sediment data available, but the area will be subject to considerable development in the next 10 years. Select an appropriate diameter and gradient circular sewer. Take Manning's n as 0.012.

Solution

Solve Manning's equation with S_0 = 1:300, and assume the sewer will run full to obtain the required pipe diameter:

$$D = \left(\frac{1.5 \times 0.012 \times 4^{5/3}}{\pi \times (1/300)^{1/2}} \right)^2 = 1.000 \, \text{m}$$

The resulting velocity, v, is as follows:

$$v = \frac{1.5 \times 4}{\pi \times 1^2} = 1.9 \, \text{m/s}$$

Assume the sediment loading category is high (H) due to anticipated development work, and determine the required velocity to avoid any permanent deposition (LOD).

From curve S-H-LOD: The velocity required for this diameter is not shown, but it is clearly above 2 m/s, which is greater than that available. Within the available gradient of 1/300, the sewer cannot be designed to carry high sediment loading without deposition.

Therefore, allow up to 2% deposition, and determine the required velocity.

From curve S-H-2%: required velocity = 1.35 m/s which is <1.9 m/s and thus achievable.

Note: this velocity is in excess of typical minimum full-bore velocities (e.g., 1 m/s).

---

mainly on data from laboratory pipe-flume experiments using single-sized, granular sediment. This is clearly a simplification of the conditions found in combined sewers (in particular), as described in the rest of this chapter. Some verification with field data has been attempted (May et al., 1996) but for these methods to become more widely used, further field research is needed (Montes et al., 2021). The benefits of using simplified rather than detailed sediment transport models for sewer sediment management have also been studied with the conclusion that a complementary approach using both methods should provide acceptable levels of uncertainty (Mannina et al., 2012).

PROBLEMS

1 Outline the various sources of sediment. Discuss ways in which sediment generation might be reduced.
2 Explain how sediment enters, moves through, and leaves an urban drainage system.
3 What are the main problems caused by sediment deposition?
4 How does deposited sediment affect the hydraulic characteristics of a sewer system?

5 Define and explain the three phases of sediment movement: entrainment, transport, and deposition.

6 Sewer sediments have been classified into five types (A–E). Compare and contrast the physical and chemical characteristics of each. What is their significance?

7 A 1.5 km long, 1.5 m diameter combined sewer suffers from sediment deposition along its whole length. Measurements reveal an average depth of 300 mm of type A sediment overlain by 20 mm of type C. A storm with a flow of 2.2 m^3/s sustained for 30 minutes completely releases the pollutants bound by the sediment bed. Assuming the incoming stormwater has negligible pollution, estimate the average concentration of BOD_5 and COD released.   [567 mg/L, 2993 mg/L]

8 Explain the three main modes of sediment transport in sewers. Under what conditions would each mode be most important?

9 Calculate the equivalent flow velocity for the sewer designed in Example 17.4 using the tractive force method.   [1.5 m/s]

10 Explain the main elements of the CIRIA sewer design procedure (for self-cleansing). What are its advantages over conventional approaches? What are its weaknesses?

11 Calculate the minimum velocity needed under CIRIA criterion 3 (τ_b = 2 N/m^2, k_b = 1.2 mm) to cleanse a 1000 mm diameter pipe running half-full. If the pipe has a sediment bed of roughness k_b = 50 mm, what will be the actual shear stress generated by this velocity?   [0.88 m/s, 6.9 N/m^2]

12 Explore the influence and importance of three variables used in the CIRIA equations. What are the implications for sewer design?

KEY SOURCES

Ackers, J.C., Butler, D., and May, R.W.P. 1996a. *Design of Sewers to Control Sediment Problems*, CIRIA R141.

Ashley, R.M., Bertrand-Krajewski, J.-L., Hvitved-Jacobsen, T., and Verbanck, M. (eds). 2005. *Solids in Sewers: Characteristics, Effects and Control of Sewer Solids and Associated Pollutants*, Scientific and Technical Report No. 14, IWA Publishers.

Fan, C.-Y. 2004. *Sewer Sediment and Control. A Management Practices Reference Guide.* US Environmental Protection Agency Report EPA/600/R-04/059.

REFERENCES

Ackers, P. 1991. Sediment aspects of drainage and outfall design. *Proceedings of the International Conference on Environmental Hydraulics, Hong Kong*, 215–230.

Ackers, P. and White, W.R. 1973. Sediment transport: New approach and analysis. *American Society of Civil Engineers Journal of Hydraulic Engineering*, 99(HY11), 2041–2060.

Ackers, J.C., Butler D., John, S., and May, R.W.P. 1996b. Self-cleansing sewer design: The CIRIA Procedure. *Proceedings of Seventh International Conference on Urban Storm Drainage*, Hannover, September, 875–880.

Arthur, S., Ashley, R., Tait, S., and Nalluri, C. 1999. Sediment transport in sewers—A step towards the design of sewers to control sediment problems. *Proceedings of Institution of Civil Engineers, Water, Maritime and Energy*, 136, March, 9–19.

Ashley, R.M. and Crabtree, R.W. 1992. Sediment origins, deposition and build-up in combined sewer systems. *Water Science and Technology*, 25(8), 1–12.

Ashley, R.M. and Verbanck, M.A. 1996. Mechanics of sewer sediment erosion and transport. *Journal of Hydraulics Research*, 34(6), 753–769.

Ashley, R.M., Arthur, S., Coghlan, B.P., and McGregor, I. 1994. Fluid sediment in combined sewers. *Water Science and Technology, 29(1–2)*, 113–123.

Bertrand-Krajewski, J.-L., Briat, P.M., and Scrivener, O. 1993. Sewer sediment production and transport modelling: A literature review. *Journal of Hydraulic Research, 31(4)*, 435–460.

Binnie and Partners and Hydraulics Research. 1987. *Sediment Movement in Combined Sewerage and Storm-Water Drainage Systems*, CIRIA PR1, London.

Bizier, P. (ed). 2007. *Gravity Sanitary Sewer Design and Construction*, ASCE Manual and Report No. 60, WEF Manual No. FD-5.

Butler, D., May, R.W.P., and Ackers, J.C. 1996a. Sediment transport in sewers—Part 1: Background. *Proceedings of Institution of Civil Engineers, Water, Maritime and Energy, 118*, June, 103–112.

Butler, D., May, R.W.P., and Ackers, J.C. 1996b. Sediment transport in sewers—Part 2: Design. *Proceedings of Institution of Civil Engineers, Water, Maritime and Energy, 118*, June, 113–120.

Chen, Y., Shi, X., Jin, X., and Jin, P. 2022. Characteristics of overflow pollution from combined sewer sediment: formation, contribution, regulation. *Chemosphere*, 298, 134254.

Crabtree, R.W. 1989. Sediments in sewers. *Journal of the Institution of Water and Environmental Management, 3*, 569–578.

Crabtree, R.W., Ashley, R.M., and Saul, A.J. 1991. *Review of Research into Sediments in Sewers and Ancillary Structures*. Report No. FR0205, Foundation for Water Research.

Fouada, M., Kishka, A., and Fadela, A. 2014. The effects of scum on sewer flows. *Urban Water Journal, 11(5)*, 405–413.

Larrarte, F., Hemmerle, N., Lebouc, L., and Riochet, B. 2017. Additional elements regarding the muddy layer in combined sewers. *Urban Water Journal, 14(8)*, 862–867.

Macke, E. 1982. *About Sedimentation at Low Concentrations in Partly Filled Pipes*, Mitteilungen, Leichtweiss—Institut für Wasserbau der Technischen Universität Braunschweig, Heft 76.

Mannina, G., Schellart, A.N.A., Tait, S., and Viviani, G. 2012. Uncertainty in sewer sediment deposit modelling: Detailed vs simplified modelling approaches. *Physics and Chemistry of the Earth, 42–44*, 11–20.

May, R.W.P. 1993. *Sediment Transport in Pipes and Sewers with Deposited Beds*, Report SR 320, HR Wallingford.

May, R.W.P., Ackers, J.C., Butler, D., and John, S. 1996. Development of design methodology for self-cleansing sewers. *Water Science and Technology, 33(9)*, 195–205.

Merritt, L. 2009. Tractive force design for sanitary sewer self-cleansing. *Journal of Environmental Engineering, 135(12)*, 1338–1347.

Merritt, L. B. and Enfinger, K.L. 2019. Tractive Force: A Key to Solids Transport in Gravity Flow Drainage Pipes. *Pipelines 2019: Multidisciplinary Topics, Utility Engineering, and Surveying* (eds J. W. Heidrick and M.S. Mihm), ASCE.

Montes, C., Kapelan, Z. and Saldarriaga, J. 2021. Predicting non-deposition sediment transport in sewer pipes using Random forest. *Water Research, 189*, 116639.

Nalluri, C. and Alvarez, E.M. 1992. The influence of cohesion on sediment behaviour. *Water Science and Technology, 25(8)*, 151–164.

Raudkivi, A.J. 1998. *Loose Boundary Hydraulics*, CRC Press.

Safari, M.J.S. and Shirzad, A. 2019. Self-cleansing design of sewers: Definition of the optimum deposited bed thickness. *Water Environment Research, 91*, 407–416.

Verbanck, M.A. 1995. Capturing and releasing settleable solids—The significance of dense undercurrents in combined sewer flows. *Water Science and Technology, 31(7)*, 85–93.

Chapter 18

Operation and maintenance

18.1 INTRODUCTION

Operation and maintenance (O&M) plays a key role in ensuring the sustainable performance of any urban drainage system. There is a temptation to assume that if there are no immediate problems, there are no O&M needs. However, drainage systems corrode, erode, clog, collapse, and ultimately deteriorate to the point of failure and beyond. So, *planned* O&M is essential for all systems.

This chapter provides a comprehensive overview of the subject beginning in Section 18.2 with an analysis of the drivers affecting the need for O&M and the strategies that can be deployed. Sewer location and inspection methods are described in Section 18.3 and sewer monitoring discussed in Section 18.4. Sewer blockages are highlighted in Section 18.5 and the various cleaning techniques to deal with them are explained in Section 18.6. An urban drainage system consists of more than just pipes, and maintenance of ancillary structures is covered in Section 18.7 and SuDS maintenance introduced in Section 18.8. Section 18.9 covers the vital topic of health and safety, with gas generation and its control discussed in the final section (Section 18.10).

18.2 MAINTENANCE STRATEGIES

There are several reasons why the maintenance of urban drainage systems is so important.

18.2.1 Public health

Maintenance of public health is paramount (see Chapter 1), and the continuing good functioning of the system can help to achieve it. In addition, the system should not cause a nuisance (e.g., odour) or a health hazard (e.g., rodents) to either its users or its operators.

18.2.2 Asset management

All systems were costly to construct and would be even more costly to replace. High priority must, therefore, be given to maintaining the physical integrity of the assets.

18.2.3 Maintain hydraulic capacity

A primary function of maintenance is to preserve the as-built hydraulic capacity of the system. This will minimise the possibility of wastewater backing up into properties or widespread

DOI: 10.1201/9781003408635-18

surface flooding. This can be done by cleaning and ensuring, as far as is practicable, that the system is watertight or has controlled water levels.

18.2.4 Minimise pollution and maintain public health and safety

Two key functions of urban drainage systems are to minimise environmental pollution and maintain public health and safety. Maintenance has a role in reducing the frequency of discharge of piped systems (especially combined sewers) into the environment, and in avoiding conditions in all drainage systems that cause excessive buildup of pollutants.

18.2.5 Minimise disruption

The water industry is judged by its regulators and customers on the efficiency with which it deals with O&M. Disruption to the general public should therefore be minimised.

Various degrees of sophistication can be built into maintenance strategies, but there are two main categories: reactive and planned.

18.2.6 Reactive maintenance

In reactive maintenance, problems are dealt with on a corrective basis as they arise (i.e., after failure): the *firefighting* approach. This approach will always be required to a certain extent, as problems and emergencies are bound to occur from time to time in every urban drainage system. However, reactive maintenance cannot reduce the number of system failures. To achieve this, a planned approach is needed.

18.2.7 Planned maintenance

In planned maintenance, potential problems are dealt with prior to failure. Unlike reactive maintenance, planned maintenance is proactive and has the objective of reducing the frequency or risk of failure. Central to planned maintenance is a comprehensive monitoring programme and analysis of existing data.

Planned maintenance is not the same as *routine* or *preventive* maintenance (operations at standard intervals, regardless of need), but involves identifying elements that require maintenance and then determining the optimum frequency of attention.

18.2.8 Operational functions

The major O&M functions are

- Location and inspection
- Cleaning and blockage clearance of pipes
- Chemical dosing
- Rodent baiting
- Ancillary equipment maintenance
- SuDS maintenance
- Fabric rehabilitation

The first six of these functions are described in this chapter. The seventh topic of rehabilitation is considered in Chapter 16.

Maintenance of sewer networks holds some specific challenges, even when compared with other industries. These include the following:

- Geographical size (e.g., dispersion and length of pipework)
- Location (e.g., on private land and under highways)
- Variability and frequency of issues (e.g., can occur with no clear cause and influenced by external factors such as rainfall, gully cleaning, and people's behaviours)
- Physical size of assets (e.g., access, confined spaces)
- Aggressive environment (e.g., hazardous gases)

18.2.9 Maintenance scheduling

Maintenance schedules are often based on operator experience and knowledge, customer complaints, or factors such as age, material, or operational conditions that are believed to have adverse effects on the state of the assets. This may result in biased data and inefficient inspection programmes making extrapolation from available data to a whole system extremely uncertain (Tscheikner-Gratl et al., 2019). However, Draude et al. (2022) report on a methodology for the optimisation of maintenance schedules resulting in the clear ordering of tasks for crews based on minimising total maintenance cost and maintenance gangs' travel times and maximising a job's priority score. Daily productivity of maintenance teams improved by 26% when applied to a case study.

18.3 SEWER LOCATION AND INSPECTION

Sewer location and inspection methods are routine tools of the maintenance engineer. Basic methods for locating sewers and manholes have been available for many years, but these have been improved and newer methods introduced. Inspection, in particular, has been revolutionised by the introduction of remote surveillance equipment.

18.3.1 Applications

The main applications of sewer location and inspection are

- Periodic inspection to assess the condition of existing sewers, chambers, and ancillaries including penstocks and screens (planned maintenance)
- Crisis inspection to investigate emergency conditions or the cause of repeated problems along a particular sewer length (reactive maintenance)
- Inspection of workmanship and structural condition of new sewers before "adoption" – see Chapter 6 (quality control)
- Investigation to understand the location and position of assets and connections for repair, replacement, or rehabilitation

18.3.2 Frequency

It is impossible to accurately estimate the rate of deterioration of a specific sewer length from a single survey. This is particularly true of insidious problems such as sulphide attacks (see Section 18.10.3). The most reliable way to monitor a sewer's condition is to carry out a series of inspections at given intervals such as those recommended in Table 18.1. The

Table 18.1 Internal condition grade transition probabilities

Internal condition grade at time t	Internal condition grade at time $t + \Delta t$				
	1	2	3	4	5
1	0.77	0.23	0	0	0
2	0	0.83	0.17	0	0
3	0	0	0.90	0.10	0
4	0	0	0	0.95	0.05
5	0	0	0	0	1

Source: Based on Savić, D. et al. 2006. *Proceedings of the Institution of Civil Engineers Water Management*, 159(2), 111–118.

Table 18.2 Recommended inspection frequencies for, and extent of, critical sewers

Internal condition grade[b]	Frequency (year)		Extent
	Category A[a]	Category B	%
5	n/a	n/a	2
4	n/a	5	8
3	3	15	13
2	5	20	17
1	10	20	60

Source: Adapted from Sewers and Water Mains Committee. 1991. *A Guide to Sewerage Operational Practices*, Water Services Association/Foundation for Water Research; and National Audit Office. 2004. *Out of Sight – Not Out of Mind. OFWAT and the Public Sewer Network in England and Wales*. Report by the Comptroller and Auditor General. HC 161 Session 2003–2004.

[a] SRM guidance (step 3 Risk Assessment) defines internal condition grades ranging from "collapse" (5) to "acceptable structural condition" (1).

[b] Category A critical sewers are those where collapse repair costs would be highest. Category B sewers, although less critical, would still have substantial collapse costs.

level of inspection chosen reflects an attempt to balance the probability of failure with its consequences.

Repeat inspection data can be used as the basis for deterioration modelling which is typically deterministic, statistical or AI-based, and Hawari et al. (2020) give a good overview of the strengths and weaknesses of each approach. An example of a statistical model is given in Table 18.2 that was prepared based on repeat closed-circuit television (CCTV) data on a UK sewer network to establish the probability of deterioration from one condition grade to another (Savić et al., 2006). For example, there is a 77% probability that on subsequent inspection an initially grade 1 sewer will remain in that condition. There is a 23% probability that it will deteriorate to grade 2. The data showed no cases of a greater than one step "fall" in grade. Generally, it is most likely that a sewer at any condition grade will stay at that grade. The data in Table 18.1 assume there have been no interventions between intervening inspections.

Regression methods have also been successfully used to determine the probability of failure of individual pipes (e.g., Ahmadi et al. 2014; Fuchs-Hanusch et al. 2015).

18.3.3 Locational survey

The first steps in a maintenance plan may include checking the accuracy and completeness of existing records of the system and then initiating a survey on the parts of the system where there is doubt. Before an inspection of any kind can take place, it is necessary to locate the manholes and thereby determine the route of all the sewers in the system under consideration.

Manhole location is usually straightforward, although a metal detector may be required if it is suspected that a cover has been buried. The position and level (cover, soffit, and invert) of each manhole can be determined using standard land surveying techniques. This procedure can be substantially speeded up using GPS (global positioning system) technology, allowing positional data to be logged on-site in seconds.

Techniques available to determine the route of a sewer range from the simple to the sophisticated. Visual inspection of flow directions in manholes is sometimes sufficient, and if not, dye tracing can be carried out. Electronic tracing is also common (Figure 18.1). A probe or "sonde" that emits radio signals is pushed, rodded, jetted, or floated along the sewer and its progress is tracked from the surface using a handheld receiver. Using this approach, sewers up to 15 m deep can be traced to a claimed accuracy of ± 10%. Interference from signals generated by other buried metallic assets can, however, cause problems.

Ground-penetrating radar is frequently used to trace the sewer and other buried asset locations. While it is not completely accurate and relies on user interpretation, it provides a non-intrusive alternative to trial holes.

18.3.4 Manual inspection

Manual inspection is only used in exceptional circumstances and only if inspection cannot be done another way (Rolfe and Butler, 1990). Clearly, health and safety issues are paramount with this approach, and these are covered in Section 18.9.

If such inspection *is* necessary, a survey begins at one manhole and works along the length of the sewer in increments (typically of 1 m) to the next (Figure 18.2). Information can be gathered on mortar loss displaced/missing bricks, cracks, sewer shape, connections, silt/debris, etc. Paper-based methods have traditionally been used to record this information, but portable data loggers or handheld computers with appropriate software are now widely used.

Progress is relatively slow using this procedure (200–400 m/d); it is costly and dangerous. However, high-quality information can be obtained.

18.3.5 Closed-circuit television (CCTV)

In this method, a small TV camera incorporating a light source is propelled through the sewer, and the images are relayed to the surface for viewing and recording (see Figure 18.3).

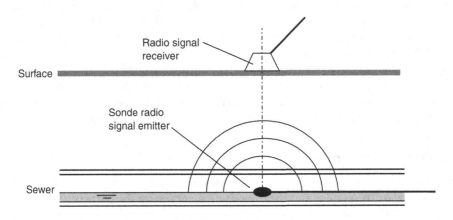

Figure 18.1 Electronic tracing of sewers.

CCTV is a popular choice for sewer inspection, not least because internal investigation of sewer systems can be carried out quickly, and with minimal disruption, avoiding lengthy shutdowns and unnecessary excavation (Rolfe and Butler, 1990). This method is particularly useful in environments that are too small or hazardous for people to enter. CCTV can also be used to locate and define the cause of a known condition or defect, and to help establish a plan of maintenance. Progress is relatively rapid with typical rates of 400–800 m/d.

Figure 18.2 Manual inspection in a confined space. (Courtesy of Scottish Water.)

Figure 18.3 CCTV image of a sewer in poor condition. (Courtesy of Telespec Ltd., Guildford.)

The method is commonly used in pipes from 100 to 1500+ mm in diameter, but it is less effective in larger pipes, due to increased lighting requirements and difficulty in achieving adequate resolution of detail with high subject-to-camera distances.

18.3.5.1 Propulsion

Generally, for sewers of <150 mm diameter, the camera is located on the end of a line and can be either pushed through or winched between manholes. For sewers of diameters greater than 150 mm, it is usual for the camera to be mounted on a self-propelled, remote-controlled tractor. The tractor speed varies according to the size of the wheels fitted and will be greater in larger sewers, with typical speeds ranging from 0.1 to 0.2 m/s.

18.3.5.2 Camera operation

The basic camera technology used in sewer inspection came originally from broadcast TV. Development has centred on gathering the maximum amount of information at a reasonable cost in very difficult and dirty conditions. Cameras are typically of the charge coupling device (CCD), solid-state type, housed in strong waterproof cases. These have high sensitivity and can be used for surveying very large sewers (over 3 m in diameter).

Lamps attached to the front or sides of the camera head provide lighting. Lenses of several focal lengths are available, including zooms, and focusing can either be pre-set or remotely controlled by the operator. The most useful view for a CCTV camera is usually axially along the sewer, but there are specific occasions (e.g., looking up a house connection or at specific problems) when a lateral view is preferable. Several techniques are available for achieving this, including a wide-angle lens, pan and tilt equipment, and an electronic "fish eye." A typical CCTV camera is illustrated in Figure 18.3.

18.3.5.3 Manhole zoom cameras (MZCs)

Instead of using a movable CCTV camera, it is possible to carry out sewer inspection with a stationary camera with a high-powered, remotely controlled zooming function and powerful lighting. The camera is lowered into a manhole on a telescopic pole and then used to zoom

Figure 18.4 A pan-and-tilt in-sewer CCTV camera. (Courtesy of Telespec Ltd., Guildford.)

in up to 40 m on either side of the manhole, capturing images and video. This enables rapid inspections (2000–3000 m/d) to be carried out over a limited distance from a manhole, making it a cost-effective screening tool before deploying other inspection techniques.

18.3.5.4 Limitations

Whilst CCTV cameras are an excellent way to inspect sewers, they are not without their limitations. There are two key areas of concern (Tscheikner-Gratl et al., 2019):

- Limited type of information acquired
  - Only a snapshot of the condition without information on the deterioration process or cause
  - No direct information on the actual hydraulic capacity of the inspected conduits
  - No information on the material properties and the geometry to estimate structural strength and stability
  - No information about the quantity of in- and exfiltration
  - Only prior defined (e.g., BS EN 13508-2:2003+A1:2011) defects are reported
- Low accuracy and repeatability
  - Relatively large percentage of false negatives and false positives in defect identification, due to the dependence on human observation of images (but see next section)
  - Numerical classification, representing good to poor condition, may vary between individual inspectors up to two steps

18.3.5.5 Automation

Open-source datasets of CCTV images are increasingly becoming available. These provide essential training data for the emerging automation techniques mentioned in the next paragraph. As an example, the Ofwat Innovation Challenge: Artificial Intelligence and Sewers Case Study contains a library of 27,262 labelled sewer CCTV images (available after free registration at https://ukwir2021.my.site.com/spring/s/).

Research on the application of automated image processing techniques using pattern recognition and machine learning (discussed in Section 21.2.2) to minimise the "human factor" shows great promise (with claimed accuracies of up to 90%), particularly in the ability to reduce the percentage of false negatives (e.g., Myrans et al. 2019; Yin et al., 2020). However, the issue of correct classification remains a challenge because standard procedures depend on semantic descriptions rather than rigorously defined quantifiable measures (Tscheikner-Gratl et al., 2019).

18.3.6 Other techniques

18.3.6.1 Laser profiling

Ring laser profiling projects a laser line onto the internal circumference of the pipe being inspected. This is viewed and the image is captured by a CCTV camera. The resulting image can be measured to provide a reasonably accurate profile of the inside of the pipe. This shows any ovality, encrustation, loss of wall material due to corrosion, etc. Repeated surveys make it possible to determine the extent to which a pipeline has deformed and therefore the type of renovation or replacement technology that might be employed to solve any problems caused.

Recent developments report an increase in accuracy due to the application of camera movement compensation methods which have been tested under laboratory conditions

(Clemens et al. 2015). The technology is promising as it provides detailed information on the 3D geometry of pipes that may be used to quantify material loss, deformation and the dimensions of obstacles and intruding lateral connections.

18.3.6.2 LiDAR scanning

LiDAR surveys are carried out by projecting a laser and measuring the time it takes for it to reach a target and reflect back to the sensor. A LiDAR survey collects continuous 2D cross sections of the pipe wall and by compiling these cross sections a high-resolution 3D model can be created. LiDAR is typically used to measure features and defects above the sewer flow level such as ovality, deformations, offset joints, and lateral sizes. It can be used in conjunction with other sensors such as CCTV.

18.3.6.3 Sonar

Before the advent of in-sewer sonar technology, it was only possible to inspect relatively empty sewers. In some cases, this required the effort and expense of over-pumping. Sonar techniques can acquire an image of the profile of a liquid-filled sewer, without the need for a light source. The sonar head is controlled from a surface processor and scans through 360° in discrete increments providing information on deposited sediment depths, FOG accumulation and pipe deformation (see Figure 18.5). Data derived from the acoustic signal can be displayed on a colour monitor and recorded. Sonar can also be combined with conventional methods in TISCIT (Totally Integrated Sonar and CCTV Inspection Technique) systems to provide the comprehensive inspection of large-diameter pipes (Thomson and Grada, 2004). A laser profiler is also included for the above-water level survey in some systems.

18.3.6.4 Infrared

Another approach to in-sewer inspection is thermal imagery, in which an infrared camera is used to gather and focus emitted black-body radiation and convert it into a form visible to the human eye. Again, no external light source is necessary. This technique is more limited in application but can be used to inspect for infiltration, based on the assumption that the wastewater and groundwater are at different temperatures.

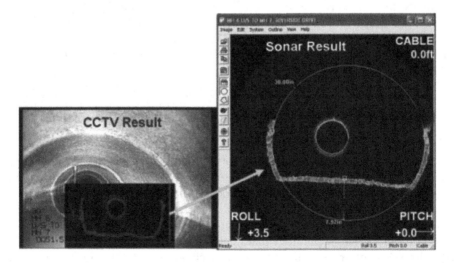

Figure 18.5 Sonar image of a half-submerged sewer with a significant sediment bed. (Courtesy of AECOM Water.)

18.3.6.5 Acoustics

Where local and rapid inspections are necessary, such as identifying where a pipe may be partially blocked, acoustic sensors can detect anomalies using a technique known as acoustic pulse reflectometry (Duan et al., 2015). Trials of the SewerBatt® technology have demonstrated a 90% accuracy compared with CCTV, and OPEX savings between 17% and 29% (Isle Utilities, 2013). Plihal et al. (2016) demonstrate the added benefits of using acoustic sensing in combination with MZCs.

18.3.6.6 UAVs

Unmanned Aerial Vehicles (UAVs) or drones are being increasingly used within the water industry. One innovative application is for sewer inspection. For example, Scottish Water and its delivery partners have developed a bespoke carbon fibre drone which carries a CCTV camera for visual inspection and a LiDAR scanner for measurement. They are typically deployed in larger, deeper or otherwise inaccessible sewers to replace manual inspection (see Figure 18.5).

18.3.7 Data storage and management

CCTV images and video are recorded in digital format and typically transferred from operator to client over the Internet, and can be viewed almost immediately after a survey has been completed. This enables multiple sewer lengths to be rapidly processed, facilitating the pinpointing and review of defects.

Sewer surveys generate a great deal of data that require careful and systematic handling. Software packages are available to aid in data management, allowing defects, coded in a standardised way (WRc, 2013) at relevant locations in the pipe or manhole, to be stored in an easily accessible database. Such information can then be used to assess the structural condition of the system systematically. Most packages produce data files in a format compatible with simulation models. It is common for databases to be part of GIS software, which allows spatial information to be held and graphically displayed. Information on many services can be held on different "layers." Sewer record databases can be linked with CAD packages to allow rapid production of drawings. Software also exists that will manage all of the above, analyse results semi-automatically and produce manhole survey reports.

18.4 SEWER MONITORING

18.4.1 Flow surveys

A typical monitoring point for a sewer flow survey is in a manhole, with the data-logging equipment located near the top of the manhole, and a sensor in the sewer that monitors depth and velocity. Some sensors enable depth and velocity to be measured from the same point. For example, the sensor is positioned on the invert and is wedge-shaped to minimise the buildup of solids (see Figure 18.7). It measures depth using a pressure transducer, and velocity using a *Doppler shift* system. An ultrasonic signal is transmitted against the oncoming flow and is reflected back by solid particles or air bubbles. The reflected signal changes in frequency, and the magnitude of the change is proportional to the particle velocity, thus giving water velocity. These give reasonable results and are widely used. Accuracy is in the order of ±10% on flow rate under ideal conditions. Battery life depends on monitoring frequency but should be in excess of 3 months when logging at a rate of 2 minutes.

Figure 18.6 Sewer inspection drone in a large, deep sewer. (Courtesy of Scottish Water)

An alternative method of detecting depth is a top–down ultrasonic sensor, positioned above the water surface (also see the next section). These may be fitted along with the sensor in the invert, or alternatively used to monitor depth only.

Velocity may also be measured using an electromagnetic meter (where a current is induced in a conductor moving across a magnetic field that is directly proportional to the speed of movement of the conductor).

Figure 18.8 shows a plot of velocity against depth for a large number of readings at a flow survey monitoring point in a sewer system. This can be used to check the data for consistency. A similar *scattergraph* is formed by plotting the log flow rate against log depth.

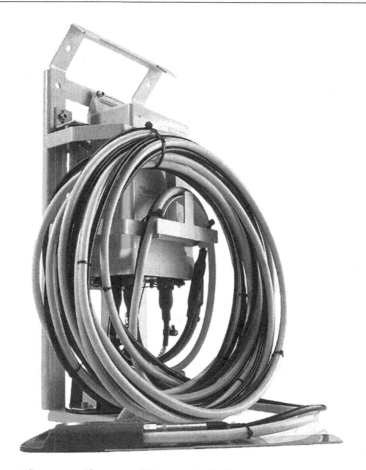

Figure 18.7 Ultrasonic flow meter. (Courtesy of Detectronics Ltd.)

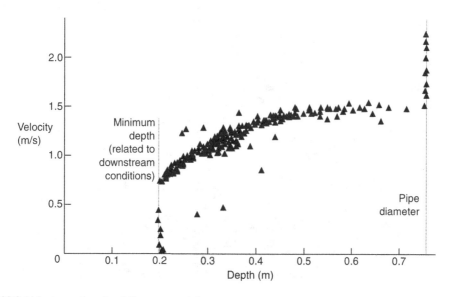

Figure 18.8 Velocity against depth "scattergraph."

18.4.2 CSO monitors

There is a continuing trend to monitor CSOs and their performance to ensure and demonstrate compliance with regulatory permits. As flow rate or total overflow volume is difficult and expensive to monitor at weir locations, spill event time and duration are more commonly monitored by the use of level sensors. The approach adopted in the UK is to fit event duration monitors (EDMs) to all CSOs giving a record of the number and duration of spill events. EDM devices are of various types, each with its own strengths and weaknesses including (CIWEM, 2021)

- Ultrasonic sensors
- Radar sensors
- Pressure transducers
- Conductivity probes
- Float or tilt switches

The key requirement for EDM equipment is that it accurately and reliably monitors and logs information about spill events. The choice of a specific sensor will also depend on the location and configuration of the monitoring location. Figure 18.9 shows a radar level sensor EDM positioned within a CSO chamber, with the overflow weir shown on the left of the image. Measurements are typically made at 2- or 15-minute intervals, depending on the significance of the discharges, and the number of spills logged. Depth measurement accuracy is typically ±5 mm. Flow measurement provides transparent data to alert stakeholders, demonstrates whether overflows are meeting their permits and helps identify where an investigation, and ultimately further enhancements, may be required. The Chartered Institution of Water and Environmental Management's (CIWEM, 2021) good practice guide provides further detail on their practical application.

Figure 18.9 CSO event duration monitor (EDM). (Courtesy of Vega Controls Ltd.)

Future enhancements will be to report on discharges from overflows in near-real time (within one hour) with the data showing where the discharge to the environment happened, when it started and when it ended (Defra, 2023). Parameters to be monitored will be dissolved oxygen, temperature and pH, turbidity and ammonia. There are calls for a wider suite to be monitored including FIOs (NEPC, 2024).

Measuring spill frequency and duration does not give a direct determination of the flow volume or pollutant load discharged from a CSO event. Plans are currently being drawn up to monitor receiving water quality upstream and downstream of key discharge points (Defra, 2023). The aim is to provide clear evidence to stakeholders (including the public) on whether improvement schemes are achieving the required outcomes, and where further upgrades may be required.

18.4.3 Smart manholes

An emerging trend is the rollout of extensive networks of monitors within sewer networks. These sensors are similar to EDMs, typically located at manholes and used to measure flow depth. The monitors relay information to a control centre where machine learning algorithms (see Section 21.2.2) can be used to distinguish between "normal" and "abnormal" flow depths in both wet and dry weather conditions, and proactively detect blockage buildup. The location can then be prioritised for cleaning (see Section 18.6) and investigation into the potential source of the problem.

18.5 SEWER BLOCKAGE

18.5.1 Definition

Blockages are defined as full or partial restrictions within the sewer that affect normal operational performance. Some are caused by inappropriate or unintended inputs to the system and some by the design, construction, or condition of the system. The causes are often interrelated. Most of the problems are more acute in smaller pipes but can escalate and affect larger sewers. The effect ranges from partial loss of capacity to complete stoppage, causing local backing up of wastewater and/or stormwater as a minimum and more widespread sewer flooding and pollution incidents in serious cases. This is a significant problem, and in 2023 there were over 270,000 sewer blockages reported in England and Wales with clearance costs of £200 million per annum. Some commentators even suggest that blockages cause the majority of sewer flooding of domestic properties and CSO spills.

18.5.2 Causes

18.5.2.1 Gross solids/sanitary waste (input)

Many cases of blockage are associated with gross solids / sanitary waste (see Section 2.3.2), especially those with non-biodegradable content or made of non-woven fabrics such as wet wipes. However, it is often difficult to establish if the solids have caused the blockage or if they are just associated with a problem generated by a system defect.

18.5.2.2 FOG/scale (input)

Solidified fats, oils, and grease (FOG) are often associated with non-domestic properties, particularly restaurants. High-temperature dishwashers often move the FOG from the premises, only for it to cool and solidify further downstream causing a loss of hydraulic capacity. However, this reasoning may be oversimplified with the biodegradation of the oil content and water hardness contributing to the buildup of FOG (Williams et al., 2012). In exceptional cases, the buildup can be so severe that "fatbergs" occur, causing a major blockage (Figure 18.10). Wall scale or encrustation, sometimes found in hard water areas, causes similar problems. FOG can also encourage sediment buildup in all sewers, and cause problems at pumping stations, fine mesh screens, and wastewater treatment plants (Ducoste et al., 2009). Anecdotal evidence suggests that 50% or more of sewer cleaning activity is associated with FOG problems.

18.5.2.3 Sediment (input)

Sediment is defined as any settleable particulate material that may, under certain conditions, form bed deposits in sewers and associated hydraulic structures. It is normally associated with large, flat sewers. Sedimentation rarely completely chokes the pipe, but it can still have a significant impact on capacity. Chapter 17 discusses the issues associated with sediment deposition in further detail.

Figure 18.10 Brick sewer fatberg. (Courtesy of Andy Brierley.)

18.5.2.4 Defects (system)

System defects are often the catalyst for blockage formation. These can range from poor design, such as severe changes in sewer direction, to poor construction practice (see the following sections). Also included in this category are defects that have arisen over the course of time such as displaced joints, pipe material (e.g., pitch fibre), and localised adverse gradients.

18.5.2.5 Traps (system)

Traps on smaller upstream sewers can be particularly problematic. These were designed to prevent odour from coming back up the system and are typically found on systems serving older properties. However, today they often run dry causing the very odours they were designed to eliminate, and they can also become blocked. It is common for the traps to be removed where they are found to be problematic.

18.5.2.6 Intruding laterals (system)

Intruding laterals or other connections are common defects resulting from poor construction practices (see Chapter 16). The intrusion reduces the cross-sectional area of the pipe and can also initiate further blockage.

18.5.2.7 Tree roots (system)

Sewers are susceptible to the intrusion of tree roots, which seek out moist conditions. The roots themselves are a nuisance, both in retarding the flow and causing similar problems to intruding laterals.

18.5.3 Prediction and detection

Given the frequency of occurrence and the severity of the consequences of blockages, there is a compelling operational need to predict where and when a blockage might occur. In practice, this means trying to assess which locations have a higher probability of occurrence so that critical or sensitive ones can be prioritised. Typically, due to the paucity and quality of data at smaller diameters, prediction is based on pipe groups or classes rather than individual sewers. Predictions for larger sewers and critical assets are typically based on thorough inspections.

Savić et al. (2006) and Ugarelli et al. (2010) have shown how blockage rates can be related to key asset data. They found blockage rate to be inversely related to sewer diameter, gradient, and length and varies between internal condition grades (defined in Table 18.2), based on 10,300 blockage events in UK sewers. Still, care needs to be taken in generalising results due to the significant influence of the local context (Rodríguez et al., 2012).

Statistical models are often used to help estimate the level (and cost) of interventions across an asset base. Erskine et al. (2014) describe such an approach to estimate the failure frequencies of a water company's assets using a Bayesian tuner. Bailey et al. (2015) present an alternative approach using decision trees to develop blockage prediction models and highlight key variables including property density, sewer length, and flow velocity. Xie et al. (2017) used a Cox proportional hazards modelling approach to identify an even wider range of explanatory variables for blockage risk in vitrified clay pipes.

Rosin et al. (2022) outline the development of an event detection system to automatically and continuously monitor the sewer network for blockages close to CSOs in near real time. The system does this by applying an evolutionary artificial neural network (see Section 21.3) discrepancy-based analysis and statistical trend-based analysis based on level sensor data installed at most CSOs. The methodology was demonstrated to detect blockage events quickly and reliably, with a low number of false alarms.

18.5.4 Blockage control

Most sewer blockages are caused by gross solids via the WC and/or FOG via the kitchen sink. FOG control is considered separately in Section 18.6.4. In terms of gross solids, products that do not disintegrate to a sufficient degree in the sewer network are a particular problem with practitioners estimating that 50–80% of blockages are caused by or associated with wet wipes. Several approaches are available to help control unsuitable solids entering the system.

18.5.4.1 Flushability standards

This approach consists of developing and implementing a flushability testing standard which all items intended to be flushed should have to achieve. In the UK, the relevant standard is "Fine to Flush" (Water UK, 2019) which specifies seven individual tests which need to be passed. For a product to qualify for certification, its intended use should mean it will likely become contaminated with faeces or other bodily waste. This ensures that only products that are used in association with the WC can be marked with the "Fine to Flush" kite mark (ruling out, for example, floor cleaning wipes or cosmetic removal wipes).

The key for this approach to be effective is to work with manufacturers and retailers of flushable products to ensure products go through the certification process. Over 100 products have been certified. Unfortunately, the product labelling apparently caused confusion among consumers in that it suggests people can dispose of as many of these products as they like into the WC. This led to the scheme being terminated in 2024, with customer engagement being the preferred ongoing approach.

18.5.4.2 Customer awareness campaigns

Customer awareness campaigns are a non-technical approach to reducing the problems caused by gross solids in sewers emphasising cultural and behavioural aspects. Their aim is to create a change in public disposal practices to reduce disposal by the waterborne route and increase it by the solid waste route. The disposal decision is a very personal one. Each individual selects the method, in the privacy of the bathroom: the item is either dropped in the WC and flushed away, or placed in a bin for subsequent removal via the solid waste system.

There have been many public awareness campaigns drawing attention to the operational and environmental problems caused by waterborne disposal. Ashley et al. (1999) investigated the advantages of encouraging a change in disposal practice and showed that such a change is both effective and improves the sustainability of the system. Sharp (2017) emphasised the great need for *sympathetic* and *sustained* engagement with the public. Clarkson et al. (2023) provide a useful summary of learnings and recommendations from UK awareness campaigns. Water UK's current campaign "Bin the Wipe" focuses on one product and has a simple, unambiguous message, "When you use a wet wipe, put it in the bin, not down the loo."

Experience indicates that campaigns can have some success but are labour-intensive and need to be repeated every few years.

18.5.4.3 Operational techniques

Thes following operational techniques to deal with blockages are covered elsewhere in the chapter:

- Prediction (Section 18.5.3)
- Detection (Section 18.5.3)
- Removal (Section 18.6)

18.6 SEWER CLEANING

18.6.1 Objectives

Sewer cleaning is carried out:

- Proactively, to remove sediment, FOG, and other blocked material in order to restore hydraulic capacity and limit pollutant accumulation
- Reactively, to deal with blockages or offensive odours
- To permit sewer inspection
- To aid sewer repair/renovation

18.6.2 Cleaning techniques

A number of cleaning techniques and methods are in use, depending particularly on location and severity, including rodding, winching, jetting, flushing, and hand excavation. A combination of more than one method may well be used in any particular locality.

18.6.2.1 Rodding or boring

Rodding or boring is primarily a manual procedure, in which short flexible rods are screwed together and then inserted into the blocked sewer. The principal action is the physical contact of the tools, although the compression of air by plungers and dense brushes can contribute. Other approaches include the use of semi-rigid, coiled glass-reinforced plastic rods supplied in continuous lengths on a reel. The procedure can be mechanised.

This technique is limited to small-diameter pipes (≤ 225 mm) at shallow depths (≤ 2.0 m) and must be close to the access point (≤ 20 m). It is particularly well suited to dislodging blockages and roots. Care must be taken not to damage the existing pipework, especially on more brittle pipe materials.

18.6.2.2 Winching or dragging

Winching or dragging is a technique involving the use of purpose-shaped buckets that are dragged through the sewer collecting sediment, which is emptied out at a manhole. Although the winch can be manually operated, power-driven devices are normally used.

The procedure is capable of removing most materials, even in large pipes, but is most effective in sewers up to 900 mm in diameter, and up to 50% silted. Care has to be taken so that damage is not caused to the sewer fabric.

18.6.2.3 Jetting

Jetting is a widely used technique that relies on the ability of an applied high-pressure (100–350 bar) stream of water to dislodge material from sewer inverts and walls and transport it down the sewer for subsequent removal. Water under these high pressures is fed through a hose to a nozzle containing a rosette of jets sited in such a way that the majority of the flow is ejected in the opposite direction to the flow in the hose. The jets propel the hose through the sewer, eroding the settled deposits in the process. A range of nozzles is available to cope with specific situations. Modern combination units incorporate vacuum or air-displacement lifting equipment to remove the material, as well as to dislodge it without the need for people entry.

Jetting is a versatile and efficient procedure for removing a wide range of materials and is widely used in practice.

18.6.2.4 Flushing

Flushing is a technique in which short-duration waves of liquid are introduced or created so as to scour the sediment into suspension and, hence, transport it downstream. Waves may be induced by

- Automatic siphons at the heads of sewers
- Dam and release using blow boards/gates
- Hydrant and hose
- Mobile water tanker

Of these, the first is rarely used, and the second is used in bespoke circumstances when flushing is particularly required. All merely move the dislodged material downstream and do not remove it from the system.

18.6.2.5 Hand excavation

Historically, large-diameter sewers were cleared by manually digging out deposited material. Labourers entered the sewer and shovelled sediment into skips that were transported to the surface for emptying. The method is limited in application to larger-size pipes (> 900 mm) and has significant health and safety implications. It is used only in exceptional circumstances.

18.6.2.6 Invert traps

These are not widely used, but they can significantly assist cleaning in problem locations by creating a preferred location to remove silt and sediment. Invert grit traps are intended to intercept sediments travelling as bed-load in combined sewers. Invert traps in existence are typically large rectangular chambers, which although effective at trapping sediments, also collect other near-bed and suspended solids. The performance of these chambers can be improved by the introduction of a cover with an open slot across the width of the invert designed to collect heavier, inorganic sediment only (Buxton et al., 2001).

Alternatives to traps include proprietary treatment systems, such as hydrodynamic and vortex separators, that help to capture silts, sediments, and other pollutants (Woods-Ballard et al., 2016). These are especially important for SuDS and particularly where geo-cellular structures are used.

18.6.2.7 Gully pots

Gully pots are provided to limit surface sediments entering the urban drainage system. Cleaning is routinely carried out by vacuum lifting equipment on mobile tankers. Cleaning frequency varies from place to place but is normally carried out once or twice a year. Gully sensors are coming onto the market which can monitor sediment depths and alert when cleaning is needed. However, these are not in widespread use. The efficiency of cleaning varies but is around 70%.

Pots not only trap and retain the sediment for which they are provided, but also other pollutants including oil. Problems arise during dry periods when the retained sediment is degraded anaerobically, allowing NH_4, chemical oxygen demand (COD), and dissolved metals to build up in the retained liquor. At the next storm event, the liquor and fine and dissolved solids will be mixed and entrained into the flow, adding significantly to the pollutant load (Langeveld et al., 2016; Wei et al., 2021).

18.6.3 Comparison of methods

No one method of cleaning is superior to the others on all occasions; each has its advantages and disadvantages. These are summarised in Table 18.3. The *Manual of Drain and Sewer Cleaning* (WRc, 2020) provides more detail and best practice guidance.

18.6.4 FOG control

A range of approaches is available to control FOG in sewers, including at-source methods, recycling, chemical treatment, and bio-augmentation.

Table 18.3 Relative performance of sewer cleaning techniques

Topic	Rodding	Winching	Jetting	Flushing
Sewer size (mm):				
<375	Good	Fair	Good	Good
450–900	Poor	Good	Fair	Fair
Maximum cleansing distance (m)	25	100	100	50
Number of manholes required	1	2	1	1
Dislodging materials:				
Invert	Fair	Good	Good	Good
Walls	Fair	Fair	Good	Poor
Joints	Fair	Fair	Good	Poor
Materials encountered:				
Silt	Fair	Fair	Good	Good
Sand/gravel	Poor	Good	Good	Good
Rocks	Poor	Good	Fair	Poor
Grease	Fair	Fair	Fair	Poor
Roots	Good	Good	Fair	Poor
Material removed?	No	Yes	Yes	No
Damage potential	Low	Medium	High	Low
Flooding potential?	No	No	No	Yes

Source: Adapted from Lester, J. and Gale, J.C. 1979. *The Public Health Engineer*, 7(3), July, 121–127.

18.6.4.1 At source

The most effective strategy to manage FOG is to employ measures to reduce its inappropriate disposal or capture it before entry into sewer systems. Soft measures include user education (Mattsson et al., 2015), published good practice guides, and provision of authoritative advice. Gelder and Grist (2015) report successes in such approaches with the most substantial reductions in blockages occurring during and after intensive and targeted approaches. Hard measures include the installation and proper maintenance of grease traps and separators (Sultana et al., 2023). These are typically fitted online, immediately after the sink outlet, and are designed to achieve physical separation of the FOG. Such approaches may be targeted at commercial and domestic consumers.

18.6.4.2 Recycling

A more sustainable option is to collect and recycle the FOG for beneficial incorporation into animal feeds and into soaps and lubricants for industrial use, and for biodiesel production. Recycling options are now being made available to domestic customers as part of kerbside collection arrangements. Collected oils are converted into various fuel sources, such as bio-diesel for vehicles.

18.6.4.3 Chemical treatment

A variety of chemical compounds can be used to treat FOG-contaminated wastewater, either before discharge to the sewer or by being introduced directly. Coagulants such as aluminium sulphate, ferric chloride, and lime can be used to dissipate surface charges and break the fat emulsion. Flocculants can be used separately or in combination to enhance settlement.

Many surfactant-based products are available, such as soaps, detergents, and degreasers, that when mixed with FOG-containing water surround the FOG molecules and facilitate emulsification, particularly when water temperature is high.

Caustics and solvents may be used to "cut" FOG. These products essentially raise the water pH and thereby increase the solubility of non-polar substances (grease) in water. However, chemicals such as these are corrosive and can potentially damage the sewer fabric.

Enzymes such as lipase degrade triglycerides, cleaving the fatty acids off the glycerol backbone of fats, thus increasing the solubility of greases. Gary and Sneddon (1999) investigated the effectiveness of enzymes on samples obtained from restaurant grease traps. They found that enzymes without surfactants performed better than those with surfactants. The effectiveness of any particular enzyme is dependent on attributes such as the type of grease, its pH, and the grease trap influent flow rate.

A significant drawback of all of these approaches is that they typically only offer a temporary solution in terms of clearing sewers because the FOG molecules are not completely broken down and can reappear and accumulate downstream where conditions return to "normal."

18.6.4.4 Bio-augmentation

These technologies consist of introducing quantities of naturally occurring or engineered microorganisms (mainly bacteria), with or without supporting media, to break down the FOG molecules. Bio-augmentation products can be broadly classified into two groups, based on the type of microorganism used: spore-forming or vegetative bacteria. Drinkwater et al. (2015) outline a protocol for the development of biological dosing into the sewer and pumping stations. Laboratory tests can indicate the effectiveness of compounds in a controlled environment, but the success is very dependent on the on-site conditions. Where manufacturers and water company operatives work together, the results of dosing have been very positive. Bio-augmentation has also been trialled for gully pot sediments but could not be recommended as a replacement for or in addition to mechanical cleaning (Scott, 2012).

18.7 ANCILLARY AND NETWORK EQUIPMENT MAINTENANCE

Many sewer networks contain some form of equipment. These may include the following:

- Pumps with valves plus air valves and washouts on rising mains
- Dosing equipment to deal with hydrogen sulphide in the network
- Combined sewer overflows with screens (static and mechanical) and automatic/remote cleaning
- Passive or dynamic flow controls, such as automated penstocks or auto-opening and auto-closing sluices
- Non-return valves may be present to prevent flooding

Maintenance is required to ensure the performance is not compromised, to maximise the asset life, and to avoid sudden and catastrophic failure. Each piece of equipment has a bespoke maintenance regime, including specific tasks and frequencies. These often link to supplier guarantees relating to performance. Sensors and monitors can provide critical information to support the maintenance regime. For example, at pumping stations, condition-based monitoring (e.g., vibration) can help to evaluate the pump performance and whether there is any deterioration.

18.8 SUDS MAINTENANCE

Table 18.4 summarises the main types of maintenance needed for various SuDS components. These consist, in particular, of sediment management (removal and safe disposal) and vegetation management, and arguably require a wider range of skills than standard pipe

Table 18.4 Common SuDS maintenance requirements

Practice	Forebay cleanout and disposal	Pruning	Removing/mowing vegetation	Inspect outlet structure	Unclog surface	Sediment removal and disposal
Retention/detention pond	X		X	X		X
Constructed wetland	X		X	X		
Bioretention	(X)	X		X	X	X
Green roof		X	X	X		
Infiltration device					X	X
Pervious pavement					X	X
Filter strip and swale			X			
Rainwater harvesting	X					

Source: Adapted from Blecken, G.T. et al. 2017. *Urban Water Journal*, 14(3), 278–290.

O&M. More detailed advice on individual SuDS components is given in Chapter 11, and further general advice can be found in *The SuDS Manual* (Woods-Ballard et al., 2016).

It is recommended that owners of developments with SuDS should be provided with an owner's manual and model agreements for SuDS – legal documents that can be used as the basis of arrangements for maintenance, and these have been published by CIRIA (Shaffer et al., 2004).

18.9 HEALTH AND SAFETY

It is important to appreciate the health and safety hazards involved with sewer operation. Many of the practices to be carried out are covered in the UK by the Health and Safety at Work Act 1974. In addition, the sewer environment is classified as a "confined space" under the Confined Spaces Regulations (HSE, 1997). Examples are sewers, manholes, storage tanks, and pump sumps. Specific safety advice is available for sewage workers (HSE, 2003). Clients, designers, and contractors also have specific responsibilities under the CDM, Construction (Design and Management) Regulations (www.hse.gov.uk/construction/cdm/2015/index.htm) that relate to the safe operation of assets once built or refurbished.

18.9.1 Atmospheric hazards

Atmospheric hazards are probably the most dangerous: explosive or flammable gases may develop at any time. Anaerobic biological decomposition of wastewater yields methane, which is lighter than air. Petrol vapour is heavier than air and tends to form pockets at the invert of the sewer. Factories should hold trade discharge licences and report the disposal of dangerous chemicals, but they may not report accidental spills or deliberately negligent acts.

The most commonly occurring toxic gas within the sewer is hydrogen sulphide. It is a flammable and very poisonous gas that has a distinctive smell. It is particularly dangerous to workers in sewers because the ability of a person to smell the gas diminishes with exposure and as the concentration increases. The amount of breathable oxygen in a sewer can be reduced or even eliminated by displacement by a heavier gas. If there is no breathable oxygen in a sewer, the life expectancy of a person entering is approximately 3 minutes.

To provide adequate ventilation, the access manhole cover and those upstream and downstream need to be lifted well before an inspection. Those entering the sewer must use a gas detector. Because of the short life expectancy in unfavourable conditions, the current best practice is to ensure that gangs have the appropriate personal protective equipment (PPE) to self-rescue, rather than relying on fire brigade assistance. For certain activities, it can be appropriate to have rescue teams on standby. If a dangerous atmosphere exists within a person-entry sewer, the inspection can be carried out using breathing apparatus.

18.9.2 Physical injury

Physical injury is an ever-present hazard in the sewer environment. It can result from falls down manholes or in the sewer, and the dropping, throwing, or misuse of equipment. The risk of drowning must not be underestimated, whether in the residual flow within the sewer or due to a sudden rise in water level after rainfall. This can be avoided by maintaining close contact with the surface at all times.

Acids and other chemicals may find their way into the sewer system and could cause burns, so appropriate PPE should be worn at all times.

18.9.3 Infectious diseases

Infection from tetanus, hepatitis B, or leptospirosis (Weil's disease) is a potential risk that should be planned for by ensuring workers are under medical supervision and inoculated as appropriate. Discarded needles found in sewers should be avoided. Rodents (such as rats) and insects can also be a health hazard. Rodent control is covered in Section 18.9.5.

18.9.4 Safety equipment

If the sewer is to be entered for inspection or any other maintenance work, suitable PPE must be worn. This includes waterproof clothing, heavy waders, gloves, safety harnesses, eye protection, and hard hats, as well as radios, safety lamps, and, if necessary, breathing equipment (escape or full). Portable gas detectors should always be used to test for toxic gases. Two-way radios have been used to keep in contact with members of the team on the surface but may be ineffective in certain conditions; visual and vocal contact with a surface member is preferable. The surface team must have access to regular weather forecasts and knowledge of trade discharges.

18.9.5 Rodent control

The key concern over rat infestation in sewer systems is their potential link to disease transmission to humans (Keeling and Gilligan, 2000). The extent of the problem is difficult to judge, but there is objective evidence to suggest that sewer-rodent populations have declined in recent years (Channon et al., 2000). As with other operational functions described in this chapter, control is either (most commonly) reactive or proactive. The reactive approach is to respond to complaints or sightings by baiting in the locality. A proactive approach consists of planned random or semi-random baiting across the network. A protocol including elements of both approaches was developed by Ashley et al. (2003).

18.10 GAS GENERATION AND CONTROL

A side effect of conveying wastewater in pipes is the generation and release of sewer gases. The most troublesome of these are generated under anaerobic conditions and comprise a range of volatile organic compounds and reduced sulphur compounds such as mercaptans. Sewerage odour is typically dominated by hydrogen sulphide. Hydrogen sulphide can cause corrosion of concrete, metal, and electrical equipment. Particularly susceptible locations are points of turbulence following long retention times (e.g., backdrop manholes, wet wells of pumping stations, and outlets of wastewater rising mains). Problems are the most serious in hot, arid climates (Ayoub et al., 2004) but will also manifest in temperate climates such as in the UK.

In addition, sewer gases, when fugitive, can cause odour nuisance (leading to customer complaints) and endanger sewer workers (as already described).

18.10.1 Mechanisms

Wastewater naturally contains sulphur as inorganic sulphate and organic sulphur compounds. The sulphate is usually derived from the mineral content of the municipal water supply or from saline groundwater infiltration. Organic sulphur compounds are present in excreta and household detergents, and in much higher concentrations in some industrial effluents such as from the leather, brewing, and paper industries.

Bacterial activity quickly depletes any dissolved oxygen that is present, and septicity can easily develop. Under anaerobic conditions, complex organic substances are reduced to form volatile fatty acids resulting in a drop in pH. *Desulfovibrio* bacteria in pipe biofilms and sediment reduce organic sulphur compounds and sulphates (SO_4^{2-}) to sulphides (S^{2-}):

$$SO_4^{2-} + C, H, O, N, P, S \rightarrow S^{2-} + H_2O + CO_2 \tag{18.1}$$

Gaseous, molecular hydrogen sulphide (H_2S) results from a reaction with hydrogen ions in the water and is pH-dependent, with more being formed in acidic conditions:

$$S^{2-} + 2H^+ \leftrightarrow H_2S \tag{18.2}$$

In pipes flowing under gravity, H_2S escaping into the atmosphere from solution in the wastewater tends to rise and accumulate in condensation water on the soffit of the pipe (see Figure 18.11). There it is oxidised by *thiobacillus* bacteria to form sulphuric acid:

$$H_2S + 2O_2 \rightarrow H_2SO_4 \tag{18.3}$$

Sulphuric acid can cause serious damage to pipe materials through consuming alkalinity in concrete or direct attack on metal.

18.10.2 Favourable conditions

The most favourable conditions for the production of hydrogen sulphide are as follows (Newcombe et al., 1979):

- Wastewater with a large proportion of trade wastes with substantial sulphide or organic sulphur contents
- Wastewater with a relatively high sulphate concentration

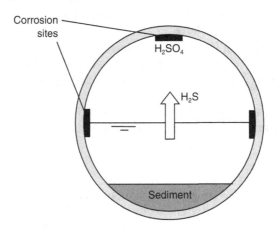

Figure 18.11 Corrosion in sewers.

- Wastewater with a low pH – the lower the pH value, the greater the proportion of molecular hydrogen sulphide
- High rate of oxygen demand to rapidly consume available dissolved oxygen
- High wastewater temperature, which accelerates biological activity
- Retention of the wastewater flow under anaerobic conditions (e.g., long sewers with flat gradients, long rising mains, and large pump wet wells)
- Low wastewater velocity that decreases the rate of oxygen reaeration and increases sedimentation

Factors tending to increase the emission of hydrogen sulphide are as follows:

- High concentrations of molecular hydrogen sulphide in the wastewater
- High wastewater velocity or turbulence
- High relative velocity and turbulence in the head space *above* the flow
- Clean wastewater surfaces with respect to oil films, surfactants, etc.

18.10.3 Sulphide buildup

Pomeroy and Parkhurst (1977) have suggested a formula that produces an index Z that is broadly indicative of the conditions under which sulphide may be formed in gravity sewers under 600 mm in diameter. The Z formula gives the following:

$$Z = \frac{3(EBOD)}{S_o^{\frac{1}{2}} Q^{\frac{1}{3}}} \frac{P}{B} \qquad (18.4)$$

where $EBOD$ is the effective biochemical oxygen demand, $BOD_5 = BOD_5 \times 1.07^{(T-20)}$ (mg/L), T is the wastewater temperature (°C), S_o is the sewer gradient (m/100 m), Q is the flow rate (L/s), P is the wetted perimeter (m), and B is the Flow width (m).

The formula contains factors representing the main influences on sulphide generation with $EBOD$ accounting for the influence of both temperature and (indirectly) sulphate content of the wastewater. Wastewater pH is assumed to be in the range 7–8. The value of Z and its interpretation are given in Table 18.5 (see also Example 18.1). The Z formula has been widely used in practice (e.g., Spooner, 2001).

Other studies (e.g., Hvitved-Jacobsen et al., 1999) indicate the particular importance of the biodegradability of the organic matter in the wastewater that is not captured in the BOD_5 test only. Vollertsen (2006) proposes a simple modification to the Z formula (Z'), such that

$$Z' = Z\left[1 + 10\left(\frac{BOD_5}{COD} - 0.47\right)\right] \qquad (18.5)$$

where COD is the wastewater COD (mg/L).

Table 18.5 Likelihood of sulphide development

Z	*Prevalence of sulphide*
<5000	Rarely present
~7500	Low concentrations likely
~10,000	May cause odour and corrosion problems
~15,000	Frequent problems with odour and significant corrosion problems

---

EXAMPLE 18.1

A 500 mm diameter concrete sewer (n = 0.012) has been laid at a gradient of 0.1%. At average flow, the pipe runs half full of wastewater. If the wastewater has a BOD_5 of 500 mg/L and the summer temperature is 30°C, estimate the likelihood that hydrogen sulphide will be generated.

Solution

Geometric properties of flow:

Area of flow, $A = \pi D^2/8 = 0.098$ m^2

Wetted perimeter, $P = \pi D/2 = 0.785$ m

Flow width, $B = D = 0.5$ m

Flow rate, Q (Equation 7.23):

$$Q = \frac{A}{n} R^{\frac{2}{3}} S_o^{\frac{1}{2}} = \frac{0.098}{0.012} \left(\frac{0.098}{0.785}\right)^{\frac{2}{3}} 0.001^{\frac{1}{2}} = 64.6 \text{L} / \text{s}$$

$EBOD = BOD_5 \times 1.07^{(T-20)} = 500 \times 1.07^{10} = 984$ mg/L.
From Equation 18.4,

$$Z = \frac{3(EBOD)}{S_o^{\frac{1}{2}} Q^{\frac{1}{3}}} \frac{P}{B} = \frac{3 \times 984}{0.1^{\frac{1}{2}} 64.6^{\frac{1}{3}}} \frac{0.785}{0.5} = 3653$$

This value of Z implies sulphide is unlikely to be present.

18.10.4 H$_2$S control

A wide range of techniques is available to control the generation or emission of hydrogen sulphide.

18.10.4.1 Sewerage detail design

Probably the most effective way of controlling the generation and emission of sewer gases is to design the sewer system carefully in the first place. In particular, self-cleansing velocities should be designed to occur regularly. Dead spots in manholes and other structures should be avoided.

Where sulphide generation cannot be avoided, perhaps because of high ambient temperatures, sewer design should avoid excessive turbulence.

Concrete is a material susceptible to H$_2$SO$_4$ attack. Boon (1992) suggests that using calcareous aggregates rather than quartz aggregates could considerably extend the life of a concrete sewer. Alternatively, pipes protected by thin epoxy resin coatings have been used.

Vitrified clay or plastic pipes have been shown to be more resistant to corrosion by sulphuric acid. These should be specified at vulnerable locations.

18.10.4.2 Ventilation

The need for good ventilation was mentioned in Chapter 6. It has several benefits, including the following:

- Helping to maintain aerobic conditions and hence preventing sulphate reduction
- Stripping any gases from the atmosphere
- Oxidising any H_2S in the atmosphere
- Reducing pipe soffit condensation

The latter point is particularly important – H_2S does not cause corrosion under dry conditions, as bacterial oxidation requires moisture.

18.10.4.3 Aeration

Dissolved oxygen in the form of air, or directly as molecular oxygen, has been widely used to treat rising mains. The injection point could be at the inlet end of the main with the aim being to maintain aerobic conditions to "toggle off" sulphate reduction from the obligate anaerobes. Alternatively, treatment could be at the discharge point to oxidise already formed sulphides, but this is a relatively slow reaction. Both methods will have relatively high capital and running costs. Dosing a nitrate salt, typically calcium nitrate, at the upstream location also has a similar effect of suppressing the sulphate-reducing bacteria and is considered to be best practice.

18.10.4.4 Oxidation

The addition of chlorine or hypochlorite oxidises any sulphides present and temporarily halts bio-activity, thus preventing further sulphide generation. Experience suggests that such treatment is only moderately effective, probably because wastewater has a high chlorine demand. Continuous addition of high doses of chlorine would be prohibitively expensive. A popular alternative is to use a stabilised solution of chlorine dioxide, which is an oxidant but does not produce harmful chlorinated organic residual compounds. Potassium permanganate, available as liquid or granules, can also be used and can yield a large reduction in the total odour potential of the sewage.

18.10.4.5 Chemical addition

Ferric(III) salts can be dosed to precipitate existing sulphides as insoluble ferrous(II) sulphide. Typically, dose rates are kept low to ensure that only minimal, and often immeasurable, additional suspended solids are created. At rates of 1–2 mg/L, the dose is not high enough to give a coagulant effect and generate solids. Iron dosing packages are readily available with compatible pumps and vessels enabling appropriate control and storage. Iron can be dosed downstream to lock up the H_2S as an insoluble precipitate. This is of particular value if nitrate dosing cannot be used upstream to switch off the septicity. Iron is limited to eliminating dissolved sulphide and does not react with other odourants.

Another possible strategy is the addition of magnesium hydroxide as a slurry to the wastewater. The effect is to increase the pH of the wastewater, thus reducing the proportion of the sulphide present as H_2S.

Cen et al. (2023) give a comprehensive and critical review of chemicals used in sewer systems, including dosing strategies.

18.10.5 Greenhouse gases

The amount of greenhouse gases (GHGs) emitted by a system or organisation is a measure of its impact on the world's climate. The critical importance of this and its implications for urban drainage have already been discussed in detail in Section 4.6.

One of the main ways that an organisation or utility's emissions are accounted for is to classify them within three different "scopes" (www.ghgprotocol.org). Scope 1 concerns those direct emissions that are owned or controlled by an organisation, whereas scope 2 and 3 indirect emissions are a consequence of their activities but occur from sources not owned or controlled by them. The Scope 1 GHGs emitted from urban drainage systems are caused by the degradation of organic matter and nitrogen present in wastewater. It produces two types of emissions: methane (CH_4) and nitrous oxide (N_2O). The former has a global warming potential about 30 times higher than carbon dioxide over a 100-year timescale and the latter 300 times.

The generation and emission of methane occurs in some sewers under anaerobic conditions and is particularly associated with pumped systems. However, estimating its magnitude and importance is difficult with reliable figures hard to find. One Australian study of three rising mains found they contributed around 18% of the total GHG emissions from an urban wastewater system (Foley et al., 2011). N_2O has been detected in a few field-scale sewer sampling campaigns (Short et al., 2014), but its contribution to the overall GHG inventory is still unclear.

Scope 2 (i.e., caused indirectly, coming from energy providers) GHG emissions from urban drainage systems are largely associated with the energy used for pumping. This has been discussed in Section 15.8. Scope 3 emissions refer to those caused by an organisation's suppliers or customers. Section 19.2.3 has further details on carbon costs and accounting.

PROBLEMS

1 Describe the main operation and maintenance functions in an urban drainage system and the reasons for carrying them out. What are the particular challenges to be met?

2 Explain the differences between reactive and planned maintenance. What are the benefits of combining both approaches?

3 Compare and contrast the main types of sewer inspection technology. What would be your main selection criteria?

4 What causes sewers to block?

5 Describe the main types of sewer cleaning equipment. Discuss their main areas of application and effectiveness.

6 Fats, oils, and grease (FOG) can cause serious operational problems. What are these, and what are suitable methods of eliminating these problems? What are "fatbergs," and what causes them?

7 Explain the main differences between the maintenance needs of piped sewer systems and SuDS.

8 Explain the dangers to the health and safety of sewer workers, and detail good working practices.

9 "Hydrogen sulphide causes serious corrosion of urban drainage systems." Discuss the validity of this statement in terms of how, when, and where the gas is formed.

10 Measurements in an existing sewer reveal the following data: diameter = 300 mm, depth of flow = 240 mm, mean velocity = 0.75 m/s, *BOD* = 750 mg/L, and mean

temperature = 30°C. Calculate Z to assess the likelihood of H_2S generation. Assume Manning's n is 0.012. [7600]

11 List the main methods to control hydrogen sulphide problems and discuss their relative merits.

12 What GHG emissions are associated with urban drainage systems and what are their implications?

KEY SOURCES

Cherqui, F., Clemens-Meyer, F., Tscheikner-Gratl, F. and van Duin, B. Eds. 2024. *Asset Management of Urban Drainage Systems. If anything exciting happens, we've done it wrong!* IWA Publishing.

Hvitved-Jacobsen, T., Vollertsen, J. and Nielsen, A.H. 2013. *Sewer Processes—Microbial and Chemical Process Engineering of Sewer Networks*, 2nd ed., CRC Press.

Jiang, G., Sun, J., Sharma, K.R., and Yuan, Z. 2015. Corrosion and odor management in sewer systems. *Current Opinion in Biotechnology, 33*, 192–197.

Tscheikner-Gratl, F., Caradot, N., Cherqui, F., Leitão, J.P., Ahmadi, M., Langeveld, J.G., Le Gat, Y., Scholten, L., Roghani, B., Rodríguez, J.P. Lepot, M., Stegeman, B., Heinrichsen, A., Kropp, I., Kerres, K., do Almeida, M. C., Bach, P.M., Moy de Vitry, M. Sá Marques, A., Simões, N.E., Rouault, P., Hernandez, N., Torres, A., Werey, C., Rulleau, B. and Clemens, F. 2019. Sewer asset management – state of the art and research needs. *Urban Water Journal, 16*(9), 662–675,

WEF/ASCE. 2022. *Urban Stormwater Controls Operation and Maintenance*. WEF Manual of Practice No. 39, ASCE/EWRI Manuals and reports on Engineering Practice No. 153.

REFERENCES

Ahmadi, M., Cherqui, F., De Massiac, J.C. and Le Gauffre, P. 2014. Influence of available data on sewer inspection program efficiency. *Urban Water Journal, 11*, 641–656.

Ashley, R.M., Channon, D., Blackwood, D., and Smith, H. 2003. A new model for sewer rodent control. *International Pest Control, 45*(4), 183–188.

Ashley, R.M., Souter, N., Butler, D., Davies, J., Dunkerley, J., and Hendry, S. 1999. Assessment of the sustainability of alternatives for the disposal of domestic sanitary waste. *Water Science and Technology, 39*(5), 251–258.

Ayoub, G.M., Azar, N., El Fadel, M., and Hamad, B. 2004. Assessment of hydrogen sulphide corrosion of cementitious sewer pipes: A case study. *Urban Water Journal, 1*(1), 39–53.

Bailey, J., Keedwell, E., Djordjevic, S., Kapelan, Z., Burton, C., and Harris, E. 2015. Predictive risk modelling of real-world wastewater network incidents, 13th Computer and Control for Water Industry Conference, CCWI 2015. *Procedia Engineering, 119*, 1288–1298.

Blecken, G.T., Hunt III, W.F., Al-Rubaei, A.M., Viklander, M., and Lord, W.G. 2017. Stormwater control measure (SCM) maintenance considerations to ensure designed functionality. *Urban Water Journal, 14*(3), 278–290.

Boon, A. 1992. Septicity in sewers: Causes, consequences and containment. *Journal of the Institution of Water and Environmental Management, 6*, February, 79–90.

BS EN 13508-2:2003+A1:2011 *Investigation and assessment of drain and sewer systems outside buildings – Part 2: Visual inspection coding system*

Buxton, A., Tait, S., Stovin, V., and Saul, A.J. 2001. The performance of engineered invert traps in the management of sediments in combined sewer systems. *Novatech 2001, Proceedings of the Fourth International Conference on Innovative Technologies in Urban Drainage*, Lyon, France, pp. 989–996.

Cen, X., Li, J., Jiang, G. and Zheng, M. 2023. A critical review of chemical uses in urban sewer systems. *Water Research*, 240, 120108.

Channon, D., Cole, L., and Cole, M. 2000. A long-term study of *Rattus norvegicus* in the London Borough of Enfield using baiting returns as an indicator of sewer population levels. *Epidemiology and Infection*, 125, 441–445.

Chartered Institution of Water and Environmental Management (CIWEM). 2021. *Event Duration Monitoring. Good Practice Guide*, Version 3.1. Urban Drainage Group. ciwem.org

Clarkson, T., Digman, C., Maclagan, O. and Patel, S. 2023. *Learning and recommendations form customer behaviour campaigns on blockage reduction*. Report No. 23/SW/01/28, UKWIR. https://ukwir. org/learning-and-recommendations-from-customer-behaviour-campaigns-on-blockage-reduction

Clemens, F. H. L. R., Stanić, N., Van der Schoot, W., Langeveld, J.G. and Lepot, M. 2015. Uncertainties associated with laser profiling of concrete sewer pipes for the quantification of the interior geometry. *Structure and Infrastructure Engineering*, 11, 1218–1239.

Defra. 2023. *Consultation on Continuous Water Quality Monitoring and Event Duration Monitoring. Implementing s.81 and s.82 of the Environment Act 2021 and associated technical guidance*. v 1.0. Department for Environment, Food & Rural Affairs.

Draude, S. Keedwell, E., Kapelan, Z. and Hiscock, R. 2022 Multi-objective optimisation of sewer maintenance scheduling. *Journal of Hydroinformatics*, 24(3), 574.

Drinkwater, A., Moy, F., and Villa, R. 2015. *Fats, Oils and Greases (FOG): Where Are We and Where We Could Be: The Feasibility of Biological Dosing into Sewer Systems and the Development/Specification of a Protocol*, Report No. 15/SW/01/14, UKWIR. https://ukwir. org/reports/15-SW-01-14/129666/Fats-Oils-and-Greases-FOG--Where-we-are-and-where-we-could-be--The-feasibility-of-biological-dosing-into-sewer-systems-and-the-development--specification-of-a-protocol

Duan, W., Kirby, R., Prisutova, J. and Horoshenkov, K.V. 2015. On the use of power reflection ratio and phase change to determine the geometry of a blockage in a pipe. *Applied Acoustics*, 87, 190–197.

Ducoste, J.J., Keener, K.M., and Groninger, J.W. 2009. *Fats, Roots, Oils, and Grease (FROG) in Centralized and Decentralized System*, WERF Report 03-CTS-16T.

Erskine, A., Watson, T., O'Hagan, A., Ledgar, S., and Redfearn, D. 2014. Using a negative binomial regression model with a Bayesian tuner to estimate failure probability for sewerage infrastructure. *Journal of Infrastructure Systems*, 20(1).

Foley J., Yuan Z., Keller J., Senante E., Chandran K., Willis J., Shah A., van Loosdrecht M. and van Voorthuizen E. 2011. *N_2O and CH_4 Emission from Wastewater Collection and Treatment Systems: State of the Science Report*, Report 29, Global Water Research Coalition.

Fuchs-Hanusch, D., Günther, M. Möderl, M. and Muschalla, D. 2015. Cause and effect oriented sewer degradation evaluation to support scheduled inspection planning. *Water Science and Technology*, 72, 1176–1183.

Gary, E., and Sneddon, J. 1999. Determination of the effect of enzymes in a grease trap. *Microchemical Journal*, 61, 53–57.

Gelder, P., and Grist, A. 2015. *Fats, Oils and Greases (FOG): Where Are We and Where Could We Be?* UKWIR, Report No. 15/SW/01/13. https://ukwir.org/reports/15-SW-01-13/129665/Fats-Oils-and-Greases-FOG--Where-we-are-and-where-we-could-be

Hawari, A., Alkadour, F., Elmasry, M. and Zayed, T. 2020. A state of the art review on condition assessment models developed for sewer pipelines. *Engineering Applications of Artificial Intelligence*, 93, 103721.

Health and Safety Executive (HSE). 1997. *Safe Work in Confined Spaces. Approved Code of Practice, Regulations and Guidance*, HSE Books.

Health and Safety Executive (HSE). 2003. *Working with Sewage. The Health Hazards*. A Guide for Employees. Pocket Guide INDG197.

Hvitved-Jacobsen, T., Vollertsen, J., and Tanaka, N. 1999. An integrated aerobic/anaerobic approach for production of sulphide formation in sewers. *Water Science and Technology*, 41(6), 107–116.

Isle Utilities. 2013. *Case Study - Collaborative Trial: Sewerbatt Acoustic Sensing Technology*. www. isleutilities.com/news/case-study-collaborative-trial-sewerbatt-acoustic-sensing-technology

Keeling, M.J. and Gilligan, C.A. 2000. Bubonic plague: A metapopulation model of zoonosis. *Proceedings of the Royal Society*, 267, 2219–2230.

Langeveld, J.G., Liefting, H.J., Schilperoot, R.P.S., Hof, A., Baars, I.B.E., Nijhof, H. 2016. Stormwater management strategies: Source control versus end of pipe. *Proceedings of Ninth International Conference on Urban Water*, Novatech, Lyon, France.

Lester, J. and Gale, J.C. 1979. Sewer cleansing techniques. *The Public Health Engineer*, 7(3), July, 121–127.

Mattsson, J., Hedstrom, A., Ashley, R.M., and Viklander, M. 2015. Impacts and managerial implications for sewer systems due to recent changes to inputs in domestic wastewater—A review. *Journal of Environmental Management*, 161, 188–197.

Myrans, J., Everson, R. and Kapelan, Z. 2019.Automated detection of fault types in CCTV sewer surveys. *Journal of Hydroinformatics*, 21.1, 153–163.

National Audit Office. 2004. *Out of Sight – Not Out of Mind. OFWAT and the Public Sewer Network in England and Wales*. Report by the Comptroller and Auditor General. HC 161 Session 2003–2004.

National Engineering Policy Centre (NEPC). 2024. *Testing the waters. Priorities for mitigating health risks from wastewater pollution*, Royal Academy of Engineering. https://nepc.raeng.org.uk/testing-the-waters

Newcombe, S., Skellett, C.F., and Boon, A.G. 1979. An appraisal of the use of oxygen to treat sewage in a rising main. *Water Pollution Control*, 78(4), 474–504.

Plihal, H., Kretschmer, F., Bin Ali, M.T., See, C.H. Romanova, A. Horoshenkov, K.V. Ertl, T. 2016. A novel method for rapid inspection of sewer networks: Combining acoustic and optical Means. *Urban Water Journal*, 13, 3–14

Pomeroy, R.D. and Parkhurst, J.D. 1977. The forecasting of sulphide buildup rates in sewers. *Progress in Water Technology*, 9, 621–628.

Rodríguez, J.P., McIntyre, N., Díaz-Granados, M., and Maksimović, Č. 2012. A database and model to support proactive management of sediment-related sewer blockages. *Water Research*, 46(15), 4571–4586.

Rolfe, S. and Butler, D. 1990. A review of current sewer inspection techniques. *Municipal Engineer*, 7, August, 193–207.

Rosin, T.R., Kapelan, Z., Keedwell, E., and Romano, M. 2022. Near real-time detection of blockages in the proximity of combined sewer overflows using evolutionary ANNs and statistical process control. *Journal of Hydroinformatics*, 24(2), 259–273.

Savić, D., Giustolisi, O., Berardi, L., Shepherd, W., Djordjevic, S., and Saul, A. 2006. Modelling sewer failure by evolutionary computing. *Proceedings of the Institution of Civil Engineers Water Management*, 159(2), 111–118.

Scott, K.M. 2012. *Investigating Sustainable Solutions for Roadside Gully Pot Management*, PhD thesis, University of Hull.

Shaffer, P., Elliott, C., Reed, J., Holmes, J., and Ward, M. 2004. *Model Agreements for Sustainable Water Management Systems, Model Agreements for SUDS*, Report, C625 CIRIA, London, UK.

Sharp, L. 2017. *Reconnecting People and Water. Public Engagement and Sustainable Urban Water Management*, Earthscan.

Short, M. D., Daikeler, A., Peters, G. M., Mann, K., Ashbolt, N. J., Stuetz, R. M. and Peirson, W. L. 2014. Municipal gravity sewers: an unrecognised source of nitrous oxide. *Science of the Total Environment*, 468–469, 211– 218.

Spooner, S.B. 2001. Integrated monitoring and modelling of odours in sewerage systems. *Journal of the Chartered Institution of Water and Environmental Management*, 15, July/August, 217–222.

Sultana, N., Roddick, F., Gao, L., Guo, M., and Pramanik, B.K. 2023. Understanding the properties of fat, oil, and grease and their removal using grease interceptors. *Water Research* 225, 119141.

Thomson, J. and Grada, L. 2004. *An Examination of Innovative Methods Used in the Inspection of Wastewater Collection Systems (CD)*, WERF Report 01-CTS-7.

Ugarelli, R., Venkatesh, G., Brattebo, H., Di Federico, V., and Saegrov, S. 2010. Historical analysis of blockages in wastewater pipelines in Oslo and diagnosis of causative pipeline characteristics. *Urban Water Journal*, 7(6), 335–343.

Vollertsen, J. 2006. Description and validation of structural condition. Chapter 2 in *CARE-S. Computer Aided Rehabilitation of Sewer and Storm Water Networks* (ed. S. Saegrov), IWA Publishing.

Williams, J.B., Clarkson, C., Mant, C., Drinkwater, A., and May, E. 2012. Fat, oil and grease deposits in sewers: Characterization of deposits and formation mechanisms. *Water Research*, 46, 6319–6328.

Woods-Ballard, B., Wilson, S., Udale-Clark, H., Illman, S., Scott, T., Ashley, R., and Kellagher, R. 2016. *The SUDS Manual*, v. 5, CIRIA C753, London.

Water Research Centre (WRc). 2013. *Manual of Sewer Condition Classification*, 5th edn. Swindon.

Water Research Centre (WRc). 2020. *Manual of Drain and Sewer Cleaning*. Swindon.

Wei, H., Muthanna, T.M., Lunday, L. and Viklander, M. 2021. An assessment of gully pot sediment behaviour under current and potential rainfall conditions. *Journal of Environmental Management*, 282, 111911.

Water UK. 2019. *Fine to Flush: Specification for a Testing Methodology to Determine Whether a Product is Suitable for Disposal Through a Drain or Sewer System*. Water Industry Specification, WIS 4-02-06.

Xie, Q., Bharat, C., Nazim Khan, R., Best, A. and Hodkiewicz, M. 2017. Cox propoertinal hazards modelling of blockage risk in vitrified clay wastewater pipes. *Urban Water Journal*, 14(7), 669–675.

Yin, X., Chan, Y., Bouferguene, A., Zaman, H., Al-Hussein, M. and Kurach, L. 2020. *Automation in Construction*, 109, 102967.

Chapter 19

Planning

19.1 WHAT IS DRAINAGE PLANNING?

19.1.1 Need and development

Urban drainage assets are of great economic, environmental, and social value. They may be wholly owned and managed by a single organisation or, as in the UK, by multiple organisations with split roles and responsibilities. The governance of urban drainage and water management more broadly is highly location-dependent; however, the OECD (2015) has outlined 12 principles for good water governance that underpin the basis for good planning. These are grouped into three broad categories concerning enhancing: effectiveness, efficiency, and trust and engagement.

Urban drainage systems in the UK as we currently know them started to be constructed over 150 years ago (as outlined in Chapter 1). As our urban areas have grown and evolved, the systems have grown. However, this has not meant wholesale sewer system capacity increases or upgrades. Indeed, overall the assets themselves have become older and subject to failure or a reduction in their "asset health" (United Utilities, 2021). This could be due to material failure, poor quality of construction, or as a result of external influences such as ground movement. Other causes of failure could be a change in loading that might result in structural collapse, or the effect of particular asset management practices.

Drainage planning is essential to determine the investment requirements (how much and when) to cope with the situation described above and provide fit-for-purpose, "future-proof" systems. The remainder of this section (19.1) sets out what is involved in drainage planning. Section 19.2 describes how drainage plans are created, and Section 19.3 considers drainage plans in the context of the broader systems with which they interact.

Current, emerging and future risks to be addressed in the planning process include future development, an increase in rainfall extremes over time (see Section 4.6), changes in legislative requirements, pressure from the public, and the need to maintain the assets already in place (Chapter 18). Planning critically supports capital and operational investment choices, which are made strategically in the short-term (e.g., yearly), medium-term (e.g., 5 years), and long-term (e.g., 25 years), and tactically, on a more frequent basis (day-to-day). Investment aims to ensure the current function and performance of the assets are maintained (through capital and operational funding) or enhanced to improve the level of service, for example, to meet new legislative requirements.

Whilst urban drainage planning has always been based on the skill and experience of drainage engineers, the first nationally agreed and internationally benchmarked guidance in the UK was developed and issued by WRc in 1983 as the *Sewerage Rehabilitation Manual* v1 (SRM) and *Sewers for Adoption* (by the National Water Council and WRc) in 1981. These documents provided the cornerstone for drainage planning (capacity and asset health) in existing and new developments, respectively.

Drainage planning has evolved through technical advances and the requirements set by regulators and governments. Technical advances and ways of working, including drainage modelling and the quality, coverage, temporal availability, analysis and management of asset data have created the ability to plan and make better investment choices. The influences on investment choices have also evolved. Where traditionally there was a focus on flooding, receiving water quality and asset health, today there is additional focus on the service that customers or communities receive, development impacts, coping with uncertainty, and the multiple benefits that drainage choices may create.

19.1.2 Frameworks in practice

Frameworks to support drainage planning have also evolved to match new requirements, new ideas and technological advances. The SRM itself has evolved through successive versions. Other guidance and frameworks have emerged over the last 20 years as indicated in Table 19.1.

Where an organisation is responsible for the management of urban drainage, there are legislative and regulatory requirements (economic, environmental, and social) that the organisation must comply with. In the UK, there are regulators in place who oversee organisations' performance and are responsible for holding them to account, with the power to impose penalties if such performance is not met. These organisations may be in private or public ownership but are independent of local or national government. In other countries, such as the USA and Australia, municipalities are typically responsible for urban drainage (Digman et al., 2020), with regulations (national and local) enforced by government agencies (e.g., the US Environmental Protection Agency).

19.1.3 Wider asset management

Drainage planning does not occur in isolation; it forms part of wider asset management thinking, planning and application. This may be within a particular organisation, or within parallel organisations with different functions, requirements, and responsibilities. To be successful, asset management principles need to be applied by the organisation to its practices including investment choices and interventions, its interactions with external organisations, and the communities it serves. The aim is to maximise the value of the assets (current and/or future).

There are many approaches to good asset management. ISO 55001: 2014 provides a framework for an asset management system and is designed to standardise effective practices. The aim is to have the activities and structures in place within an organisation that enable costs and risks to be managed, to maximise value for the organisation and those it serves.

The Institute of Asset Management outlines 10 asset management topics in its *The Pathway to Excellence in Asset Management* (Institute of Asset Management, 2021). These are

1. Purpose and context
2. Leadership and governance
3. Organisation and people
4. Strategy and planning
5. Asset management decision-making
6. Information management
7. Life cycle delivery
8. Risk management
9. Review and continual improvement
10. Values and outcomes

Table 19.1 Evolution of documents supporting urban drainage planning and their purpose

Framework	Period in use	Overview/purpose of the documents
Sewers for Adoption. Updated as Sewerage Sector Guidance in England (Water UK, 2023)	1981–	Guidance for the design and construction of foul and surface water sewers for new developments to ensure that they can be adopted by water companies. Versions created for individual nations, for example, Scotland and Northern Ireland. Latest England version includes the use of SuDS designated as sewers (Chapter 6).
Sewerage Rehabilitation Manual (SRM). Updated as Sewerage Risk Management (WRc, 2023).	1983–	In two parts. Part 1 sets out an approach to sewer risk management including a framework to evaluate performance and plan for investment in urban drainage assets (Section 16.8). This guidance underpinned the development of Drainage Area Plans for a sewer network or part of a sewer network. Part 2 contains a sewer renovation design guide which includes guidance on understanding failures and how to rectify them in piped systems (Section 16.9).
Urban Pollution Management Manual (FWR, 2019)	1994–	Provides good practice guidance for the water industry and regulators on the impact, assessment, and management of urban wet weather discharges to rivers and seas (Section 2.6.2).
Capital Maintenance Planning – A Common Framework (UKWIR, 2002)	2002–2011	Methodologies to support organisations understand capital maintenance needs and plan for future interventions, for example, based on asset age, and health, and determine the expenditure required to manage the asset performance.
Surface Water Management Action Planning (Defra, 2018)	2010–	Framework and guidance for local authorities to coordinate, plan, and lead activities relating to flood risk from surface water. Includes flooding from storm sewers and combined sewers, watercourses, highway drainage, and flows on the surface.
Drainage Strategy Framework (Environment Agency / Ofwat, 2013)	2013–	A framework for water companies to create a drainage strategy based on six principles: partnership, uncertainty, risk, whole life costs and benefits, live process, and innovative and sustainable solutions.
Storm Overflow Assessment Framework (Environment Agency / Water UK, 2018)	2018–	Assessment framework targeting overflows that have a high spill frequency to evaluate if they cause environmental harm. Considers effect on river water quality by intermittent sewer discharges.
Drainage and Wastewater Management Plans (Water UK, 2019)	2019–	A framework to evaluate risks (present and future) and assess proposed interventions to manage risks related to drainage and treatment works. Includes short-, medium-, and long-term epoch planning scenarios. Covers capacity and asset health requirements.
Water Cycle Studies (Environment Agency, 2021)	2021–	Guidance to support practitioners plan water supply, wastewater, and flood risk for major development sites, or as part of statutory development plans.

Within each group, there is a range of subjects, many of which interact and influence others. Elements and interactions that fundamentally affect good urban drainage planning are listed below.

- Information management (group 6) provides the underlying structure for holding asset data, how it can be accessed, and how it can be utilised. Data quality and coverage underpins the decisions made (group 5) in asset strategy and planning (group 4) and during the delivery lifecycle (group 7) where assets are created, operated, maintained, renewed and/or disposed of, to deliver the values and outcomes desired by the organisation
- The purpose and context (group 1) sets the direction of values and outcomes (group 10) to underpin the strategy and planning (group 4) for the short and long term, whilst influencing an organisation's approach to risk appetite and management (group 8), controlled by the actions of leadership and approach to governance (group 2)

Continuous improvement, and the mechanisms to undertake and apply the learning to make changes within the organisation, must also be present (group 9).

Urban drainage assets will not only be affected by decisions made under each group category but also where they connect with, or can be influenced by, other assets, which also have decisions made to manage their performance. Examples of these interactions include the following:

- A wastewater treatment plant (WTP) and its capacity will influence choices upstream in the catchment with respect to the retention and release of flows after storm events
- A pumping station, where the performance deteriorates over time (e.g., impellor wear), will cause a decrease in flows pumped forward resulting in upstream impacts of overflow spills and flooding
- The construction of new flood defences that raise a river's top water level (to prevent flooding) will result in a sewer outfall being able to discharge less and may cause flooding in the drainage network
- The connection of a new development where all flows are slowly drained into a sewer or where foul flows only are drained and stormwater is harvested will significantly change volumes discharged during or after storm events

With multiple assets and multiple outcomes to achieve, the asset management approach within an organisation must balance competing priorities. Decisions will be made based on costs and benefits as well as how funding has been agreed and allocated, for example, to specific environmental improvement programmes. Organisations may also be governed by financial accounting rules that steer how their agreed funding can be invested, for example, following financial reporting standards (www.ifrs.org).

The management of assets that interact, especially in urban drainage, may not be the responsibility of one organisation. Therefore, the interaction between the organisations, how they undertake asset management and the effects this has on others need to be understood. Understanding the effect should influence plans developed and deployed, from the creation of assets through to their operation and management.

For example, in England, a local authority may construct roadside SuDS that drain eventually to a sewer. The roadside SuDS may also be maintained and operated by the local authority, but the performance of these assets will affect the performance of the sewers, influencing the outcomes achieved by the sewerage company. Furthermore, the timing of when actions or interventions take place affects the asset performance. This highlights how two organisations need to work together to create an overall outcome. Decisions made in isolation can result in a poor customer, community, or environmental impact.

19.2 CREATING DRAINAGE PLANS

19.2.1 Elements that steer drainage plans

19.2.1.1 Cost

Here we set out the basic terms used for defining costs in the context of drainage planning. These are applied in later sections: to the inclusion of costs in developing investment choices (Section 19.2.1.4), and to considering approaches to evaluating costs (Section 19.2.3).

The term *Totex* (total expenditure, Equation 19.1) applies to an asset or group of assets that brings together *Capex* (capital expenditure) and *Opex* (operational expenditure) over their long-term operating life. This is also referred to as a whole life cost. The term *Totex* may be used to describe the investment into new assets or upgrades to existing assets to meet new regulatory requirements or improve the overall urban drainage system performance to a higher standard than before. *Totex* can also be used to describe the cost of a large investment plan, for example, over a 5-year period, and the investment required to deliver such a plan.

$$Totex = Capex + Opex \qquad (19.1)$$

The term *Botex* refers to base expenditure (Equation 19.2). These are the costs to keep the current assets running and operating as expected. *Botex* is the *Opex* to run the existing assets and the *Base Capex* (the capital maintenance costs that maintain the long-term capability of the assets).

$$Botex = Opex + Base\ Capex \qquad (19.2)$$

19.2.1.2 Carbon

Carbon is a common shorthand description of six greenhouse gases (GHG) listed in the Kyoto Protocol (United Nations, 1998). These GHG emissions are often expressed as equivalent quantities (in terms of global warming potential) of carbon dioxide (CO_2e) (UKWIR 2012). *Capital carbon* relates to the GHG emissions that occur from assets that are created, refurbished, or demolished. *Embodied carbon* represents the emissions related to the creation (including energy used) of products and materials (UKWIR, 2023). *Operational carbon* relates to the GHG emissions linked to operation and maintenance activities. As with cost, the total carbon for a project or plan can be developed and is referred to as *Totex Carbon*.

19.2.1.3 Benefits

Benefits link to the performance of assets and the resulting capital or operational expenditure that creates a change in performance and outcomes. Benefits may be qualitative (an impact might be scored, for example, if it will make a system more resilient), quantitative (a measured impact, for example, is the number of people who benefit or the number of properties where flooding will not occur) or monetised (a monetary value is placed on the benefit and its size). Traditionally, benefits from urban drainage systems relate to a reduction in flooding and pollution, and to creating cleaner waterbodies where a discharge takes place. However, it is now common to consider the wider environmental and societal benefits that accrue, especially when considering the use of SuDS that are nature-based. These benefits are discussed further in Section 11.3.12.

19.2.1.4 Level of service and risk management

Organisations typically make commitments to manage assets to achieve a certain performance. To do this, they must understand how assets will perform, the implications when they do not, and the actions to take. Organisations may target attaining a level of service, for example, no flooding or surcharge for a particular type of event, as advocated in Sewerage Sector Guidance (Water UK, 2023), where no exceedance from the sewer should occur below a 1 in 30-year return period rainfall event (discussed in Section 10.2.1).

The risks of such occurrences now and in the future have to be understood (discussed in Section 19.2.2) to enable investment choices to be outlined and agreed upon. Some levels of service may be more aspirational within the existing drainage network, and the risk (both probability and consequence) has to be understood.

Typically, to manage risk,

- Strategy will set the direction based on the vision and purpose.
- Planning will outline in greater detail what needs to be done where and when. This depends on the types of interventions – Capex and Opex – and the size of the interventions (major capital through to minor capital). These become more detailed as the drainage plans are enacted, and designs are developed.
- Tactical planning provides the local level detail where the decision-making is made on a day-to-day basis.
- Decision-making processes and systems are put in place to review the cost and benefit of projects or programmes of work, with the change in risk updated following interventions.

Figure 19.1 visualises how evaluation approaches and strategic and tactical plans may be developed within an asset management framework and overall drainage strategy to address key risks encompassed by legislative and regulatory requirements. As risks are evaluated, planned for and addressed, the risks change. These risks, to be addressed as part of an approach in Sewerage Risk Management (WRc 2023), include the following.

- Environmental: relates to discharges from sewer overflows, WTPs, stormwater outfalls as well as the sewers when the capacity is exceeded
- Hydraulic capacity: relates to the drainage network capacity under different conditions and how performance can change when interacting with other assets and systems
- Condition and structural: relates to the asset health of the drainage network that leads to operational issues (e.g., sewer blockage, potential for collapses) and hydraulic issues (e.g., siltation reducing the conveyance capacity, infiltration increasing flows beyond planned)
- External risks: relates to activities that are not within the control of the drainage network owner (e.g., disposal of inappropriate material to the sewer, misconnections)

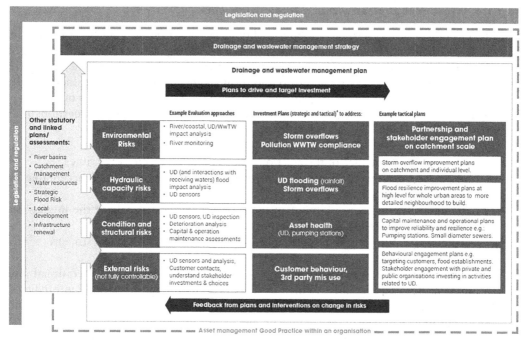

Notes: * Plans may target assets or outcomes or both to drive service and performance. Urban drainage (UD). Wastewater Treatment Works (WwTW).

Figure 19.1 Examples of the risks addressed through planning (strategic and tactical) and potential wider interactions.

19.2.1.5 Legal and regulatory requirements

National and local governments may pass or amend legislation that directly affects the requirement of an organisation to achieve a level of performance or improvement. Such legislation may relate to critical infrastructure planning, requirements to improve the aquatic environment or a change in how water and urban drainage are managed, including an organisation's responsibilities. For example, legislation may drive the reduction in the frequency of CSO spills due to concern about their impact on public health and the ecological health of receiving waters. The Environment Act 2021 provided greater clarity on the requirements of the 1994 Urban Wastewater Treatment Regulations and led to the creation of the Storm Overflows Discharge Reduction Plan in England (discussed in Section 13.4.2).

Organisations such as water service providers in England, Wales, and Scotland are regulated and measured on how they achieve performance targets, typically related to an outcome. Regulatory frameworks may create performance commitments, with financial or reputational rewards or penalties depending on success in meeting the targets. For example, a performance commitment may target internal and external flooding. Whilst a sewer may flood, if it does not cause internal flooding (inside a property) or external flooding (within the curtilage of a property), it is not recorded towards a performance commitment (Ofwat, 2018). These descriptions and ways to measure performance typically change over time.

19.2.1.6 Investment choices within a Totex framework

Investment choices in the short to long term should be based on an understanding of the current, and estimation of the future asset performance. Future performance could be related to "today" or to future epochs such as 10 years ahead, or an end-point such as 2050. These choices relate back to the priorities of the organisation and the people they serve. Increasingly, the public and society influence such priorities. Regulations also drive the types of asset performance that have to be considered when creating drainage plans. Typical questions that might need to be answered include the following:

- How will the performance of the asset or groups of assets change over time?
- Is the performance acceptable now, and what level of performance is acceptable or desirable in the future?
- To change the performance now or in the future, what is the scale of the investment needed, when is it needed, and what types of interventions are required?
- How certain are we that the performance can be achieved with the investment?

To answer these questions and others, choices have to be made related to operational and capital interventions considered in a Totex environment, along with the resulting predicted benefits. Table 19.2 shows options within a Totex hierarchy that can be blended together and deployed to support investment decisions. The hierarchy has the most preferable option first (eliminate), with fabrication last. Drainage plans therefore provide the platform for these decisions, and these plans can range from more targeted and delivery focused short-term plans to more strategic long-term plans. The following sections briefly outline examples of why investment planning is undertaken. Example 19.1 illustrates the application of a blended Totex approach and consideration of the options of when to implement it.

Whilst Totex refers to cost, understanding capital and operational carbon also drives decision-making. Frameworks have been developed to help practitioners target project and programme options further. Table 19.3 outlines a framework that brings thinking from PAS 2080 (BSI, 2023) and highlights the trade-offs between different competing priorities (when they do not always align) and how they can be brought together.

19.2.1.7 Maximising existing asset life and value

While much focus in drainage planning is on the need for new assets to drive improvements or the amendment of existing ones to cope with flooding and pollution, for example, there is also a need to maximise the life and value of existing assets. Such investment is aimed at preventing the deterioration of an asset and requires an understanding of the optimum time to intervene.

Table 19.2 Totex hierarchy options that should be blended together to develop optimum solutions

Totex hierarchy option	Option description
Eliminate	Remove the root cause of the principal threat or pressure
Collaborate	Partner with stakeholders to develop mutually beneficial (and funded) solutions
Operate	Operate and maintain assets and systems differently, driven via data
Invigorate	Leverage existing asset capabilities or enhance headroom
Fabricate	Construct new assets, on a "designed to operate" basis, using efficient construction approaches

Planning may use models to estimate the cost and which epoch interventions will be introduced, or create the framework for real-time data to determine the optimum intervention timing, for example, when pump efficiency reaches a point where replacement, refurbishment, or change of operation is required. Planning enables reactive, pre-emptive, and proactive interventions to take place in a managed way (as discussed in Section 18.2). The goal is to avoid an unacceptable level of performance occurring, for example, a brick sewer collapsing that leads to significant investment to recover the asset's health and performance.

19.2.1.8 Responding to public pressure

Public opinion can also influence planning when individual members or groups of the public call for change (e.g., due to environmental harm (Court of Justice of the European Union,

Table 19.3 Framework incorporating Totex options and a carbon reduction hierarchy

Totex options	Carbon reduction hierarchy		
	Avoid carbon use	Switch mindset / method	Improve solutions / other areas
Eliminate	Evaluate basic needs and explore alternatives using network and catchment systems-level approaches	Adopt adaptive planning and apply thresholds. Switch to performance-based standards	Improve asset data, smarter performance monitoring and control
Operate	Avoid introducing new demand for fossil fuel use, chemicals, etc.	Replace fossil fuel use with renewables and support the circular economy	Improve operational regimes to improve efficiency
Invigorate	Avoid new construction by reusing or refurbishing existing assets	Integrated nature-based solutions with existing asset systems using risk-based approaches	Apply circular economy principles to maximise by-product recovery
Fabricate	Avoid designs that depend on carbon-intensive technologies, materials, consumables	Switch to alternative technologies, leaner designs, low carbon materials, more off-site fabrication	Use construction technologies that avoid waste, minimise materials and energy use on site

Source: Johnson, A. 2023. *Institute of Water Magazine*, Autumn, 2023.

EXAMPLE 19.1

An overflow at a pumping station close to the WTP, but separated from it by a railway line, is spilling too frequently and causing environmental harm in the waterbody. The overflow has a screen although it is undersized for the peak 5-year return period flow requirement. At the pumping station, the pumps occasionally do not pump forward the required flow rate. The permit for the overflow is incorrect with a clear error in the pass-forward flow requirement.

Develop a plan to meet the short-term target of permit compliance and the longer-term target of reducing environmental harm.

Solution

Figure 19.2 shows various Totex options that can be combined in different ways to meet the short- and long-term targets. The plan introduces a level of adaptation depending on outcomes and decisions. Options considered as part of the plan are outlined in Table 19.4.

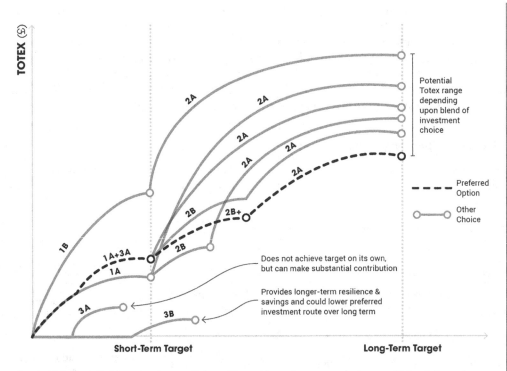

Figure 19.2 Range of interventions utilising a Totex hierarchy approach that could be delivered over the short to long term to achieve targets (Example 19.1).

Table 19.4 Summary of Totex options

Option	Summary	E	C	O	I	F
1A	Correct the permit and update the screen.	✓	✓			✓
1B	Provide an initial increase in storage to reduce the frequency of spills and meet pass-forward flow requirements.					✓
2A	Major grey infrastructure investment to increase pass-forward flow at the pumping station, and tank storage at the overflow.					✓
2B	Surface water management targeting the removal of flow upstream from highways, large individual landowners (schools, hospitals, commercial shopping locations), and houses.	✓	✓			✓
2B+	Active system control utilised in combination with surface water management.	✓		✓		✓
3A	Monitoring of the pumping station and change of impellors to improve the handling of solids that come from the screen. Operational response put in place to react.			✓	✓	
3B	Customer behaviour targeted upstream to discourage the disposal of sanitary products and wet wipes to reduce the impact on the screen and pumping station, with the potential to remove the need to upsize the screen.	✓	✓			

E = Eliminate, C = Collaborate, O = Operate, I = Invigorate, F = Fabricate

2012)). There are times when a localised problem leads to complaints and attention that the organisation wishes to address and a strategic decision to invest is made even when not necessarily meeting a typical cost–benefit threshold (e.g., frequent nuisance flooding of a highway).

19.2.2 Understanding asset performance

19.2.2.1 Current and future performance quantification

Within a plan, it is important to understand a baseline position and how internal and external factors may change performance over time. Planning for a potential future performance level is critical as it influences choices made today, including types and scales of investment.

Each organisation has to balance up the level of granularity, precision, and confidence in each area of interest when developing a plan, and the cost to achieve this versus the benefit gained. At a drainage planning level for a catchment draining to a WTP, or a number of catchments that may discharge to a river length, the interactions between different systems and components are important. However, they are unlikely to contain the detail that enables an intervention design to be automatically created from the plan without further work.

Organisations typically have data and tools at their disposal to understand current and estimated future performance to underpin the development of a plan. At times, these tools (often in the form of models) need to be created or updated and maintained to provide an accepted level of confidence in performance estimates. The amount of data (both static and live) available has significantly increased over the last 5–10 years and provides a good platform for plans and for understanding current performance. However, the quality of the data, its robustness, and level of validation, are important factors informing the level of confidence in conclusions drawn (Section 21.5).

Current performance is influenced by factors at the time and therefore can vary day-to-day. In the case of drainage, performance is influenced by both uncontrollable and controllable inputs. Uncontrollable examples include rainfall and sewer solids (gross solids and FOG). Controllable factors (within limits) are influenced by ongoing investment choices and interventions deployed. An example of this is the management of the sewer flows if a diversion (such as over-pumping) is required to enable sewer rehabilitation (Chapter 16) or cleansing (Chapter 18).

Data provides insight based on what has happened and can be used to estimate future performance. Understanding future performance relies on the use of tools and models which are typically statistical or deterministic. Questions that may have to be answered to create investment plans include the following:

- What will be the potential performance of the system,
  - o If a certain type of rainfall event occurs that exceeds system capacity, causing flooding?
  - o For typical rainfall over a year or longer, for example, on discharges to receiving waters?
- How will assets change over time?

Statistical models are defined and discussed in Section 21.1. Examples of statistical models related to asset health are deterioration models (Erskine et al., 2014) that estimate how sewers may degrade over time based on a range of factors. The models can be used to create an investment plan for a range of deterioration scenarios and the level of service or asset health level an organisation wishes to maintain over time. The model could include such factors as the asset material, age, surrounding environment such as ground conditions, and historical data on its condition.

Deterministic models (Chapter 20) can be used to understand performance for a range of scenarios allowing for different influencing factors and predictions (e.g., climate change and its impact on rainfall and temperature on drainage systems and receiving waters). Such models may vary in detail, quality, and accuracy and will depend on the intended purpose. Models used for drainage planning will need to balance the level of confidence required to support investment choices with the level of effort in time and cost to create, simulate, and analyse for different performance requirements. For example, a 1D–2D coupled drainage model (see Section 20.10.3) may be used to estimate the flooding impact for a range of rainfall events and durations, but could be cumbersome and slow to estimate the discharges from CSOs when simulating a 10- or 25-year rainfall time series.

19.2.2.2 Critical factors

When developing a plan it is necessary to determine the future factors and time periods to be considered. These may be based on different scenarios that require different approaches to replicate or influence performance. As such, some scenarios and their related factors may be included within deterministic modelling (e.g., rainfall, growth) and others considered in the long-term influence on cost and benefit, adopting more qualitative approaches (Environment Agency, 2018). Ofwat (2022) proposed examples of various scenarios to be applied as part of long-term drainage and wider asset management planning including the following:

- climate (e.g., changes in rainfall, temperature)
- technology (e.g., Internet of Things, nature-based solutions, behaviours)
- demand (e.g., growth, water consumption, urban creep, regulations, and standards)
- ambition (e.g., scale of environmental improvement)

Table 19.5 outlines critical factors and their influence on drainage planning. With such a wide range of factors, careful consideration is required of the tools and models, the level of confidence in their quantification, the approach to risk, and how the outputs will be used. For example, will the factors be used to create an adaptable and flexible plan, are they required to show the range of futures, or are they needed to select a preferable future that sits within plausible, possible, or probable futures (Lyons, 2018)? Further details about scenarios and futures are given in Section 24.9.

19.2.3 Evaluating costs, carbon and benefits of plans, programmes and projects

Costs and targeted benefits have been long-standing elements of decision-making. For example, the cost to reduce the likelihood of a property flooding internally may have a value based on willingness to pay for the prevention of flooding and so the number of properties protected versus the cost of the scheme could be evaluated. More recently, the inclusion of the amount of carbon (capital and operational) and the wider benefits have become more important in decision-making. An introduction to costs and carbon and benefits is given in Sections 19.2.1.1–19.2.1.3.

Business planning over longer time horizons to determine system-wide performance improvements will be at a high level, whereas tactical planning will be more granular with increasing levels of detail and certainty moving through the asset lifecycle (AACE, 2020).

Table 19.5 Critical factors that affect drainage planning

Factor	Influence on drainage planning
Area being drained (runoff)	Coupled with rainfall, influences the amount of rainfall-runoff entering the drainage system by inflow and infiltration (Section 3.4). Can include hard paved and permeable surfaces.
Asset health	As assets age, they become worn and require remedial action. Mechanical, electrical, instrumentation, control, and automation assets may require more frequent replacement, refurbishment or operational management. Other assets such as pipes and concrete will have longer asset lives; however, elements such as pipe seals may fail earlier. SuDS are also likely to require a different approach to ensure their asset health remains within an "operational window." Decisions on health and its influence on performance are critical in medium- and long-term Totex plans (Chapter 18).
Behaviours	Affects assumptions regarding urban creep, how solids may be disposed of (and hence leads to blockages in sewers causing pollution and flooding). Affects the uptake of property-level interventions.
Growth	Causes an increase in dry weather flow (DWF) in sewer systems that can reduce headroom and have an impact on fixed controls such as those at pumping stations. Local impacts may be seen where new development occurs or there is a significant change of use. Potential uncertainty on when such a development may take place.
Non-water Infrastructure	Wider plans for other infrastructure renewal (e.g., highways and footpaths) and regeneration plans (e.g., social housing improvements) may influence choices for urban drainage interventions: what can be done and when.
Population	Causes an increase in DWF and load (solids and dissolved pollutants) in sewer systems that can reduce headroom and have an impact on fixed controls such as those at pumping stations.
Public pressure	The public has had a growing influence on environmental planning and outcomes. Considering elements of current practice that may be influenced, or expectations changed, enables scenarios to be tested.
Rainfall and dry periods	Affects predictions of the scale of flooding or impact on receiving waters. Dry periods influence the buildup of pollution, affecting the magnitude of the first foul flush (Section 13.3.2), the impact on discharges, and the performance of screens at overflows and WTP inlet works.
Regulatory shifts	Changes in legislation and regulation (e.g., environmental, funding) will directly influence the strategy, the plans to meet requirements, and the behaviours of an organisation.
Temperature	Particularly relevant for future epochs, changing the impact that urban drainage discharges have on the receiving water by affecting water quality parameters' decay rates.
Urban Creep	Affects the area of runoff, which is unchecked and often unknown. Contributes to increased storm response to the sewer with flooding and water quality impacts (Section 10.3.3).
Water infrastructure	Proposals for water resources and distribution have long-term implications for rainwater and wastewater reuse as part of planning, linked to per capita consumption.
Water use	Affects DWF and headroom. Changes in the future and for new developments affect future decisions in combination with growth and population (Chapter 3).
WTP infrastructure	WTPs will typically have a numerical or descriptive permit against which performance is assessed. Management and upgrades planned to meet current or future requirements affect and influence decisions that can be made in the urban drainage network.

19.2.3.1 Determining costs

Costs may consider the capital cost only, operational cost only or the total expenditure / whole life cost (where costs are discounted into the future (HM Treasury, 2023)). Determining costs for conventional infrastructure used in urban drainage such as pipes, pumps, and screens is well-established with supporting material (e.g., Spon's Civil Engineering and Highway Works Price Book, AECOM (2023)) and provides a good understanding of unit costs for different contexts.

The costs for SuDS and nature-based solutions are less certain for capital and operational expenditure. However, databases have been collated internationally (The Water Research Foundation, 2020), and tools to support SuDS costs, carbon management and benefits estimation are being developed (Digman et al., 2024). SuDS costs are discussed further in Section 11.6.4.

19.2.3.2 Determining carbon emissions

Managing whole life carbon, typically applying the principles and requirements of the PAS 2080 framework (BSI, 2023), across each stage of the asset cycle should be undertaken, and, therefore, this forms an important part of contemporary drainage planning. Estimating carbon follows a similar approach to costs, with capital, operational, and whole life carbon of schemes taken into account. Avoided carbon emissions can also be estimated and may be treated as a benefit.

UKWIR have created a Carbon Accounting Workbook (UKWIR, 2023) which provides guidelines for UK water service providers to estimate the carbon embodied in constructing and maintaining assets (UKWIR, 2012). The workbook is regularly updated and enhanced, including changes in carbon accounting practices (UKWIR, 2023). Carbon emission data has also been collated, such as in CESMM4 (ICE, 2019), for different materials and activities in a similar way to costs. These carbon values for high-level planning are likely to include all-in values for raw material collection, transportation, manufacturing, construction/installation, and operation and maintenance. As more detailed plans and projects emerge, the granularity and individual assessment of activities will improve.

19.2.3.3 Determining benefits

Benefits may be quite specific, for example, improvement of a waterbody or reduction of flood risk to a property. Flood risk (introduced in Section 12.4) can be assessed using annualised damages (FHRC, 2024). Benefits may also consider the wider outcomes of a solution and can be grouped into sub-areas often with a monetary value. These can be categorised using frameworks typically created for specific contexts, such as

- Triple bottom line: assesses the economic, environmental, and social benefits
- Ecosystem services: considers the direct and indirect contributions to ecosystems in four groups – provisioning, regulating, cultural, and supporting
- Multi-capitals (Capitals Coalition, 2022): groups benefits or the value from them into "capitals" such as natural, social, human, manufactured, intellectual, and financial

Various guidance and tools exist to establish the value of the benefits. These frameworks may place a monetary value to enable a cost–benefit analysis to be undertaken, or may use a scoring system. Examples of frameworks and methods deployed include the following.

- Ciriabest (CIRIA, 2024): enables the benefits of solutions using blue-green and grey infrastructure to be estimated, and schemes compared (Horton et al., 2019). Creates a monetary value presented in an ecosystem services framework and a five-capitals framework. Used to supplement other frameworks and appraisals (see Section 11.3.12)
- Value toolkit developed by the Construction Innovation Hub (2022): adopts a four-capitals model to be applied through different stages of lifecycle delivery (from needs to operation)
- Water Industry National Environment Programme (Defra/Ofwat, 2022): outlines steps to develop and evaluate proposals, including the benefits, in a structured manner to meet the statutory and non-statutory requirements of government priorities in England

Operational and capital maintenance expenditure benefits, especially linked to tactical planning, were often very targeted, focusing only on those that the activities promote. However, it is becoming more common to consider wider benefits to inform decision-making. For example, to reduce internal and external sewer flooding to properties as a result of a blockage in a local sewer, the benefit is the avoided incident occurring to residents, including disruption and impact on well-being. A further benefit will be a reduction in reactive operational response and no customer contact.

Organisations that deliver large and multiple programmes of work are likely to develop and use their own frameworks to enable decisions to be made, linked to their plans for current and future investment (e.g., Yorkshire Water, 2021). Their decision-making tools enable a large portfolio of risks, needs, solutions, and outcomes to be assessed and prioritised.

19.2.3.4 Utilising a cost–carbon–benefit framework

Drainage plans are typically structured to support decision-making including how costs, carbon and benefits, etc., are established over different timeframes. During planning, the level of detail needed will be different to that required when a particular programme or project is being designed. Appraisals are required in drainage plans to enable decisions to be made on what, why and when to invest, and the level of benefit expected. Appraisal guidance is well developed with documented steps to create a business case for investment (e.g., The Green Book (HM Treasury, 2023) and Cost Benefit Analysis and the Environment (OECD, 2018)).

A drainage plan will normally consider the current and future risks and work within a framework that balances different planning epochs. Within a plan, risks, interventions, and outcomes have to be prioritised within funding boundaries and mechanisms, for example, balancing the willingness and ability for the plan to be funded and paid for (such as by customers through bills or taxation through the government).

A cost–carbon–benefit framework is typically applied to make decisions on which risks are addressed. This may be at a local level, or at a country/governmental level. There may also be aspects that determine whether investment is appropriate such as utilising BATNEEC (Best Available Technology Not Entailing Excessive Costs) (Sorrell, 2002) or BTKNEEC (Best Technical Knowledge Not Exceeding Excessive Costs) principles. These approaches typically include costs and benefit, but also consider affordability.

Frameworks may consider costs, carbon, and benefits in different ways. The carbon (monetary) values (current and forecast) can be calculated for different schemes or programmes and compared to demonstrate the level of benefit and as part of a wider benefit appraisal. In the last decade, there has been a growing recognition of wider benefits that can also be included and accounted for (if funding rules allow). For example, it is now common to consider and make decisions on investments that might balance least cost versus most cost–carbon beneficial.

19.2.4 Engaging with stakeholders

Urban drainage, as confirmed in this chapter, does not sit in isolation. It has an impact on communities and the environment and on the planning, delivery, operation, and maintenance of other forms of infrastructure. There is a need therefore for plans to be developed in collaboration and consultation with stakeholders. Different stakeholders have different interests, therefore planning stakeholder engagement is crucial from the outset. Thought should be given to how stakeholders' views and plans will influence and integrate with the drainage plan.

Plans are often consulted on by the lead organisations. This may be on a range of aspects including the following:

- Strategic intent, priorities for the organisations and when they expect the priorities to be delivered
- Gaining knowledge and evidence to inform the plan by understanding the needs and risks better
- Understanding how proposals may be received and viewed, including a willingness to pay or value placed on the plan or elements of the plan

Stakeholders that may be engaged and consulted include the following:

- Those with legal roles and responsibilities with whom the plan may interact from a drainage perspective
- Those with legal roles and responsibilities who may influence the plan such as planning authorities and other infrastructure providers
- Community and special interest groups
- Business and the general public

There is a shift in how plans are created, with an increasing emphasis on developing them collaboratively where more than one organisation is responsible for drainage, or heavily influences, or is influenced by the choices a plan proposes. Such an approach is critical for adaptive planning and is discussed further in the next section.

19.3 DRAINAGE PLANNING IN THE BROADER CONTEXT

19.3.1 Applying systems thinking

Systems thinking is not a new concept and has been applied in urban drainage for many years but in an ad-hoc and not always deliberate way. Essentially it is how we think (and therefore act and plan) in a way which joins up disciplines, infrastructure and organisations. As the interconnectivity of human-made and natural systems becomes more complex, how we make sense of these interactions and plan for them becomes more important (Whitcomb et al., 2020). Complexity is often a function of the historical evolution of the water and wider infrastructure planning and management systems.

Within urban drainage, planning for operational and capital investment should consider how individual assets interact with those around them, maybe kilometres upstream and downstream. More obvious systems understanding is seen through the use of sewer hydraulic simulation models (Chapter 20) that demonstrate how the flows in the system behave and change when interventions are put in place. For example, the control of a pumping station will influence the deposition and movement of solids and sediments upstream, the management of solids in a wet well, and pump performance and efficiency. The change in control of a CSO may increase flows downstream or increase surcharge depths that could lead to flooding.

Figure 19.3 outlines how urban drainage interacts with a range of other systems including infrastructure, the natural environment, the built environment, communities, and regulation. Some consequences of these interactions are given below:

- The control and management of discharges from CSOs helps to improve the health of rivers and seas
- Retrofitting urban drainage often requires engagement and permissions from third parties such as highway authorities, spatial planning authorities or private landowners, and can be subject to planning legislation
- Future mobility choices (e.g., shared mobility, active travel) will influence the amount of highway required for vehicle movement and parking (ITF, 2017) especially considering 15–30% of urban areas are typically used for parking (WSP & Farrells, 2016). Such space influences both the volume of runoff and the availability of space to build retrofit stormwater management measures (Chapter 11)

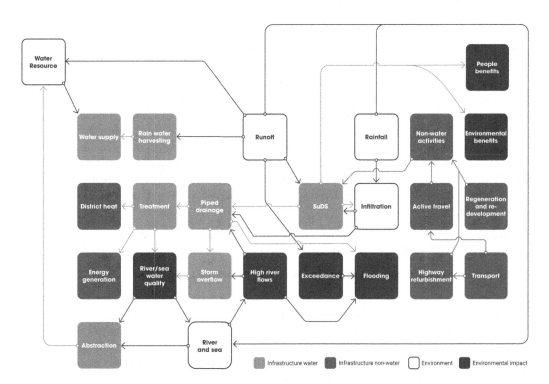

Figure 19.3 Examples of different systems and how they interact with each other.

- The regeneration and redevelopment of urban areas can enable stormwater to be managed more efficiently and at a lower cost than retrofitting at a later date. For example, social housing improvements involving the replacement of roofs, facias, gutters, and downpipes could also include the introduction of household water management features such as slow-draining water butts, planters, and rain gardens (Chapter 11)
- Water reuse supports a reduction in potable water demand linked to water resources strategy and planning (Section 24.4.1)
- The quantity of wet wipes / sanitary solids disposed of via the sewer system influences the frequency of blockages in small-diameter pipes leading to flooding and affecting the performance of screens at CSOs and at inlet works (Section 18.5)
- The frequency of road gully cleansing can influence the quantity of silts and sediments entering the urban drainage system; these have the potential to be deposited and reduce system performance or be discharged directly to waterbodies impacting water quality (Section 18.6.2.7)
- The volume of stormwater mixed with wastewater retained in the sewer system may result in more sludge and biosolids created at a WTP

19.3.2 Integrated planning

The recognition of interconnectivity provides evidence as to why there is a call and shift towards integrated water management (IWM) and IWM planning. This is not straightforward, with different interpretations by regulators and those looking to plan and deliver schemes (Bromwich et al., 2022); however, understanding and collaboration are fundamental to its success. In some cities, this has been in place for some time, such as in Melbourne, Australia (Digman et al., 2020). Here, an initial focus on stormwater management and water-sensitive urban design shifted to a broader agenda as a result of issues faced by the city including drought.

In the UK, plans and pilot studies for the implementation of integrated urban drainage have been in place for some time (Balmforth et al., 2009). However, the more formal and progressive creation of partnerships has taken time, and in some cases has been a direct result of major flooding from extreme rainfall. Other partnerships are forming where it is recognised that more can be delivered together, and the opportunities to achieve this are available by joining up funding from water and non-water sources. Examples of non-statutory voluntary partnerships that enable integrated planning include the Glasgow Strategic Drainage Partnership (MGSDP, 2016), the Northumbrian Integrated Drainage Partnership (Water UK, 2021), Living With Water (livingwithwater.co.uk), and the Greater Manchester Integrated Water Management Plan (GMCA, United Utilities and Environment Agency 2023).

19.3.3 Creating adaptive plans to align investment choices through collaboration

19.3.3.1 Connecting opportunities from different organisations

Considering how systems and their planning and investment cycles interact provides the opportunity to be more efficient, increase the value of investments, and reduce disruption. For example, if there is the need to reduce flows into a combined sewer network and a highway refurbishment scheme is planned, then aligning the timing of the elements of the work will be beneficial to all parties and the surrounding community. This concept of applying flexibility in how and when to deliver schemes to meet joined-up needs can be thought of as part of adaptive planning.

Mapping systems to define interactions enables an understanding of what can be achieved now and in the future. Different investment periods may mean projects will be at different stages (from a risk or need, through to detailed design, or ready to start on site). Understanding where these projects are located, their value, the benefit they are looking to create, and their delivery timescales and regulatory requirements enables potential alignment opportunities to be established or challenged. An example of this is in Greater Manchester (GMCA, United Utilities and Environment Agency 2023) where key partners have come together (the Combined Authority which includes the transport authority, the environmental regulator and the water company) to co-create a programme of schemes to develop and deliver together. The plan goes further: to develop the capacity and capability of people in the region.

The Northumbrian Integrated Drainage Partnership brings together the flooding responsibilities of different partners where they overlap spatially. A joint investigation funded by multiple parties typically takes place and, where appropriate, a business case is created that satisfies the needs of the different funding organisations. The result has been the delivery of over 21 schemes (at the end of 2023) reducing flood risk to over 6000 properties (Daughety, 2024). Flexibility in delivery has been achieved through staging the delivery of schemes, such as at Killingworth and Longbenton (susdrain, 2018, Lewis, 2019) which provided protection to over 3500 properties (Box 19.1).

BOX 19.1   Creating and implementing a drainage master plan in Killingworth and Longbenton (KLB)

This collaborative scheme protects communities from flooding while stepping away from traditional practices. Esh-Stantec co-created the £5.2m partnership scheme at KLB with Northumbrian Water (NWG), the Environment Agency, and North Tyneside Council.

The work started in 2011 with NWG creating an integrated drainage model which predicted flood risk below and above ground. Significant effort was placed on developing the integrated model to facilitate the accurate replication of flooding to give confidence in the ultimate solution and the effect it would have. This enabled engineers to work with the partners to develop a drainage master plan and a strategy to inform the design that reduces the flood risk to 3500 homes. The strategy from the outset looked to measure the potential benefits to the community. The flooding was complex, coming from the watercourses and sewers as well as flows that cannot enter either of them.

The scheme introduced SuDS across the catchment, utilising a range of solutions including the following:

- Providing upstream attenuation by reprofiling a watercourse and controlling flows
- Diverting a watercourse from the combined sewer network to another watercourse
- Creating extra capacity to manage the diverted flow from the watercourse into a new detention basin and amphitheatre at a high school
- Lowering the normal lake water level, disconnecting its outfall from the combined sewer system and creating an exceedance basin to cope with increased flows.

The design adopted resilient principles (Section 24.7) including future-proofing for climatic changes, and providing redundancy at low cost through "exceedance" approaches that could quickly recover to their primary designed function. Landscape architects led the design to fit with the local environment and create additional amenity, ecological, health, biodiversity, educational, and recreational benefits. This included novel floating islands that create habitat and improve water quality, adopting learning from the Netherlands, planted by local schoolchildren. The scheme resulted in multiple benefits exceeding £30m calculated using the ciriabest tool (Section 11.3.12) to inform the partnership funding application and £4m saving against traditional approaches. The other benefits included improved water quality by reducing storm overflow spills and flow to treatment and created capacity to support regeneration in the area.

The project was delivered over a 4-year period to maximise opportunities when they arose (e.g., reprogramming of the scheme to align with the new school built at Longbenton). Construction finished in autumn 2019. Aligning the timing of opportunities is a critical element in creating savings and delivering a strategy. The work led to partnership agreements for appropriate organisations to own and maintain the created SuDS and is a leading demonstration of how to implement schemes successfully over time.

19.3.3.2 Creating aligned plans in the future

Aligning investment plans is not necessarily straightforward when each system or sector has well-established funding cycles and routes to agree to funding. However, the creation of integrated plans that are cognisant of others' requirements is likely to be a key enabler.

Creating visibility of plans, risks, needs, and intent on what to deliver, where and when is a key enabler to unlocking collaborative working, funding, and delivery. The aim of the Greater Manchester IWMP (GMCA, United Utilities and Environment Agency 2023), introduced in the previous section, is to align future planning, with delivery post-2030 in mind. Creating the platforms and frameworks to enable the sharing and hence visibility within a partnership will be fundamental.

In England and Wales, future iterations of the Drainage and Wastewater Management Plans, provide an opportunity to consider this and to create a framework that supports such an approach, across sectors.

PROBLEMS

1 Why is urban drainage planning important and how does a strategic approach inform tactical planning and implementation?
2 What are the core components of an urban drainage plan?
3 What role does asset management play in influencing urban drainage planning?
4 What are the potential benefits that urban drainage interventions can bring and what approaches can you use to assess those benefits, especially those beyond traditional benefits such as flooding and water quality?
5 How can the Totex and Carbon Reduction Hierarchy influence strategic and tactical decision-making when developing and implementing plans on a programme and project level?
6 What are the key aspects that influence adaptive planning in urban drainage? In Example 19.1, consider and outline what the key interactions are between the different options and pathways.

7 What stakeholders could be involved and what roles could they have in urban drainage planning and implementation?

8 When adopting a systems thinking approach, how can this influence what an urban drainage plan looks like and how such a plan is developed?

KEY SOURCES

Institute of Asset Management. 2021. *The Pathway to Excellence in Asset Management. IAM Maturity Scale & Guidance*. https://uk2.theiam.org/media/4244/iam-the-pathway-to-excellence-in-asset-management-preview.pdf

Water UK. 2019. *A framework for the production of Drainage and Wastewater Management Plans*. https://www.water.org.uk/news-views-publications/publications/framework-production-drainage-and-wastewater-management-plans

WRc. 2023. *Sewerage Risk Management*. http://srm.wrcplc.co.uk

REFERENCES

AACE. 2020. *Cost Estimate Classification System – As Applied in Engineering, Procurement, and Construction for the Building and General Construction Industries*, AACE International Recommend Practice No. 18R-97, https://web.aacei.org/docs/default-source/toc/toc_18r-97.pdf?sfvrsn=4

AECOM. (Eds.) 2023. *Spon's Civil Engineering and Highway Works Price Book*, 3th Edition, Routledge.

Balmforth, D., Digman, C.J., Shaffer, P. and Butler, D. 2009. Research Framework – The *Implementation of Integrated Urban Drainage*. Environment Agency Science Report, SC070064.

Bromwich, B., Crilly, D. and Banerjee, J. 2022. Water governance and system coordination across diverse risk-management cultures. *Water International*, 47(7), 1037–1047.

BSI. 2023. PAS 2080 *Carbon Management in Buildings and Infrastructure*, 2nd Edition.

Capitals Coalition. 2022. *A Navigation Through Value Accounting Methods*. https://capitalscoalition.org/publication/a-navigation-through-value-accounting-methods/

CIRIA. 2024. *ciriabest. (CIRIA'S Benefits Estimation Tool)*. https://www.susdrain.org/resources/best.html

Construction Innovation Hub. 2022. *Value Toolkit Overview*. https://constructioninnovationhub.org.uk/media/2dobmihh/20220927_hub_valuetoolkit_overview_interactive.pdf

Court of Justice of the European Union. 2012. *Judgment Of The Court (First Chamber) 18 October 2012 (Failure of a Member State to fulfil obligations — Pollution and nuisance — Urban waste water treatment — Directive 91/271/EEC — Articles 3, 4 and 10 — Annex I(A) and (B)) In Case C 301/10*. https://curia.europa.eu/juris/document/document.jsf?docid=128650&doclang=EN

Daughety, A. 2024. *Personal Communication*.

Defra. 2018. *Surface Water Management. An Action Plan*. https://assets.publishing.service.gov.uk/media/5b4c7ab840f0b6184ba3645a/surface-water-management-action-plan-july-2018.pdf

Defra/Ofwat. 2022. *Guidance: Water Industry National Environment Programme (WINEP) Methodology*. https://www.gov.uk/government/publications/developing-the-environmental-resilience-and-flood-risk-actions-for-the-price-review-2024/water-industry-national-environment-programme-winep-methodology

Digman, C.J., Gorton, E., Horton, A., McMullan, J., Morris, W., Riisnaes, S. and Woods-Ballard, B. 2024. *Improved Understanding of Retrofit SUDS Whole Life Cost, Carbon and Delivery to Enable Deployment*, UKWIR Report 24/SW/01/29.

Digman, C.J., McCloy, A., Mclarnon, C., Shaffer, P. and Szczyrba, M. 2020. *When is Surface Water Removal and Management the Most Cost-effective Solution?* UKWIR Report 20/SW/01/20. https://ukwir.org/when-is-surface-water-removal-the-most-cost-beneficial-solution-1

Environment Agency. 2018. *Accounting for Adaptive Capacity in FCERM Options Appraisal*. https:// www.gov.uk/flood-and-coastal-erosion-risk-management-research-reports/accounting-for-adaptive-capacity-in-fcerm-options-appraisal

Environment Agency. 2021. *Water Cycle Studies. Guidance*. https://www.gov.uk/guidance/water-cycle-studies

Environment Agency / Ofwat. 2013. *Drainage Strategy Framework*. https://www.ofwat.gov.uk/wp-content/uploads/2015/12/rpt_com201305drainagestrategy1.pdf

Environment Agency / Water UK. 2018. *Storm Overflow Assessment Framework*. https://www.water.org.uk/wp-content/uploads/2018/12/SOAF.pdf

Erskine, A., Watson, T., O'Hagan, A., Ledgar, S., and Redfearn, D. 2014. Using a negative binomial regression model with a Bayesian tuner to estimate failure probability for sewerage infrastructure, *Journal of Infrastructure Systems*, 20(1).

FHRC. 2024. *Flood and Coastal Erosion Risk Management Handbook for Economical Appraisal* https://www.mcm-online.co.uk/handbook

FWR. 2019. *Urban Pollution Management Manual*, 3rd editon. http://www.fwr.org/UPM3/

GMCA, United Utilities and Environment Agency. 2023. *Enhancing Life through Water – Integrated Water Management Plan (Draft Plan)*. https://democracy.greatermanchester-ca.gov.uk/documents/s27343/10A%20Integrated%20Water%20Management%20Plan.pdf

HM Treasury. 2023. *Guidance. The Green Book*. https://www.gov.uk/government/publications/the-green-book-appraisal-and-evaluation-in-central-government/the-green-book-2020

Horton, B., Digman, C.J., Ashley, R.M. and McMullan, J. 2019. *B£ST Guidance – Guidance to Assess the Benefits of Blue and Green Infrastructure using B£ST*, CIRIA Report W047b, Release version 5.

ICE. 2019. *CESMM4 Revised. Civil Engineering Standard Method of Measurement*. ICE Publishing.

ISO 55001: 2014. *Asset Management – Management Systems*.

ITF. 2017. *Shared Mobility Simulations for Helsinki*, OECD. https://www.itf-oecd.org/sites/default/files/docs/shared-mobility-simulations-helsinki.pdf

Johnson, A. 2023. Moving from trade-offs to synergies in the water sector, *Institute of Water Magazine*, Autumn, 2023.

Lewis, G. 2019. Killingworth & Longbenton Water Management Ph 3 (2019) - A partnership approach delivering a sustainable flooding solution with benefits to the wider community. *UK Water Projects Online*, https://waterprojectsonline.com/case-studies/killingworth-ph3-2019/

Lyons, G. 2018. *Handling Uncertainty in Transport Planning and Decision Making - Report of a Roundtable Discussion held in London on 20 July*, https://uwe-repository.worktribe.com/output/861803

MGSDP. 2016. *The Metropolitan Glasgow Strategic Drainage Partnership – Surface Water Management Masterplan Summary and Actions*, https://www.mgsdp.org/CHttpHandler.ashx?id=38023&p=0

National Water Council and WRc. 1981. *Sewers for Adoption*, 1st Edition.

OECD. 2015. *Principles on Water Governance*, https://www.oecd.org/cfe/regionaldevelopment/OECD-Principles-on-Water-Governance-en.pdf

OECD. 2018. Cost Benefit Analysis and the Environment – Further Developments and Policy Use, https://www.oecd.org/governance/cost-benefit-analysis-and-the-environment-9789264085169-en.htm

Ofwat. 2018. *Reporting Guidance – Sewer Flooding*, https://www.ofwat.gov.uk/wp-content/uploads/2018/03/Reporting-guidance-sewer-flooding.pdf

Ofwat. 2022. *PR24 and beyond: Final guidance on long-term delivery strategies*. https://www.ofwat.gov.uk/publication/pr24-and-beyond-final-guidance-on-long-term-delivery-strategies/

Sorrell, S. 2002. The meaning of BATNEEC: interpreting excessive costs in UK industrial pollution regulation, *Journal of Environmental Policy and Planning*, 4(1), 23–40.

susdrain 2018 *Killingworth and Longbenton, Surface Water Management Scheme, North Tyneside*, https://www.susdrain.org/case-studies/pdfs/suds_awards/019_18_04_30_susdrain_suds_awards_killingworth_and_longbenton_sw_mgt_scheme.pdf

The Water Research Foundation. 2020. *International Stormwater BMP Database: 2020 Summary Statistics* https://www.waterrf.org/system/files/resource/2020-11/DRPT-4968_0.pdf

UKWIR. 2002. *Capital Maintenance Planning – a Common Framework. Volume 1: Overview*. Report 02/RG/05/3, https://ukwir.org/eng/reports/02-RG-05-3/66808/Capital-Maintenance-Planning-A-Common-Framework-Volume-1-Overview

UKWIR. 2012. *A Framework for Accounting for Embodied Carbon in Water Industry Assets*, Report number 12/CL/01/15, https://ukwir.org/reports/12-CL-01-15/66617/A-Framework-for-Accounting-for-Embodied-Carbon-in-Water-Industry-Assets

UKWIR. 2023. *Workbook for Estimating Operational GHG Emissions – CAW Version 17*, Report 23/CL/01/36.

United Utilities. 2021. *Asset Health in the Water Sector. Proposal for a Framework*, https://www.ofwat.gov.uk/wp-content/uploads/2021/04/United-Utilities-Asset-Health-Framework-Future-Ideas-Lab.pdf

United Nations. 1998. *Kyoto Protocol to the United Nations Framework Convention on Climate Change*.

Water UK. 2021. *Working Together to Improve Drainage and Environmental Water Quality. An Overview of Drainage and Wastewater Management Plans*, https://www.water.org.uk/wp-content/uploads/2021/10/Working_Together_an_overview_of_Drainage_and_Wastewater_Management_Plans.pdf

Water UK. 2023. *Sewerage Sector Guidance*. https://www.water.org.uk/sewerage-sector-guidance-approved-documents

Whitcomb, C., Davidz, H. and Groesser, S. (Eds.) 2020. *Systems Thinking*, MDPI Book.

WSP and Farrells. 2016. *Making Better Places: Autonomous Vehicles and Future Opportunities*, https://www.wsp.com/-/media/Sector/Global/Document/Making-better-places.pdf

Yorkshire Water. 2021. *Our Contribution to Yorkshire. A Review of our Impact and Public Value*, https://www.yorkshirewater.com/media/022dl5eo/our-contribution-to-yorkshire-main-report-february-2021.pdf

Chapter 20

Modelling in practice

20.1 MODELS IN URBAN DRAINAGE ENGINEERING

There are two chapters in this book on modelling for urban drainage. This chapter (Chapter 20) deals with modelling within engineering practice and research that is based on established representations of hydrological, hydraulic and water quality processes. Chapter 21 covers innovative approaches to modelling.

There are three stages to the current chapter. We begin by considering *what models are for*. Section 20.1 is an introduction to models in urban drainage engineering, and Section 20.2 considers urban drainage models in context. The processes we are aiming to represent are summarised in Section 20.3 for flow modelling, and in Section 20.4 for quality. We then consider *how models are used*. In Section 20.5, we look at how the type and complexity of models should be matched to the aims of the study. Section 20.6 concentrates on setting up and validating flow models, and Section 20.7 concentrates on quality models. We finish by looking at *how models work*, at how they can actually represent the physical processes involved. Section 20.8 considers modelling of flow in sewers, Section 20.9 modelling of SuDS, Section 20.10 modelling of flooding, and Section 20.11 modelling of quality. We do not go beyond illustrating some basic principles. It would not be realistic to cover the theory underpinning all available software, and in any case, every commercial package is different and most are changing rapidly.

20.1.1 Use of models

The purpose of models in urban drainage engineering is to represent the system and its response to different conditions in order to answer questions about it, such as why it does or does not perform as expected now or in the future, and how to improve it. In this context, the system could be a conventional sewered system, a sewered system that included SuDS or a drainage system based on SuDS only.

Urban drainage modelling software has two distinct uses: for *design* (of new and retrofitted systems) and for *simulation* (of existing systems). In design, the physical details of a proposed drainage system are determined so the system will behave satisfactorily when exposed to specific conditions. Design software effectively provides automation of some of the design processes defined in Chapters 9, 10, 11, 13, and 14, typically combined with land imagery, 3D visualisation of system and performance, design process management, CAD, and costing.

In a simulation, the physical details of the system already exist, and the user is interested in how the system responds to particular conditions (in terms of flow rate and depth, the extent of surcharge and surface flooding, and scale and impact of pollution). The aim is usually to investigate how the system performs, understand the causes, and find out if the system needs to be improved. The model acts as a planning aid. Most of this chapter focuses on the use of modelling software for simulation.

DOI: 10.1201/9781003408635-20

Apart from hydrological, hydraulic, and quality aspects, some urban drainage modelling packages include the ability to replicate real-time interventions and operational rules and receive live data streams (as considered in Section 22.4). They have versatile data interfaces with geographical information system (GIS) compatibility, graphical representation of the system and its response to storms, and the potential to interact with design tools in an automated manner.

20.1.2 Model types

Physically based computer simulation models are based on accepted mathematical relationships between physical parameters, and all involve some element of simplification. No model covers every raindrop or every variation in the catchment surface. In deterministic models, one combination of input data will always give the same output, and randomness is not accounted for. The fact that these models include simplifications and ignore random effects, combined with the uncertainty associated with input data and with field measurements, means that it would be a reckless or naive modeller who stated, "the results predict the exact behaviour of the system." That modeller would be far wiser to recommend the results as being *indicative*.

For most sewer simulations that are concerned with hydrological and hydraulic aspects, deterministic models have become the standard tools. Where necessary, sensitivity analysis is undertaken to explore the impact of uncertainty in model parameters. This is explored further in Section 21.7.

Various other approaches also exist to model urban drainage systems. These include empirical, neural network, conceptual, and stochastic methods. These are discussed in further detail in Section 21.2.

20.1.3 Modelling aims

In many parts of this chapter, we give separate coverage to the aspects of urban drainage modelling that relate specifically to flow in pipes, to SuDS, to surface flooding and to quality. However, in many actual catchment studies, the modelling of these components is *integrated*. This integration is also likely to include conditions in the *receiving waters* (e.g., river flow). Clearly, outflows from a drainage system will affect flow in the receiving water, and water levels in the receiving water will affect flow in the drainage system and the likelihood of surface flooding. The main commercially available simulation packages offer fully integrated modelling of flow in the drainage system and in the receiving waters.

Levels of complexity in drainage models are defined and discussed in Section 20.5. In all modelling work, the approach must be suited to the aims. The level of complexity should be that needed to achieve the aims of the study, and no higher. For example, this may involve balancing computational time or survey demands with the level of detail specified for the model.

20.2 URBAN DRAINAGE MODELS IN CONTEXT

20.2.1 The modeller

Although sewer system modelling involves the application of mathematical methods, most commentators stress the fact that success is heavily reliant on the skill, experience, and judgement – even intuition – of the human beings who set up and run the models.

Osborne et al. (1996), comparing US- and UK-based runoff models and software in a large-scale application, comment that "the two approaches turn out to be similar and a good engineer can get sensible results using either approach. However an inexperienced engineer will probably get bad results using either approach." A German comparison of sewer flow and quality models (Russ, 1999) concluded that "the reliability and achievable accuracy depend more on the user's qualification, experience and care than on the performance of the model."

While there is software available that can fit variables to measured data, how the model will be used and its purpose require the modeller to fully understand the implications of their decisions that can be detrimental to its application. First, this is relevant when small storm events are used to verify or calibrate a system without realising the consequence of decisions made when larger events are simulated and some predictions become unreliable. Second, when understanding the cause of a problem and how to address it, a model auto-calibrated without justification of the variables linked to the reality within a catchment area, limits or prevents the modeller from achieving the model's purpose.

20.2.2 Confidence in the model

Models are built or updated to meet specific objectives. Whatever the purpose is, it is important to understand the level of confidence in the model to be able to replicate the system's performance. A model may not be built to the same level of detail across the whole of the catchment. This may result in differing levels of confidence across the model. Confidence can be assessed by reference to the data used to build the model, particularly in the data used to validate it (see Section 20.6.4), such as measured from short-term flow surveys, long-term data records, and historical flooding records.

Water service providers have many drainage systems modelled to varying levels of detail and for various purposes. It is common that an existing model may be used for a purpose different to that for which it was originally built. Therefore, assessing whether it is appropriate to use the model, the level of confidence the modeller has in its use, and whether further upgrades are necessary is a critical step.

20.2.3 Model use

Models are built, maintained, and upgraded for numerous purposes. They may be used by a single stakeholder or by multiple stakeholders when working in partnership. They can be broken down into the following four broad categories.

- Understanding system performance now and in the future: this is typically carried out either to understand a particular problem or to predict what may occur in the future and therefore plan for it. Some models form part of wider understanding and planning, as discussed in Chapter 19.
- Developing (capital) solutions to solve a problem (based on the understanding of the system performance): here, the model can be used to identify thresholds or triggers for when an intervention becomes necessary, and test alternative approaches to solving or managing a given need. It is a tool in the design development process and is best used when integrated with design and not viewed as a separate function.
- Developing and testing operational and maintenance solutions: it is becoming more common to improve drainage system performance through targeted and timely operational interventions. A model enables the testing of the solution and importantly indicates what may happen if such interventions are not put in place.
- Forecasting and real-time "live" running to support and make dynamic system interventions: models can be run in real time and with forecast data to enable a "what if" to be determined, and dynamic decisions to be taken, such as when to turn on critical pumps or hold flows back in a drainage network.

The Chartered Institution of Water and Environmental Management's *Code of Practice for the Hydraulic Modelling of Urban Drainage Systems* (CIWEM, 2017) and *Integrated Urban Drainage Modelling Guide* (CIWEM, 2021) provide a comprehensive overview of the use of models.

20.3 ELEMENTS OF URBAN DRAINAGE MODELS

20.3.1 Overview of the components

A physically based, deterministic model of sewer flow must represent the *inputs* (rainfall, infiltration, and wastewater flow) and convert them into the information that is needed: flow rate and depth within the system and at its outlets (e.g., at overflows or through flooding). These outlets can become inputs into other models (e.g., assessing the impact of overflow discharges). The model carries out this conversion by representing (mathematically) the main physical processes that take place. The model must, therefore, be reasonably comprehensive: we could not expect to leave out an important process and still produce accurate results. In order to represent processes in a physically based mathematical form, a good level of scientific knowledge about the processes is needed. Therefore, urban drainage flow models are based on a body of research information about runoff, pipe flow, and so on. However, as has already been pointed out, the model is also bound to have varying levels of detail with some elements of simplification.

At a very general level, therefore, three factors greatly influence the accuracy, confidence, and usefulness of the simulations by a particular physically based modelling package: the comprehensiveness of the model, the reliability and completeness of the scientific knowledge on which it is based, and the appropriateness of the simplifications it contains.

The word *model* tends to be used in a number of ways. We are referring throughout to mathematical, computer-based models. The mathematical representation of each process can be termed a *model*, as can the combination of all the processes (into a software package). However, these models are simply tools ready to do a job: simulation of a flow in a particular catchment. To do this job, a great deal of data is required about, for example, the surfaces of the catchment and the network of sewers. These data can improve the confidence in the simulations by reducing the uncertainty (often through physical surveys). The specific application of a software package to a particular catchment requires great effort in checking, calibration, and verification, as considered later.

Typically, a model of the minor drainage system contains *links* and *nodes*. The links are generally pipes, in which the model must represent the relationship between the main hydraulic properties: cross-sectional size, gradient, roughness, flow rate, and depth (links are also used to represent pumps and other features). The nodes are generally manholes, at which there may be additional head losses and changes of level (and are also used to represent storage). In addition to these primary building blocks, the package must also be capable of simulating conditions in more specialised ancillary structures: including tanks and combined sewer overflows (CSOs).

The emphasis throughout is on unsteady conditions: on variations of flow rate and depth with time. A crucial element in a flow-modelling package is the way in which it simulates these unsteady conditions. Common methods are presented in Section 20.8.

We now consider the main physical processes that must be represented in a software package for flow modelling.

20.3.2 Rainfall

The model will be used to find the response of the catchment and the drainage system to particular rainfall patterns. Straightforward examples would be simple constant rainfall or, more realistically, rainfall with a particular storm profile (variation of rain intensity with time). This would be generated for a specified return period using the types of relationship between intensity, duration, and frequency considered in Section 4.3. To model the operation of CSOs, or storage facilities that need to be emptied during dry periods, a typical sequence of wet and dry periods would be studied using time series rainfall stochastically generated for the specific location (see Section 4.5) or observed using rain gauge and/or radar data. Spatial variation of rainfall is an important consideration in large catchments. During model verification (described more fully in Section 20.6.4), rain gauge records or radar rainfall data are used as rainfall input to the model, and the flow simulations by the software package are compared with flows measured in the system.

Issues relating to climate change have been considered in Section 4.6.

20.3.3 Rainfall to runoff

The conversion of rainfall into "runoff," water destined to find its way into the sewer system, is a highly complex process. There are many reasons for rainwater not to become stormwater in the sewer. It may, for example, soak into the ground (even on "impervious" surfaces, via cracks), may form puddles and later evaporate, or may be caught in the leaves of a tree. There is an obvious distinction between water that falls on a roof or a road and that which falls in an undrained garden; but where, for example, there is a grass strip beside a pavement, some water falling on the grass may run off onto the pavement and enter the sewer, whereas some water falling on the pavement may run onto the grass and infiltrate. Once the ground is saturated though, it can become runoff again and enter the drainage system. The methods of representing these processes have been considered in Section 5.2.

20.3.4 Overland flow direct from runoff

The two main considerations here are the amount of rainwater that will enter the sewer and how much time it will take to get there. The amount is typically dealt with in the conversion to runoff. Clearly, the extent to which water entering from one area will overlap with that entering from another will have a significant effect on the way the flow in the sewer builds up with time. Again, the physical processes are highly complex, with many ways in which surface irregularities can affect the flow. Overland flow is usually represented in a simplified form, and methods have been given in Section 5.3. We refer here to the initial flow of runoff. Flow over the surface in flood conditions is considered in Section 20.3.7.

20.3.5 Dry weather flow

In a combined sewer system, the stormwater joins the flow of wastewater in the sewer. Realistic simulation of wastewater generation is an important function in a flow-modelling package and can form a significant proportion of the flow in larger catchments. Methods are presented in Sections 3.2 and 3.3 outlining the wastewater components.

20.3.6 Infiltration

Another significant component of dry weather flow is infiltration, that is ingress of water that finds its way from the soil into the drainage system. In piped systems, this is typically through cracks and holes formed where the fabric has deteriorated (either over time or through

Table 20.1 Detail on the elements of a flow model

Topic	Chapter/Section
Rainfall	4
Rainfall to runoff	5.2
Overland flow direct from runoff	5.3
Dry weather flow	3.2, 3.3, 9.3.2
Infiltration	3.4, 9.3.4
Surface flooding	12, 20.3.7

damage). Modelling often considers different types of infiltration that may enter the drainage system in response to rainfall at different times. The largest tends to be groundwater infiltration, which occurs as a result of longer-term rainfall patterns that increase the groundwater level to a point where it can enter the drainage system. In coastal regions, this may also have tidal influences and may show a tidal pattern. Non-dry weather infiltration may occur as water percolates through the surface during and shortly after rainfall and finds its way into the drainage system. However, this is hard to distinguish from runoff flows.

Further details on infiltration and its representation have been given in Section 3.4. Validation is only really possible with short- and long-term monitoring, with an emphasis on the long term. Care is needed when applying these models to ensure extrapolation during large events does not yield unreliable results.

20.3.7 Surface flooding

The five main processes described above must be included in a software package for flow modelling when the capacity of the (minor) sewer system is not exceeded. When the capacity *is* exceeded, a sixth element may become very important in modelling the hydraulic response of a drainage system: surface flooding and flood routing.

The main concepts of surface water flooding are introduced in Chapter 12, and approaches to including surface flooding and routing in urban drainage models are described in Section 20.10.

The output of this modelling effort generally comes in the form of simulated variations of flow rate and depth with time at chosen points within the sewer system and at outlets from it. These may be manholes or where greater detail is added, gullies. There is usually particular interest in the ability of the sewer system to cope with the simulated flows, and thus on the extent of possible pipe surcharge or surface flooding.

Table 20.1 summarises the parts of this text that contain more detail on each of the elements of a flow model.

20.4 WATER QUALITY MODELLING

Sewer systems can be direct contributors of pollution to waterbodies. Stormwater outfalls can discharge pollution, as discussed in Section 5.4; serious short-term pollution can arise from CSOs, as considered in Section 2.5 and Section 13.3. The effects can be counteracted by approaches such as reducing the amount of stormwater entering the system, building additional storage within the system, rehabilitating or replacing CSO structures, or treating the discharge. In order to design these measures to achieve high standards, and to evaluate them fully, it is necessary to have information on the rate at which pollutants flow into the structures from the sewer system, and how the pollutants are distributed in the dry weather and storm flow. This information is provided by models of *quality* (distribution of pollutants) in sewer flow.

Physically based deterministic quality models are now widely available in practice, but it should be recognised that the accuracy of such models is not as high as flow models because of the complex physical, chemical, and biological processes involved. Willems (2008), for example, found that the uncertainty in the results contributed by the water quality part of a sewer model was an order of magnitude higher than that for the flow component. That said, with careful and expert calibration against measured data, these models can be a powerful tool to estimate sewer impacts on receiving waters and so develop appropriate solutions.

This section discusses some of the requirements that are common to all methods of sewer quality modelling.

20.4.1 The processes to be modelled

The main aim of a sewer-quality model is to simulate the variation of concentration of pollutants with time at chosen points in a sewer system. These simulations will be used to understand the impact on the receiving waters and where needed improve the performance of the system, for example, by aiding the design of CSOs to enhance the retention of pollutants in the sewer system. It is also common to monitor and model the receiving waterbody, as the individual discharge alone does not demonstrate its impact.

The main quality parameters, modelled separately (directly or applied as standard concentrations), will be the standard determinands described in Section 2.3, including suspended solids, oxygen demand (biochemical oxygen demand [BOD] or chemical oxygen demand [COD]), ammonia, dissolved oxygen (DO), and others depending on the model.

Pollutants find their way into combined sewers from two main sources: wastewater and the catchment surface. Once in the system, the material may move unchanged with the flow in the sewers, be transformed, or become deposited. Deposited pollutants may subsequently be re-entrained or eroded, usually in response to an increase in flow.

The main elements of the system that influence the quality of the sewer flow are indicated schematically in Figure 20.1 (for a combined sewer).

20.4.2 Wastewater inflow

The flow rate and the concentration of pollutants vary with time in dry weather flow in a fairly repeatable daily (diurnal) pattern (see Sections 3.2.3 and 9.3.1). The variation is related to patterns of human behaviour, and there tend to be differences between weekdays and weekends. Industrial and commercial flows may be present, in addition to domestic contributions, and these can lead to significant differences in pollutant load and profile. Infiltration can be a significant fraction of dry weather flow (Sections 3.4 and 9.3.4) and tends to dilute the wastewater.

20.4.3 Catchment surface

In periods of dry weather, there is a buildup of pollutants on roads, roofs, and so on, and these are washed into the drainage system by the next rainfall (Section 5.4.2). In general terms, the amount entering the drainage system depends on the quantity of material accumulated on the catchment surface, intensity of the rain, intervention points to catch pollutants (such as gullies), level of maintenance, and nature of the overland flow.

20.4.4 Gully pots

As with catchment surfaces, pollutants tend to build up in gully pots in dry weather. Material remaining in the liquor is regularly washed into the drainage system, even by minor storms. Only high-return-period storms are thought to disturb previously deposited, heavier solids.

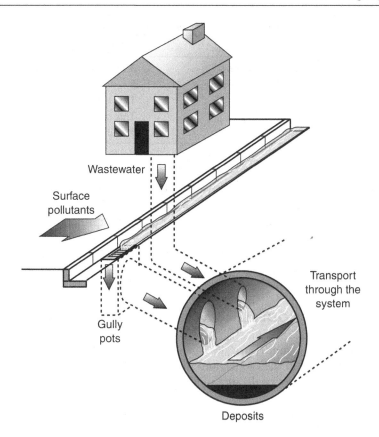

Figure 20.1 Main elements that influence sewer-flow quality (combined sewer).

20.4.5 Transport through the system

Once they have been carried into the pipe system, and provided they are not attached to solids that are deposited, pollutants are transported by the moving liquid. It is generally assumed that the pollutants move at the mean liquid velocity, though there are cases when this might be inappropriate, as considered later.

20.4.6 Pipe and tank deposits

At low flows in sewers, especially during the night, solids (and pollutants attached to them) may settle and form deposits. At higher dry weather flows or during storms, they may be re-eroded, releasing suspended solids and dissolved pollutants into the flow. The resulting increase in pollutant concentration depends on the characteristics of the system, of the catchment, of the dry weather and storm flows, and the antecedent dry period. As is clear from Chapter 17, the deposition and erosion of sediments is a complex subject, with many different types of sediment and associated pollutants. Flow patterns in tanks and large trunk sewers also lead to deposition, and the subsequent erosion can have a significant effect on quality.

The mechanisms by which pollutants are introduced to the flow, and are subsequently transported with it, are heavily dependent on hydraulic conditions. Therefore, physically based deterministic quality models need to be based on hydraulic models of the sewer system. The main deterministic quality models in current use are based on established flow models. The accuracy of quality simulations is strongly affected by the accuracy of the hydraulic simulation on which they were based.

Table 20.2 Chapters relevant to quality modelling

Topic	Chapter/Section
Wastewater inflow	3
Catchment surface	5
Sediment	17
Flow modelling	This chapter
First foul flush	13.3.3

In some systems, a significant feature in the variation of sewer-flow quality with time is the first flush in early storm flows, which may contain particularly high pollutant loads. This effect is considered in Section 13.3.2. It is important that it is represented in a model of sewer-flow quality.

Table 20.2 lists the parts of this book that deal with the processes that should be represented in a sewer quality model.

20.5 MATCHING THE MODELLING APPROACH TO THE AIMS

The various purposes of a model are described in Section 20.2.3. The purpose relates to the type of detail required in the model, and ultimately to the cost of the exercise. CIWEM (2017) defines three categories of detail a model may have

- *Limited detail/simplified*: used to provide a limited indication of performance and generate flows for other parts of the model
- *Planning/general purpose*: used to understand comparative risk across an area
- *High level of detail*: appropriate where detailed investigations and design are required

It is not only the level of detail that is important, but also the quality of the data that sit behind this, and the level of data validation and collection that help to improve the confidence in any model output.

One level of complexity relates to modelling in one or two dimensions. A one-dimensional model represents system performance longitudinally: along a sewer or watercourse. A two-dimensional model also represents performance laterally. CIWEM (2021, p12) identifies the following categories.

- 1D: one dimension (for example a sewer and/or watercourse model)
- 2D: two dimensions (for example a pluvial runoff and overland flow model)
- 1D–2D: a coupled one-dimension and two-dimension model (for example with sewers and watercourses modelled in 1D but coupled with a 2D mesh to model overland flow)

This is covered in more detail in Section 20.10.3.

It is likely that the aims of a particular modelling exercise have been derived from the planning work that has been carried out, as considered in Chapter 19. Section 19.2.2 has discussed the critical decisions involved and how they affect the modelling of the system.

Of course, good drainage planning is not just about modelling. Indeed, it could be argued that there is often a case for using modelling *less* and giving more emphasis to understanding the results and using them to make informed decisions in order to bring genuine benefits.

20.6 SETTING UP AND VALIDATING A SYSTEM MODEL

20.6.1 Input data

A flow model requires a physical definition of the drainage system (at the level of detail defined above), information about the catchment for runoff calculations, and information about specific inflows. In a conventional system in which pipes run between manholes, this generally requires the type of data listed in Table 20.3. Particular attention is given to ancillaries (CSOs, pumping stations, syphons, etc.) as they form controls in the network and require representation in the model. All manholes and pipes have reference codes (see Section 6.3.1).

If a package is used for design, the input data are for the proposed system, and the model simulates the response of that system to specified rainfall conditions.

If a package is used to model an existing drainage system, the input data are that which exists for the system. If the system has poor records, it may be necessary for extensive sewer survey work to be carried out. This is expensive, but good sewer records are of great value, not just for modelling. Significant effort may be needed to define the catchment area data, especially where there is incomplete information on the connection of particular properties or sub-areas to the sewer system. Data are typically held and processed within GISs.

The level of detail implied in Table 20.3, combined with the discussion of appropriate levels of complexity in Section 20.5, highlights the potential danger of overcomplicating models. There may sometimes be a need to simplify components to focus, in effect, on representing the hydraulics, not the data.

The model simulates the response of the system (represented by the previously provided data) to specified rainfall patterns. The rainfall data may take a number of forms:

- For verification or calibration, using rain gauge or rainfall radar records
- For simulation, analysis, and design, sets of synthetic storms (Section 4.4), and or time-series rainfall (Section 4.5).

20.6.2 Model testing

Initial checks on a catchment model are needed to make sure that the model is behaving satisfactorily in mathematical terms and to eliminate obvious mistakes in the input data. It is necessary to check against instability for a variety of extreme conditions. In addition, the overall volume entering the system, or part of the system, should be compared with the overall volume leaving.

20.6.3 Flow surveys

To create a successful working model of an existing system, data are needed on how the catchment actually responds to particular rainfall conditions. Flow monitoring gives records of the hydraulic performance of the system, and the conditions (mainly rainfall) that produced that performance. If that rainfall is used as input to the model, a comparison between the resulting simulation and the flow survey data should tell us how much confidence we have in our model.

The various approaches to rainfall measurement are covered in Section 4.2. The flow data may come from short-term surveys or from long-term monitoring.

In short-term surveys, in-sewer measurement is carried out at selected sites, and an appropriate selection of sites is essential. There are a number of considerations. The number of sites must suit the purpose of the model to enable the required level of confidence to be

Table 20.3 Data requirements for flow models

Element	Data type
Manholes	Reference number
	Location (map reference)
	Ground level
	Storage volume
	Head-loss parameter
Pipes	Reference number
	Connectivity information
	Length
	Shape
	Size
	Roughness
	Invert level
Catchment	Total area contributing
	Pervious/impervious areas
	Slope of ground
	Soil data
	Flooded areas
Gullies	Location
	Connectivity
Inflows	Population-derived dry weather flow
	Infiltration
	Industrial inflows/trade effluent
	Overland flows
	Land drainage
SuDS features	Location
	Characteristics
	Connection to sewers
Ancillary structures	Location
	Data to define hydraulic performance
CSOs	Geometry
Storage tanks	Inflow/outflow arrangements
	Screens
	Geometry
Pumping stations	Trigger and control levels
	Pump characteristics
Outfalls	Hydraulic characteristics
Receiving waters	Boundary conditions, for example, water levels, tidal influence
	Location
	Type, for example, freshwater, marine, bathing water
	Relevant output from the watercourse model (within an integrated model)

achieved; their location must allow comparisons with the simulation at locations of particular interest. Critically though, they must also be appropriate from a hydraulic and a practical point of view: with safe and adequate access, and without excessive disruption to the flow, or deposited sediment. Further details can be found in Section 18.4.1.

Long-term or permanent monitoring takes place at the wastewater treatment plant (WTP) and sometimes at ancillaries (CSOs and pumping stations). In addition, some water companies have fitted thousands of network sensors (measuring depth) within the sewer system connected via telemetry (see Section 18.4.3). These data provide a more complete picture of the catchment response over time to rainfall across a range of conditions, seasonally and annually. Whereas short-term flow surveys may use a higher density of rain gauges, long-term monitoring typically requires good rainfall radar supported with a smaller number of rain gauges to enable validation.

In practice, there may be no shortage of in-sewer data in terms of the *quantity* of data; but *quality* of data must remain the priority. This relates to the discussion of model complexity in Section 20.6.1.

20.6.4 Model calibration/verification against measured flow data

It is useful to reflect on some of the terminology used during the model development and building process, particularly the distinction between the terms *calibration* and *verification*. During the calibration of a model, the most appropriate values of the various model parameters are sought. In verification, the model parameters and their values have now been established, and input is run through the model to produce results that are compared with known conditions (e.g., flow survey data). In this way, the agreement between computed and observed values can be verified. The verification is, strictly speaking, only valid for that particular location and only over the range of the available data.

In theory, if a model is deterministic, it does not need calibration. If all the input parameter values are accurate and the physics of the processes are simulated sufficiently well, accurate results should be produced without calibration. In practice, however, many of the input parameters are not or cannot be accurately ascertained and the physics is only approximated, thereby making it necessary to resort to default values, which may not be representative of the site in question.

Price and Osborne (1986) defined verification as the art of demonstrating that a model, which incorporates previously calibrated sub-models, correctly represents the reality of the particular system being studied. They see calibration as occurring only during model development, whereby the physical phenomena represented by the various sub-models are tested, under varied conditions, on many catchments to ensure approximation to observed data with sufficient accuracy. Verification is then carried out by the model user (the drainage engineer), primarily to demonstrate the physical details of the sewer system are correctly incorporated in the model. Figure 20.2 is a plot of actual and simulated hydrographs produced during verification of a flow model.

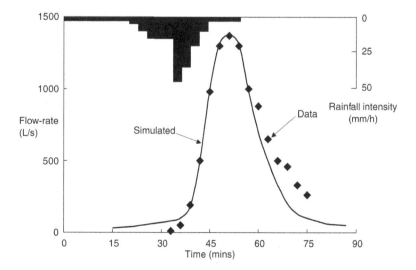

Figure 20.2 Model verification: hydrograph.

When calibration as defined above is carried out, Vojinovic and Solomatine (2006) identify a number of possible approaches. It may be a trial-and-error process in which parameters are adjusted "according to the modeller's judgement" (and should be based on data/other evidence relating to fundamental principles and mechanisms, not just to get a "better fit with observed data") in order to provide a better match with observed data. Alternatively, this adjustment process may be built into the modelling software (see also Wangwongwiraj et al., 2004), or the software may be based on a number of different models whose outputs can be weighted to improve the match. Comparing these methods with each other and with what we have called verification, Vojinovic and Solomatine (2006) found that all of the approaches will "incur a degree of uncertainty." They warned that "making decisions on a single model output without the consideration of uncertainties and risk involved is very dangerous and misleading." In particular, changing values without understanding the "engineering consequences" can result in a tool that cannot be readily used to understand problems or test certain types of solutions.

Further guidance on the practice of verifying a model against measured data and historical observations is provided by the CIWEM Urban Drainage Group's *Code of Practice for the Hydraulic Modelling of Urban Drainage Systems* (CIWEM, 2017). Calibration of integrated urban wastewater system models is covered in Section 21.6 in the next chapter.

20.6.5 Evaluating confidence

Evaluating model confidence provides a measure of the suitability of using the model against its intended purpose. This evaluation can be through various ways and ranges from simple to complex and qualitative to quantitative. CIWEM (2017) considers the confidence in various aspects of the model components, recognising their importance and contribution to defining the quality of the model. This includes

- Asset data – considers where the data come from, how much has been surveyed/confirmed, and the quality of the survey method.
- Sub-catchment – relates to the areas defined including impermeable area, method of area assessment and amount of data collected, confirmation of populations, and input data.
- Flow survey – considers the quality and extent of the flow survey and over what period. Includes both short-term, long-term, and permanent survey collection.
- Dry weather verification – considers the flow parameter predictions within an upper and lower bound tolerance.
- Storm verification – concerns the ability to represent the flow, depths, volumes, shape, and timing of flows on individual and full period events.
- Historical verification – considers the performance of the network against known flooding and longer-term data sets (such as monitors at works, pumping stations, and event duration monitors at overflows)

The CIWEM guidance also provides a framework to enable model confidence to be assessed, and those who commission the building or the use of a model can use it to meet their own defined needs.

20.6.6 Documentation

Setting up and developing a sewer system model is an investment in the gathering of information. Some of the benefits may be physical, in the form of improved system performance as a result of successful rehabilitation work. However, many of the benefits remain as information: data on the properties of the system, understanding of its response to particular conditions, and the potential to predict its performance under new conditions. The value of this information is related to the quality and completeness of the documentation of all stages of model development. The conventions used in the documentation depend on the type of model and on the standard practice within the organisation carrying out the modelling.

20.7 USING WATER QUALITY MODELS

20.7.1 Model applications

Water quality models are needed when detailed knowledge is required of the pollutants entering, moving through, and (crucially) leaving the sewer system. The main applications for quality models are therefore studies of

- Intermittent and continuous discharges to receiving waters (river, estuary, marine)
- Loads discharged to WTPs
- Sediment transport and deposition in sewer systems

Of these three, the first application is the most common, and this is often carried out in association with a model representation of the receiving water. Therefore, it is appropriate to consider the two components together, where similar processes take place and similar methods apply.

20.7.2 Types of sewer quality models

The *UPM Manual 3* (FWR, 2019) recommends the use of two types of sewer quality models, detailed and simple, depending on the application.

Detailed models are deemed to be appropriate where the network is complex, flat, there is interaction with the WTP, or there are significant trade discharges. A verified flow *and* water quality model is used, typically with default values for water quality parameters. Here, event mean spill concentrations may be applied to the discharges or time-varying concentrations depending on the receiving waterbody modelling approach. Simple models are normally only appropriate in small catchments with little interaction, and flows are controlled, enabling representation of the system through a tank model (Section 14.4) or simplified sewer model.

20.7.3 Types and complexity of river quality models

Four levels of model complexity are proposed in the *Storm Overflow Assessment Framework* (Environment Agency, 2018). The choice of level is derived from the complexity of the problem. The appropriate river model is defined as follows.

Level 1 – Statistical distribution used to represent upstream river flow; hydraulics modelled as a simplified channel with uniform steady conditions; simplified water quality model representing the main oxygen demand processes using default values

Level 2 – River flow time series used to represent upstream river flow, otherwise as level 1

Level 3 – Calibrated flow routing model, with representation of advective pollutant transport

Level 4 – Calibrated hydrodynamic model, with calibrated advection/dispersion model

20.7.4 Planning an integrated quality study

An integrated study of the drainage system and receiving waters provides a greater understanding of the causes of pollution and how to address it. Before collecting data and creating models, the following components should be considered. A river context is used as a common example.

20.7.4.1 Drivers

The initial step is to understand why the study is being completed. For example, is there already a known issue and river quality target to achieve under the WFD or is the failure not understood? Is it high-level planning or a scheme to be approved by the regulator? These considerations drive the level of data collection, study approach, and model detail required.

20.7.4.2 Catchment boundary

The catchment boundary needs to be derived to establish river flows. The same is undertaken for the urban drainage catchment to understand potential interactions.

20.7.4.3 Waterbodies

To determine the level of modelling and calibration effort, it is important to understand which waterbodies require assessment. For example, it may be a whole river network or just a downstream reach. Key locations to identify for data collection and calibration purposes will be where there are permanent sampling locations and downstream of any thought to be key discharges.

20.7.4.4 River modelling approach

The approach to river flow modelling in part depends on the accuracy required but predominantly depends on the physical characteristics of the watercourses. Most studies tend to be in free-flowing reaches, and so a routing model will suffice. If there are backwater effects or slow-flowing water, then it would be appropriate to use a 1D approach. For particularly deep or wide waterbodies, then it may be necessary to use a 2D or 3D approach to capture the effects of stratification or streaming. It is possible to use a combination of approaches.

20.7.4.5 Point source discharges

Reviewing sewer models and GIS data will help CSOs, WTPs, and surface water outfalls to be identified. These can be compared to the river catchment boundary to define the extent of upstream modelling required. Typically, a sewer model is used to derive overflow pollutant discharges, whereas external distributions are applied to stormwater and final effluent discharges.

20.7.4.6 Rainfall and evaporation

As a minimum, representative historical rainfall data are needed that cover the calibration and verification periods used in the study. However, hydrological models require multi-year records due to the subsurface processes affecting catchment wetness (see Section 4.5).

During analysis, a 10-year time series is typically used, preferably historical rainfall but synthetic series can also be used. Where the catchment is large or there are changes in elevation, often more than one series are applied.

20.7.5 Input data and model calibration/verification

The concept of verifying a pre-calibrated quality model is a less realistic proposition than for a flow model (as presented earlier in the chapter). Variations and uncertainties are much larger, making it harder to transfer experience or default values from one catchment to another, even though apparently similar. Hence, a major issue is that these models require a great deal of data (either in the sewer or in the river). A typical deterministic quality model requires the type of input data specified in Table 20.4.

The decision to collect water quality data in the sewer and receiving waterbody depends on the nature of the study, the proposed modelling approach, and the outcomes to achieve. In some cases, despite the caveats above, the relative expense of gathering full catchment-specific data means that quality modellers may have to rely on default data.

It is common to consider dry weather flow first in this process. Typically, a starting point is to calibrate against observed WTP influent data to validate or change the default profiles. The simulations are then compared with any measured data, and default values are replaced by amending the inputs to the dry weather pollutants, in particular, from trade discharges, until the model is deemed to be calibrated. Fixed calibration criteria are not typically specified, but the relative magnitude and timing of the observed response should be represented by the model (Wastewater Planning Users Group [WaPUG], 2006). Figure 20.3 shows actual and simulated dry weather flow quality data used in the verification of a model. In the waterbody, comparisons are undertaken and compared against spot samples taken monthly (with a random date and time) and any continuous monitoring by the environmental regulator. This generally provides a dry weather river calibration, typically over a 3-year period, and accounts for the background water quality, final effluent, and direct trade discharges. The process is then applied to storm flow.

Ideally at this stage, a comprehensive sensitivity analysis should be carried out to capture the inherent uncertainties in the process (e.g., Schellart et al., 2008) and to explicitly estimate the statistical validity of the results. In practice, this may not be undertaken due to model run times, the complexity of river models, and financial constraints. This often leads to more conservative default setups being used.

Table 20.4 Input data for a quality model

Wastewater	For a typical day in the week and at the weekend:
	• Variation of flow with time
	• Variation of pollutant concentrations with time
	• Trade discharges
Stormwater	For two or three separate rainfall events:
	• Variation of flow with time
	• Variation of pollutant concentrations with time
Surface pollutants	Land use characteristics
	Particulate characteristics
Deposits	Initial sediment depth
	Characteristics of sediment

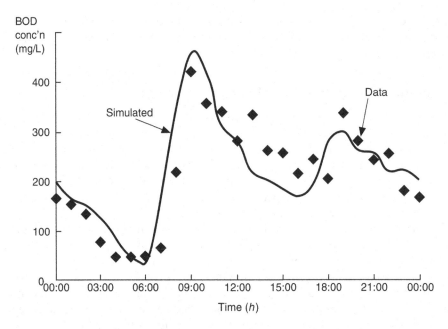

Figure 20.3 Model verification: pollutograph (dry weather flow).

20.7.6 Data collection in sewers and receiving waters

Taking samples from the flow at particular locations in the system and analysing them for pollutant concentrations is an important operation in the development of quality models for specific systems. Samples from the dry weather flow may be needed to provide catchment-specific input data, and samples from both dry weather and storm flow are needed for calibration/verification.

Other catchment-specific data collection will normally be needed. This may include taking samples of particulate material on the catchment surface and measuring the depth and properties of the sediment layer in pipes. Most importantly, though, it is often the data collected in the receiving waterbody that enable model validation through physical sampling and continuous monitoring (e.g., sondes), such as for dissolved oxygen and temperature for dry and wet weather. A high-frequency spot sample may also be undertaken to capture the diurnal effects in dry weather and the effect of CSOs and surface water discharges in storms. In the UK, permanent in-river sampling stations are currently being rolled out to monitor receiving water quality upstream and downstream of key discharge points (see Section 18.4.2).

There is guidance on the collection of field data to support quality models in the *UPM Manual 3* (FWR, 2019) and the *Guide to the Quality Modelling of Sewer Systems* (WaPUG, 2006).

20.8 MODELLING FLOW IN SEWERS

The remaining sections of this chapter deal with ways that physical processes can be represented in models. The aim is to provide a fuller understanding of how models do their job. Of course, engineers involved in drainage modelling may not feel the need to have detailed knowledge of the underpinning representations of physical processes in the software they use. However, we hope that those seeking a fuller understanding of how models work will find this material useful.

We present, as illustrations, some basic principles relevant to drainage modelling. We are not aiming to explain how all models work or to cover all aspects of relevant theory.

20.8.1 Representing unsteady flow

Wastewater flow varies with the time of day, and during storm conditions, inflow to the sewer system can vary dramatically with time. The representation of unsteady (time-varying) flow is an important component in a sewer-flow software package.

In unsteady flow in a part-full pipe, there is a far more complex relationship between depth and flow rate than there is in steady flow (described in Section 7.4). Also, as a storm wave moves through a sewer system, it *attenuates* (spreads out and the peak reduces) and it *translates* (moves along). The relationship between flow rate (or depth) and time cannot be accurately predicted without taking this effect into account. In addition, accurate simulation of unsteady flow may save on the over-design that might result from assuming that waves did not change shape. (If you are used to associating the word *wave* with an effect that lasts for a few seconds, remember that increases in flow in sewer systems resulting from rainfall, that is, storm waves, may last many hours.)

There are a number of methods available for analysing unsteady conditions in a sewer system. Some are based on approximations, and others on attempts to give a full theoretical treatment of the physics of the flow. The main methods are for free surface flows. Adjustments for surcharged pipes are considered later.

20.8.2 The Saint-Venant equations

For gradually varied unsteady flow in open channels (including part-full pipes), the full one-dimensional theoretical treatment leads to a pair of equations usually referred to as "the Saint-Venant equations" after A.J.C. Barré de Saint-Venant, who first published them in the middle of the nineteenth century. A clear derivation of these equations is available in Chow (1959).

There are two equations: a dynamic equation and a continuity equation. The dynamic equation can be written as follows:

$$S_f = S_o - \frac{\partial y}{\partial x} - \frac{v}{g}\cdot\frac{\partial v}{\partial x} - \frac{1}{g}\cdot\frac{\partial v}{\partial t}$$

$$\underset{1}{\rightarrow}|$$

$$\underset{2}{\rightarrow}|$$ (20.1)

$$\underset{3}{\rightarrow}|$$

where y is the flow depth (m), v is the velocity (m/s), x is the distance (m), t is the time (s), S_o is the bed slope (–), and S_f is the friction slope (–).

In this form, the components that make up the equation can be identified. The part marked "1" above includes no variations with distance or time and applies to uniform steady conditions. Taking the terms up to "2" includes variations with distance but not with time, and applies to nonuniform steady conditions. The whole equation "3" also includes variation with time and applies to nonuniform unsteady conditions (the ones that interest us here).

Equation ((20.1)) is commonly presented in terms of flow rate rather than velocity and is given together with the continuity equation:

$$\frac{\partial Q}{\partial t} + \frac{\partial}{\partial x}\left(\frac{Q^2}{A}\right) + gA\frac{\partial A}{\partial x} - gA\left(S_o - S_f\right) = 0 \tag{20.2}$$

$$B\frac{\partial y}{\partial t} + \frac{\partial Q}{\partial x} = 0 \tag{20.3}$$

where Q is the flow rate (m³/s), A is the area of the flow cross section (m²), and B is the water surface width (m).

The Saint-Venant equations are valid in situations where the following assumptions are appropriate:

- The pressure distribution is hydrostatic.
- The sewer bed slope is so small that the flow depth measured vertically is almost the same as that normal to the bed.
- The velocity distribution at a channel cross section is uniform.
- The channel is prismatic.
- Friction losses estimated by steady flow equations (see Section 7.3.2) are valid in unsteady flow.
- Lateral flow is negligible.

20.8.3 Simplifications of the full equations

Equation ((20.1)), the dynamic equation, includes terms for the bed slope, the friction slope, the variation of water depth, and the variation of flow rate with distance and time. Some of these terms may be more significant than others, giving opportunities for simplifying the equations.

The greatest simplification is to assume that most of the terms in Equation ((20.1)) can be ignored and reduce it simply to

$$S_o = S_f \tag{20.4}$$

This is the equivalent of ignoring all but part 1 of Equation ((20.1)) and implies that the relationship between flow rate and depth is the same as it would be in steady uniform flow (as in Section 7.4).

Combining Equations 20.3 and 20.4 gives

$$\frac{\partial Q}{\partial t} + c\frac{\partial Q}{\partial x} = 0 \tag{20.5}$$

The wave, called a *kinematic wave*, does not attenuate but translates at the wave speed, c.

A less extreme simplification involves ignoring just the variation of flow rate with time (the "unsteady" effects), that is using Equation ((20.1)) up to "2". The equivalent of Equation 20.5 is now

$$\frac{\partial Q}{\partial t} + c\frac{\partial Q}{\partial x} = D\frac{\partial^2 Q}{\partial x^2} \tag{20.6}$$

This *diffusion wave* travels at the same speed, c, as the kinematic wave but, as a result of the term on the right-hand side of the equation, is subject to diffusion. The diffusion coefficient, D, regulates the attenuation of the wave as it propagates downstream. For simplicity, c and D are usually regarded as constant although they vary. Table 20.5 gives the range of hydraulic conditions accounted for in the methods.

20.8.4 Numerical methods of solution

The main equations derived in the last two sections are partial differential equations since Q (or v) and y are functions of both distance (x) and time (t). The most common method of solution is using finite differences, involving dividing distance and time into small, discrete steps. This can be represented on a schematic two-dimensional grid showing x and t together, as in Figure 20.4.

In Figure 20.4, the distance step is shown as Δx and the time step as Δt. The points where the lines intersect are *calculation nodes*. The nodes marked with a circle are the distance nodes for time = 0 (say, the start of the calculation). Flow conditions for these are likely to be known (e.g., baseflow all along the pipe before the beginning of storm flow). The nodes marked with a square are the time nodes for distance = 0 (say, the upstream end of the pipe). Flow conditions for these may also be known (e.g., inflow of stormwater varying with time).

Table 20.5 Hydraulic conditions accounted for by simplifications of wave equations

Accounts for	Kinematic wave (1)	Diffusion wave (2)	Dynamic wave (3)
Wave translation	✓	✓	✓
Backwater	✗	✓	✓
Wave attenuation	✗	✓	✓
Flow acceleration	✗	✗	✓

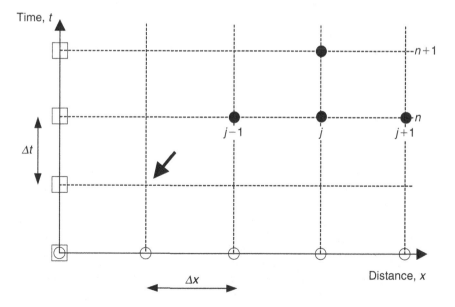

Figure 20.4 Numerical solution of flow equations: x–t grid.

To calculate the conditions at the node marked with the arrow using a kinematic wave model, we can use the known values at neighbouring nodes. Once this calculation is complete, it can be repeated for successive distance nodes (the nodes to the right of the arrow on the same horizontal line). When we have calculated conditions at all distance nodes for that time step, we can proceed to the next time step (the horizontal line above the one we have dealt with).

We have to take into account the *boundary conditions* – the hydraulic conditions at the limits of our system, for example, the flow rate at the inlet and the water level at the pipe outlet.

There are many ways in which finite difference calculations can be carried out. In Figure 20.4 the jth distance node and the nth time node are marked.

Suppose we are solving Equation 20.5:

$$\frac{\partial Q}{\partial t} + c\frac{\partial Q}{\partial x} = 0$$

We write the value of Q at the jth distance node and the nth time node as Q_j^n.

Suppose we have reached the stage where we wish to calculate Q_j^{n+1}. If we assume that Q varies linearly within each distance and time step, we can write

$$\frac{\partial Q}{\partial t} \text{ as } \frac{Q_j^{n+1} - Q_j^n}{\Delta t}$$

and

$$\frac{\partial Q}{\partial x} \text{ as } \frac{Q_{j+1}^n - Q_j^n}{\Delta x} (\text{forward difference})$$

or

$$\frac{Q_j^n - Q_{j-1}^n}{\Delta x} (\text{backward difference})$$

or

$$\frac{Q_{j+1}^n - Q_{j-1}^n}{2\Delta x} (\text{central difference})$$

If we substitute these types of expression into Equation 20.5, it can be rearranged with the unknown, Q_j^{n+1}, by itself on the left-hand side, and solved directly. In solutions where this is the case, the method is described as *explicit*.

There are many different explicit finite difference methods, some incorporating "half-steps," for example, the intermediate calculation of $Q_{j+(1/2)}^{n+(1/2)}$, to improve accuracy and stability.

More complex formulations, in which the unknown value appears on both sides of the finite difference equation, are described as *implicit*. Even though each set of equations is more difficult to solve, implicit methods can be used with longer time steps than explicit methods,

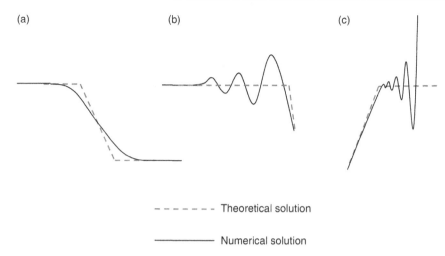

(a)   (b)   (c)

– – – – – – – – Theoretical solution

——————— Numerical solution

Figure 20.5 (a) Numerical diffusion; (b) numerical oscillation; (c) instability.

and therefore often have the advantage of being more computationally efficient. More information on these methods can be found in the texts recommended at the end of this section.

These approaches to numerical solutions can suffer from various anomalies, especially when the input data contain rapid changes. Two common problems are illustrated in Figure 20.5. Figure 20.5a shows *numerical diffusion*, where values are smoothed out in what should be a zone of rapid change. Figure 20.5b shows *numerical oscillation*: small fluctuations at points of change. These problems arise because the method requires variations that are actually continuous to be treated as a series of linear steps.

The problems are overcome by the selection of appropriate solution methods and suitable time and distance steps. Explicit schemes usually have to satisfy the *Courant condition* to maintain stability:

$$\frac{\Delta x}{\Delta t} \geq c \tag{20.7}$$

Figure 20.5c shows instability, in which errors introduced by the finite difference method are amplified as the calculation proceeds. This may cause the solution to go completely out of control. A common cause is the use of a time step that is too long.

More information on numerical methods of solving partial differential equations can be found in a number of books, some highly specialised. Accessible introductions to the subject are given by Chadwick et al. (2021), Koutitas (1983), Vreugdenhil (1989), and Yen (1986).

20.8.5 Surcharge

It is common for pipes in drainage systems to occasionally experience surcharge – to run as full pipes under pressure rather than open channels with a free surface (see Section 7.4.5) and indeed some pipes such as inverted syphons (Section 8.4) are always surcharged and can be modelled as such.

The methods described so far in this section are for free-surface flows. In a surcharged pipe, Equation 20.3, for example, would present problems; the terms B (water surface width)

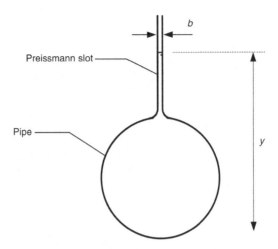

Figure 20.6 Preissmann slot.

and y (flow depth) would be meaningless: B is equal to zero, and y is always equal to pipe diameter regardless of flow rate.

A concept that allows Equations 20.2 and 20.3 to continue to be applied in surcharge conditions is the *Preissmann slot* (Yen, 1986). An imaginary slot (Figure 20.6) is introduced above the pipe, which allows y to exceed the pipe diameter and give the effect of pressurised flow.

The width of the slot b is calculated precisely to suit the conditions and must not be so wide that it has a significant effect on continuity. For a circular pipe of diameter D, this is given by

$$b = \frac{\pi D^2 g}{4c^2} \qquad (20.8)$$

where c is the speed of a wave in a pressure pipe (m/s).

In large piped systems, care is needed as the Preissmann slot can introduce significant storage that is not physically present. This can affect model accuracy and result in an under-prediction of flooding in surcharge conditions or inaccurately represent a storage calculation.

20.9 MODELLING SUDS

SuDS components are typically included in the main commercial sewer modelling packages, and a wide range of research models and tools has also been developed. All provide a representation of component hydrologic and hydraulic performance, but other issues such as water quality and economic impacts are less comprehensively covered (Elliott and Trowsdale, 2007; Lerer et al., 2015). Caution is advised in evaluating model results as already complex models now have the added burden of representing (perhaps poorly characterised) two- or three-dimensional components and flow patterns, depending on the level of effort and detail undertaken during modelling.

Compared with linear pipe systems, SuDS provision has a spatial element to it, and greater knowledge is required about the local urban context in order to specify appropriate components. This has given rise to the development of mapping tools (typically GIS-based), which can support option choice and decision-making. For example, Makropoulos et al. (2001), Ellis and Viavattene (2014), and Voskamp and Van de Ven (2014) describe tools that seek to prioritise the placement of SuDS components and quantify the impact on flood risk. Ferrans et al. (2022) give a comprehensive review of SuDS modelling approaches and the use of models in creating decision support systems for SuDS.

As an example of the inclusion of SuDS in a sewer modelling package, we consider SWMM, the Storm Water Management Model of the US EPA (EPA, 2023). This is one of the most widely used drainage modelling packages worldwide; it is open-source public software and free to use. SuDS are referred to as low-impact development (LID) controls.

> SWMM allows engineers and planners to represent combinations of green infrastructure practices as low impact development (LID) controls to determine their effectiveness in managing runoff

> (EPA, 2023).

SWMM can explicitly model eight types of SuDS components: green roofs, rooftop discon-nection, water butts, rain gardens, infiltration trenches, swales, pervious pavements, and bio-retention areas. (For information on components see Section 11.2.) SuDS components can be included in sub-catchments that also contain conventional drainage or can be located within a sub-catchment drained entirely by SuDS.

Kaykhosravi et al. (2018) compared and evaluated 11 drainage packages that included the modelling of SuDS. SWMM compared well and was recommended for preliminary and detailed SuDS design. Specifically for green roofs, Jeffers et al. (2022) found good agreement between experimental data for green roof performance and simulations using SWMM.

20.10 MODELLING FLOODING

An important role of sewer flow models is to represent the performance of the system (both minor *and* major) during extreme rainfall events. Under those conditions, the system will initially surcharge and then may well "spill" onto the urban surface: *exceedance flow* (see Section 12.2.2). In this section, we discuss and evaluate both the conventional approach to modelling this phenomenon and the many newer methods developed. Of course, the model will only predict flooding where it has a point to escape, and this should always be con-sidered when comparing model predictions to reality (e.g., if the property drainage is not included in the model, this might be the lowest point of escape and hence different than that of the model).

A key component is discussed first: the site terrain model (DTM/DEM).

20.10.1 DTM/DEMs

Commercially available flow models have a GIS interface that is intimately linked to a rep-resentation of the "height characteristics" of the urban catchment surface. There is some confusion over the terms used, but in urban flood modelling a digital terrain model (DTM) is taken to be the topographic model of the ground surface only. A digital elevation model (DEM) is based on the underlying DTM but also includes elevation data of buildings and other prominent features. A DEM forms the underlying foundation for any analysis of sur-face flooding.

In the UK, a variety of data sources is available to derive suitable DTM/DEMs ranging from basic land surveys to airborne-generated Light Detection and Ranging (LiDAR). Those now typically used are summarised in Table 20.6.

Where multiple sources of topographical data are brought together, attention must be paid to the risk that there may be differing physical levels for the same point in the ground model, and this may affect flood predictions. This can be critical in determining, for example, whether or not a property would flood at a certain flood depth.

20.10.2 Virtual flood cones

Models, simulating in one dimension only use virtual flood cones or reservoirs on top of each node to represent the flood flows once they have left the underground piped system in which they are temporarily stored (see Figure 20.7). Typically, the water is allowed to flow back to the underground system as long as the capacity allows, or it can be lost from the system altogether. The cone is only loosely related to the surrounding topography (by its user-defined geometry), so it can only give a rough measure of flow depth (particularly important when

Table 20.6 Available sources of DTM/DEM data

Land survey
- The oldest and most widely used method of obtaining topographical data for a specific site
- The most accurate form of data collection with accuracy up to ±1 cm

GIS sewer records
- Include some form of manhole geo-referencing and a relative ground level
- Used to produce an initial catchment DTM using thematic mapping capabilities within GIS

Light Detection and Ranging (LiDAR)
- A process that uses an aeroplane/helicopter mounted with an optically based scanner
- Typically available on a horizontal grid between 0.25 and 2 m
- Vertical accuracy typically ranges from ±5 to 15 cm (although some older data sets may be less accurate)

Source: Adapted from Hale (2004).

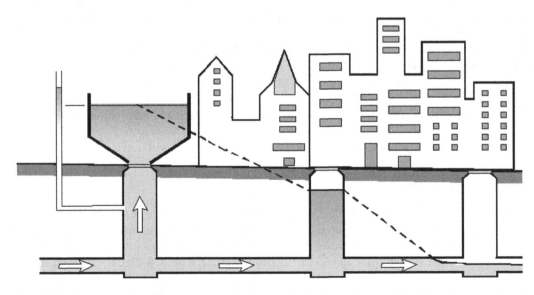

Figure 20.7 Conventional approach to sewer flood modelling using a virtual cone. (After Maksimović and Prodanović (2001)).

evaluating flood damage – see Section 11.4). Although the interaction between the flood flows and pipe flows is captured to some extent, this is of limited realism. Care is needed in defining the cone size, as too small a cone can create a substantial head and force more flow through the system than in reality, for example. However, this approach is quick and functional and does give an indication of surface flood volume. Where manholes cannot flood, they can be sealed in a model, and this can affect the hydraulic profile in larger return period events.

20.10.3 One-dimensional–two-dimensional (1D–2D) coupled models

As the name suggests, 1D–2D models consist of the tight coupling of 1D pipe flow models with 2D surface flow models. In 2D models, the solution domain (the whole DTM) is discretised as a coordinate system of nodes (i.e., grid points), where each point is represented by spatial coordinates (X, Y, Z). The flood routing model typically also solves a version of the Saint-Venant equations. Both models share two unknown variables, namely, velocity (or flow rate) and water level, with the difference that in the 1D case, the velocity has only one component, while in the 2D case, it has two orthogonal components (X, Y).

The continuity equations for network nodes, energy equations for nodes and pipe/channel ends, and the complete Saint-Venant equations for flow in pipes and surface channels are simultaneously solved. They can represent the interaction of the major and minor systems both in terms of outfalls linked to a natural major system (e.g., watercourse, tidal) and water exiting and returning to the minor system (Figure 20.8). Results are expressed as standard sewer outflow hydrographs and exit flood volumes, but also the volume of flood flow remaining on the surface in ponds and flood levels in flooded areas.

1D–2D models, especially when coupled, are considered to be the most accurate representations of urban surface flooding currently available but achieve this accuracy at the expense of high computational burden both in terms of time and data requirements (Bamford et al., 2008).

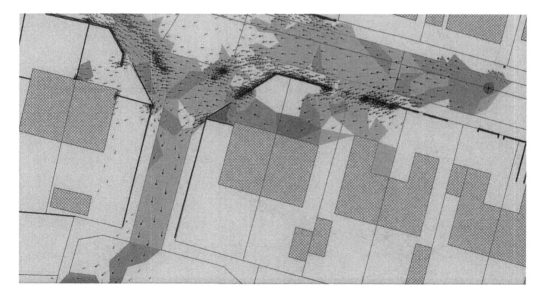

Figure 20.8 Example of a 1D–2D model with water escaping from the drainage system, flowing overland, and ponding around properties. Darker colour represents an increase in water depth, and thick lines indicate physical barriers such as walls that direct flow.

Guo et al. (2021) present a review of approaches to urban surface water flood modelling, including the effectiveness of hydrodynamic models based on shallow water equations with varying levels of complexity.

20.10.4 Rapid flood spreading models

Rapid flood spreading models take the volume of flow generated by a pipe-flow model and distribute it realistically but rapidly over the catchment surface. The earliest versions of this type of model automatically delineated the individual flooding sub-catchments. This was carried out by establishing a grid on the catchment. Then, by comparing the relative elevation of each grid intersection, it was possible to compute flow direction across every cell. Knowing this for each cell helped the lowest (pit) cells to be identified and the boundary between each pit or pond to be found. It was also established whether or not each of the pits contained a sewer node (manhole) and where it was located in the pond. Individual properties were then assigned to the grid cells. Once complete, flood volume could be quickly and reliably distributed across the catchment (Ryu, 2008). The advantage of this approach is its relative ease of use and therefore speed. Flooding is assigned to local low points in the catchment and can also cross sub-catchment boundaries. However, it does not represent any flow movement and hence has no time component.

More recent approaches use 2D cellular automata (CA)–based models that employ simple transition rules and a weight-based system rather than solving the full shallow water equations. These do allow the representation of flow movement. The CA technique provides a versatile method for modelling complex physical systems using simple operations (Wolfram, 1984) allowing a significant increase in computational speed. Guidolin et al. (2016) report on this type of flood spreading model WCA2D that is capable of simulating water depth and velocity variables with reasonably good agreement against a benchmark 1D–2D model (Figure 20.9), but using only a fraction of the computational time and memory. In the case of a real-world example, the model run times were up to eight times faster.

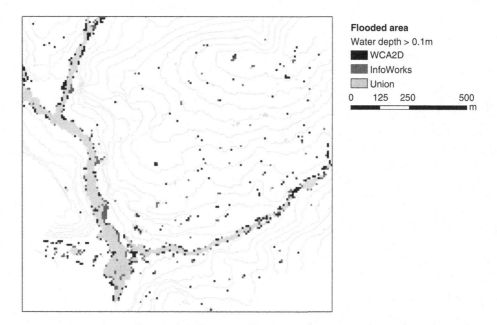

Figure 20.9 Area predicted as flooded by benchmark model only (dark grey pixel), by WCA2D only (black pixel), and by both models (light grey pixel) at 8 m resolution. (Courtesy Dr. Albert Chen.)

20.11 MODELLING QUALITY

20.11.1 Advection/dispersion

Transport of pollutants with the flow is normally represented by one of two alternative forms of equation:

$$\frac{\partial c}{\partial t} + v \frac{\partial c}{\partial x} = 0 \tag{20.9}$$

or

$$\frac{\partial c}{\partial t} + v \frac{\partial c}{\partial x} = \frac{\partial}{\partial x}\left[D \frac{\partial x}{\partial x} \right] \tag{20.10}$$

where x is the distance (m), t is the time (s), c is the concentration of pollutant (kg/m³), v is the mean velocity of flow (m/s), and D is the longitudinal dispersion coefficient (–).

Equation 20.9 represents *advection*, the movement of pollutants at the mean velocity of flow. Equation 20.10 additionally includes *dispersion*, the spreading out of pollutants relative to mean velocity. These two mechanisms are illustrated in Figure 20.10.

Figure 20.10a shows a slug of pollutant in a pipe (equally distributed horizontally and vertically), which is moved along at the mean flow velocity without spreading out: *advection only*. Figure 20.10b shows a similar slug of pollutant that is moved along the pipe at the mean flow velocity and at the same time spreads out: *advection* and *dispersion*.

Nearly all practical examples of sewer flow are strongly dominated by advection. For this reason, not all sewer quality models include dispersion.

Numerical methods for solving these equations are similar to those outlined in Section 20.8.4.

20.11.2 Completely mixed "tank"

An alternative to Equations 20.9 and 20.10 is to treat each pipe length as a conceptual tank in which pollutants are fully mixed with the flow. The governing equation is

$$\frac{d(Sc)}{dt} = Q_1 c_1 - Q_0 c_0 \tag{20.11}$$

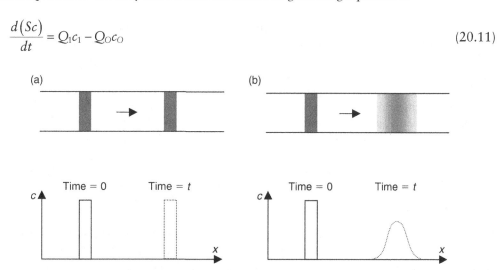

Figure 20.10 (a) Advection; (b) advection and dispersion.

where c, c_I, and c_O are concentrations in the pipe, at the inlet to pipe, and at the outlet (kg/m^3), respectively; Q_I and Q_O are flows at the inlet to pipe and at the outlet (m^3/s), respectively; and S is the volume of liquid in the pipe length (m^3).

This equation can be solved to give progressive estimates of concentration in each pipe as a whole. It does not include a distance or a velocity term and is therefore not capable of explicitly modelling the progress of pollutants at mean velocity.

20.11.3 Sediment transport

Pollutants transported in the body of the flow by advection and dispersion are either dissolved or suspended. Pollutants in solution remain in that form whatever the flow regime, although they may be transformed by biochemical processes (discussed in more detail in the next section). Pollutants in suspension may, however, be affected by the flow regime. At low flows, they may concentrate in a near-bed layer or deposit to form a sediment bed, and at high flows, they may re-erode. Larger, heavier solids may never (or rarely) achieve suspension, yet be transported as bed-load. The complex hydraulics of sediment movement is discussed in Chapter 17.

All sewer quality models have at least some representation of the movement of solid-associated pollutants through the system in terms of

- Mechanics: entrainment, transport, and deposition
- Sediment bed
- Solid attachment

20.11.3.1 Mechanics

A relatively straightforward way to model the entrainment, transport, and deposition of pollutants is by using one of the sediment transport equations that predict the volumetric sediment carrying capacity of the flow c_v (e.g., Macke's equation for suspended solids transport, Equation 17.3). Thus, at each time step, for an incoming pollutant concentration c, if

- $c < c_v$: all the incoming pollutant is transported, and if deposited sediment is available this may be eroded up to the carrying capacity c_v.
- $c > c_v$: only c_v / c of the incoming pollutant is transported. The remainder is subject to deposition.

The most straightforward approach is to consider just one type of solid. In principle, many solid-size fractions could be represented, although considerable data would be needed for calibration/verification. Representing two fractions (coarse and fine) or more can improve model performance, although would require more data to calibrate a quality model.

We saw in Section 17.4 that sediment can travel in suspension and as bed-load. Some models aim for a more realistic representation by using equations for total sediment transport. However, the mechanics of both types of transport processes can be represented separately.

20.11.3.2 Sediment bed

In the appropriate hydraulic conditions, pollutants will move out of suspension and become deposited on the pipe invert. Most models allow for the development of a sediment bed. At its simplest, this can be considered to have no impact on the hydraulics of the flow. A more realistic representation will include loss of flow area and increased hydraulic roughness that can be communicated to the flow model. A further elaboration could be to include the hydraulic effects of the sediment bed forms (Section 17.5).

The simplest structure of the bed is a simple, single layer of sediment that is supplied by deposition and removed by erosion. The settling velocity of the solid may be used to fix the rate of deposition. Erosion may be controlled by the sediment transport capacity of the flow (as previously mentioned) or by a defined critical shear stress. A more refined approach is to represent two layers; one containing stored (type A) sediment, and the other, erodible type C deposits (see Section 17.5).

Time-dependent effects such as consolidation and cohesion have also been introduced in a simple form in some models. Such software packages also include the ability to transport, deposit, and erode different sediment fractions.

20.11.3.3 Solid attachment

Pollutants may be modelled in two forms: dissolved and solid-attached, for the reason mentioned at the beginning of this section. Potency factors f are specified (Huber, 1986), which simply relate the concentration of solid-attached pollutant (c_s) to solid concentration c, as $c_s = fc$. An example for a pollutant n is given in Equation 20.12:

$$K_{Pn} = C1\left(IMKP - C2\right)^{C3} + C4 \tag{20.12}$$

where K_{Pn} is pollutant n's potency factor, $IMKP$ is the maximum rainfall intensity over a 5-minute period (mm/h), and $C1$–$C4$ are coefficients varying for each pollutant.

20.11.4 Gross solids

A basis for predicting the behaviour of gross solids is described in Section 9.5. This information has been used by Butler et al. (2003) to develop a model of gross solid transport that was subsequently developed to link directly with commercial hydraulic modelling software to understand the drainage network and its hydraulic performance (Spence et al., 2016). The model can support the assessment of engineered solutions, including the design of storage facilities and the development of efficient screening systems at CSOs and WTPs.

The model represents two aspects of gross solid transport: advection (movement with the flow, but not necessarily at the mean water velocity) and deposition/erosion. The fact that gross solids in actual sewers possess an almost infinite variety of properties, sizes, and states of physical degradation means that simplification is needed. The model represents this range by a finite number of distinct solid *types*. For each solid type, the model requires advection and deposition properties: the values of α and β in Equation 9.8, and the critical value of velocity and depth for deposition as described in Section 9.5.1.

The computational basis of the gross solids model is to "track" the progress of individual solids or groups of solids through the system. At any point in distance and time, the mean flow velocity is known from the hydraulic model. This can be converted to solid velocity using the relationship specified for the particular solid type. The velocity of a solid in any instantaneous position is thus known, and this can be used to progressively track its movement through the system. If depth or velocity over any section decreases to below the value specified as causing deposition, the progress of solids in the section is halted until the value is again exceeded.

The model includes a simplified model of solids transport in the smaller pipes in the upstream parts of the catchment (where not specifically modelled in the drainage model), and another of solids behaviour in CSO structures, to create a comprehensive model of gross solids loadings throughout a combined sewer system (Digman et al., 2002; Spence et al., 2016). The work has also included the collection of an extensive set of field data at three sites for comparison with model simulations demonstrating its ability to predict the first foul flush of solids.

20.11.5 Modelling pollutant transformation

In a gravity sewer, the main transformation processes act within or between the atmosphere, the wastewater itself, the biofilm attached to the pipe wall, and the sediment bed (see Figure 20.11). In a pressure sewer, there is no atmospheric phase and the biofilm is distributed around the pipe perimeter.

Of particular importance are those processes associated with the biodegradation of organic material. These are caused by microorganisms occurring either on the pipe wall as a biofilm or in suspension in the wastewater. The biofilm will be more influential in smaller pipes and suspended biomass in larger sewers. These are aerobic processes, requiring the presence of adequate dissolved oxygen (DO) levels. Thus, parameters such as BOD or COD to represent the organic material and DO to represent the toxic state of the wastewater need to be modelled. Anaerobic processes are not normally modelled in detail (but see Section 18.10).

All sewer quality models have at least some representation of pollutant transformation through the system. This can range from simplistic to sophisticated, as follows:

1. Conservative pollutants
2. Simple decay expressions
3. Complex processes approach

20.11.6 Conservative pollutants

Conservative pollutants are those that are not affected by any chemical or biochemical transformation processes. Pollutant concentration may still vary due to the processes of advection and dispersion.

Some sewer quality models omit representations of pollutant degradation or biochemical interactions. The justification for this is not that all pollutants *are* conservative, but that the processes are relatively insignificant. This may be a reasonable assumption for short-retention

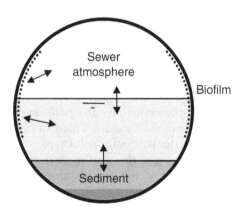

Figure 20.11 In-sewer pollutant interactions.

systems but will be inaccurate in systems with, for example, long, large-diameter trunk sewers of well-aerated outfalls.

20.11.7 Simple decay expressions

A second simplified approach is to model the transformation of individual pollutants using a simplified, summary model of the reactions. A first-order decay model is a common example of this, where

$$\frac{dc}{dt} = -kc \tag{20.13}$$

where c is the pollutant concentration (g/m³) and k is the rate constant (h⁻¹).

Thus, the pollutant concentration decreases with time and temperature, such that

$$k_T = k_{20}\theta^{T-20} \tag{20.14}$$

where k_T is the rate constant at T °C (h⁻¹), k_{20} is the rate constant at 20°C (h⁻¹), and θ is the Arrhenius temperature correction factor (–).

This approach ignores any interactions that may occur between the various substances (e.g., DO and BOD).

20.11.8 Complex processes approach

Current software packages offer the facility to model the complex processes discussed above in more detail by representing the oxygen balance of the flow and its various components. An example of the types and methods of representing these is given below.

20.11.8.1 Oxygen balance

DO in the flow results from a balance between oxygen supplied by aeration from the atmosphere and that consumed by the microorganisms in the wastewater and biofilm. Processes in the sediment bed may also exert an additional oxygen demand. This can be represented as an *oxygen balance* as follows:

$$\frac{dc_0}{dt} = K_{LA}\left(c_{0,S} - c_0\right) - \left(r_w + r_b + r_s\right) \tag{20.15}$$

where $c_{0,S}$ is the saturation dissolved oxygen concentration (g/m³), c_0 is the actual dissolved oxygen concentration (g/m³), K_{LA} is the volumetric reaeration coefficient (h⁻¹), r_w is the oxygen consumption rate in the bulk water (g/m³.h), r_b is the oxygen consumption rate in the biofilm (g/m³.h), and r_s is the oxygen consumption rate in the sediment (g/m³.h).

20.11.8.2 Reaeration

Reaeration is a naturally occurring process of diffusion. Oxygen in the atmosphere is dissolved into the liquid up to saturation levels that depend mainly on temperature. Aeration may be increased by turbulence caused by manhole backdrops or pumps. Pomeroy and

Parkhurst (1973) have derived an empirical relationship for reaeration in sewers based on a number of hydraulic parameters:

$$K_{LA} = 0.96\left(1 + 0.17\left(\frac{v^2}{gd_m}\right)\right)\gamma\left(S_f v\right)^{3/8}\frac{1}{d_m} \tag{20.16}$$

where v is the mean velocity (m/s), d_m is the hydraulic mean depth (m), γ is the temperature correction factor (1.00 at 20°C), and S_f is the hydraulic gradient (–).

20.11.8.3 Oxygen consumption in the bulk flow

The oxygen consumption of wastewater (also known as the oxygen uptake rate) varies with the age and temperature of the wastewater but (provided conditions are aerobic) is independent of oxygen concentration. Typical values are 1–4 mg/L.h.

20.11.8.4 Oxygen consumption in the biofilm

Oxygen consumption by biofilms is a complex phenomenon affected by substrate and oxygen availability, among other factors. Pomeroy and Parkhurst (1973) expressed consumption empirically as follows:

$$r_b = 5.3\left(S_f v\right)^{1/2}\frac{c_0}{R} \tag{20.17}$$

where R is the hydraulic radius (m).

20.11.8.5 Oxygen consumption in the sediment

Anaerobic processes in the sediment bed will produce oxygen-demanding by-products resulting in a *sediment oxygen demand* (SOD). This is exerted when the sediment bed is eroded.

PROBLEMS

1 What is meant by "a physically based deterministic model of flow in a sewer system"? What should such a model cover? To what uses might it be put?
2 "There is often a case for using modelling *less*." What point is really being made here?
3 Substitute in Equation 20.5 using finite difference forms proposed in Section 20.8.4 (with forward differences for both terms). Rearrange to give Q_j^{n+1} on the left-hand side.
4 Explain why *calibration* and *verification* mean different things (in the context of sewer-flow models). How is verification carried out in practice?
5 "Models are always wrong." So what is the point of using them (in urban drainage)?
6 Explain why quality modelling of a sewer system is more difficult than flow modelling.
7 Describe the physical processes that should be covered in a deterministic sewer quality model.
8 Classify and describe the various approaches to pollutant transformation modelling. What are their relative merits?

9 "Attempting deterministic modelling of sewer-flow quality is a waste of time." Some experts seriously hold this view. What do *you* think?

10 Describe alternatives to physically based deterministic models for sewer-flow quality. What are their advantages and disadvantages?

11 What are the main steps to be carried out in an integrated study?

KEY SOURCES

Chadwick, A.J., Morfett, J.C., and Borthwick, M. 2021. *Hydraulics in Civil and Environmental Engineering*, 6th edn, Spon Press.

Chartered Institution of Water and Environmental Management (CIWEM). 2017. *Code of Practice for the Hydraulic Modelling of Urban Drainage Systems*, www.ciwem.org/special-interest-groups/urban-drainage-group

Chartered Institution of Water and Environmental Management (CIWEM). 2021. *Integrated Urban Drainage Modelling Guide*, www.ciwem.org/special-interest-groups/urban-drainage-group

REFERENCES

Bamford, T.B., Digman, C.J., Balmforth, D.J., Waller, S., and Hunter, N. 2008. Modelling flood risk—An evaluation of different methods. *WaPUG Autumn Conference*, Blackpool.

Butler, D., Davies, J.W., Jefferies, C., and Schütze, M. 2003. Gross solid transport in sewers. *Proceedings of the Institution of Civil Engineers, Water and Maritime Engineering*, 156(WM2), 175–183.

Chow, V.T. 1959. *Open-Channel Hydraulics*, McGraw-Hill.

Digman, C.J., Littlewood, K., Butler, D., Spence, K., Balmforth, D.J., Davies, J.W., and Schütze, M. 2002. A model to predict the temporal distribution of gross solids loadings in combined sewerage systems. *Global Solutions for Urban Drainage: Proceedings of the Ninth International Conference on Urban Drainage*, Portland, Oregon, on CD-ROM.

Elliott, A.H. and Trowsdale, S.A. 2007. A review of models for low impact urban stormwater drainage. *Environmental Modelling and Software*, 22, 394–405.

Ellis, J.B. and Viavattene, C. 2014. Sustainable urban drainage system modeling for managing urban surface water flood risk. *CLEAN-Soil, Air, Water*, 42(2), 153–159.

Environment Agency 2018. *Storm Overflow Assessment Framework*. Environment Agency.

EPA 2023. https://www.epa.gov/water-research/storm-water-management-model-swmm

Ferrans, P., Torres, M.N., Temprano, J. and Sánchez, J.P.R. 2022. Sustainable urban drainage system (SUDS) modelling supporting decision-making: a systematic quantitative review. *Science of the Total Environment*, 806 (2022) 150447.

Foundation for Water Research (FWR). 2019. *Urban Pollution Management Manual*, 3rd edn.

Guidolin, M., Chen, A.S., Ghimire, B., Keedwell, E.C., Djordjevic, S., and Savic D.A. 2016. A weighted cellular automata 2D inundation model for rapid flood analysis. *Environmental Modelling and Software*, 84, 378–394.

Guo, K., Guan, M. and Yu, D. 2021. Urban surface water flood modelling – a comprehensive review of current models and future challenges. *Hydrology and Earth System Sciences*, 25, 2843–2860.

Hale, J.N. 2004. Urban flood routing in the UK … the next step? Proceedings of UDM04 Sixth International Conference on Urban Drainage Modelling, Dresden, Germany, August, 97–104.

Huber, W.C. 1986. Deterministic modeling of urban runoff quality, in *Urban Runoff Pollution* (eds H.C. Torno, J. Marsalek and M. Desbordes), NATO ASI Series G: Ecological Sciences—Vol. *10*, Springer-Verlag.

Jeffers, S., Garner, B., Hidalgo, D., Daoularis, D. and Warmerdam, O. 2022. Insights into green roof modelling using SWMM LID controls for detention-based designs. *Journal of Water Management Modeling*, 52, C484.

Kaykhosravi, S., Khan, U.T. and Jadidi, A. 2018. A comprehensive review of low impact development models for research, conceptual, preliminary and detailed design applications. *Water, 10*, 1541.

Koutitas, C.G. 1983. *Elements of Computational Hydraulics*, Pentech.

Lerer, S.M., Arnbjerg-Nielsen, K., and Mikkelsen, P.S. 2015. A mapping of tools for informing water sensitive urban design planning decisions—Questions, aspects and context sensitivity. *Water, 7*, 993–1012.

Makropoulos, C., Butler, D., and Maksimovic, C. 2001. GIS-supported stormwater source control implementation and urban flood risk mitigation, in *Advances in Urban Stormwater and Agricultural Runoff Source Controls* (ed. J. Marsalek), Kluwer, 95–105.

Maksimović, C. and Prodanović, D. 2001. Modelling of urban flooding—Breakthrough or recycling of outdated concepts? Urban Drainage Modeling: *Proceedings of the Speciality Symposium of the EWRI/ASCE World Water and Environmental Resources Congress*, Orlando, Florida, May.

Osborne, M., Dumont, J., and Martin, N. 1996. Beckton and Crossness catchment modelling: Hydrologic modelling—Two routes to the same answer. *Proceedings of the Seventh International Conference on Urban Storm Drainage, 1*, Hannover, September, 443–448.

Pomeroy, R.D. and Parkhurst, J.D. 1973. Self purification in sewers. *Advances in Water Pollution Research. Proceedings of the Sixth International Conference*, Jerusalem, Pergamon Press, 291–308.

Price, R.K. and Osborne, M. 1986. Verification of sewer simulation models, in *Urban Drainage Modelling* (eds C. Maksimović and M. Radojkovic), Pergamon Press.

Russ, H.-J. 1999. Reliability of sewer flow quality models—Results of a North Rhine-Westphalian comparison. *Water Science and Technology, 39*(9), 73–80.

Ryu, J. 2008. *Decision Support for Sewer Flood Risk Management*, Unpublished PhD thesis, Imperial College, University of London.

Schellart, A.N.A., Tait, S.J., Ashley, R.M., Farrar, D., and Hanson, D. 2008. Uncertainty in deterministic predictions of flow quality modelling and receiving water impact. *Proceedings of the 11th International Conference on Urban Drainage*, Edinburgh, September, on CD-ROM.

Spence, K.J., Digman, C., Balmforth, D., Houldsworth, J., Saul, A., and Meadowcroft, J. 2016. Gross solids from combined sewers in dry weather and storms, elucidating production, storage and social factors. *Urban Water Journal, 13*(8), 773–789.

Vojinovic, Z. and Solomatine, D.P. 2006. Evaluation of different approaches to calibration in the context of urban drainage modelling. *Proceedings of the Seventh International Conference on Urban Drainage Modelling and Fourth International Conference on Water Sensitive Urban Design*, Melbourne, May, 2, 651–660.

Voskamp, I. and Van de Ven, F. 2014. Planning support system for climate adaptation: Composing effective sets of blue-green measures to reduce urban vulnerability to extreme weather events. *Building and Environment, 83*, 159–167.

Vreugdenhil, C.B. 1989. *Computation Hydraulics: An Introduction*, Springer-Verlag.

Wangwongwiraj, N., Schlütter, F., and Mark, O. 2004. Principles and practical aspects of an automatic calibration procedure for urban rainfall-runoff models. *Urban Water Journal, 1*(3), 199–208.

Wastewater Planning Users Group (WaPUG). 2006. *Guide to the Quality Modelling of Sewer Systems, v 1.0. Wastewater Planning Users Guide*, https://www.ciwem.org/assets/pdf/Special%20Interest%20Groups/Urban%20Drainage%20Group/Guide-to-the-Quality-Modelling-of-Sewer-Systems.pdf

Willems, P. 2008. Quantification and relative comparison of different types of uncertainties in sewer water quality modeling. *Water Research, 42*(13), 3539–3551.

Wolfram, S. 1984. Cellular automata as models of complexity, *Nature, 311*, 419–424.

Yen, B.C. 1986. Hydraulics of sewers, in *Advances in Hydroscience*, Vol. *14* (ed. B.C. Yen), Academic Press.

Chapter 21

Innovations in modelling

21.1 INTRODUCTION

This chapter introduces some of the more novel concepts in urban drainage modelling. First, we discuss models that are not physically based and rely more on data and less on fundamental physical equations (Section 21.2), including emerging artificial intelligence approaches. As such, they also rely on good model calibration (discussed in Section 21.6) to make sure that these relatively simple models accurately approximate reality. In Section 21.3, we briefly outline the concept and methods of optimisation and discuss its application in urban drainage. We also introduce the idea that a single model may not be able to accurately (or sufficiently) describe the complete urban water system and suggest two novel ways to address this problem: integrated models (Section 21.4) and standards for coupling a series of different models (Section 21.5). How uncertain all these models are and what we can do about it are the focus of Section 21.7. Also discussed (Section 21.8) are how virtual case studies (VCSs) can be used to draw generalised conclusions about engineering interventions, and trends in the visualisation of model results (Section 21.9).

21.2 ALTERNATIVE, NON-PHYSICALLY BASED, MODELLING APPROACHES

This section summarises approaches to modelling that can be considered alternative to more traditional, physically based models. They can be used for modelling flow as well as quality, and their basic premise is that they rely more on data or simplified conceptual descriptions of the drainage system and less on fundamental physical equations.

Every urban drainage model receives input data and produces output data. So, for any given system, the modeller needs to collect data on the actual, real-world, system behaviour. These data typically consist of recorded historical events (e.g., rainfall), and consequences observed as a result of these events (e.g., runoff, possibly resulting in urban flooding). To reproduce the relationship between events driving the physical processes and their consequences within a model, the events become an *input* and the consequences an *output*. The model converts input into output using a set of mathematical procedures sometimes called *transfer functions*, representing the relationship between events and their consequences. The key difference between physically based models (discussed in previous chapters) and the alternative approaches discussed in this chapter, is that physically based models attempt to describe the system in detail, by creating a mathematical equivalent of each major, relevant, physical process: rainfall to runoff, transport of pollutants, sediment deposition in sewers, and so on.

DOI: 10.1201/9781003408635-21

On the contrary, alternative approaches "simply" map *known*, historical, inputs onto outputs, by finding statistical or other relationships between them. They then capture these relationships in a transfer function and use this function to calculate outputs for new (real or hypothetical) inputs. These transfer functions do not have a direct physical interpretation – we cannot point to an equation and say "This is where rainfall is converted to runoff." The formulation and potential success of the model have a mathematical explanation, but not a physical one. The model is a *black box*. It follows that a disadvantage of such models is that they cannot, for example, be used to look at options to upgrade a sewer system, since the physical properties of the system are not contained in the model. Remember, the model is exclusively based on relationships between known (existing, historical) inputs and outputs: if the new sewer system is not there yet in the records, there is no relationship to capture! They are, however, very quick, and as such are ideal for real-time applications, such as early warning systems for urban floods (see Section 22.5). Some of the most well-known categories of "alternative" models are briefly presented in the following sections.

21.2.1 Empirical models

An empirical model is based on observation rather than theory. It usually represents the real system by simple relationships (e.g., a polynomial data-fitting equation) that rely for their accuracy on parameters that are calibrated using observed data (e.g., using a least squares optimisation method). However, despite their simplicity (or perhaps because of it), empirical models have their role in urban drainage. For example, regression models have been used to assess pollutant loads in urban runoff (Maniquiz et al., 2010) but also to provide flow estimations for wastewater systems, where historical records were missing (Brito et al., 2017).

21.2.2 Machine Learning Models and AI

Machine learning (ML) is a subset of artificial intelligence (AI) that uses algorithms to allow computers to perform tasks without being explicitly programmed for those tasks. Instead, they learn patterns and make decisions based on data. While AI is a broad field that seeks to create machines capable of mimicking human intelligence, machine learning is specifically concerned with algorithms that improve through experience. In other words, all machine learning is AI, but not all AI is machine learning. In the past decade, ML models have begun to be widely used in urban drainage and urban water systems more generally, especially in three main areas (Kwon and Kim, 2021): drainage system operation (i.e., real-time control), risk assessment and management (e.g., flood prediction and response measures deployment) and drainage infrastructure maintenance (e.g., sewer replacement and rehabilitation planning).

Artificial neural networks (ANNs) are certainly the most well-known and often used forms of machine learning in urban drainage applications. In ANNs, inputs are treated as "signals" which are passed between artificial "neurons" arranged in layers. Each neuron receives signals from a number of "upstream" neurons, applies a weighting to each input, and then applies a transfer function before outputting signals to "downstream" neurons, passing the modified signal from layer to layer. The network is then trained (by changing the weights associated with each neuron) to reproduce the relationships in training data sets, matching known inputs to known output, provided by the modeller. Given enough training on good data, the network can "learn" how to make useful predictions when provided with new inputs, without the need for hard to assess physical parameters or complex flow equations.

Although ANNs have been around for a long time (e.g., they have been used successfully in sewer modelling (Loke et al., 1996) as well as in real-time applications in sewer modelling for control (Darsono and Labadie, 2007) for more than 15 years), two recent developments in computer science, *Deep Learning* (DL) and *Reinforcement Learning* (RL) have significantly enhanced the capabilities and use cases for ML in a wide range of engineering applications, including urban drainage.

Deep Learning (DL) is a subset of ML that utilises neural networks with many layers (hence "deep") to analyse various forms of data. Deep Learning models are particularly powerful for tasks where large amounts of data are available. As big data technologies (see Chapter 22) increase the amount of data available for urban drainage, DL is becoming more popular (Fu et al., 2022) as more complex neural network architectures gain momentum. Two representatives of this trend are the Long Short-Term Memory (LSTM; Hochreiter and Schmidhuber, 1997) neural networks, and Convolutional Neural Networks (CNNs; LeCun et al., 2015). LSTM networks, belonging in the general category of recurrent neural networks, incorporate mechanisms that captures efficiently long-term temporal dependencies, which typically appear in sequential data such as time-series (e.g., rainfall), speech, and text. On the other hand, CNNs have found wide applicability for the efficient processing of grid-like data such as images and videos.

Reinforcement Learning (RL) is an area of ML where an "agent" (e.g., an urban drainage control algorithm) learns how to "behave" in an environment by taking actions and receiving rewards or penalties. The goal is to find a strategy that will maximise the agent's cumulative reward over time. This approach allows algorithms to "learn by doing" (initially through modelling, but eventually through real-time monitoring of the impact of their actions in a "live" drainage system) and continuously improve. Perhaps more importantly, the pursuit of longer-term strategy (e.g., minimisation of urban flooding or overall river pollution) allows the algorithm to make small "sacrifices" (e.g., allowing a few CSO spills at a suitable time) to ensure longer-term objectives are better met. In other words, the algorithm learns and decides (eventually) like an experienced engineer. Recent examples of using RL to prevent urban flooding (Bowes et al., 2021) including CSO management (Zhang et al., 2023) demonstrate the significant potential of this approach.

Ever-increasing computational power has enabled the rise of ensemble ML algorithms and approaches, which combine the predictions from multiple individual models. Such an approach is bootstrap aggregation (known as "bagging"), which entails the individual training of multiple models, under certain variations (either on the structure of the model or the training dataset used), to provide an aggregated prediction. A key representative of this category is the Random Forest algorithm (Breiman, 2001), which provides predictions based on a batch of individually trained decision trees. This algorithm was recently implemented to provide predictions on the optimal settings of actuators in a large urban drainage system (van der Werf et al., 2023). Another ensemble method is "boosting," which entails the sequential building of models, where each new model in the sequence tries to improve the previous one (Friedman, 2002).

21.2.3 Conceptual or meta-models

In a conceptual model, the physical system is represented by highly simplified "concepts," for example, the representation of the physics of pipe flow by a simple tank system. Detailed treatment of individual processes is replaced by overall global representation. A good example of this approach is the open-source CITY DRAIN model developed by Achleitner et al. (2007).

21.2.4 Stochastic models

A stochastic model includes randomness. Unlike a deterministic model, it does not necessarily give the same output for the same input. Simple empirical models may be designed to be stochastic, and this is appropriate since the measured data on which the empirical model is based are certain to contain random elements. A complex, physically based drainage or urban flood risk assessment model can also have stochastic characteristics by introducing random influences on some inputs (usually rainfall) and showing the effect on the output. The output from a stochastic model will not be a single answer but a range of answers, possibly represented by a mean and standard deviation. For example, Efstratiadis et al. (2022) employ stochastic methods to account for three important sources of uncertainty in urban flooding: how storms change over time; how prior soil moisture affects flooding and how runoff propagation depends on flood event intensity. The stochastically generated rainfall events and soil conditions are then used to drive an integrated urban drainage model (see also the next section) to calculate a range of flood hazard and resulting risk values with different probabilities of occurrence to allow for more informed decisions on potential measures and levels of protection.

21.3 OPTIMISATION

Optimisation is a collection of mathematical principles and methods used to solve quantitative problems in many disciplines, including urban drainage. The optimisation process typically has three elements: an *objective function* or functions, a collection of *decision variables* and a set of *constraints*. An optimisation model aims to find the value of its decision variables which results in the best outcome for its objective function(s) without violating the constraints. If a problem has more than one objective (i.e., is multi-objective), additional functions can be considered in the process. The goal is to minimise or maximise the objective function(s) while satisfying all the constraints.

In practical terms, "the best" outcome is an optimal or near-optimal solution with low computational effort. Effort is typically measured in terms of the computation time and/or the computer memory that is consumed by the method. For many optimisation methods, there is a trade-off between solution quality and computational effort.

Many different optimisation approaches have been developed. These range from simple Monte Carlo methods (Freni et al., 2008), controlled random search (Butler and Schütze, 2005), and evolutionary algorithms (Muschalla, 2008), all the way to multi-objective algorithms such as the Pareto preference ordering (Khu and Madsen, 2005), genetic algorithms (Schütze et al., 2001), surrogate-enhanced evolutionary annealing simplex algorithms (Tsoukalas et al., 2016), and more. Shishegar et al. (2018) give a comprehensive overview of optimisation methods used in the urban drainage field.

The main applications in urban drainage are the following:

- System design (Section 10.2.2)
- System operation (Section 22.4.4)
- Model calibration (Section 21.1)

In each of these cases, an optimisation method will typically "call" a simulation model many times to compute the decision variables in the search for the best objective function(s). This imposes a high computational burden. Typical approaches to alleviate this can be classified into four main categories:

- Parallel computing (e.g., Cheng et al., 2005; Dias et al., 2013)
- Computationally efficient optimisation algorithms (e.g., Tolson et al., 2009; Tolson and Shoemaker, 2007)
- Strategies to opportunistically avoid (expensive) model evaluations (e.g., Razavi et al., 2010)
- Surrogate modelling techniques (e.g., Forrester and Keane, 2009; Razavi et al., 2012; Tsoukalas and Makropoulos, 2014)

The latter approach is considered further in Section 21.6.1.

21.4 INTEGRATED MODELLING

Conventional practice has been to operate and, therefore, model the various components of the engineered urban wastewater cycle (water treatment, distribution, sewerage and storm drainage, wastewater treatment, and environmental compartments) in isolation. Each component has been engineered to meet the needs of its users and the environment but with little feedback or cross-reference to other components. The importance of more combined management of the various urban water system components is now recognised by both researchers and practitioners. This move towards ever more *integration* (1) considers all parts, or components, of the system; (2) involves water conservation and diverse fit-for-purpose water supplies; (3) works at a range of scales (both central and decentralised); and (4) allows the establishment of links with other environmental cycles (e.g., energy and nutrients) (Bach et al., 2014; Brown et al., 2009; Mitchell, 2004).

21.4.1 Why do we need integrated models?

The need for an integrated model of the system is similar to the need for a model of any of the individual elements. In addition, representing and understanding the urban system as a whole potentially allows better, more cost-effective solutions to be engineered.

The urban wastewater system as a whole contains numerous elements that can be utilised to prevent water pollution. For example, there is little point in "wasting" storage volume in tanks and pipes, or treatment capacity, on weak wastewater. If the available capacity could primarily be used to capture the most polluted flows, the potential for pollutant discharge reduction can be increased. Storage other than in the urban drainage system may also be exploited (e.g., the time lag of processes in the WTP and the receiving water). If the WTP and CSOs discharge into the same receiving water, carefully timed release of effluents could minimise overall pollutant discharges.

Calibrating and verifying such large integrated models can be challenging because of the need for large data sets from diverse parts of the system. Muschalla et al. (2008) provide advice on the procedures to be adopted, and Solvi (2007) describes in detail the calibration and verification of an integrated model case study.

21.4.2 Defining and classifying the level of integration

The formulation of a unique definition for integration and classification format for urban water systems modelling is rather a hard task due to the complexity of the latter and the subjectivity of what can be considered as "integrated." Rauch et al. (2005) suggested that thinking about "degrees of integration" is more useful than simply using the term *integration*.

Bach et al. (2014) have suggested that to move towards integrated modelling in urban water systems, we need to consider three key points:

- Modelling of a multitude of components (biophysical, economic, and beyond) and interactions between them
- Consideration of acute, chronic, and delayed impacts of water quantity and quality processes in long simulation periods
- Ability to see both local processes and the global "big picture" to better inform decision-making, policies, or scientific knowledge

Bach et al. (2014) also suggested four different integration levels to categorise urban water models according to the physical and institutional system delineation, model complexity, and differences between their development philosophies.

- *Integrated component-based models (ICBMs)*: ICBMs encompass "whole-of-subsystem" or "plant-wide" integration and focus on the integration of components within the local urban water subsystem (e.g., coupling several treatment processes within a wastewater treatment plant).
- *Integrated urban drainage models (IUDMs)*: IUDMs focus on the integration of the subsystems of urban drainage (wastewater and/or stormwater). They are also known as *integrated urban wastewater systems*. These are some of the most well-known forms of integrated models that are typically used to study and evaluate upgrade options for a local WTP, non-pipe decentralised technologies in stormwater management, ways of reducing CSO emissions, or showing the combined impact of different parts of the drainage stream on receiving waters.
- *Integrated urban water cycle models (IUWCMs)*: At this level of integration, the IUDMs and IWSMs are combined into one modelling framework. This integration is typically used in strategic planning and conceptual assessment of sustainable solutions (e.g., combinations of central and decentralised technologies) in urban water management, rather than devising detailed control strategies. Such a model is, for example, UWOT, which is described briefly in the next section.
- *Integrated urban water system models (IUWSMs)*: This is the highest level of integration in urban water systems that takes into account and draws links among environmental, social, and economic aspects of water-related issues. An inter- and multi-disciplinary approach is adopted to study and model the total urban water cycle. For an example of a complete socio-technical modelling of the urban water system, see Makropoulos (2017).

21.4.3 Methods and tools for integrated system modelling and design

Irrespective of which *level* of integration the modeller is looking to achieve, there are, typically, two different approaches to employ: The first is to use a single software package with which to build the whole integrated model, and the second is to integrate existing models (built using different software packages) into one model "chain." Extensive reviews of these modelling approaches can be found in Elliott and Trowsdale (2007) and in Bach et al. (2014).

Typical examples of the first approach include *SIMBA#* (IFAK, 2023) and *WEST* (DHI, 2023), while combined modelling of water and wastewater flows can also be performed with

CITY DRAIN (Achleitner et al., 2007; Burger et al., 2010) and *UrbanCycle* (Hardy et al., 2005). *UrbanBEATS* (Bach, 2012), *DAnCE4Water* (Rauch et al., 2012), *UWOT* (Rozos and Makropoulos, 2013), and CityWat (Dobson and Mijic, 2020) also support a higher integration level by allowing the modelling of additional aspects of the urban water cycle (i.e., social, economics, energy, etc.). UWOT is presented in Box 21.1 as an example.

BOX 21.1   The urban water optioneering tool (UWOT)

The UWOT model (Makropoulos 2017; Rozos and Makropoulos, 2013) enables the simulation of the three main components of the urban water cycle (i.e., water supply, wastewater disposal, and stormwater drainage) in an integrated framework. Unlike other typical urban water models that simulate actual water flows, UWOT adopts an alternative approach based on the generation, aggregation, and transmission of demand signals, starting from household water appliances and moving towards the source. This demand-oriented conceptualisation approach has the advantage of directly representing the principal purpose of infrastructure, which is to serve the need for water supply and wastewater disposal. Figure 21.1 presents a simplified example of the water flows from the abstraction, through the transmission network to the treatment plant, and then to the distribution network (left panel), along with the equivalent representation in UWOT (right panel). The part inside the ellipse describes the external water system, including abstractions, transmission, and treatment of raw water, whereas the lower part of the panel shows the generation of the demand and the disposal of wastewater and stormwater to waterbodies (the internal water system).

UWOT distinguishes between two signal types: *push* and *pull* signals. The push signals express a need to dispose of a specific volume of water (e.g., the output of a washing machine). The pull

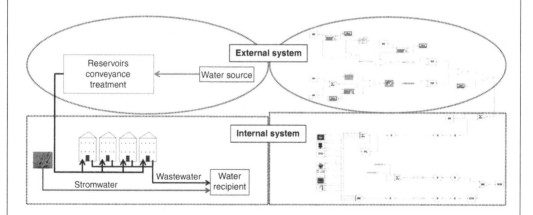

Figure 21.1 Left panel: representation of the urban water cycle. Right panel: source-to-tap modelling of the complete urban water cycle with UWOT in Matlab Simulink. The external water supply system is presented inside ellipses, while the water consumption and the drainage system are shown inside rectangles. (After Rozos, E. and Makropoulos, C. 2013. *Journal of Environmental Modelling and Software*, 41, 139–150.)

signals express a demand for a specific volume of water (e.g., the water required for the operation of a washing machine). Pull signals do not bear a qualitative characterisation because water that covers a demand is assumed to meet the quality standards imposed by regulations. Push signals are characterised by a qualitative value that can express any preselected water quality parameter.

The information required by UWOT to simulate the operation of water and wastewater components (water consumption, required energy, BOD values, capital and operational cost, etc.) is stored in a database that is called a *technology library*. The library is populated with information obtained from surveys, practitioner manuals, scientific publications, laboratory pilot tests, or even other models (see below), comprising both central and distributed management technologies. The user can define system components (from appliances to reservoirs and from WTPs to CSOs) and their topologies (connections between components) via an advanced (drag and drop) graphical user interface. UWOT has been successfully applied to simulate the urban water cycle at a wide range of spatial and temporal scales, from a single household, all the way to city-wide scales. UWOT has also been coupled with other models, such as Agent-Based Models (Koutiva and Makropoulos, 2016) to assess how water demand management affects water system reliability as well as with water resource management and water distribution models (Nikolopoulos et al., 2022) within a stochastic framework (see Section 21.2.4) to assess the resilience of integrated urban water systems at large scales.

21.4.4 Benchmarks

The International Water Association (IWA) has produced a series of Benchmark Simulation Models (e.g., BSM1, BSM2) for wastewater treatment plants which consist of a predefined plant layout, process models, sensor and actuator models, influent wastewater characteristics and evaluation criteria (Jeppsson et al., 2013). The aim of these benchmarks is to allow a more objective comparison of different system modifications and control strategies at WTPs, and they have been widely used in industry and academia. These models have been extended "outside the fence" to incorporate the urban catchment and sewer system. Saagi et al. (2016) describe such a benchmark model and show its applicability and capabilities within integrated urban wastewater system modelling.

21.5 COUPLING MODEL TO OTHER MODELS: EMERGING STANDARDS

As discussed earlier, an alternative approach to building the integrated model within one software package is to link up existing models, built using different software packages into a model chain. To be able to do this, one needs to make sure that the outputs of one model (e.g., the rainfall-runoff urban hydrology model) become an input for the next "downstream" model in the chain (e.g., the hydraulic model of the drainage system), and so on, all the way down to the recipient waterbody. The typical way of accomplishing this model link is to write bespoke code (sometime termed *gluing routines*). This approach, however, has a serious disadvantage: one needs to change these routines every time a new model is linked in the chain, or indeed a new variable is being passed from model to model.

A promising effort to address this issue has been the development of an open model-coupling standard, the *Open Modelling Interface (OpenMI)*, which allows for consistent and efficient integration of different models at runtime, currently at its 2.0 version (Harpham et al., 2019). OpenMI-compliant models can be easily combined into integrated modelling systems without the need for bespoke glueing routines. OpenMI v.2.0 is an official Open Geospatial Consortium (OGC) standard (see OGC's page on OpenMI 2.0) and the OGC holds the definition of this ratified version of the standard. The OpenMI Network supports the approach (https://publicwiki.deltares.nl/display/OPENMI/Home), and several studies have been conducted across disciplines using the standard (see Knapen et al., 2013).

More integrated models require, however, more raw data, and the need for standardisation not only of the data exchanged between models, but also of water observations that serve as inputs to these models is emerging as an important issue. A noteworthy development in standardisation, aiming to provide a common data exchange format for hydrological time series, and therefore facilitate the exchange of such data sets across different models and information systems is *WaterML*. WaterML 2.0 is an open standard and information model for the encoding, representation, and communication of water observation data. It has been designed and developed by many national and international organisations from the public and private sectors, within the OGC (Open Geospatial Consortium, 2014).

Of course, interoperability between data sources and models, as well as between models themselves is not a challenge exclusive to urban drainage engineers. As more "smart" applications emerge (see also Chapter 22) across disciplines, generic (i.e., not water-specific) interoperability standards are also emerging, promising to link real-time data from the field to databases, analytics, and models in near-real time. One of the most promising is *FIWARE* (www.fiware.org). FIWARE is an open-source interoperability framework designed for the development of smart applications across various sectors, such as Smart Cities and the Internet of Things (IoT). The framework offers modular components, known as "generic enablers," to aid in tasks ranging from data connectivity to processing, making it a popular choice for smart solution projects, particularly in Europe. Water applications of FIWARE are also beginning to emerge (e.g., Sweetapple et al., 2022).

21.6 CALIBRATION ASPECTS FOR INTEGRATED URBAN WASTEWATER SYSTEMS

As discussed in Chapter 20, the majority of urban drainage applications require the calibration of the parameters of the model. In the formal calibration problem, the internal properties of the model, either physical or conceptual, are unknown and have to be fine-tuned by optimising (in this case minimising) the departures of the simulated responses of the model at hand against the observed, real-life system responses. Therefore, calibration is a particular case of optimisation as discussed in Section 21.3.

The calibration of urban drainage models is, however, characterised by several difficulties, the most significant of which are

- The high computational burden of drainage models, which makes calibration a highly time-consuming process
- The vast amount of data and measurements that are required to capture the nature of each component of the urban water cycle
- The fact that calibration procedures may lead to equally good combinations of parameters, reduces the confidence in model predictions. This concept is associated with calibration uncertainty and is also known as *equifinality*.

These difficulties are further exacerbated in the calibration of integrated models (Rauch et al., 2002; Jamali et al., 2020). In this especially challenging case, three different calibration approaches have been used (Bach et al., 2014):

- Calibration/optimisation of the whole integrated model at once (often a very difficult and sometimes impossible task)
- Initial calibration/optimisation of the upstream models, gradually adding the next immediate downstream component
- Calibration/optimisation of individual sub-models separately before integrating them. In this case, of course, it is not possible to assess the effect of remaining errors in the sub-models after integration.

21.6.1 Computational approaches

In the computational procedure of calibration, the simulation runtime required by the drainage model is by far the most time-consuming element, and the one that imposes a practical barrier to optimisation: typically, several thousand runs of the model, each driven by very long (often stochastically generated) time series, are required by the most powerful optimisation algorithms (of the evolutionary kind, such as genetic algorithms) to reach a good (calibrated) parameter set. If, however, each model run needs several minutes to complete, then the whole calibration procedure may well cost days or even weeks of simulation time. This is clearly impractical.

Surrogate models (also termed *meta-models* or *response surface models*), introduced in 21.3.3, are particularly useful as they can be applied to virtually any optimisation problem, including the urban drainage calibration variety. The key concept of this technique is to generate (inexpensive) meta-models that are accurate in a certain region (i.e., in a potentially optimal area) of the search space and use these meta-models in combination with the more expensive complete models to intelligently guide the optimisation algorithm (Couckuyt et al., 2013). Typical surrogate-based optimisation algorithms are the Multistart Local Metric Stochastic RBF (Regis and Shoemaker, 2007), DYnamic COordinate Search-Multistart Local Metric Stochastic RBF (Regis and Shoemaker, 2013), and Surrogate-Enhanced Evolutionary Annealing Simplex Algorithm (Tsoukalas et al., 2016).

The key benefit of these algorithms is that they can be used to successfully calibrate complex (even integrated) models using very long time series and driven by stochastic inputs (such as rainfall). This allows us to perform otherwise computationally impractical models, including formal uncertainty analysis of (calibrated) model parameters and model results. A discussion of the process of uncertainty analysis in urban drainage is the subject of the following section.

21.7 UNCERTAINTY ANALYSIS

Urban drainage models, like any other mathematical models, provide a simplified representation of reality and subsequently are characterised by an inherent uncertainty. The total elimination of uncertainty is not possible to achieve, and hence, an uncertainty analysis is of high importance to quantify the level of reliability of the models, providing a robust basis for their application in practice (e.g., to provide a level of confidence for a model used for risk analysis). Extensive reviews on uncertainty analysis in urban drainage models can be found in Deletic et al. (2012), Dotto et al. (2012), and Freni et al. (2009).

21.7.1 Types of uncertainty in urban drainage modelling

The total uncertainty in model results is typically attributed to three main sources:

1. Model input uncertainty: derives from the imperfect or infrequent measurement of data used in model calibration and validation, and as initial or boundary conditions (e.g., systematic and/or random errors)
2. Parameter uncertainty: defined as the uncertainty in the model parameter vector and is related to the processes and data used in model calibration. This uncertainty is influenced by the
   a. Accuracy and availability of observations (e.g., measurement errors in both inputs and outputs)
   b. Appropriateness of calibration data sets (e.g., use of non-representative calibration data)
   c. Efficiency and appropriateness of calibration/optimisation algorithms
   d. Appropriateness of objective functions
3. Model structural uncertainty: also known as "model error," is related to the ability of the model to represent the real mechanism. This error can be further discerned into conceptualisation error (i.e., scale issues, omission or simplification of key processes), ill-posed equations, and poor identification of boundary conditions and numerical methods.

The above sources of uncertainty are interlinked, and their decomposition into individual components is rather a hard task requiring assumptions on the statistical properties of the individual errors. Figure 21.2 graphically presents the main sources of uncertainty along with their interconnections.

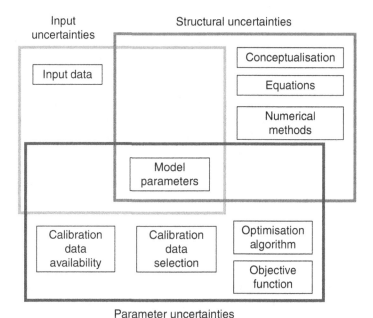

Figure 21.2 The main sources of uncertainties in urban drainage models and interlinks between them. (Adapted from Deletic, A. et al. 2012. *Physics and Chemistry of the Earth*, 42–44, 3–10.)

21.7.2 Uncertainty analysis in urban drainage modelling

The complexity and non-linearity of drainage models do not allow for the analysis, estimation, and propagation of uncertainties via analytical methods. This process typically involves probability theory, Monte Carlo (MC) sampling techniques, and formal or informal Bayesian statistical methods.

The most popular, Monte Carlo–based uncertainty assessment method is the Generalised Likelihood Uncertainty Estimator (GLUE), developed by Beven and Binley (1992). This approach rejects the idea of an *optimum model* or parameter set in favour of the *equifinality concept* that assumes that many parameter sets or model structures may have an equal likelihood of producing acceptable simulations.

In GLUE, different candidate model structures are identified along with the parameters that most affect the output. A large number of parameters are generated by prior probability distributions using an MC-type sampling process. The models are run for each generated parameter set, and the model outputs are compared to the available calibration data. A quantitative measure is then used to assess the performance of each model, based on the model residuals. Different formal or informal likelihood measures or combinations of them can be used. The models and their corresponding parameter sets that provide a likelihood measure that reaches a minimum threshold are retained and rescaled so that their cumulative probability sum is equal to 1. The uncertainty bands of model predictions are then calculated.

The implementation of the above general methodology requires the modeller to make subjective decisions at each step of the process (Beven, 2012):

- The model or models to include in the analysis
- The feasible range for each parameter value
- The sampling strategy for the parameter sets
- An appropriate likelihood measure or set of measures
- The conditions for rejecting models

The main advantage of the GLUE methodology is the flexibility to account for different sources of uncertainty along with the fact that there is no need for an assumption of the error distribution functions. For instance, by considering different parameter sets or model structures, one can explicitly assess parameter uncertainty or structural uncertainty, respectively. By using different input and output data, the effect of measurement errors can be estimated. Further to the high computational burden imposed by the great number of required simulations, the method has been mainly criticised for the use of informal likelihood measures for statistical inference and for the choice of the threshold of some likelihood measure above which models are considered acceptable.

A comparative analysis of the use of GLUE and other methods for parameter uncertainty analysis in urban drainage models is provided by Dotto et al. (2012).

21.8 VIRTUAL CASE STUDIES

The concept of benchmark systems was discussed earlier in this chapter (Section 21.4.4) where engineering solutions or interventions can be trialled on a standard system and so compared with other interventions. However, drawing conclusions from the results of individual investigations (even on a standard) is questionable and may result in erroneous conclusions that are not applicable to other systems. Ideally, solutions should be trialled on many different case studies but due to data and time constraints, this may not be possible. VCSs can provide a useful answer to this and enable a wider analysis without the need for extensive case study data.

One method of VCS generation is presented by Moderl et al. (2009), which uses an adapted Galton–Watson branching process to generate multiple distinct dendritic sewer system layouts, and designs the components according to standard values. Other methods to generate spatial layouts of water infrastructure networks include graph theory algorithms (Duque et al. 2020) and agent-based models (Urich et al. 2010). An example application is given in Section 24.7.

21.9 VISUALISING THE RESULTS OF MODELS

Visualisation of model results is crucial in water management, in general, and urban drainage, in particular. Presenting complex data in an intuitive manner ensures that both professionals and stakeholders can understand, interpret, and act upon the findings. With the advancement of technology, there has been a shift from traditional 2D visualisations to more immersive experiences.

Historically, 2D graphical representations, such as contour maps indicating water flow or bar charts showing rainfall intensities, have been the standard approach to results visualisation. They provide a straightforward view of data across a spatial dimension and are still used extensively in literally all the commercial and most of the research models available to date.

More recent applications involve 3D visualisation, where drainage systems, water flow, and flood risk can be represented in a real-world context (Zhi et al., 2019; Wang et al., 2019). These provide depth, making it easier to grasp the physical scale and relationships of different elements.

Both 2D and 3D visualisations are part of a model's user interface (UI) – the point of direct interaction between the user and the software. When designing this interaction, traditionally emphasis was placed on function and clarity, ensuring that inputs could be accurately fed into the model and results could be clearly read. However, as technology has evolved and user expectations grew, there has been a significant shift towards user experience (UX). This evolution moves beyond just the interface, encompassing the entirety of a user's interaction with a model, from ease of use to the satisfaction of outcomes. While a clean and intuitive UI remains crucial, modern computational models are now designed with the holistic "user journey" in mind. Elements like guided workflows, interactive visualisations, and real-time feedback loops are integrated to make the modelling process more intuitive and engaging. This shift not only makes models more user-friendly but also increases their adoption rate. By prioritising UX, developers ensure that water professionals can focus on the intricacies of their work rather than grappling with software, leading to more efficient and effective water management solutions.

21.9.1 Emerging trends in visualisation: immersive and playful

Two of the most recent, promising developments in model results visualisation are *Augmented* and *Virtual Reality*, sometimes collectively called *Mixed Reality*.

Augmented Reality (AR) superimposes digital information on the real world. In urban drainage, AR can overlay model results onto a physical space using devices such as smartphones or AR glasses. For instance, an engineer can point a device at a stormwater drain and see predicted flow rates, flood extents, upcoming maintenance needs, or potential blockages (see Puertas et al. (2020), for an example of an AR facility to run and visualise the results of numerical flood models, or Schall et al. (2010) for the description of software that helps engineers in the field to visualise underground infrastructure, like sewerage pipes). Figure 21.3 displays an AR mobile application (named "CirculAR"; Katika et al., 2022) used to improve citizen awareness of and engagement with water innovations.

Virtual Reality (VR) provides a fully "immersive" experience. Professionals can "walk" through a digital representation of the drainage system, observe water flow in real-time simulations, or even experience flood scenarios in a controlled environment. This depth of interaction enhances understanding and aids in decision-making.

As these more immersive ways of interacting with model results gain ground due to rapid technological progress in terms of both hardware and software, new more playful ways of interacting with models are also emerging. A noteworthy mention is the rise of so-called,

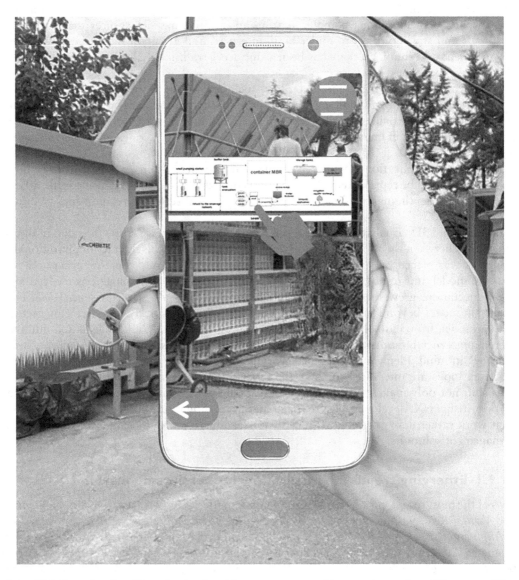

Figure 21.3 Example of an AR application for citizen engagement with water innovations. (Courtesy of Tina Katika.)

"Serious Games" (games designed for purposes other than just entertainment). In the realm of water management, these games offer interactive platforms where "players" can engage with water systems, make management decisions, and witness outcomes. They provide an educational tool, making, for example, complex urban drainage concepts accessible to a broader audience, from policymakers to the general public (see, for example, Khoury et al., 2023 for an example of a serious game that brings the concept of water in the circular economy closer to stakeholders). By gamifying the experience, users are not just passive observers; they're at least in principle, active participants, which fosters a deeper understanding of the trade-offs inherent in urban drainage and an increased ownership of co-created results (from new flood infrastructure design to CSO spill thresholds). Serious games do not even have to be computer-based (see, for example, Figure 21.4). However, a word of caution: the current

Figure 21.4 Board and associated paraphernalia of SuDSbury serious game. Reproduced with permission from: SuDSbury: A serious game to support the adoption of sustainable drainage solutions, Nguyena, et al., Urban Water Journal © 2023 The Author(s). Published by Informa UK Limited, trading as Taylor & Francis Group.

state of serious games design for water management is far from ideal (see Mittal et al., 2022 for a critical assessment of their contribution to decision-making). A lot of work in this field still needs to be done.

Nevertheless, the visualisation of model results in urban water applications, in general, and urban drainage, in particular, has seen transformative advancements. From static 2D maps to immersive AR and VR experiences, the field is continually evolving, ensuring that complex data is both accessible and actionable (Makropoulos and Savic, 2019). With tools like serious games, the realm of urban water management is also reaching out, beyond professionals, promoting widespread understanding and collaborative solutions co-created with policymakers and societal actors.

PROBLEMS

1 List the main alternative modelling approaches, and explain their main characteristics.
2 Define the integration of urban models, and discuss their main principles.
3 List the main parts of the urban wastewater system and consider ways in which they interact with each other. What are the potential benefits of integrated system control?
4 What is the difference between single and multi-objective modelling? What benefits can be gained from the latter?
5 Describe the main standards for model and data integration.
6 What is calibration of a model? What are the main difficulties in the calibration of integrated urban models?
7 List and describe the typical approaches to address the high computational burden of urban water models.
8 What are the main sources of uncertainty in urban drainage modelling? Explain a suitable method to assess them.
9 What is a virtual case study and why would you use one or more of them?
10 What is the difference between Augmented and Virtual Reality?
11 What is the difference between a model and a (serious) game?

KEY SOURCES

Bach, P.M., Rauch, W., Mikkelsen, P.S., McCarthy, D.T., and Deletic, A. 2014. A critical review of integrated urban water modelling—Urban drainage and beyond. *Environmental Modelling and Software*, 54, 88–107.

Deletic, A., Dotto, C.B.S., Mccarthy, D.T., Kleidorfer, M., Freni, G., Mannina, G. 2012. Assessing uncertainties in urban drainage models. *Physics and Chemistry of the Earth*, 42–44, 3–10.

Makropoulos, C. and Savic, D.A. 2019. Urban hydroinformatics: Past, present and future. *Water*, 11(10), 1959.dotto

Razavi, S., Tolson, B.A., and Burn, D.H. 2012. Review of surrogate modeling in water resources. *Water Resources Research*, 48(7), W07401.

Rozos, E. and Makropoulos, C. 2013. Source to tap urban water cycle modelling. *Journal of Environmental Modelling and Software*, 41, 139–150.

Wang, Q., Zhou, Q., Lei, X. and Savić, D.A. 2018. Comparison of multiobjective optimization methods applied to urban drainage adaptation problems. *Journal of Water Resources Planning and Management*, 144(11), 04018070

REFERENCES

Achleitner, S., Möderl, M., and Rauch, W. 2007. CITY DRAIN © An open source approach for simulation of integrated urban drainage systems. *Environmental Modelling and Software*, 22(8), 1184–1195.

Bach, P.M. 2012. *UrbanBEATS: An Exploratory Model for Strategic Planning of Urban Water Infrastructure* (Online). Melbourne. www.urbanbeatsmodel.com

Beven, K. 2012. *Rainfall-Runoff Modelling: The Primer*, 2nd edn, John Wiley and Sons, Ltd, Chichester, UK.

Beven, K. and Binley, A. 1992. The future of distributed models: Model calibration and uncertainty prediction. *Hydrological Processes*, 6, 279–298.

Bowes, B.D., Tavakoli, A., Wang, C., Heydarian, A., Behl, M., Beling, P.A. and Goodall, J.L. 2021. Flood mitigation in coastal urban catchments using real-time stormwater infrastructure control and reinforcement learning. *Journal of Hydroinformatics*, 23(3), 529–547.

Brito, R.S., Almeida, M.C. and S. Matos. 2017. Estimating flow data in urban drainage using partial least squares regression. *Urban Water Journal*, 14(5), 467–474.

Breiman, L., 2001. Random Forests. *Machine Learning*, 45, 5–32.

Brown, R.R., Keath, N., and Wong, T.H.F. 2009. Urban water management in cities: Historical, current and future regimes. *Water Science Technology*, 59(5), 847–855.

Burger, G., Fach, S., Kinzel, H., and Rauch, W. 2010. Parallel computing in conceptual sewer simulations. *Water Science Technology*, 61(2), 283–291.

Butler, D. and Schütze, M. 2005. Integrating simulation models with a view to optimal control of urban wastewater systems. *Environmental Modelling and Software*, 20, 415–426.

Cheng, C.-T., Wu, X.-Y., and Chau, K.W. 2005. Multiple criteria rainfall–runoff model calibration using a parallel genetic algorithm in a cluster of computers. *Hydrological Sciences Journal*, 50(6), 1069–1087.

Couckuyt, I., Deschrijver, D., and Dhaene, T. 2013. Fast calculation of multiobjective probability of improvement and expected improvement criteria for Pareto optimization. *Journal of Global Optimization*, 60(3), 575–594.

Darsono, S. and Labadie, J.W. 2007. Neural-optimal control algorithm for real-time regulation of in-line storage in combined sewer systems. *Environmental Modelling and Software*, 22(9), 1349–1361.

DHI. 2023. *Wastewater treatment process modelling at its finest*. Available from: https://www.mikepoweredbydhi.com/products/west

Dias, B.H., Tomim, M.A., Marcato, A.L.M., Ramos, T.P., Brandi, R.B.S., Junior, I.C.D.S., and Filho, J.A.P. 2013. Parallel computing applied to the stochastic dynamic programming for long term operation planning of hydrothermal power systems. *European Journal of Operational Research*, 229(1), 212–222.

Dobson, B. and Mijic, A. 2020. Protecting rivers by integrating supply-wastewater infrastructure planning and coordinating operational decisions. *Environmental Research Letters*, 15, 114025.

Dotto, C.B.S., Mannina, G., Kleidorfer, M., Vezzaro, L., Henrichs, M., Mccarthy, D.T. 2012. Comparison of different uncertainty techniques in urban stormwater quantity and quality modelling. *Water Research*, 46(8), 2545–2558.

Duque, N., Duque, D., Aguilar, A., Saldarriaga, J., 2020. Sewer network layout Selection and hydraulic design using a mathematical optimization framework. *Water*, 12 (12), 3337.

Efstratiadis, A., Dimas, P., Pouliasis, G., Tsoukalas, I., Kossieris, P., Bellos, V., Sakki, G.K., Makropoulos, C. and Michas, S. 2022. Revisiting flood hazard assessment practices under a hybrid stochastic simulation framework. *Water*, 14(3), 457.

Elliott, A.H. and Trowsdale, S.A. 2007. A review of models for low impact urban stormwater drainage. *Environmental Modelling and Software*, 22, 394–405.

Forrester, A. and Keane, A. 2009. Recent advances in surrogate-based optimization. *Progress in Aerospace Sciences*, 45(1–3), 50–79.

Freni, G., Mannina, G., and Viviani, G. 2009. Uncertainty assessment of an integrated urban drainage model. *Journal of Hydrology, 373*(3–4), 392–404.

Friedman, J.H., 2002. Stochastic gradient boosting. *Computational Statistics & Data Analysis, 38,* 367–378.

Freni, G., Maglionico, M., Mannina, G., and Viviani, G. 2008. Comparison between a detailed and a simplified integrated model for the assessment of urban drainage environmental impact on an ephemeral river. *Urban Water Journal, 5*(2), 87–96.

Fu, G., Jin, Y., Sun, S., Yuan, Z. and Butler, D., 2022. The role of deep learning in urban water management: A critical review. *Water Research, 223,* 118973.

Hardy, M.J., Kuczera, G., and Coombes, P.J. 2005. Integrated urban water cycle management: The UrbanCycle model. *Water Science Technology, 52*(9), 1–9.

Harpham, Q.K., Hughes, A. and Moore, R.V. 2019. Introductory overview: The OpenMI 2.0 standard for integrating numerical models. *Environmental Modelling & Software, 122,* 104549.

Hochreiter, S., and Schmidhuber, Jürgen. 1997. Long short-term memory. *Neural Computation, 9*(8), 1735–1780.

Institut für Automation und Kommunikation (IFAK). 2023. *SIMBA# - Simulation software for water and wastewater,* www.ifak.eu/en/produkte/simba

Jamali, B., Bach, P.M. and Deletic, A., 2020. Rainwater harvesting for urban flood management–An integrated modelling framework. *Water Research, 171,* 115372.

Jeppsson, U., Alex, J., Batstone, D.J., Benedetti, Comas, J., Copp, J.B., Corominas, L., Flores-Alsina, X., Gernaey, K.V., Nopens, I., Pons, M-N, Rodríguez-Roda, I., Rosen, C., Steyer, J-P., Vanrolleghem, P.A., Volcke, E.I.P. and Vrecko, D. Benchmark simulation models, quo vadis. 2013. *Water Science and Technology.* 68(1), 1–15.

Katika, T., Karaseitanidis, I., Tsiakou, D., Makropoulos, C., & Amditis, A. 2022. Augmented reality (AR) supporting citizen engagement in circular economy. *Circular Economy & Sustainability, 2*(3), 1077–1104.

Khu, S.T. and Madsen, H. 2005. Multiobjective calibration with Pareto preference ordering: An application to rainfall-runoff model calibration. *Water Resources Research, 41,* 1–14.

Khoury, M., Evans, B., Chen, O., Chen, A.S., Vamvakeridou-Lyroudia, L., Savic, D.A., Djordjevic, S., Bouziotas, D., Makropoulos, C. and Mustafee, N., 2023. NEXTGEN: A serious game showcasing circular economy in the urban water cycle. *Journal of Cleaner Production, 391,* 136000.

Knapen, R., Janssen, S., Roosenschoon, O., Verweij, P., de Winter, W., Uiterwijk, M. and Wien, J.E. 2013. Evaluating OpenMI as a model integration platform across disciplines. *Environmental Modelling & Software, 39,* 274–282.

Koutiva, I. and Makropoulos, C. 2016. Modelling domestic water demand: An agent based approach. *Environmental Modelling & Software, 79,* 35–54.

Kwon, S.H. and Kim, J.H., 2021. Machine learning and urban drainage systems: state-of-the-art review. *Water, 13*(24), 3545.

LeCun, Yann, Yoshua Bengio, and Geoffrey Hinton. Deep learning. 2015. *Nature, 521*(7553), 436–444.

Loke, E., Warnaars, E.A., Jacobsen, P., Nelen, F., and Almeida, M. 1996. Problems in urban storm drainage addressed by artificial neural networks. *Proceedings of the Seventh International Conference on Urban Storm Drainage, 3,* Hannover, September, 1581–1586.

Makropoulos, C. 2017. Thinking platforms for smarter urban water systems: Fusing technical and socio-economic models and tools. *Geological Society, London, Special Publications, 408(1),* 201–219.

Maniquiz, M.C., Lee, S. and Kim, L.H., 2010. Multiple linear regression models of urban runoff pollutant load and event mean concentration considering rainfall variables. *Journal of Environmental Sciences, 22*(6), 946–952.

Mittal, A., Scholten, L. and Kapelan, Z. 2022. A review of serious games for urban water management decisions: current gaps and future research directions. *Water Research, 215,* 118217.

Mitchell, V.G. 2004. Integrated urban water management, a review of current Australian. *Technical Report of the Australian Water Conservation and Reuse Research Program,* a joint initiative of CSIRO and Australian Water Association, Melbourne, Australia.

Moderl, M., Butler, D. and Rauch, W. 2009. A stochastic approach for automatic generation of urban drainage systems. *Water Science and Technology, 59*(6), 1137–1143.

Muschalla, D. 2008. Optimization of integrated urban wastewater systems using multi-objective evolution strategies. *Urban Water Journal*, 5(1), 57–65.

Muschalla, D., Schütze, M., Schroeder, K., Bach, M., Blumensaat, K., Klepiszewski, K., 2008. The HSG guideline document for modelling integrated urban wastewater systems. *Proceedings of the 11th International Conference on Urban Drainage*, Edinburgh, September, on CD-ROM.

Nikolopoulos, D., Kossieris, P., Tsoukalas, I. and Makropoulos, C. 2022. Stress-testing framework for urban water systems: A source to tap approach for stochastic resilience assessment. *Water*, 14(2), 154.

Nguyena, J., Mittala, A., Kapelan, Z. and Scholten, L. 2023. SuDSbury: A serious game to support the adoption of sustainable drainage solutions. *Urban Water Journal*, 21(2), 204–218.

Open Geospatial Consortium (OGC). 2014. *SensorML: Model and XML Encoding Standard*, Mike Botts.

Puertas, J., Hernández-Ibáñez, L., Cea, L., Regueiro-Picallo, M., Barneche-Naya, V. and Varela-García, F.A., 2020. An augmented reality facility to run hybrid physical-numerical flood models. *Water*, 12(11), 3290.

Rauch, W., Bertrand-Krajewski, J., Krebs, P., Mark, O., Schilling, W., Schütze, M., and Vanrolleghem, P.A. 2002. Deterministic modelling of integrated urban drainage systems. *Water Science Technology*, 45(3), 81–94.

Rauch, W., Seggelke, K., Brown, R., and Krebs, P. 2005. Integrated approaches in urban storm drainage: Where do we stand? *Environmental Management*, 35(4), 396–409.

Rauch, W., Bach, P.M., Brown, R., Deletic, A., Ferguson, B., De Haan, J. 2012. Modelling transition in urban drainage management. *Ninth International Conference on Urban Drainage Modelling*. Belgrade, Serbia.

Razavi, S., Tolson, B.A., Matott, L.S., Thomson, N.R., MacLean, A., and Seglenieks, F.R. 2010. Reducing the computational cost of automatic calibration through model preemption. *Water Resources Research*, 46(11), W11523. doi:10.1029/2009WR008957

Regis, R.G. and Shoemaker, C.A. 2007. A stochastic radial basis function method for the global optimization of expensive functions. *INFORMS Journal on Computing*, 19(4), 497–509.

Regis, R.G. and Shoemaker, C.A. 2013. Combining radial basis function surrogates and dynamic coordinate search in high-dimensional expensive black-box optimization. *Engineering Optimization*, 45(5), 529–555.

Rozos, E. and Makropoulos, C. 2013. Source to tap urban water cycle modelling. *Journal of Environmental Modelling and Software*, 41, 139–150.

Saagi, R., Flores-Alsina, X., Fu, G., Butler, D., Gernaey, K.V. and Jeppsson, U. 2016. Catchment and sewer network simulation model to benchmark control strategies within urban wastewater systems. *Environmental Software and Modelling*, 78, 16–30.

Schall, G., Schmalstieg, D. and Junghanns, S., 2010. VIDENTE-3D visualization of underground infrastructure using handheld augmented reality. In *GeoHydroinformatics: Integrating GIS and Water Engineering*, 207–219, CRC Press.

Schütze, M., Butler, D., and Beck, M.B. 2001. Parameter optimisation of real-time control strategies for urban wastewater systems. *Water Science Technology*, 43(7), 139–146.

Shishegar, S., Duchesne, S. and Pelletier, G. 2018. Optimization methods applied to stormwater management problems: a review. *Urban Water Journal*, 15(3), 276–286.

Solvi, A.M. 2007. *Modelling the Sewer-Treatment-Urban River System in View of the EU Water Framework Directive*, PhD thesis, Ghent University, Belgium.

Sweetapple, C., Salomons, E., Le Gall, F., Abid, A., Vamvakeridou-Lyroudia, L., Chen, A.S. and van den Broeke, J. 2022. Integrating Epanet and FIWARE for Development of Water Distribution System Digital Twins. In *Advances in Hydroinformatics: Models for Complex and Global Water Issues—Practices and Expectations* (1081–1086). Springer Nature Singapore.

Tolson, B.A. and Shoemaker, C.A. 2007. Dynamically dimensioned search algorithm for computationally efficient watershed model calibration. *Water Resources Research*, 43(1), WR004723.

Tolson, B.A., Asadzadeh, M., Maier, H.R., and Zecchin, A. 2009. Hybrid discrete dynamically dimensioned search (HD-DDS) algorithm for water distribution system design optimization. *Water Resources Research*, 45(12), W12416.

Tsoukalas, I. and Makropoulos, C. 2014. A surrogate based optimization approach for the development of uncertainty-aware reservoir operational rules: The case of nestos hydrosystem. *Water Resources Management*, 29, 4719–4734.

Tsoukalas, I., Kossieris, P., Efstratiadis, A., and Makropoulos, C. 2016. Surrogate-enhanced evolutionary annealing simplex algorithm for effective and efficient optimization of water resources problems on a budget. *Environmental Modelling and Software*, 77, 122–142.

Urich, C., Sitzenfrei, R., Moderl, M. and Rauch, W., 2010. An agent-based approach, for generating virtual sewer systems. *Water Science and Technology*, 62(5), 1090–1097.

Wang, C., Hou, J., Miller, D., Brown, I. and Jiang, Y. 2019. Flood risk management in sponge cities: The role of integrated simulation and 3D visualization. *International Journal of Disaster Risk Reduction*, 39, 101139.

van der Werf, J. A., Kapelan, Z., & Langeveld, J. G. 2023. HAPPy to control: A heuristic And predictive policy to control large urban drainage systems. *Water Resources Research*, 59, e2022WR033854.

Zhang, Z., Tian, W. and Liao, Z. 2023. Towards coordinated and robust real-time control: A decentralized approach for combined sewer overflow and urban flooding reduction based on multi-agent reinforcement learning. *Water Research*, 229, 119498.

Zhi, G., Liao, Z., Tian, W., Wang, X. and Chen, J., 2019. A 3D dynamic visualization method coupled with an urban drainage model. *Journal of Hydrology*, 577, 123988.

Chapter 22

Smart systems

22.1 INTRODUCTION

Recent technological innovations and developments in information and communication technologies (ICTs), provide new perspectives towards a more dynamic, real-time, and "smart" management of urban drainage infrastructure. This "smartness" emerges from a combination of new real-time *data*, captured through novel *sensors*, which are analysed by new *models* running increasingly on the Cloud to provide real-time services, such as real-time control (RTC) or early warning (Makropoulos and Savic, 2019). This increasingly tight integration between physical assets, such as pipes and pumps in the field, and our models simulating, predicting, and controlling their performance through a continuous stream of bi-directional real-time data exchange, is giving rise to a novel, but promising concept: that of *digital twins*. At the same time, it also gives rise to new risks for smart water systems, as these become "cyber-physical" infrastructures rather than purely physical ones.

This chapter starts (Section 22.2) with an account of the main new data sources for urban drainage applications (including Internet of Things [IoT] devices such as Smart Sensors, Earth Observation Systems, and also Citizen Observatories). It then briefly presents the concept of a digital twin (Section 22.3) discussing what this could mean for urban drainage. It then proceeds with an account of two key applications of smart urban drainage: RTC and early warning. Section 22.4 presents RTC for urban drainage systems – well understood, but not yet widely practised. Here, we suggest that technological and methodological developments now make it possible to model and optimise the effect of control on the urban wastewater system as a whole. Then the chapter describes the role of early warning systems (EWSs) as a key tool in the management of water-related hazards (Section 22.5), allowing us to prevent loss of life and limit property and infrastructure damage by predicting storm events in advance and warning vulnerable communities. Especially in the urban setting, EWSs deal with (pluvial) flash floods and (fluvial) surface water/river floods, as well as coastal floods and storm-induced landslides, mud flows, and debris flows. Last, we discuss emerging risks from this move towards smarter systems, specifically addressing the issue of *cyber-physical security* of water infrastructures, and by providing examples of relevant work to shield our systems from harm (Section 22.6).

22.2 NEW DATA SOURCES

The evolution of urban drainage has long been intertwined with the progression of technology. Historically, models and systems depended on periodic field measurements and rudimentary monitoring tools, limiting real-time assessment and timely response. Fast forward

DOI: 10.1201/9781003408635-22

to the twenty-first century, a transformative era for data has emerged, thanks to the IoT, smart meters, and advanced sensors. These technologies are shaping the way we understand, design, and maintain urban drainage systems.

IoT refers to the network of physical devices embedded with sensors, software, and other technologies to connect and exchange data over the Internet. In the realm of urban drainage, IoT devices can collect a myriad of data points, from water flow rates to pollutant concentrations, enabling instantaneous feedback and adaptive system adjustments. Today's sensors can detect everything from water quality parameters like pH, turbidity, and contaminant concentrations to physical parameters like water level and flow velocity. When strategically placed throughout an urban drainage network, these sensors can provide a holistic view of the system's performance and indicate areas of potential concern. As these sensors become cheaper and more reliable, they afford a much-increased granularity in terms of both spatial and temporal coverage. A case in point is the so-called smart meter. Smart meters are advanced measurement tools that provide real-time data on water usage, flow rates, and even potential blockages. They are installed at household and network levels and can revolutionise urban drainage by monitoring water consumption and wastewater generation patterns at different timescales (Kossieris and Makropoulos, 2018) in real time, thus facilitating RTC (see Section 22.4) of the integrated urban water system. For example, Oberascher et al. (2021) present a way to improve urban drainage performance through advanced control of IoT-based micro storages in the form of "smart" rain butts (see also Section 11.2.1.3). They report that combined sewer overflow (CSO) volume can be reduced using this smarter, IoT-driven approach, "by between 7 and 67% depending on rain characteristics [...] and an applied control strategy."

22.2.1 Earth observation and soft sensors

Beyond sensors deployed in the field (such as those described above), new data streams are also coming into play, in the form of Earth Observation (EO) datasets. These are generated by sensors onboard satellites such as Landsat, Sentinel, and others, which capture high-resolution imagery and multi-spectral data of the Earth's surface. Of particular interest for urban drainage are EO datasets of

- land use and land cover: these help with the identification of impermeable surfaces, green areas, and waterbodies – information which is essential to estimate runoff coefficients and to model water flow patterns within urban areas
- topographic information: used to derive Digital Elevation Models (DEMs) which can help in delineating catchments and identifying natural flow paths, aiding flood prediction and drainage infrastructure planning
- surface temperatures, soil moisture conditions and rainfall estimates: all of which are crucial parameters for urban drainage models.

These new EO-derived datasets can be combined with field measurements to develop "Soft Sensors." Soft (as opposed to "hard," i.e., hardware-based) sensors are simply mathematical models that integrate data from various sources (like EO datasets, IoT devices, and physical sensors), processing this information to generate estimates of unmeasured (or unmeasurable) variables at different spatial or temporal scales. For example, Palmitessa et al., (2021) present a soft-sensor scheme that assimilates field observations into an ensemble of 1D hydrodynamic models to validate physical sensor measurements with high confidence.

Using soft sensors that combine the (relatively few) field observations we typically have in an urban catchment with regularly updated high-resolution information we always have, through satellites for the entire urban catchment, our models can capture and integrate much more of the available information and become more trustworthy.

22.2.2 Crowdsourcing and citizen observatories

Today's wide use of smart mobile communication devices (such as smartphones) along with the advent of new ICT services (i.e., greater capabilities of sharing, storing, processing, and displaying vast amounts of data and information on the fly) allows for an active engagement of citizens, beyond scientists and water professionals, in water management. In the "human observatories" framework, citizens act as "sensors" that collect and share valuable information and observations, providing up-to-date awareness of water infrastructures and environmental problems, such as urban flooding (Wehn et al., 2015).

In an urban setting, citizens-as-sensor technologies find wide applicability, especially in flood risk management. A characteristic example of such a technology is the PEARL Detective smartphone application (Figure 22.1) which enables users to share information about flood incidents with the authorities.

The user can submit, through the smartphone application, text reports and other information (including photos) on a flood-related event (e.g., increase in river water level, overtopping) or on situations that may cause problems during a future event (e.g., damaged pipes, obstructed inlets). The system supports two-way communication, also informing the users about flood events that are forecasted in their area.

Many water service providers have opened new digital communication channels to enable customers to report potential problems. For example, a customer can report a potential pollution incident using WhatsApp or SMS messaging, including location and photographic evidence. Such approaches are increasingly playing an important role in shaping smart urban drainage systems of the future, creating a more direct link between infrastructure, authorities/companies, and citizens/customers.

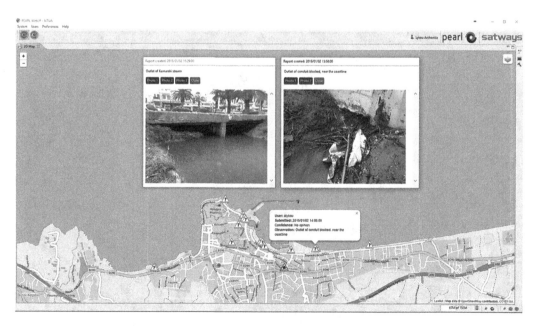

Figure 22.1 The PEARL Detective app applied for flood protection in the city of Rethymno. (Reproduced from http://www.pearl-fp7.eu/, courtesy of Christos Makropoulos.)

22.3 DIGITAL TWINS

A "digital twin" is a dynamic, digital replica of a physical system or process, which accurately mimics and simulates its real-world counterpart (see Figure 22.2). Essentially, a digital twin changes as its real-world counterpart changes and vice versa. In the context of urban drainage, this means creating a comprehensive digital model of the entire drainage network, infrastructure, and the environment in which it operates, and continuously updating it with real-time data sourced from diverse sources, such as IoT devices and EO datasets. Clearly, such a radical interaction between different systems suggests the need for reliable open interoperability standards, such as FIWARE (see Section 21.5).

Unlike traditional models, digital twins evolve with time, adapting to changes in the real-world system. With IoT and EO, digital twins receive constant data feeds, enabling them to reflect current conditions accurately. Perhaps more importantly, professionals can interact with digital twins, simulating different scenarios and testing possible interventions in the "real system" before deploying them in the field, bringing planning and operation closer together (Pedersen et al., 2021).

By being constantly updated with data from sensors and devices, digital twins can provide an ideal infrastructure for RTC of drainage infrastructures. For instance, if sensors detect a surge in water flow rates, the digital twin can simulate potential outcomes, guiding controllers to adjust floodgates or pump operations instantly (see Section 22.4). An early example of such an application, towards a digital twin of an urban drainage system, can be seen in Bartos and Kerkez (2021) where the authors show that by fusing data from the field into an urban drainage model, running in real time, forecasting of stormwater depths and flows is significantly improved. Digital twins can also integrate weather forecasts, river flow rates, and current protection infrastructure states to predict potential flooding scenarios. This predictive capability provides authorities with early warning (see Section 22.5).

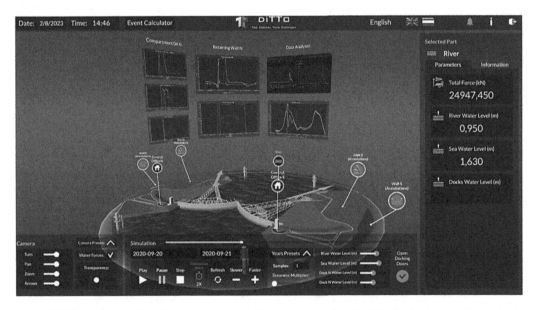

Figure 22.2 Example of a digital twin prototype of the storm surge barrier Maeslantkering, the Netherlands. (Courtesy of Luc Ponsioen.)

22.4 REAL-TIME CONTROL

Most urban drainage systems are still managed in an essentially passive mode. That is, fixed elements are provided to operate in one way only without the opportunity for intervention of any kind. Combined sewer systems, for example, were conventionally designed (Chapter 13) to handle flows from storms of high return periods (low frequency) and, therefore, contain large amounts of storage volume (even without dedicated stormwater detention). For most storm events, this storage is not fully utilised. So there should be significant scope to improve the performance of the system by actively or smartly managing this storage in response to changing input (e.g., rainfall) and output (e.g., flooding, overflows, interaction with the wastewater treatment plant (WTP)).

22.4.1 Definition

An urban drainage system can be considered to operate under RTC when process data (state information), which is currently monitored in the system, is used to operate flow regulators during the actual process. Thus, system information (e.g., rainfall, water level) is continuously collected and processed and is used to make decisions about the operation of devices (e.g., pumps, weirs) in the system in real time to limit the occurrence of adverse effects. This is also sometimes called *active control*, *active system control* or *active system management*.

The ultimate target in the application of RTC in urban drainage systems is the reduction of the risk of flood events and/or sewer overflows while minimising the need to increase the capacity of the existing system (García et al., 2015). Other applications include smoothing the profile of flow into the WTP and protecting the WTP during storm conditions from process overload.

22.4.2 Equipment

The main hardware elements to be found in RTC systems are

- *Sensors* that monitor the ongoing process
- *Regulators* that manipulate the process
- *Controllers* that activate the regulators
- *Data transmission systems* that carry measured data from sensors to controllers and signals from controllers to regulators

Together, these four elements form the *control loop* that is common to all RTC systems (considered in Section 22.4.3.2). More details on RTC components can be found in Campisano et al. (2013).

22.4.2.1 Sensors

Although there is a large number of sensor types, only a few are suitable for RTC of urban drainage systems. These include rain gauges, sewer water level gauges, flow rate gauges, and water quality sensors. In general, requirements include long-term resistance to water and chemicals, the suitability for continuous recording, remote data transmission, long life or continuous energy source, robustness, reliability and low maintenance.

Water level measuring devices are the most commonly used sensors. They are used to determine the state of storage facilities, to evaluate flow depths into pipes, and to monitor the discharges through regulators such as a gate orifice. See Section 18.4 for more details on water level sensors.

Flow rate estimates can be obtained from weirs and gates via level-flow converting relationships. More accurate measurements can be provided via electromagnetic and ultrasonic flow rate meters.

Technological improvements are making sensors more durable in the difficult environment of drainage systems, enabling the measurement of various quality parameters, such as temperature, pH, turbidity, suspended solids, organic, carbon, chemical oxygen demand, conductivity, ammonia, nitrate, and total nitrogen. The use of water quality sensors, although showing considerable potential, remains limited in the RTC framework.

Rain gauges along with radar stations are typically used to deliver information on the spatio-temporal distribution of rainfall over large catchments. Rainfall and flow forecasts from radar measurements are used to configure the control strategy (Fuchs and Beeneken, 2005).

22.4.2.2 Regulators

Sewer flow regulators, also known as actuators, are the components of the RTC system that are used to control water flows and levels. They include a variety of pumps (constant or variable speed), moveable weirs, penstocks, gates, inflatable dams, valves, and flow splitters (see Chapters 8 and 15). Chemical dosing devices and aeration devices can also be used to adjust the quality parameters of the flow.

22.4.2.3 Controllers

A controller is required for every regulator. It accepts an input signal and adjusts the regulator accordingly. The most common types of digital controllers are programmable logic controllers (PLCs) or remote terminal units (RTUs). Controllers can be broadly classified into two categories: continuous and discrete. The most common method of discrete control is the two-point controller which has only two positions: on/off or open/closed. An example of two-point control is shown in Figure 22.3 for a simple pumping station. Here, the pump switches off at a low level and on at a high level as shown.

The disadvantage of two-point control is that the frequency of switching can become excessive. Three-point controllers have been developed to overcome this problem, and they are typically used for regulators such as automatic penstocks and moveable weirs. The most commonly used controller for continuously variable regulator settings is the *PID* (*Proportional-Integral-Derivative*) *controller*. Simplified versions (P, PI, and PD) are also available.

To prevent adverse effects in the case of controller breakdown or data transmission failures, emergency strategies should be planned and implemented on-site.

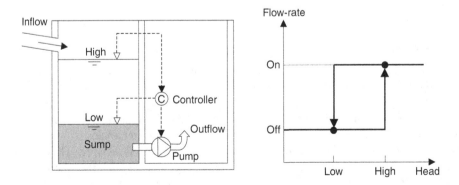

Figure 22.3 Two-point control at a pumping station.

22.4.2.4 Data transmission systems

The RTC components described above can be distributed over the entire drainage system, and hence a data transmission (telemetry) system is required to ensure communication among them. Data transmission is either wired or wireless and should be fast and safe. Non-infrastructure locations typically use wired data transmission whilst network/infrastructure sensors more often rely on mobile-phone communications (e.g., 4G). This is relevant as wireless systems can be less reliable and the risk of temporary losses of signal must be considered as part of the RTC design.

22.4.3 Control

22.4.3.1 Classification

RTC systems can be broadly classified as local, global, or integrated. Under local control, regulators use process measurements taken directly at the regulator site (e.g., by a float) but are not remotely manipulated from a control centre, even if operational data are centrally acquired. An example is an automatic penstock being operated in relation to the water level monitored by an ultrasonic level monitor.

In global RTC systems, regulators are operated in a coordinated way by a central computer with respect to process measurements throughout the entire system. This central computer is often part of a control system, also known as a SCADA system (supervisory control and data acquisition) that monitors and controls the main components of the drainage system. For example, the mobilisation of upstream and downstream storage volumes is linked in order to avoid emptying the upper tank into the already-full lower tank. In large and complex drainage systems, both local and global RTC systems can co-exist.

The newest class (and least proven in practice) is that of integrated control where control can be influenced by process measurements derived from systems outside the sewer system (e.g., WTP, receiving water). Integrated modelling was discussed in Chapter 21.

22.4.3.2 Control loop

The *control loop* is the basic element of any RTC system. In this loop, measured values of the controlled variable are compared with a set of reference values, known as *set-points*. The outcome of this comparison then determines how the variable will be adjusted. Two main types of control loops can be distinguished (see Figure 22.4): feedback and feed forward.

- Feedback control: Control commands are actuated depending on the measured deviation of the controlled process from the set-points. Unless there is a deviation, the feedback controller is not actuated.
- Feed forward control: This anticipates the immediate future values of these deviations using a model of the process controlled and activates the controls ahead of time to compensate. Its accuracy, therefore, depends on the effectiveness of the model.

22.4.3.3 Control strategy

A *control strategy*, also known as a *control algorithm*, can be defined as either the time sequence of all regulator set-points or the set of control rules in a RTC system. Control can either be performed *offline*, via static rules, or *online* via rules that are continuously updated depending on fast computer forecasts of the system state (*model predictive control* [MPC]). The simplest strategy is to keep the set-points constant, but time-varying set-points arguably

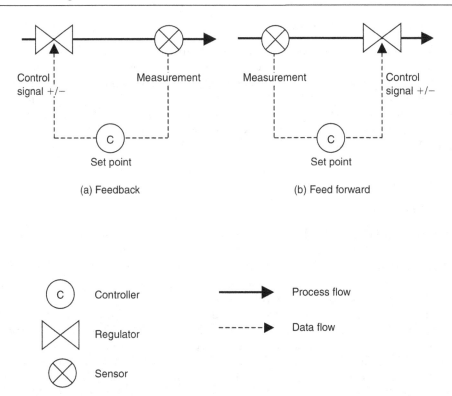

Figure 22.4 Control loop.

give better performance, allowing the system to react to the non-regular transient storm events to which it is subjected.

A great variety of methods has been proposed for the development of control strategies (García et al., 2015; Ocampo-Martinez, 2010; Schütze et al., 2004, Sun et al., 2021). In general, control strategies can be a product of either a heuristic approach or a more rigorous optimisation procedure. Heuristic control algorithms are based on the experience of the operations personnel to derive the initial set-points. Then, multiple simulation runs can be performed to improve the performance of the controllers. General control strategies are

- Preferential upstream storage: Wastewater/stormwater is retained first in the upper reaches of the network to reduce downstream flooding impacts.
- Preferential downstream storage: Wastewater/stormwater is retained first in the lower reaches of the network to minimise upstream CSO impacts.
- Balanced storage: The various storage elements are evenly filled throughout the catchment.

Through this process, *decision matrices* containing control actions for all possible combinations of the systems inflow and state variables can be produced *a priori* (offline). The control actions are presented in the form of "if … then … else" rules, where the first part of the statement describes the current state of the system. Relatively simple rules can be amended to improve system performance further. Although this type of algorithm is simple to implement and understand, it requires high levels of expertise from the staff and good knowledge of the system's behaviour. Fuzzy logic has also been used to obtain control values according to fuzzy rules (Klepiszewski and Schmitt, 2002; Nikoli et al., 2010).

Mathematical optimisation algorithms have also been used to derive optimum control rules on the basis of specific criteria (see Section 21.3). Operational objectives are translated into the minimisation or maximisation of an objective function subject to (physical) constraints. For example, a simple RTC optimisation procedure would be to minimise the sum of all CSO discharge volumes V_i over a time horizon t_i to t_f as shown in Equation (22.1):

$$\sum_{i=t_i}^{t_f} V_i \rightarrow \min \tag{22.1}$$

The aim is to identify the optimum solution and, therefore, the "best" performance of the system, given all constraints and assumptions. Some commonly used criteria are as follows (Joseph-Duran et al., 2014):

- Minimisation of flooding in streets/buildings
- Maximisation of the treated wastewater in the system
- Minimisation of operational costs (pump stations and WTPs)
- Minimisation of the water pollution released to the environment

Further to the *a priori* specified control strategies, periodical determination of the control rules depending on the current state of the system can also be implemented. In *online optimisation*, a number of possible control actions are evaluated, at predefined time steps (e.g., 5 minutes), and the most beneficial (on the basis of certain criteria) is applied. In the MPC method, a mathematical model of the process is frequently updated with current measurements, generating a sequence of future control actions within a predefined, finite-time, horizon (García et al., 2015). The sequence of actions is derived via an optimisation procedure that uses the predicted outputs of previous steps and the set of system constraints (operational, technical, etc.). The MPC approach has the advantage of anticipating the system's response to forthcoming rainfall events and the capacity to capture the complex non-linear system dynamics. However, the high computational effort required to evaluate the impact of different control actions in a drainage system can be an impediment to implementation particularly in terms of connection to existing telemetry systems. To alleviate this burden, simplified drainage models, balancing accuracy and complexity, can be used (see Section 21.6.1). Features of these models are discussed in García et al. (2015) and Antonio and Gutierrez (2014). The application of MPC methods in urban drainage systems is discussed in Marinaki and Papageorgiou (2005) and Ocampo-Martinez (2010).

Information collection and interpretation is a vital part of the processing of any control strategy. Historical data are very useful but forecast information can be even more valuable to allow the system to become ready for the expected load. Thus, possible sources of information are as follows:

- Flow, level, and quality measurements in upstream sewers: System reaction must be within the time of flow.
- Rainfall measurements and results from rainfall/runoff models: Available reaction time is extended to the time of concentration of the catchment.
- Rainfall forecasts: These gain additional time depending on the forecast lead time.

22.4.4 Integrated control

With the advent of integrated modelling tools that can represent the dynamic hydraulic and water quality processes within the entire system, opportunities are created for controlling the performance of the system as a whole (see also Section 21.4). In this section, we focus on *integrated control*, with the objective of minimising detrimental impacts on receiving waters, which is characterised by two aspects:

- *Integration of objectives*: control objectives within one sub-system may be based on criteria measured in other sub-systems (e.g., operation of pumps in the drainage system directed at minimising oxygen depletion in the receiving waterbody).
- *Integration of information*: control decisions taken in one sub-system may be based on information about the state of other sub-systems (e.g., operation of pumps in the drainage system based on WTP effluent data).

RTC of the components of the integrated system, therefore, takes into account the state of the whole system when utilising the control devices available in order to reach operational objectives defined for any location in the system. Table 22.1 summarises some of the measures and objectives of control, as well as methods applied to determine control strategies.

Figure 22.5 shows diagrammatically how individual sub-systems can be placed in an integrated framework to include control and optimisation. Here, an integrated model has been applied to a catchment, and various degrees of control are simulated. These include a base case with local control only, an optimised variation of this with fixed set-points, and an example of simple integrated control. The performance of each of the control scenarios is optimised to allow fair comparison.

Figure 22.6 illustrates the effects of these scenarios in terms of reaching a single objective, namely, (1) the duration (percentage of run time) of the oxygen concentration below a 4 mg/L threshold value at any location in the receiving water (a river in this case) or (2) the minimum DO concentration in the river during the simulated time period. It can be seen that control with optimised fixed set-points leads to improved performance over the base case. Further improvement is achieved for the integrated control scenario.

22.4.4.1 Multi-objective control

In the case of *multi-objective control*, more than one operational objective is considered. A number of studies (e.g., Fu et al., 2008; Newman et al., 2014; Wang et al., 2018) have

Table 22.1 Components of control of the urban wastewater system

Sub-system	Devices	Objectives	Decision-finding methods
Sewer system	• Pumps • Weirs • Gates	• Prevention of flooding • CSO reduction (frequency, volumes, loads)	• Heuristics, experience • Self-learning expert system • Offline optimisation
Treatment plant	• Weirs, gates • Return sludge rate • Waste sludge rate • Aeration	• Equalisation of flows • Maintenance of effluent standards • Process maintenance • Improved water quality • Flood protection	• Online optimisation • Model predictive control • Application of control theory
Receiving river	• Weirs • Gates		

Source: Schütze et al. (1999).

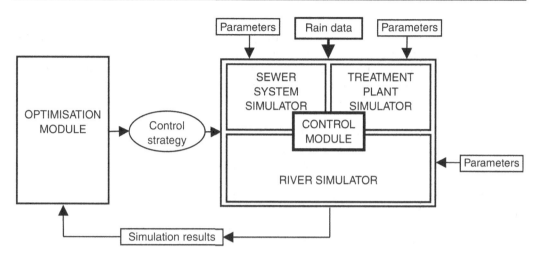

Figure 22.5 Overview of the "Integrated Simulation and Optimisation Tool." (Reproduced from Schütze et al. (1999), with permission of publishers Pergamon Press and copyright holders IAWQ.)

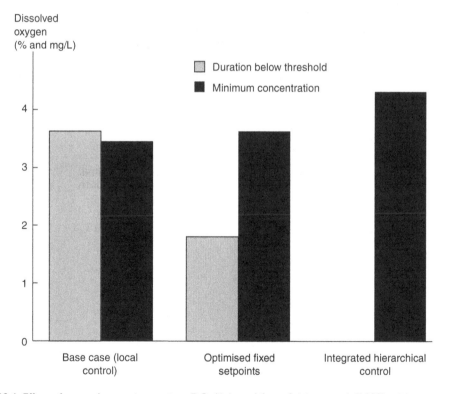

Figure 22.6 Effect of control scenario on river DO. (Adapted from Schütze et al. (1999), with permission of publishers Pergamon Press and copyright holders IAWQ.)

used multi-objective optimisation techniques to develop control strategies for the integrated urban wastewater system to establish what the possible trade-offs are between key parameters such as water quality determinands, energy consumption, and cost. By way of example, Figure 22.7 shows the output of a study on an integrated system where both river water dissolved oxygen (DO) concentration and ammonium concentration are used as optimisation objectives for many different control strategies, using a *genetic algorithm* as the optimiser. A clear trade-off is evident between the two parameters; an increase in one implies a decrease in the other (a *Pareto front*). In other words, it is not possible to both fully maximise river DO concentration and at the same time minimise ammonium concentration.

In practice, feasible solutions on the Pareto front can be reduced to a great extent by only considering those that comply with river water quality standards (the area shaded grey in Figure 22.7). Comparison should also be made with the non-optimised base case indicating the improvement that can be achieved in terms of both DO and ammonium concentrations.

Other optimisation methods have been trialled with some success and both Lund et al. (2020) and Sun et al. (2020) used MPC within an integrated framework to simulate the benefits of RTC on system pollution control. Wang et al. (2018) provide a useful overview and comparison of multi-objective optimisation methods applied to urban drainage adaptation problems.

22.4.4.2 Emerging approaches

Heuristic methods, such as (offline) rule-based control and fuzzy logic control, are intuitive to end-users and effective in simple systems (Kroll et al., 2018; Mounce et al., 2020). However, it becomes increasingly difficult to obtain optimal predefined solutions for all scenarios as the system complexity grows (Schmitt et al., 2020). MPC, on the other hand, searches for

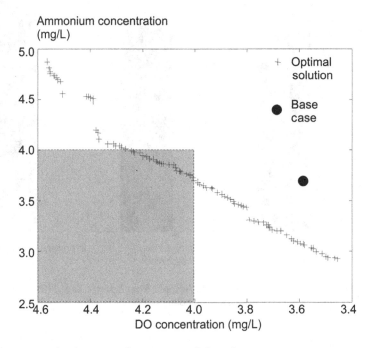

Figure 22.7 Pareto optimal solutions with respect to DO and ammonium concentration. (Adapted from Fu, G., Butler, D., and Khu, S.T. 2008. *Environmental Modelling and Software*, 23[2], 225–234.)

optimal operational solutions recursively for each control horizon to adapt to the dynamics of drainage systems (Lund et al., 2020; Luo et al., 2023). Although several studies have shown that MPC can outperform heuristic control methods in flood and CSO mitigation (Luo et al., 2022; Sun et al., 2020), implementation in practice is limited due to the heavy computational burden.

Deep reinforcement learning (DRL), introduced in Section 21.2.2, has started to be applied in the wider water system field and specifically to urban drainage (at least in terms of research studies). The idea is that, using DRL, the control policy can be trained in advance of the application period, thus avoiding the difficulties of computational burden (Bowes et al., 2021). Recent studies (e.g., Saliba et al. (2020); Tian et al. (2022)) have shown the promise of such techniques and it is expected that as ever newer approaches emerge further gains and benefits will be achieved (Salimans et al., 2017).

22.4.5 Applicability

Prior to implementing a complex and expensive RTC scheme, a comprehensive investigation should be conducted to identify the costs and benefits of deploying this solution to a particular drainage system. There are numerous indicators and criteria to help make such a decision (Beeneken et al., 2013; García et al., 2015). Perhaps the greatest single indicator of the benefit to be derived from the RTC of a particular catchment is the magnitude of the useful storage. In under-designed systems, with low storage volumes, RTC will be of little benefit. But, in an over-designed system with large storage volumes, flooding and CSO discharges will be infrequent anyway, and little additional benefit will accrue. Properly designed systems with sufficient but not excessive storage that can be "activated" should produce the best results. Appropriate distribution of storage around the catchment is also important.

A second major factor is the hydraulic load to which the network is subjected. Assuming storage volume is fixed, RTC will provide no benefit for minor storm events that are handled effectively by the passive system. For large storm events, the benefits are more limited because, once all the storage capacity is fully utilised, the only remaining option is to allow discharges to the receiving water. However, even for large events, RTC offers some potential to more evenly distribute CSO discharges and to better control the first foul flush (see Section 13.3.3). RTC is, however, most effective at reducing the frequency of CSO spills from smaller but still significant storms. These minor spills are common occurrences in conventional passive systems where controls are pre-set at the design stage (often many years earlier).

Other network characteristics that favour RTC applications are

- Spatially distributed inputs (rainfall)
- Spatially distributed storage
- Larger, flatter, more looped sewer networks
- Many controllable elements (e.g., storage tanks, pumps, overflows)

22.4.6 Benefits and drawbacks

In general, RTC has the following major benefits:

- Reduction in the risk of surface flooding by utilising the full storage of the system
- Reduction in pollution spills by detaining more wastewater within the system
- Reduction in capital costs by minimising the storage and flow-carrying capacity requirements of the system
- Reduction in operating costs by optimising pumping and maintenance costs

- Enhancement of WTP performance by balancing inflow loads and allowing the plant to operate closer to its design capacity
- Control of sewer sediments

Other benefits include the following:

- Flexibility to respond to changes in the catchment or local failure in the system
- Better understanding of the performance of the system
- Better system supervision and record-keeping

Typical RTC benefits include large reductions in CSO spills at the most sensitive locations, reduced frequencies of overflow operation (by about 50%), and reduced annual CSO volumes (by 10–20%). Meng et al. (2017) showed by modelling a case study that integrated RTC could improve river water quality by over 20% to meet the EU Water Framework Directive "good status" requirements (see Section 2.6.1.3) with a 15% reduction in costs. That said, operational experience and verified, sustainable benefits have *not* been widely reported (van Daal et al. 2017).

Drawbacks are relatively few, and reluctance to use smarter, more active systems, in general, and RTC, in particular, seems to still rest mainly on the lack of (good) operational experience. However, there are also legitimate concerns regarding the increased maintenance cost and commitment needed, sensor reliability and data quality, lack of relevant skills, and issues surrounding the risk of failure. In the UK, local control is very common, but globally controlled systems are rare. As more RTC systems go online worldwide such as in Austria (Box 22.1), Canada (Jean et al., 2022), and the Netherlands (Liefting et al., 2024), the experience gained should improve confidence and accelerate uptake in appropriate circumstances. This will be further propelled by the rising prominence of the whole smart systems agenda and industrial capacity building.

BOX 22.1 Vienna, Austria

The City of Vienna in Austria has a population of 1.8 million spread over some 260 km² and served by 2200 km of sewer system. The central part of the system is combined, with separate systems on the periphery. Large relief sewers have been constructed along the banks of the rivers Donau, Donaukanal, Wienfluss, and Liesing to minimise CSO discharges after significant rainfall events. A central RTC system has been developed that triggers control devices within the system to activate the storage within the relief sewers. The system consists of the following elements:

- Control devices to regulate water level or flow
- Measurement devices for water level, flow, and rainfall
- A SCADA system to collect the measured data, transmit set-points, and display information about the system
- A central control strategy to generate a control decision based on measured and forecast data

The system contains 25 rain gauges, 40 in-sewer flow measurement devices, and 20 water level measurement units installed at 25 sites. The data are fed into a simplified model of the sewer network. Control decisions are made online, based on "if-then" rules evaluated with the aid of fuzzy logic.

The scheme is being implemented in three phases. The first phase, consisting of the catchment on the left bank of the River Donau, went operational in 2005. Figure 22.8 shows predicted reductions of around 60% for CSO BOD and COD load. The second phase was integrated with the first and became operational in 2006. Evaluation of measured results between March and September 2006 showed that discharge volumes had been reduced by 40% compared with previous years. The third phase was due for completion in 2015 (Fuchs and Beeneken, 2005; Nowak, 2007), however, no further update on performance has been reported.

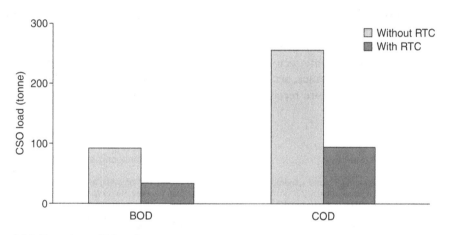

Figure 22.8 First phase CSO pollution load reduction. (Adapted from Nowak, R. 2007. *IWA Water 21,* April, 36–37.)

22.5 EARLY WARNING SYSTEMS

Among natural hazards, storms and floods are the most lethal, affecting millions and causing heavy economic losses. According to the Emergency Event Database (EM-DAT), in Europe, between 2000 and 2016, severe events affected more than 1 million people, caused around 1500 human deaths, and resulted in EUR 20 billion direct economic losses (Centre for Research on the Epidemiology of Disasters [CRED], 2016). At the same time, climate change projections show a future increase in the magnitude and frequency of extreme events (see Section 4.6). To mitigate and reduce these impacts, various risk management frameworks and plans have been issued, both in European and at national/regional levels (European Commission, 2007), aiming to improve the prevention, protection, preparedness, and emergency response to such events.

22.5.1 Definition and elements

An early warning system is defined as

> The set of capacities needed to generate and disseminate timely and meaningful warning information to enable individuals, communities and organizations threatened by a hazard to prepare and to act appropriately and in sufficient time to reduce the possibility of harm or loss"

(UNISDR, 2009).

According to the UN Office for Disaster Risk Reduction (World Meteorological Organization [WMO], 2015), an effective and people-centred EWS is compiled by the integration of four key elements: (1) *risk knowledge*, (2) *monitoring and warning service*, (3) *dissemination*, and (4) *emergency response capacity* (Figure 22.9).

The implementation of a complete EWS requires the synergy of these four components, based on an appropriate legislative and legal framework, along with the close collaboration of many agencies and actors (e.g., international bodies, national and local governments, regional institutions and organisations, communities, non-governmental organisations, science community, media, private sector). Failure of any part of the system may imply failure of the whole system. For instance, accurate warnings will have no impact if the population is not prepared or if the alerts are received but not disseminated by the agencies receiving the messages.

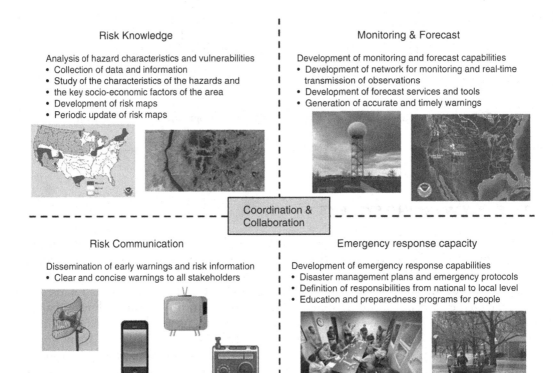

Figure 22.9 An efficient and people-centred EWS.

22.5.1.1 Risk knowledge

Risk knowledge on the one hand supports the development of an efficient EWS, and on the other, enables the development of effective preparedness measures and management plans in advance of an event. The assessment of risks is based on the combined study of hazards along with the vulnerabilities of the population and assets (e.g., agricultural production, infrastructure, and homes). In this regard, hazard characteristics (e.g., magnitude, duration, frequency, location) are analysed on the basis of key socio-economic factors (e.g., casualties, construction, and crop damages). Information on the patterns and trends of hazards and vulnerabilities can be used by the decision-makers at different levels (local, national, regional, and global) to plan risk reduction strategies, such as (WMO, 2015) (1) casualties reduction, (2) medium- and long-term planning to reduce economic losses and develop resilience, and (3) risk financing and transfer (e.g., insurance) to transfer and/or redistribute the financial impacts of disasters.

22.5.1.2 Monitoring and warning service

The two core components of an EWS are the sensor network system for continuous monitoring of hazard variables and the warning service, able to detect, forecast, and communicate the occurrence of a hazard. Both systems must reliably operate on a 24/7 basis, while their systematic maintenance and update are required to remain operational.

The monitoring network includes the sensors and the appropriate telecommunication infrastructures, for the real-time transmission of observations to the forecast system. For flood-oriented EWS, the network is typically composed of rainfall gauges, streamflow gauges, weather radars, and satellites (Hill et al., 2010). The last two are the key sources of hydrometeorological information used as input in weather prediction models, due to their ability to detect and track the evolution of clouds, providing precipitation estimates over large areas. Ground gauges provide accurate rainfall observations that are used to correct the error in radar and satellite estimates (see Section 4.2). Historic rainfall data along with discharge measurements from streamflow gauges are also used to calibrate the hydrological and flood models of the EWS.

Communication of data from the monitoring network to the forecast centre is conducted by various means. Data from local rain and runoff gauges are usually transmitted via wired or wireless Internet links, public telecommunication lines, and radio links (VHF, HF, and UHF). Observations from radar and satellite networks, operated mainly via international and national meteorological services, are typically available via Internet links and satellite communication links (VSAT). Towards the efficient implementation of a multi-hazard EWS, the World Meteorological Organization has established the Global Telecommunications System to enable the exchange of data at local, national, and international levels (Hill et al., 2010).

22.5.1.3 Dissemination

An efficient EWS should ensure that the warnings and forecast information are communicated to the decision-makers and/or directly to the general public in a timely fashion. This is achieved via a multi-channel communication system that is operated 24 hours a day, every day of the year. The warning information must be clear, concise, and in a familiar format to enable the immediate response of those at risk.

A typical typology to communicate information on the severity, timing, and level of confidence of a forecast hazard is the three-level "Ready, Set, Go" concept, adopted by many EWS in the US (Hill et al., 2010).

The dissemination system must be reliable and redundant, while it is important to provide update and cancellation capabilities.

22.5.1.4 Emergency response capacity

Once a warning is issued, the line of action must be clear and understood by all EWS stakeholders. In this regard, well-prepared and well-tested disaster management plans and emergency protocols should be developed by the appropriate authorities, defining roles and responsibilities from the national level down to the neighbourhood level. Equally crucial is the systematic education and preparedness programmes that inform the general public about the nature of the hazards, the potential risks, and options for safe behaviour. Towards this, there are many community-based disaster risk reduction programmes, such as the US Storm Ready Program (see www.stormready.noaa.gov), which aims to increase resilience to floods by establishing community-based plans, raising public awareness, and training people.

22.5.2 Forecast services

The forecast service is the core of any EWS, able to collect, process, and analyse various types of data in order to produce forecasts on the characteristics of a forthcoming flood event. Many EWS of different complexity and sophistication levels have been developed during the last decades, tailored to the special characteristics of the hazard and the focal area (Alfieri et al., 2012), as well as the interplay between different hazards (such as floods and fires (Kochilakis et al., 2016)). The spatial and temporal extent of the different water-related hazards is summarised in Figure 22.10 (top panel). The corresponding meteorological inputs that are used in the forecasting process are shown in the bottom panel, with a qualitative assessment of their skill towards the forecast lead time.

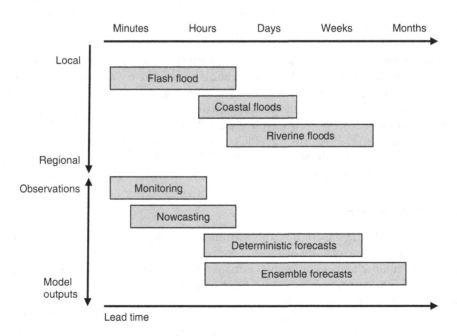

Figure 22.10 Spatio-temporal extent of water-related hazards and the corresponding meteorological input for forecasting. (Adapted from Alfieri, L. et al. 2012. *Environmental Science and Policy*, 21, 35–49.)

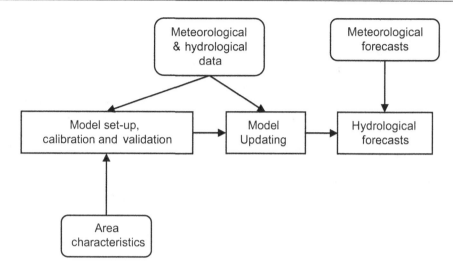

Figure 22.11 Main components, data, and tasks of forecast service. (Adapted from Arheimer et al. (2011))

The simplest forecast systems, usually employed for long-lasting river floods, use water level data from the upstream regions. If the measurements exceed predefined critical thresholds (derived from past flood observations), then the warning system is activated for downstream areas. More sophisticated EWS systems can provide greater lead times by coupling meteorological predictions with hydrological simulation models to provide runoff forecasts, which are then compared with warning levels.

In general, a complete flood forecasting system requires (Borga et al., 2011; Collier, 2007) (1) a precipitation detection system, typically consisting of remote sensing (radar, satellites) and ground rainfall and water-level gauges; (2) a regional and global weather prediction model, able to provide short-range meteorological forecasts, typically 24–48 hours ahead; and (3) hydrological and/or hydraulic models to simulate the response of the catchment, at a wide range of time scales, over predictions. These components along with the data involved in the hydrological forecasting chain are presented schematically in Figure 22.11.

Especially in the case of urban areas, the timely forecasting of flash flood events is of vital importance for the protection and awareness of the people. Such events are usually associated with heavy storms of short duration and high intensity, while the main difference between flash floods and other water-related hazards is their rapid onset (from minutes to a few hours) which restricts the opportunity for effective warning and preparation. Flash floods typically occur in streams and small catchment areas of a few hundred square kilometres, whose characteristics (e.g., steep slopes, saturated soils, extensive impervious surfaces) favour quick response to intense rainfall rates (Hapuarachchi et al., 2011).

22.5.2.1 Flow forecasting models

"The heart of any flow forecasting system is a hydrological model" (Serban and Askew, 1991). Hydrological and hydraulic models support the transformation of meteorological predictions (e.g., precipitation, air temperature, wind speed, humidity) into runoff forecasts (see Chapters 5, 7, and 10). These are generally classified into data-driven (black-box) and conceptual or physical-process models. The main representatives of the first category are the transfer function models (Sene and Tilford, 2004) and neural network models (Sahoo et al.,

2006). Forecast frameworks that couple neural network models with multiple sources of information (i.e., gauged and satellite data as well as numerical weather prediction [NWP] model forecasts) can be found in Chiang et al. (2007).

Unlike data-driven models, conceptual and physical-process hydrological models try to mathematically represent the various components of hydrological cycles. In flood forecasting, both lumped and (semi-) distributed models have been widely used (Hapuarachchi et al., 2011). Representative of the first category is the Sacramento soil moisture accounting model (SAC-SMA model) (Burnash et al., 1973), used by the US National Weather Services (NWS). Typical spatially distributed models applied in a forecasting framework include the TOP-MODEL model (Beven and Binley, 1992) and the LISFLOOD model (De Roo et al., 2000). Distributed models support simulations at finer temporal and spatial scales, and hence, they can be considered as more suitable in flash flood applications. However, the high computational burden along with the high demands for observed data may restrict their operational applicability.

The application of hydrological models on a real-time basis requires their continuous updating in order to constrain the various sources of uncertainty (i.e., input error, parameter, and structural uncertainty) at each time step of the forecast procedure. The update concerns (Serban and Askew, 1991) (1) the input variables (e.g., meteorological and hydrological predictions); (2) the state variables of the modelled system (e.g., water depths, extent of saturated and unsaturated areas, water level in reservoir and other flood protection structures); and (3) the model parameters or structure (i.e., recalibration of the model with updated data). Typically, formal (Vrugt et al., 2003) or informal (i.e., GLUE; Beven and Binley, 1992) Bayesian methodologies are used in this process.

22.5.2.2 Meteorological predictions

The efficiency of any EWS depends on lead time, where accurate weather forecasts, especially on precipitation, are available. In recent years, many different techniques have been developed to provide precipitation forecasts at different spatial extents, temporal analyses, and forecast ranges. These techniques are mostly based on NWPs and weather radar (see Section 4.2.2) and satellite estimates (Section 4.2.3), as well as on their blending.

Short-range precipitation forecasts (up to 6 hours ahead), also known as nowcasts (Golding, 2000), have until recent years been based on extrapolation techniques that use radar and satellite observations in order to capture the spatio-temporal dynamic of precipitation field and forecast the motion of a discrete set of rain objects (Alfieri et al., 2012; Borga et al., 2011; Collier, 2007). Real-time measurements from ground rain gauges are typically used to adjust the error in precipitation estimation. Forecasting extrapolation techniques can be grouped into those that use correlation techniques, track the centroid of an object, and use NWP wind advection techniques (Bowler et al., 2004). Hybrid approaches, artificial neural network applications, and Bayesian statistical techniques have also been applied (Hapuarachchi et al., 2011).

Although these types of methods have shown some success in simulating and forecasting the precipitation field, there are weaknesses in developing convection of rainfall fields in new areas and capturing cell splitting and decay that usually control heavy rain dynamics and, hence, flash floods (Borga et al., 2011). The new generation of high-resolution NWP models, with finer temporal resolution (lower than 1 hour) and spatial resolution of a few kilometres, offer the prospect of producing useful forecasts of convective storms on scales applicable to flood prediction. Further prediction accuracy can be achieved by combining NWP forecasts with radar rainfall estimation. A characteristic operational forecast system that follows this data assimilation approach is the Nimrod precipitation nowcasting system (Golding, 1998) of the UK Met Office that provides precipitation forecasts from 1 to 6 hours lead time (Collier, 2007).

To take into account the uncertain and complex nature of the atmosphere ensembles of many NWPs, known as ensemble prediction systems (EPSs), rather than single deterministic forecasts, can be used. EPSs are typically derived by perturbating the initial conditions of the model. The EPS is then imported into a hydrological model to produce a set of different possible flood predictions (Cloke and Pappenberger, 2008). With the use of ensembles, the accuracy of general weather forecasts can be improved at medium-range (Tracton and Kalnay, 1993) and at short-range (Du et al., 1997) lead times (Collier, 2007). A typical operational EPS system is the short-term ensemble prediction system (STEPS), which merges an extrapolated nowcast with different downscaled NWP model forecasts in order to provide the probability of precipitation (Bowler et al., 2004).

22.5.2.3 Flood occurrence criteria

To infer whether a rainfall (and subsequent runoff) forecast may lead to flooding, criteria and reference thresholds need to be established.

The *flow comparison method* compares the forecasted runoff with a flooding flow. In the simplest case, flooding flow is defined by an observed flooding threshold. However, the estimation of a flooding flow is a rather hard task, especially in urban areas where flood modelling is complex due to human-made interventions and constructions. As such, bank-full discharge, calculated using the uniform flow resistance formula, is typically used as a flooding threshold (Hapuarachchi et al., 2011). Despite its apparent simplicity, this method requires extensive and detailed information on the characteristics of the river or urban drainage system (i.e., Manning's roughness coefficient [see Section 7.5], channel slope, and cross-section geometry) as well as their spatial variation. Otherwise, this method is highly uncertain. Bank-flow discharges can also be derived via the statistical analysis of historical discharge data. According to Hapuarachchi et al. (2011), there is a good statistical relationship between the bank-full flow and a flow with a return period of between 1 and 2 years. Simulated runoff series from models can also be used in the case of inadequate observed data (Reed et al., 2007). Further to observed flooding thresholds, a more probabilistic-distributed approach can also be used, employing long-term simulated records to develop a flooding flow frequency curve for each cell of the grid (Reed et al., 2007).

Similarly, the *rainfall comparison method* compares the rainfall forecasts against the rainfall required to produce flooding flow in a given area. The main representative of this approach is the flash flood guidance (FFG) which is defined as the volume of rainfall of a given duration distributed uniformly over a small catchment (<300 km^2) that is just enough to cause minor flooding at the catchment outlet. If the forecasted rainfall volume is greater than the characteristic rainfall volume, then flooding in the catchment is likely to occur. FFG was developed in the US (Georgakakos, 2006) and is fully operational within the US NWS River Forecast Centers that issue FFG throughout the day for every county in their area. The determination of the FFG is conducted via a hydrological model that is run with increasing amounts of accumulated rainfall for a particular duration (e.g., 1, 3, 6 hours). The simulated runoff is plotted against the required rainfall, and the FFG is retrieved for a known threshold runoff value. US national estimates of threshold runoff have been derived by Carpenter et al. (1999). The hydrological model is updated periodically with the current soil moisture state. In the UK, the Flood Forecasting Centre (FFC) issues a Flood Guidance Statement (FGS) that provides a daily flood risk assessment for all types of natural flooding: coastal/tidal, river, groundwater, and surface water. The FGS presents risk assessment by county and unitary authority across England and Wales over 5 days.

22.6 EMERGING RISKS IN CYBER-PHYSICAL URBAN WATER SYSTEMS

As drainage and other urban water systems become more digitalized and are actively monitored and managed by a combination of sensors, actuators, SCADA systems, and embedded PLCs in real time, new challenges also emerge. The main vulnerability lies in the increased exposure of urban water infrastructures to cyber-physical attacks (CPA) due to the networking, communication, and remote-control mechanisms. This extended attack surface means that these systems are susceptible not only to traditional physical attacks, like sabotage or component destruction but also to a plethora of cyber-attacks, including Denial of Service (DoS) attacks and SQL injection threats (Taormina et al. 2017). The potential adversaries are varied, ranging from hacktivists and terrorists to disgruntled employees and state hackers.

To underline the gravity of these threats, real-world water cyber-physical systems (CPSs), despite not typically being associated with CPAs, have indeed faced several CPA incidents in recent times. Studies indicate that they are now among the most targeted critical infrastructure (ICS-CER, 2016). Recent work on identifying and modelling water systems as CPSs is starting to deliver "stress-testing platforms" (Nikolopoulos et al., 2020) that can simulate the information flow between the cyber control layer and the feedback interactions with the physical water infrastructure system that is under control (see Figure 22.12). Such platforms allow engineers and planners to develop and test multiple CPA scenarios (Moraitis et al., 2023) on various elements of the SCADA, including sensors, actuators, and PLCs, assessing the impact they have on the hydraulic response of the network and the level of service to support strategic planning and risk management.

Although urban drainage systems are lagging behind other water systems (such as water distribution networks) in digitalisation and, as such, are not, at the moment, exposed to CPAs, caution and careful risk assessment and management are certainly warranted when transitioning to more digital-twin-oriented urban water systems as discussed earlier.

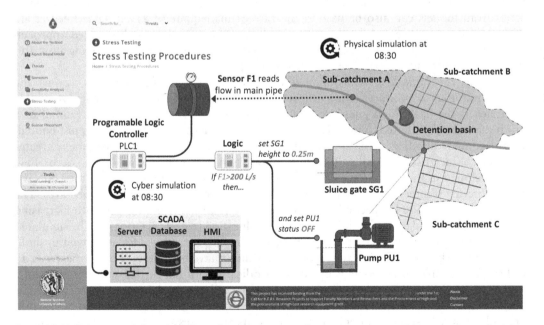

Figure 22.12 Schematic of the cyber-physical representation of a drainage system within a stress-testing platform (Based on PROCRUSTES presented in Nikolopoulos et al., 2020 and Moraitis et al., 2023)

PROBLEMS

1 What are the main sources of new data sets available to urban drainage professionals?
2 What is a Digital Twin and what are its main functions?
3 Define real-time control and describe the main hardware components needed to implement it in an urban drainage system.
4 What are the advantages and disadvantages of urban drainage RTC? In what situations is RTC likely to be beneficial?
5 Explain how you would go about developing an RTC strategy and give an example of a possible practical approach. What sources of data would you need?
6 Define an early warning system and describe the main key elements needed for its efficient implementation.
7 Describe the main components, data, and tasks that are involved in the forecast service of a typical early warning system.
8 Explain the concept of the citizen-observatories framework. What sources of data would it use?
9 What is the source of the main vulnerabilities facing cyber-physical water infrastructures and what do we need to do to manage the resulting risks?

KEY SOURCES

García, L., Barreiro-Gomez, J., Escobar, E., Téllez, D., Quijano, N., and Ocampo-Martinez, C. 2015. Modeling and real-time control of urban drainage systems: A review. *Advances in Water Resources*, 85, 120–132.

Hapuarachchi, H.A.P., Wang, Q.J., and Pagano, T.C. 2011. A review of advances in flash flood forecasting. *Hydrological Processes*, 25, 2771–2784.

Marinaki, M. and Papageorgiou, M. 2005. *Optimal Real-Time Control of Sewer Networks. Advances in Industrial Control*, Springer.

Ocampo-Martinez, C. 2010. *Model Predictive Control of Wastewater Systems, Advances in Industrial Control*, Springer, London.

World Meteorological Organization (WMO). 2015. *Synthesis of the status and trends with the development of early warning systems*. Background Paper prepared for the Global Assessment Report on Disaster Risk Reduction 2015. Geneva.

REFERENCES

Alfieri, L., Salamon, P., Pappenberger, F., Wetterhall, F., and Thielen, J. 2012. Operational early warning systems for water-related hazards in Europe. *Environmental Science and Policy*, 21, 35–49.

Antonio, L. and Gutierrez, G. 2014. On the modeling and real-time control of urban drainage systems: A survey. *11th International Conference of Hydroinformatics*.

Arheimer, B., Lindström, G., and Olsson, J. 2011. A systematic review of sensitivities in the Swedish flood-forecasting system. *Atmospheric Research*, 100, 275–284.

Bartos, M. and Kerkez, B., 2021. Pipedream: An interactive digital twin model for natural and urban drainage systems. *Environmental Modelling & Software*, 144, 105120.

Beeneken, T., Erbe, V., Messmer, A., Reder, C., Rohlfing, R., Scheer, M. 2013. Real time control (RTC) of urban drainage systems—A discussion of the additional efforts compared to conventionally operated systems. *Urban Water Journal*, 10, 293–299.

Beven, K. and Binley, A. 1992. The future of distributed models: Model calibration and uncertainty prediction. *Hydrological Processes*, 6, 279–298.

Borga, M., Anagnostou, E.N., Blöschl, G., and Creutin, J.-D. 2011. Flash flood forecasting, warning and risk management: The HYDRATE project. *Environmental Science and Policy*, 14, 834–844.

Bowes, B.D., Tavakoli, A., Wang, C., Heydarian, A., Behl, M., Beling, P.A. and Goodall, J.L. 2021. Flood mitigation in coastal urban catchments using real-time stormwater infrastructure control and reinforcement learning. *Journal of Hydroinformatics*, 23(3), 529–547.

Bowler, N.E.H., Pierce, C.E., and Seed, A. 2004. Development of a precipitation nowcasting algorithm based upon optical flow techniques. *Journal of Hydrology*, 288, 74–91.

Burnash, R.J.C., Ferral, R.L., McGuire, R.A., McGuire, R.A., and Center, U.S.J.F.-S.R.F. 1973. *A Generalized Streamflow Simulation System: Conceptual Modeling for Digital Computers*. US Department of Commerce, National Weather Service, State of California, Department of Water Resources.

Campisano, A., Cabot Ple, J., Muschalla, D., Pleau, M., and Vanrolleghem, P.A. 2013. Potential and limitations of modern equipment for real time control of urban wastewater systems. *Urban Water Journal*, 10, 300–311.

Carpenter, T.M., Sperfslage, J.A., Georgakakos, K.P., Sweeney, T., and Fread, D.L. 1999. National threshold runoff estimation utilizing GIS in support of operational flash flood warning systems. *Journal of Hydrology*, 224, 21–44.

Centre for Research on the Epidemiology of Disasters (CRED). 2016. *OFDA/CRED International Disaster Database*, Université Catholique de Louvain, Brussels, Belgium.

Chiang, Y.-M., Hsu, K.-L., Chang, F.-J., Hong, Y., and Sorooshian, S. 2007. Merging multiple precipitation sources for flash flood forecasting. *Journal of Hydrology*, 340, 183–196.

Cloke, H.L. and Pappenberger, F. 2008. Evaluating forecasts of extreme events for hydrological applications: An approach for screening unfamiliar performance measures. *Meteorological Applications*, 15, 181–197.

Collier, C.G. 2007. Flash flood forecasting: What are the limits of predictability? *Quarterly Journal of the Royal Meteorological Society*, 133, 3–23.

De Roo, A.P.J., Wesseling, C.G., and Van Deursen, W.P.A. 2000. Physically based river basin modelling within a GIS: The LISFLOOD model. *Hydrological Processes*, 14, 1981–1992.

Du, J., Mullen, S.L., and Sanders, F. 1997. Short-range ensemble forecasting of quantitative precipitation. *Monthly Weather Review*, 125, 2427–2459.

European Commission. 2007. *Directive 2007/60/EC of the European Parliament and of the Council of 23 October 2007 on the Assessment and Management of Flood Risks*, 27–34.

Fu, G., Butler, D., and Khu, S.T. 2008. Multiple objective optimal control of integrated urban wastewater systems. *Environmental Modelling and Software*, 23(2), 225–234.

Fuchs, L. and Beeneken, T. 2005. Development and implementation of a real-time control strategy for the sewer system of the city of Vienna. *Water Science and Technology*, 52, 187–194.

Georgakakos, K.P. 2006. Analytical results for operational flash flood guidance. *Journal of Hydrology*, 317, 81–103.

Golding, B. 2000. Quantitative precipitation forecasting in the UK. *Journal of Hydrology*, 239, 286–305.

Golding, B.W. 1998. Nimrod: A system for generating automated very short range forecasts. *Meteorological Applications*, 5(1), 1–16.

Hill, C., Verjee, F., and Barrett, C. 2010. *Flash flood early warning system reference guide*. University Corporation for Atmospheric Research, NCAR/UCAR.

ICS-CERT (Industrial Control Systems Cyber Emergency Response Team). 2016. *ICS-CERT year in review*. Arlington, VA: Cybersecurity and Infrastructure Security Agency.

Joseph-Duran, B., Jung, M.N., Ocampo-Martinez, C., Sager, S., and Cembrano, G. 2014. Minimization of sewage network overflow. *Water Resources Management*, 28, 41–63.

Klepiszewski, K. and Schmitt, T.G. 2002. Comparison of conventional rule based flow control with control processes based on fuzzy logic in a combined sewer system. *Water Science and Technology*, 46, 77–84.

Kochilakis, G., Poursanidis, D., Chrysoulakis, N., Varella, V., Kotroni, V., Eftychidis, G., 2016. A web based DSS for the management of floods and wildfires (FLIRE) in urban and periurban areas. *Environmental Modelling and Software*, 86, 111–115.

Kossieris, P. and Makropoulos, C., 2018. Exploring the statistical and distributional properties of residential water demand at fine time scales. *Water*, 10(10), 1481.

Kroll, S., Weemaes, M., Van Impe, J. and Willems, P. 2018. A methodology for the design of RTC strategies for combined sewer networks. *Water*, *10*(11), 1675.

Liefting, H.J., Schoester, J., Schepers, J. and Langeveld, J.G. 2024. Development and implementation of a large-scale real time control system in Rotterdam. *16th International Conference on Urban Drainage*, Delft, June.

Lund, N.S.V., Borup, M., Madsen, H., Mark, O. and Mikkelsen, P.S. 2020. CSO reduction by integrated model predictive control of stormwater inflows: a simulated proof of concept using linear surrogate models. *Water Resources Research*, *56*(8), e2019WR026272.

Luo, X., Liu, P., Cheng, L., Liu, W., Cheng, Q. and Zhou, C. 2022. Optimization of in-pipe storage capacity use in urban drainage systems with improved DP considering the time lag of flow routing. *Water Research*, *227*, 119350.

Luo, X., Liu, P., Xia, Q., Cheng, Q., Liu, W., Mai, Y., Zhou, C., Zheng, Y. and Wang, D. 2023. Machine learning-based surrogate model assisting stochastic model predictive control of urban drainage systems. *Journal of Environmental Management 346*, 118974.

Makropoulos, C. and Savic, D.A., 2019. Urban hydroinformatics: Past, present and future. *Water*, *11*(10), p. 1959.

Meng, F., Fu, G. and Butler, D. 2017. Cost-effective river water quality management using integrated real-time control technology. *Environmental Science & Technology*, *51*, 9876–9886.

Moraitis, G., Sakki, G.K., Karavokiros, G., Nikolopoulos, D., Tsoukalas, I., Kossieris, P. and Makropoulos, C. 2023. Exploring the cyber-physical threat landscape of water systems: A socio-technical modelling approach. *Water*, *15*(9), 1687.

Jean, M. È., Morin, C., Duchesne, S., Pelletier, G. and Pleau, M. 2022. Real-time model predictive and rule-based control with green infrastructures to reduce combined sewer overflows. *Water Research*, *221*, 118753.

Mounce, S., Shepherd, W., Ostojin, S., Abdel-Aal, M., Schellart, A., Shucksmith, J. and Tait, S. 2020. Optimisation of a fuzzy logic-based local real-time control system for mitigation of sewer flooding using genetic algorithms. *Journal of Hydroinformatics*, *22*(2), 281–295.

Newman, J.P., Dandy, G.C. and Maier, H.R. 2014. Multiobjective optimization of cluster-scale urban water systems investigating alternative water sources and level of decentralization. *Water Resources Research*, *50*(10), 7915–7938.

Nikoli, V., Žarko, Ć., Ivan, Ć., and Petrovi, E. 2010. Intelligent decision making in wastewater treatment plant SCADA system. *Automatic Control and Robotics*, *9*, 69–77.

Nikolopoulos, D., Moraitis, G., Bouziotas, D., Lykou, A., Karavokiros, G. and Makropoulos, C., 2020. Cyber-physical stress-testing platform for water distribution networks. *Journal of Environmental Engineering*, *146*(7), 04020061.

Nowak, R. 2007. Real-time control of Vienna's sewer system. *IWA Water 21*, April, 36–37.

Oberascher, M., Rauch, W. and Sitzenfrei, R., 2021. Efficient integration of IoT-based micro storages to improve urban drainage performance through advanced control strategies. *Water Science and Technology*, *83*(11), 2678–2690.

Palmitessa, R., Mikkelsen, P.S., Law, A.W. and Borup, M., 2021. Data assimilation in hydrodynamic models for system-wide soft sensing and sensor validation for urban drainage tunnels. *Journal of Hydroinformatics*, *23*(3), pp. 438–452.

Pedersen, A.N., Borup, M., Brink-Kjær, A., Christiansen, L.E. and Mikkelsen, P.S., 2021. Living and prototyping digital twins for urban water systems: towards multi-purpose value creation using models and sensors. *Water*, *13*(5), p. 592.

Reed, S., Schaake, J. and Zhang, Z. 2007. A distributed hydrologic model and threshold frequency-based method for flash flood forecasting at ungauged locations. *Journal of Hydrology*, *337*, 402–420.

Sahoo, G.B., Ray, C. and De Carlo, E.H. 2006. Use of neural network to predict flash flood and attendant water qualities of a mountainous stream on Oahu, Hawaii. *Journal of Hydrology*, *327*, 525–538.

Saliba, S.M., Bowes, B.D., Adams, S., Beling, P.A. and Goodall, J.L. 2020. Deep reinforcement learning with uncertain data for real-time stormwater system control and flood mitigation. *Water*, *12*(11), 3222.

Salimans, T., Ho, J., Chen, X., Sidor, S. and Sutskever, I. 2017. Evolution strategies as a scalable alternative to reinforcement learning. *arXiv preprint,t* arXiv:1703.03864.

Schütze, M., Butler, D., and Beck, M.B. 1999. Optimisation of control strategies for the urban wastewater system – An integrated approach. *Water Science and Technology, 39(9)*, 209–216.

Schütze, M., Campisano, A., Colas, H., Schilling, W., and Vanrolleghem, P.A. 2004. Real time control of urban wastewater systems—Where do we stand today? *Journal of Hydrology, 299*, 335–348.

Schmitt, Z.K., Hodges, C.C. and Dymond, R.L. 2020 Simulation and assessment of long-term stormwater basin performance under real-time control retrofits. *Urban Water Journal, 17(5)*, 467–480.

Sene, K. and Tilford, K. 2004. *Review of transfer function modelling for fluvial flood forecasting.* R&D Technical Report W5C-013/6/TR, Defra/Environment Agency, Flood and Coastal Defence R&D Program.

Serban, P. and Askew, A.J. 1991. Hydrological forecasting and updating procedures. *Hydrology for the Water Management of Large River Basins, 201*, 357–369.

Sun, C., Romero, L., Joseph-Duran, B., Meseguer, J., Muñoz, E., Guasch, R., Martinez, M., Puig, V. and Cembrano, G. 2020. Integrated pollution-based real-time control of sanitation systems. *Journal of Environmental Management, 269*, 110798

Sun, Y., Hu, X., Li, Y., Peng, Y. and Yu, Y. 2021 A framework for deriving dispatching rules of integrated urban drainage systems. *Journal of Environmental Management, 298*, 113401.

Taormina, R., Galelli, S., Tippenhauer, N.O., Salomons, E. and Ostfeld, A., 2017. Characterizing cyber-physical attacks on water distribution systems. *Journal of Water Resources Planning and Management, 143(5)*, 04017009.

Tian, W., Liao, Z., Zhang, Z., Wu, H. and Xin, K. 2022 Flooding and Overflow Mitigation Using Deep Reinforcement Learning Based on Koopman Operator of Urban Drainage Systems. *Water Resources Research, 58(7)*, e2021WR030939.

Tracton, M.S. and Kalnay, E. 1993. Operational ensemble prediction at the national meteorological center: Practical aspects. *Weather and Forecasting, 8*, 379–398.

UNISDR. 2009. *UNISDR Terminology on Disaster Risk Reduction*, International Strategy for Disaster Reduction, Geneva.

van Daal, P., Gruber, G., Langeveld, J. Muschalla, D. and Clemens, F. 2017. Performance evaluation of real time control in urban wastewater systems in practice: Review and perspective. *Environmental Software and Modelling, 95*, 90–101.

Vrugt, J.A., Gupta, H.V., Bouten, W., and Sorooshian, S. 2003. A shuffled complex evolution metropolis algorithm for optimization and uncertainty assessment of hydrologic model parameters. *Water Resources Research, 8*, 39.

Wang, Q., Zhou, Q., Lei, X. and Savić, D.A. 2018. Comparison of multiobjective optimization methods applied to urban drainage adaptation problems. *Journal of Water Resources Planning and Management, 144(11)*, 04018070.

Wehn, U., Rusca, M., Evers, J., and Lanfranchi, V. 2015. Participation in flood risk management and the potential of citizen observatories: A governance analysis. *Environmental Science and Policy, 48*, 225–236.

Chapter 23

Low-income communities

23.1 INTRODUCTION

Providing water, sanitation, and drainage for all people regardless of socio-economic status, age, gender, and mobility is one of the most important global issues of our time. Governments and international agencies have made repeated commitments to achieve universal provision of water and sanitation going back to the 1976 UN Conference on Human Settlements (Habitat I), but it has not been achieved. The UN's current initiative consists of 17 sustainable development goals (SDGs). These aim to end poverty, protect the planet, and ensure prosperity for all. One of these (SDG 6) is focussed on "Clean water and sanitation," and aims by 2030 to "achieve access to adequate and equitable sanitation and hygiene for all and end open defecation, paying special attention to the needs of women and girls and those in vulnerable situations" (www.un.org/sustainabledevelopment/water-and-sanitation/).

In 2019, UN-Water launched the SDG 6 global acceleration framework with the aim to deliver fast results, at an increased scale, towards the goal of ensuring the availability and sustainable management of water and sanitation for all by 2030, acknowledging that business as usual will not achieve the goal (www.unwater.org/our-work/sdg-6-global-acceleration-framework).

In 2023, the UN Water Conference saw the adoption of the "Water Action Agenda" consisting of the collection of *voluntary* commitments to accelerate progress (sdgs.un.org/conferences/water2023). However, following the conference, more than 100 international water experts wrote to the UN Secretary-General criticising its poor "accountability, rigour and ambition," concerned that the lack of scientific underpinning and binding agreements would fail to deliver the SDG 6 goals (Vairavamoorthy, 2023).

By 2022, over 1.5 billion people still did not have access to even basic sanitation services, such as private toilets or latrines. Of these, 419 million still defecate in the open. In 2019, poor sanitation was believed to be the main cause of over 0.5 million deaths of which children under 5 years old are the most affected group (www.who.int/news-room/fact-sheets/detail/sanitation). Although poor or absent sanitation accounts for such a high proportion of illness, debility, and death among the poor, and is a substantial cause of high infant mortality rates, the diseases actually causing death are largely preventable.

The greatest water and sanitation needs can often be found in the informal peri-urban areas where slums and shantytowns proliferate mostly in the global south, and where by 2050 some 3 billion people will be living. These are rarely given services prior to their establishment and subsequent provision places severe financial strain on already over-stretched government resources. Yet, if these basic services can be provided, the intention is that better public health will allow an "upward spiral" of social and economic development, leading to increased productivity, higher standards of living, and improved quality of life (Parkinson et al., 2007).

An important part of the upward spiral of development is the provision of other services, such as storm drainage and solid waste collection, and the organisation of community hygiene education. If these can be deployed, both water supply and sanitation services will, in turn, function more effectively, and additional health benefits will accrue. In fact, an integrated whole of municipal services (see Figure 23.1) will bring maximum benefits. The significant role (and deficiencies) of local authorities and municipalities in supplying these services to a growing population that cannot afford to pay has been highlighted (Ashipala and Armitage, 2011; Kobel and Del Mistro, 2012).

In addition to public health benefits, sanitation is also valuable in giving dignity and privacy to people, in providing a cleaner, more pleasant living environment, and in reducing the severity and impact of malnutrition. Good drainage has similar benefits and is also valuable in reducing nuisance and economic loss due to flooding.

In this chapter, the emphasis is placed on urban drainage services in the context of the social and financial constraints of low-income communities. Section 23.2 concentrates on health implications, Section 23.3 gives an overview of options available, and Section 23.3 through Section 23.7 cover, in turn, the major approaches available: on-site sanitation, off-site sanitation, storm drainage, and grey water management. Section 23.8 summarises the implications of climate change on the main sanitation approaches.

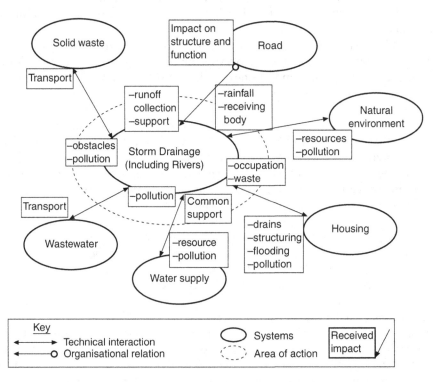

Figure 23.1 Integrated municipal services. (Reproduced from Wondimu, A. and Alfakih, E. 1998. Urban drainage in Addis Ababa (Ethiopia): Existing situation and improvement ideas. *Fourth International Conference on Developments in Urban Drainage Modelling—UDM '98*, London, 823–830, with permission of the authors.)

23.2 PUBLIC HEALTH IMPLICATIONS

23.2.1 Diseases

First in public health importance are the many faeco-oral infections where the contaminated faeces of one person are transmitted via water, hands, insects, soil, or plants to other individuals. These diseases include the well-known "waterborne" diseases such as cholera and typhoid, but also the many common diarrhoeal diseases that particularly affect young children, contributing to malnutrition and death. In fact, these diarrhoeal diseases are often the greatest cause of child mortality (Cairncross and Ouano, 1991).

The larvae of helminths like roundworms and hookworms may be transmitted from person to person when infected faeces are left on the ground. Children are particularly at risk when playing or bathing in faecally contaminated stormwater (see Figure 23.2). Some roundworm eggs (e.g., *Ascaris*) remain viable for long periods.

A further group of diseases may be classified as "water-related" rather than waterborne. For example, mosquitoes spread several different diseases. Their connection with water is that each species selects different types of waterbodies in which to breed. Malaria is the best known of these diseases and is transmitted by the female *Anopheles* mosquito. These do not usually breed in heavily polluted water but can multiply in swamps, pools, puddles, and other poorly drained areas. *Culex* mosquitoes favour polluted water (e.g., pit latrines, blocked storm drains) and are a vector for filariasis, which can lead to elephantiasis. Other mosquito species spread diseases such as dengue and yellow fever.

Another important disease is schistosomiasis (bilharzia), which can be transmitted in poorly drained urban areas (Figure 23.2). If standing stormwater becomes contaminated with egg-infected excreta, the microscopic larvae (miracidia) that hatch out can multiply in the bodies of small aquatic snails. From every infected snail, thousands of infective cercariae emerge and swim in the water. Local residents become infected through their skin when

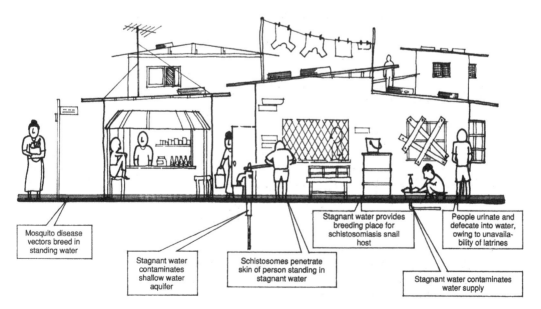

Figure 23.2 Health implications of poor drainage. (Reproduced from Cairncross, S. and Ouano, E.A.R. 1991. *Surface Water Drainage for Low-income Communities*, World Health Organization, with permission of WHO, Geneva.)

Table 23.1 Classification of diseases linked to lack or precariousness of urban drainage

Group		Disease
I	Diseases transmitted by flying vectors that can multiply in pools and wetlands	Urban yellow fever, dengue, filariasis, malaria
II	Diseases in which the etiological agent uses an intermediate aquatic host that can multiply in wetlands	Schistosomiasis
III	Diseases transmitted by direct contact with water or soil (with the presence of the hosts) – contamination is favoured by floods and wetlands	Leptospirosis
IV	Diseases transmitted by ingestion of water contaminated by etiological agents present in wetlands and floods that enter into water distribution systems	Typhoid fever (water), cholera and other diarrhoea (water)
	Diseases transmitted by direct contact with contaminated soil – contamination is favoured by floods and wetlands	Hepatitis A (water), ascaridiasis (water), trichuriasis (water), hookworm (water and soil)

Source: After Souza, C. et al. 2002. *Water21*, October, 40–41.

they stand or play in the water. Table 23.1 summarises the major diseases and their link to urban drainage.

23.2.2 Concerns

Improved water supply, sanitation, drainage, and hygiene education are important components in obstructing the transmission route of these diseases. Of course, in providing engineering works for this purpose, great care is needed to ensure that new disease transmission routes or insect breeding sites are not created inadvertently. Unfortunately, there are credible reports implicating drains as a source of infection. An emerging major concern is the spread of bacteria that have acquired antimicrobial resistance (see Section 2.3.11), which have been sampled from drainage systems (Blom, 2015).

23.3 OPTION SELECTION

23.3.1 Sanitation

Alternative types of sanitation can be conveniently classified as on-site or off-site. In on-site methods, the excreta storage/treatment is in or near the individual dwelling. Off-site systems remove the excreta from the dwelling for disposal. Additionally, systems can be designated as dry or wet, with wet systems using water to transport the excreta (see Table 23.2). Whatever type is chosen, the basic functions are the same, and they are to

- Collect the excreta
- Transport it to a suitable location and/or store it for treatment
- Treat it
- Reuse it and/or discharge it to the environment

A good sanitation system also minimises or removes health risks (Manga et al., 2022) and limits negative impacts on the environment.

Table 23.2 Classification of sanitation options

	Dry	Wet
On-site	Pit latrines	Pour-flush latrines
	VIP latrines	Septic tank systems
	Composting latrines	Aqua privies
	Communal latrines	
	Urine-diverting dry toilet	
Off-site	Bucket latrines	Conventional sewerage
	Vault latrines	Unconventional sewerage

The final choice between on- and off-site systems will normally be financial. In low-density, low-income settlements, on-site systems will almost certainly be the most cost-effective option. In higher-density areas, the feasibility of using on-site methods becomes less, and some form of sewerage may be appropriate. In northeast Brazil, for example, Sinnatamby (1986) demonstrated that unconventional (shallow) sewerage is cost-effective for population densities exceeding about 160 hd/ha. Population densities in slums and shantytowns can be much greater than this (2000 hd/ha is not uncommon).

Conventional sewerage, the most expensive option, will only be appropriate where property values are high and occupiers can pay for the full costs involved. Suitable upgrade paths also need to be considered. However, each case is different, and the merits of all the options need to be thoroughly evaluated in technical terms and by considering social, cultural, financial, and institutional factors. This may involve considerably more consultation with community members than in conventional planning and design practice. Indeed, it has been argued that co-designed and co-produced schemes involving the urban poor working with local government without external funding are the most likely to succeed in the long term (Nance and Ortolano, 2007; Satterthwaite et al., 2015).

This approach broadly matches the IWA analytical framework *Sanitation 21*, which argues that one size does not fit all, but that a properly functioning system may well have elements of both on- and off-site systems (Parkinson et al., 2014). Scott and Cotton (2020) have developed the "Sanitation Cityscape," which is a citywide sanitation planning framework that identifies the key factors of urban sanitation and locates them within three conceptual environments: Living, Service Delivery, and Enabling (see Figure 23.3). Using a set of 16 indicators and existing tools (such as Sanitation 21 and excreta flow diagrams (Peal et al. (2020)), the framework provides a rationale for targeting interventions taking into account the context and externalities. This reinforces the earlier message in the introduction to this chapter of the importance of planning and operating sanitation in conjunction with other technical services and the socio-economic milieu.

23.3.2 Storm drainage

Storm drainage options are more limited. The main classification is between "closed" systems, relying on underground pipes, "open" systems requiring open channels, and on-site options (see Table 23.3).

In most situations, conventional piped drainage will not be an option, unless it is part of a simplified system. Open channels are widely used but need to be carefully designed, constructed, and maintained. On-site options include many of the approaches discussed in Chapter 11 (SuDS), which may have application in this context.

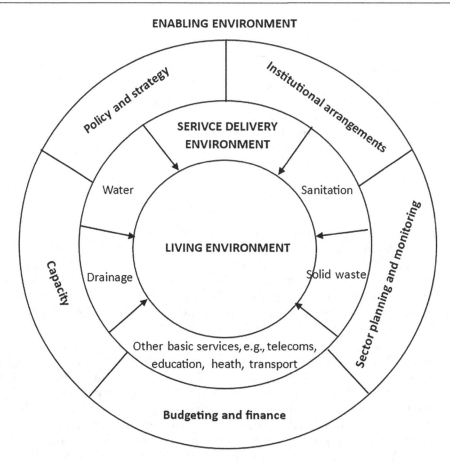

Figure 23.3 The Sanitation Cityscape conceptual framework. (Adapted from Scott, P. and Cotton, A.P. 2020. The Sanitation Cityscape – Toward a conceptual framework for integrated and citywide urban sanitation. Frontiers in Environmental Science, 8,70.)

Table 23.3 Classification of drainage options

Open	Open channels
	Road-as-drain
Closed	Conventional piped drainageDual drainage
On-site	Stormwater management

23.4 ON-SITE SANITATION

On-site disposal is widely practised in low-income communities. This can be a perfectly satisfactory urban solution (even in crowded areas) provided the plot size is large enough and sub-soil conditions are suitable for disposal of the effluent to the ground without danger of contamination of nearby water sources (wells, for example).

The following sections contain an outline of the on-site options available. Design methods and typical details can be found in Pickford (1995) and Mara (1996b).

23.4.1 Latrines

23.4.1.1 Pit latrine

The simplest form is the *pit latrine* (Figure 23.4a). This consists of a squatting hole or plate directly above a pit in the ground into which the excreta falls. The conditions within the pit are anaerobic, promoting gaseous products (mainly CO_2 and CH_4 but also some malodorous gases) that escape into the atmosphere. The excreta gradually decomposes and a solid residue accumulates in the pit bottom. Water, urine, and other liquids infiltrate into the ground through the pit walls and base. When the residue reaches about 0.5 m from the top, the latrine needs to be either cleaned out or abandoned. Pits are usually approximately 1 m in diameter (or 1 m square) and up to 3 m deep.

As the pit latrine works by allowing infiltration of liquids into the surrounding soil, it follows that there must be sufficient open space available on the individual plot, the actual size depending on the conductivity of the soil. However, pit latrines require only 1–2 m² of space, making them suitable sanitation options even in high-density areas (Reed, 1995).

23.4.1.2 VIP latrines

Ventilating pipes can be used to overcome the common complaints of bad smells and insect nuisance. The ventilated improved pit (VIP) latrine consists of a slightly offset pit with a vertical vent having fly-proof netting at its exit (see Figure 23.4b). The latrine is kept dark so any flies hatched in the pit are attracted to the light at the top of the vent and are trapped and die. The wind across the top of the vent causes low pressure and, therefore, an updraught extracting any foul odours. The pipe can also be painted black, which helps heat up the air inside, causing it to rise and ventilate the pit. The shelter needs to be well-ventilated to allow a through-flow of air.

Permanent VIP latrines can also be built with two chambers (alternating twin pit VIPs) where pits are filled and emptied alternatively. This allows safe manual emptying of "old"

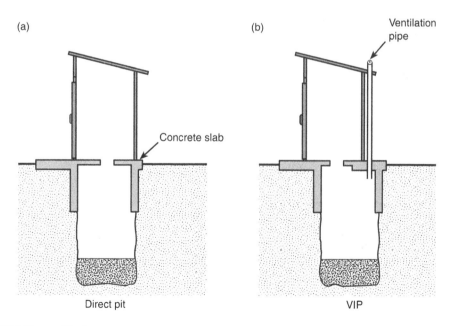

Figure 23.4 Types of latrines.

sludge, but does, however, require more of the householder than other options (Tayler, 1996). Thus, VIP latrines constitute a very useful option, even in highly built-up, low-income areas.

23.4.1.3 Pour-flush latrines

Another solution to the odour and insect problem is to provide a pan and trap with a water seal above the pit. This has the additional benefit of removing the direct line-of-sight between the user and the faeces below. Well-designed pans can be washed down with 1–3 L of water poured from a handheld vessel.

Pour-flush latrines are widely and successfully used in low-income communities where there is a nearby water source, such as a well or standpipe.

23.4.1.4 Composting latrines

Composting faeces with vegetable wastes in an enlarged pit latrine or other chamber offers another method of on-site treatment with reuse potential. These arrangements require considerable care and continuous attention, with 5 months or so needed to produce pathogen-free compost. The compost can be either sold or used locally to replace chemical fertilisers. Odours can be reduced by separating urine. Some users have shown a reluctance to handle the by-products, and there is concern over affordability.

23.4.1.5 Communal latrines

Communal latrines are acceptable in some situations (e.g., city-centre areas), although this depends on the attitudes and habits of the local people. In some places, privacy is considered so important that local people refuse to use them. Where they are acceptable, and where arrangements can be made for frequent and regular cleaning, their cost is much lower than that of individual household latrines.

23.4.2 Septic tank systems

Septic tank systems consist of a tank and, ideally, a drainage field (see Figure 23.5). The tank is a watertight, underground vessel that provides conditions suitable for the settlement, storage, and (temperature-related) anaerobic decomposition of excreta. Sludge accumulates at the bottom of the tank and has to be emptied periodically. A hard crust of solidified grease and oil forms on the surface. Wastewater is fed directly to the tank, through which it flows, and then on to the drainage field. Direct discharge to a ditch, stream, or open drain is not recommended but is a common practice in many Third World cities. The drainage field consists of a soakaway or sub-surface irrigation pipe system, which drains the effluent into the surrounding soil and provides additional treatment.

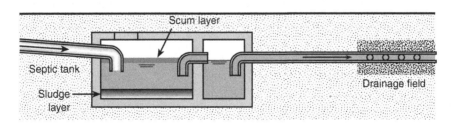

Figure 23.5 Septic tank system.

Tanks should have a minimum volume of 1 m³. The desludging interval should be short enough to ensure the tank does not become blocked but long enough to allow the benefits of anaerobic reduction in sludge volume. This will be typically 2–5 years.

Septic tanks require significantly more space than pit latrines. Depending on water use and soil conditions, this can range from 10 to 100 m² (Reed, 1995). They are also expensive to construct and operate but have proved satisfactory in low- and high-income countries alike, especially in low-density housing areas (Butler and Payne, 1995).

23.4.3 Aqua privies

An *aqua privy* consists of a latrine set over a septic tank. The squatting plate is connected to a pipe that dips below the surface of the liquor in the chamber below. Overflowing liquor infiltrates into the soil through a drainage field, and the water level in the tank is made up with small amounts of cleaning water. Regular desludging is required.

Aqua privies were popular, but problems have always been encountered in maintaining the necessary water level in the tank and with faecal fouling of the drop pipe. Both can cause odour nuisance and health hazards.

23.4.4 Urine-diverting dry toilets

A urine-diverting dry toilet operates without water and has a pan that diverts the urine away from the faeces. The urine is collected and drained from the front area of the toilet, while faeces fall through a large chute in the back. Drying material such as lime, ash, or earth is added into the same hole after defecating. These toilets facilitate the on-site reuse of urine in gardens and other small plots. The concept of this *ecosan* option is to exploit the nutrient value of the excreta for food production (see Section 24.4.2).

23.4.5 New technologies

Due in part to the lack of commercial drivers, there has been relatively little development in sanitation technologies over the years. This deficiency was noticed and is being addressed by the Bill and Melinda Gates Foundation. They began funding research in this area in 2011 as part of their "Re-invent the Toilet Challenge" and this initiative continues to support the development and commercialisation of products that:

- Remove harmful pathogens from human waste and recover valuable resources such as energy, clean water, and nutrients
- Operate "off grid" without connections to water supply and sewers, and require minimal electricity
- Cost less than US$0.05 per user per day
- Promote sustainable and profitable sanitation services and businesses that operate in poor urban settings
- Can appeal to everyone, in developed as well as developing nations

The initiative has resulted in the development of more than 25 processing components and technologies that are available for commercialisation by product and sanitation service companies (www.gatesfoundation.org/our-work/programs/global-growth-and-opportunity/ water-sanitation-and-hygiene/reinvent-the-toilet-challenge-and-expo).

One of these technologies is the Cranfield nanomembrane toilet presented in Box 23.1. Tobias et al. (2017) also discuss the early development, feasibility, and acceptance of the

BOX 23.1   Cranfield nanomembrane toilet

Figure 23.6 shows a schematic of the nanomembrane toilet. Below the toilet rim, there is a drum that is mechanically rotated by 180° to empty the contents into a settling tank below using the opening/closing action of the toilet seat. This is supported by a "swipe" to ensure complete urine/ faeces separation. Once the toilet seat is re-opened, the rotating drum is reset for the next use. Faeces are then separated in the tank below under gravity. At the base of the tank is a mechanical screw with an even pitch and 60° incline giving further solid/liquid separation and partial dewatering. Solids then enter into a convection drier sited above the combustor, which operates on a continuous basis. The residual liquid is passed through a thermally driven membrane process that produces pathogen-free water that can be collected locally through a thermal recovery process (Hanak et al., 2016). Power is provided through novel harvesting processes using thermal or chemical gradients. Development of the various sub-components of the toilet continues (Hennigs et al., 2021).

System configuration

A waterless self-contained toilet for private household of 10 people

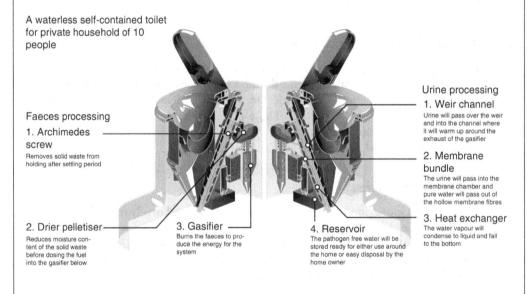

Faeces processing

1. Archimedes screw
Removes solid waste from holding after settling period

2. Drier pelletiser
Reduces moisture content of the solid waste before dosing the fuel into the gasifier below

3. Gasifier
Burns the faeces to produce the energy for the system

Urine processing
1. Weir channel
Urine will pass over the weir and into the channel where it will warm up around the exhaust of the gasifier

2. Membrane bundle
The urine will pass into the membrane chamber and pure water will pass out of the hollow membrane fibres

3. Heat exchanger
The water vapour will condense to liquid and fall to the bottom

4. Reservoir
The pathogen free water will be stored ready for either use around the home or easy disposal by the home owner

Figure 23.6 Cranfield nanomembrane toilet system. (Courtesy of Ewan McAdam.)

Blue Diversion toilet. Source separation (of urine and faeces) and on-site water recovery were found to be feasible and accepted, provided users can be convinced that the emptying service and water recovery process work reliably. Further details can be found here: www. bluediversiontoilet.com.

23.5 OFF-SITE SANITATION

23.5.1 Bucket latrines

Bucket systems are a traditional form of excreta removal. Waste is simply deposited into a container, and the "nightsoil" is collected on a daily basis. This method is still widely practised in urban areas, although it is objectionable from most points of view (including being hazardous to health), but at least it does remove excreta from the household.

23.5.2 Vault latrines

In principle, these are similar to bucket latrines except the squatting plate/seat is joined directly to a closed, watertight chamber where the excreta are deposited. Vaults need to be periodically emptied often by scoops or ladles. Objections similar to those for bucket latrines can be raised. The vault system is widely and successfully used in Japan where vacuum trucks empty the waste and transport it to be treated centrally. Similar systems connected to conventional WCs (*cesspools*) are also used in some remote properties in high-income countries. However, the need for very regular emptying makes this option expensive.

23.5.3 Conventional sewerage

As mentioned, the most expensive sanitation option is conventional sewerage. The advantages of using conventional sewerage (as described in the rest of this book) should by now be clear. However, the disadvantages for low-income communities are manifest: high cost; the need for an ample water supply; the difficulty of construction, operation, and maintenance; and the potential for serious pollution at the outfall or overflows (unless expensive wastewater treatment is proposed). In addition, there are two other practical causes for concern.

23.5.3.1 Septicity

High temperatures accelerate decomposition and limit the amount of oxygen that can be dissolved in water, leading to the rapid development of anaerobic conditions. Such conditions can give rise to H_2S production, resulting in corrosion of cementitious materials (see Chapter 18).

23.5.3.2 Blockage

Blockage can be caused or exacerbated by

- Abuse of the system through ignorance
- Use of traditional methods for anal cleansing (e.g., leaves, rags, stones, newspaper)
- Use of traditional methods for pot cleansing (e.g., sand, ash)
- Low water use
- Too few sewer connections

See Section 18.5 for a more in-depth discussion on blockage.

23.5.4 Unconventional sewerage

23.5.4.1 Simplified sewerage

Simplified sewerage (also known as *shallow* or *condominial sewerage*) is similar to conventional separate foul sewerage except it is reduced to the basics and less-conservative assumptions are used in its design. Thus, sewer diameters, depths, and gradients are reduced compared with conventional systems, and locally available materials are utilised. Hydraulic design is also similar to conventional foul sewers as described in Chapter 9 (see Table 23.4).

Table 23.4 Typical unconventional sewerage design criteria

Criteria	Simplified	Settled
Per capita flow (L/d)	100	100
Peak factor (× DWF)	2	1.5
Minimum velocity (m/s)	0.5	0.3
Maximum proposed depth of flow	0.75	1.0
Minimum pipe size (mm)	100	75
Minimum slope	1/200	–
Minimum cover (mm)[a]	350–500	350–500

[a] Depending on location.

In addition to the relaxation of the hydraulic design criteria, the following practical details are changed (Reed, 1995):

- Conventional access points are replaced by ones of smaller diameter or rodding eyes
- Access point spacing is increased
- More maintenance responsibility is taken on by residents
- Layout is amended, in particular, back-of-property collectors are used to minimise sewer length

The latter point is characteristic of the variant known as condominial *sewerage*. Stormwater drainage is needed to exclude runoff from such systems.

Simplified sewerage systems have the most obvious application in high-density, low-income areas where space is at a premium and on-site solutions are inappropriate. However, their main advantage is their increased affordability with costs found to be approximately one-third of that of conventional sewerage in an Indian context (Mara and Broome, 2008). Some of the disadvantages of conventional systems also apply to simplified systems, with the main concern being the problem of blockages; however, both laboratory work (Gormley et al., 2013) and feedback from the field indicate blockage propensity to be low (Mara, 1996).

23.5.4.2 Settled sewerage

Settled sewerage (also termed *small-bore*, *solids-free*, or *interceptor tank sewerage*) consists of small-diameter sewers connected to small tanks that collect individual household wastewater and capture much of the solid material. Solids accumulate in the tank and require periodic removal. The tank is provided to

- Settle out heavier material that normally requires relatively high self-cleansing velocities
- Settle out larger solids, grease, and scum that might potentially block smaller pipes
- Attenuate individual flow inputs to reduce the peak flow

This allows small pipes to be provided at nominal fall. Some designs (Otis and Mara, 1985) allow sections of adverse pipe gradient and pipes flowing surcharged, provided there is an overall positive hydraulic gradient. As with simplified sewerage, hydraulic design is similar to that for conventional foul sewers but using the criteria summarised in Table 23.4. However, because surcharged pipes are allowed in some systems, extra care is needed in designing the system to ensure all tanks can empty under gravity. Mara (1996) gives a worked example and further explanation. The tanks can be designed as single-chamber septic tanks (see Section 23.4.2), although there are reports of much smaller tanks being used in certain areas (Tayler, 1996).

Settled sewerage may be the best option and the most economical choice where septic tanks are already in existence. Probably the main concern about these systems is their long-term performance. How well and for how long will they operate if tank desludging is neglected? What provision, if any, has been made for proper collection and safe disposal of the sludge? At the time of writing this edition, there has still been relatively little uptake and reported positive experience of these systems.

23.6 STORM DRAINAGE

The provision of drainage for low-income communities often lags behind water supply and sanitation. This is sub-optimal because, if flooding is frequent, water supply and sanitation will be difficult to install and ineffective in operation. As has already been argued, storm drainage also fulfils a disease control function.

23.6.1 Flooding

When using conventional storm drainage in developed countries, systems are designed to run full at relatively modest storm return periods (1 or 2 years) with the knowledge and expectation that flooding will occur even less frequently (e.g., once every 5–10 years).

This situation can be contrasted with many low-income communities living with the monsoon. During the rainy season, high-intensity storms take place frequently, and drainage systems can fail very quickly due to lack of capacity and blockage, and widespread flooding occurs. As storm drains are routinely contaminated with sullage and faeces, the flooding is with dilute wastewater. However, a detailed study of some low-income householders' attitudes towards flooding is very revealing (see Box 23.2).

So, in low-income communities subject to frequent flooding, even in improved areas, local people want to know most of the very things the drainage engineer finds challenging to tell them without detailed modelling and forecasting technology. When will it flood, how long will it last, where will it flood, and how high will the water rise (Kolsky et al., 1996)?

BOX 23.2   Community attitudes in Indore, India

Flooding was ranked low in comparison to other risks and problems, such as improvements in job opportunities, provision of housing, mosquitoes, and smelly back lanes.

A major concern mentioned by residents relates to the predictability of the flooding event – even extensive inundation is bearable if expected. Interventions aimed at ameliorating the effects of flooding should try to take into account the needs of the community to understand and adapt their coping strategies, if necessary.

Residents in flat areas give equal, if not more, weight to the after-effects of rains, as compared to immediate effects. Water may stand for long periods in very flat areas. Respondents feel that this makes walking difficult and allows the breeding of mosquitoes. Faeces-contaminated mud caused by stormwater is seen as most problematic as it is also a perceived source of mosquitoes and noxious smells.

In areas of the slum improvement programme, residents had high expectations of drainage improvement when the projects were initiated and appear to have expected that flooding and inundations would cease or be reduced substantially. It is likely that the feeling that their expectations have not been met is in part an immediate sense of disappointment that flooding has not ceased altogether.

Drainage interventions would gain favour if residents understood and were clearly informed about the effects (good and bad) on the environmental risk, which they perceive as inherently a natural event. This, of course, necessitates technical personnel being able to predict the consequences of technical interventions. Such a strategy might reduce the scale of expectation (Stephens et al., 1994).

23.6.2 Open drainage

23.6.2.1 Open channels

Open drains have a number of advantages when compared to closed pipes because

- They are cheaper to build as they are simpler and shallower than closed pipes
- Blockage with refuse and washed-in sediment can be more easily monitored and safely cleaned out
- They use available heads more efficiently
- Mosquito breeding is easier to control than in closed drains

The simplest and cheapest drains are unlined channels along the roadside with water shedding from the road to the drain by positive fall. The sides of an unlined drain should not slope by more than 1:2 to ensure that they will be stable. If the gradient along the drain exceeds about 1:100, a lining will usually be required to protect the channel from damage by scouring (Figure 23.7a).

One of the keys to making open drains work in practice is to maintain them properly. Drains require cleaning at regular intervals (whatever the self-cleansing velocity specified) since material from the street inevitably gets washed, blown, or dumped in. Marais et al. (2004) estimated litter loads as high as 6000 kg/ha.yr in some South African informal settlements lacking storm drainage systems and organised refuse collection. Kolsky et al. (1996) demonstrated how the performance of the drainage system is substantially reduced by even small amounts of sediment in the system. Cleaning technique is also important. "Sweepings" from drains are traditionally left to dry in roadside piles prior to collection. Unfortunately, this material can quickly find its way back to the drain from whence it came. A compromise solution is to build channels (under the footpath, for example) with removable covers

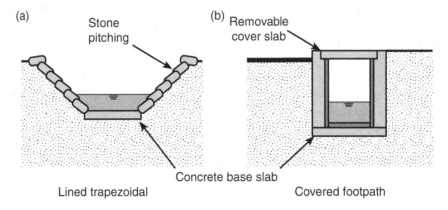

Figure 23.7 Typical open drainage channels.

(see Figure 23.7b) along their length. The covers discourage the entry of rubbish and sediment but can be removed for cleaning, if necessary.

Design can be undertaken in a similar way to that of storm sewers described in Chapter 10. The major differences lie in the roughness of the channel being used and in ensuring velocities are not so high as to damage unlined channels (see Example 23.1).

23.6.2.2 Road-as-drain

Experience suggests that blockage of drains is likely to be a recurrent problem and that frequent stormwater flow along the streets of low-income communities during the monsoon season is inevitable. Therefore, logic dictates that engineers should consider the deliberate and controlled routing of surface runoff exceedance flows over the streets of the slum (see Figure 23.8). The street surface should be depressed below the level of housing sites and not *vice versa*, as is usually the case. This may involve more extensive site grading but will assure

Figure 23.8 Road-as-drain in Indore, India.

EXAMPLE 23.1

An unlined (but vegetation-free) trapezoidal open channel (geometry shown in Figure 23.9) is to be built alongside a road with a gradient of 1:100. If the design stormwater flow is estimated at 300 L/s, calculate the depth of flow in the channel, and check its velocity. Use Manning's equation (Equation 7.23), and assume a roughness of $n = 0.025$.

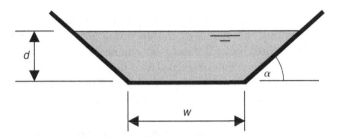

Figure 23.9 Trapezoidal channel geometry (Example 23.1).

Solution

A trapezoidal channel of bottom width $w = d/2$ and side slope 1:2 has
Cross-sectional area, A:

$$A = d(w + d\cot\alpha) = 2.5d^2$$

Hydraulic radius, R:

$$R = \frac{d(w + d\cot\alpha)}{w + 2d/\sin\alpha} = 0.5d$$

Thus,

$$Q = \frac{1}{n}2.5d^2(0.5d)^{2/3} S_o^{1/2}$$

which for $Q = 0.3$ m³/s and $S_o = 0.01$ gives $d = 0.32$ m.
Velocity, $v = Q/A = 0.3 \times 2/(5 \times 0.32^2) = 1.2$ m/s.
This velocity is unlikely to be high enough to self-cleanse and may cause erosion of the unlined channel.

more reliable drainage than any system based on conduits (open or closed) that are subject to solids deposition. Roads as drains are also easier to maintain; street sweeping is easier than drain cleaning, and residents have an interest in keeping roads reasonably clear for access (Kolsky, 1998).

This is by no means a universal panacea, however. It is most appropriate in narrow streets where heavy vehicles do not pass at all and traffic is light. It will not be appropriate on steep streets (say >5%), and, in flatter areas, the micro-topography may well prevent surface routing of all flows. Road-surfacing material needs to be carefully selected (e.g., concrete, compacted gravel, or stone) to provide erosion protection.

23.6.3 Closed drainage

23.6.3.1 Conventional drainage

The main advantage of closed drains is that they do not take up surface space. They also reduce the risk of children playing in or falling into polluted water and the possibility of vehicles damaging the drains or falling into them. Also, in principle, sediment entry to the system can be minimised with catchpits and gully pots. The main problem is blockage due to refuse or sediment that does enter, and then the difficulty of cleaning out such material. Satisfactory simplified versions of storm drainage systems have not yet been developed.

As with conventional sewerage, conventional drainage using closed drains is the most expensive option available. Even though it is the most highly engineered approach, uncritical transference of European or North American standards has littered developing countries with expensive infrastructure that does not work. Hence, piped drains should be built in low-income tropical areas only after very careful consideration of the other options.

23.6.3.2 Dual drainage

Dual major/minor drainage, as described in Chapter 12, also shows promise for low-income communities.

23.6.4 SuDS

The topics of SuDS (sustainable urban drainage systems) and stormwater management have been dealt with comprehensively in Chapter 11. However, the question to be posed here is, is such an approach viable in a low–income community context? The answer is a qualified "yes," and Reed (2004) found that SuDS are already being used, albeit informally. The problems encountered by systems in informal settlements are essentially similar to those found in an industrialised country context, but to a greater degree compounded by their "contentious, turbulent and legally uncertain nature" (Jiusto and Kenney, 2016). Strengths, weaknesses, opportunities, and threats are summarised in Table 23.5.

Table 23.5 Potential for use of SuDS in Sub-Saharan cities

Strengths	Weaknesses	Opportunities	Threats
Viable flood risk management option	Difficulty in quantifying long-term performance	Augmentation of water supply	Lower prioritisation on urban agenda (true of drainage, in general)
Addresses triple goals of controlling quantity and quality of runoff and supports diversity	Operation and maintenance may be costly and require new skills	Maybe part of infrastructure upgrading programmes	Difficulty in convincing decision-makers of new approaches
Predicted to be cheaper in the long run and can use local materials	Requires multi-stakeholder decision-making	Opportunity for more inclusive decision-making	May become health and safety threats if not maintained
More adaptive option	Relatively untested approach with a lack of system data	Support for urban agriculture	Gentrification based on SuDS may lead to the displacement of poor
Contributes to urban water cycle management	May require large open spaces of land	Jobs for relatively unskilled workers	Poor solid waste management

Source: Adapted from Mguni, P. et al. 2016. *Natural Hazards*, 82, 241–257.

23.7 GREY WATER MANAGEMENT

Grey water (or *sullage*) is defined as all the wastewater produced by a dwelling except that used in association with excreta disposal and arises from personal washing, laundry, food preparation, and the cleaning of kitchen utensils. Grey water can have a significant quantity of chemical pollutants and pathogenic microorganisms. Chapter 3 gives further details on the quantity and quality of wastewater, including this component. However, the whole issue of its management in an informal settlement context is often neglected.

It is commonplace that grey water (especially cooking and laundry sullage) is simply thrown outside the dwelling and inevitably finds its way into stagnant pools of water. These pools will ultimately overflow into open-channel stormwater drains. Even when grey water pipes are provided, they can still lead to the same stormwater drains. The resulting pollution can create a host of health and environmental impacts.

23.7.1 Options

The key guiding principles for management options are that grey water must not be allowed to (Carden et al., 2007):

- Pond on the surface
- Enter stormwater systems (including any SuDS if provided)
- Build up in the soil to such an extent that it damages it or significantly pollutes the groundwater

On- or off-site options for grey water include

- Beneficial use (e.g., irrigation)
- Treatment
- Disposal

Further discussion on these options is given by Carden et al. (2007) in the context of South Africa, Al-Mamun et al. (2009) in that of Malaysia, and Sikder et al. (2016) in Bangladesh. Oteng-Peprah et al. (2018) give a useful overview of grey water characteristics and treatment options. Grey water recycling is discussed further in Section 24.4.1.

23.8 IMPACT OF CLIMATE CHANGE

The causes and consequences of climate change have been discussed in detail in Section 4.6. However, in this section, we focus on its potential impacts on sanitation in low-income communities. These may be direct, where water is an essential part of the technology operation (e.g., sewerage) or indirect, where the capacity of the environment to attenuate the adverse effects of wastes is changed. All sanitation technologies will be vulnerable to climate change to some extent and all have some adaptive capacity. Howard and Bartram (2010) qualitatively assessed the main technologies to determine resilience. They were categorised as to whether resilience was high (resilient to most possible climate changes), medium (resilient to a significant number of possible climate changes) or low (resilient to a restricted number of climate changes). Pit latrines were considered to have relatively high resilience, septic tanks and different forms of sewerage medium resilience, and no technology was considered to

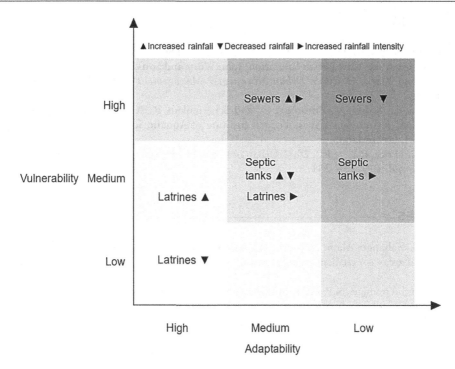

Figure 23.10 Resilience of sanitation options under climate change. (Adapted from Howard, G. and Bartram, J. 2010. Vision 2030. The resilience of water supply and sanitation in the face of climate change. Technical report. World Health Organization.)

have low resilience. Further detail on the vulnerability and adaptability of each technology group to different rainfall regimes is summarised in Figure 23.10, where a darker colour implies lower resilience.

PROBLEMS

1 Explain the importance of taking an integrated approach to municipal service provision.
2 What are the main health-related issues concerned with poor sanitation and storm drainage?
3 Classify and compare the main sanitation options available to low-income communities.
4 Classify and compare the main storm drainage options available to low-income communities.
5 Compare and contrast the various types of pit latrines.
6 Discuss what is meant by "simplified" and "settled" sewerage, and explain how they differ from conventional sewerage.
7 Assess the relative merits of open and closed storm drains.
8 For the channel cross section described in Example 23.1, what gradient is needed to reduce the flow velocity to 1 m/s to avoid erosion?   [1:140]
9 Discuss the pros and cons of using SuDS in low-income communities.
10 What is grey water, and why does it need to be managed in an informal urban setting?
11 In what ways could climate change affect the performance of sanitation systems?

KEY SOURCES

Mara, D.D. (ed.). 1996a. *Low-Cost Sewerage*, John Wiley and Sons.

Mara, D.D. 1996b. *Low-Cost Urban Sanitation*, John Wiley and Sons.

Parkinson, J. and Mark, O. 2005. *Urban Stormwater Management in Developing Countries*, IWA Publishing.

Tilley, E., Lüthi, C., Morel, A., Zurbrügg, C., and Schertenleib, R. 2008. *Compendium of Sanitation Systems and Technologies*, Swiss Federal Institute of Aquatic Science and Technology (Eawag), Dübendorf, Switzerland.

Vojinovic, Z. and Price, R.K. (eds.). 2007. Urban drainage in developing countries, Special Issue. *Urban Water Journal*, 4(3), 135–231.

REFERENCES

Al-Mamun, A., Zahangir Alam, M., Idris, A., and Sulaiman, W.N.A. 2009. Untreated sullage from residential areas—A challenge against inland water policy in Malaysia. *Pollution Research*, 28(2), 279–285.

Ashipala, N. and Armitage, N.P. 2011. Impediments to the adoption of alternative sewerage in South African urban informal settlements. *Water Science and Technology*, 64(9), 1781–1789.

Blom, K. 2015. Drainage systems, an occluded source of sanitation related outbreaks. *Archives of Public Health*, 73(8), 1–8.

Butler, D. and Payne, J.A. 1995. Septic tanks: Problems and practice. *Building and Environment*, 30(3), 419–425.

Cairncross, S. and Ouano, E.A.R. 1991. *Surface Water Drainage for Low-Income Communities*, World Health Organization.

Carden, K., Armitage, N., Sichone, O., and Winter, K. 2007. The use and disposal of greywater in the non-sewered areas of South Africa: Part 2—Greywater management options. *Water SA*, 33(4), 433–442.

Gormley, M., Mara, D.D., Jean, N., and McDougall, I. 2013. Pro-poor sewerage: Solids modelling for design optimization. *Municipal Engineer*, 166(ME1), 24–34.

Hanak, D.P., Onabanjo, T., Wagland, S.T., Patchigolla, K., Fidalgo, B. and Manovic, V. 2016. Conceptual energy and water recovery system for self-sustained nano membrane toilet. *Energy Conversion and Management*, 126, 352–361.

Hennigs, J., Ravndal, K.T., Parker, A., Collinsa, M., Jiang, Y., Koliosa, A.J. McAdam, E. Williams, L., and Tyrrel, S. 2021. Faeces–Urine separation via settling and displacement: Prototype tests for a novel non-sewered sanitation system. *Science of the Total Environment*, 753, 141881.

Howard, G. and Bartram, J. 2010. Vision 2030. *The resilience of water supply and sanitation in the face of climate change. Technical report*. World Health Organization.

Jiusto, S. and Kenney, M. 2016. Hard rain gonna fall: Strategies for sustainable urban drainage in informal settlements. *Urban Water Journal*, 13(3), 253–269.

Kobel, D. and Del Mistro R. 2012. Evaluation of non-user benefits towards improvement of water and sanitation services in informal settlement. *Urban Water Journal*, 9(5), 347–359.

Kolsky, P. 1998. *Storm Drainage. An Engineering Guide to the Low-Cost Evaluation of System Performance*, Intermediate Technology Publications.

Kolsky, P.J., Parkinson, J., and Butler, D. 1996. Third world surface water drainage: The effect of solids on performance. Chapter 14, in *Low-cost Sewerage* (ed. D.D. Mara), John Wiley and Sons.

Manga, M., Kolsky, P., Rosenboom, J.W., Ramalingam, S., Sriramajayam, L., Bartram, J., and Stewart, J. 2022. Public health performance of sanitation technologies in Tamil Nadu, India: Initial perspectives based on E. coli release. *International Journal of Hygiene and Environmental Health*, 343, 113987.

Mara, D.D. 1996. Unconventional sewerage systems: Their role in low-cost urban sanitation. Chapter 2, in *Low-Cost Sewerage* (ed. D.D. Mara), John Wiley and Sons.

Mara, D.D. and Broome, J. 2008. Sewerage: A return to basics to benefit the poor. *Proceedings of the Institution of Civil Engineers—Municipal Engineer, 161(4)*, 231–237.

Marais, M., Armitage, N., and Wise, C. 2004. The measurement and reduction of urban litter entering stormwater drainage systems: Paper 1—Quantifying the problem using the City of Cape Town as a case study. *Water SA, 30(4)*, 469–482.

Mguni, P., Herslund, L., and Jensen, M.B. 2016. Sustainable urban drainage systems: Examining the potential for green infrastructure-based stormwater management for Sub-Saharan cities. *Natural Hazards, 82*, 241–257.

Nance, E. and Ortolano, L. 2007. Community participation in urban sanitation: Experiences in northeastern Brazil. *Journal of Planning Education and Research, 26*, 284–300.

Oteng-Peprah, M., Acheampong, M.A. and deVries, N.K. 2018. Greywater characteristics, treatment systems, reuse strategies and user perception—a Review. *Water Air and Soil Pollution, 229*, 255.

Otis, R.J. and Mara, D.D. 1985. *The Design of Small Bore Sewer Systems*, TAG Technical Note. 14, The World Bank.

Parkinson, J., Tayler, K., and Mark, O. 2007. Planning and design of urban drainage systems in informal settlements in developing countries. *Urban Water Journal, 4(3)*, 137–149.

Parkinson, J., Lüthi, C., and Walther, D. 2014. *Sanitation 21: A Planning Framework for Improving City-wide Sanitation Services*, IWA Publishing.

Peal, A., Evans, B., Ahilan, S., Ban, R., Blackett, I., Hawkins, P., Schoebitz, L., Scott, R., Sleigh, A., Strande, L and Veses, O. 2020. Estimating safely managed sanitation in urban areas; Lessons learned from a global implementation of excreta-flow diagrams. *Frontiers in Environmental Science, 8*, 1.

Pickford, J. 1995. *Low-cost Sanitation. A Survey of Practical Experience*, Intermediate Technology Publications.

Reed, R.A. 1995. *Sustainable Sewerage. Guidelines for Community Schemes*, WEDC, Intermediate Technology Publications.

Reed, B. 2004. *Sustainable Urban Drainage in Low-Income Countries—A Scoping Study*, WEDC, Loughborough University.

Satterthwaite, D., Mitlin, D., and Bartlett, S. 2015. Editorial: Is it possible to reach low-income urban dwellers with good-quality sanitation? *Environment and Urbanization, 27(1)*, 3–18.

Scott, P. and Cotton, A.P. 2020. The Sanitation Cityscape – Toward a conceptual framework for integrated and citywide urban sanitation. *Frontiers in Environmental Science, 8*, 70.

Sikder, T., Hossain, Z., Pingk, P. B., Biswas, J.D., Rahman, M., Hossain, M. 2016. Development of low-cost indigenous filtration system for urban sullage: Assessment of reusability. *Future Cities and Environment, 5(2)*, 1–8.

Sinnatamby, G.S. 1986. *The Design of Shallow Sewer Systems*, United Nations Centre for Human Settlements.

Souza, C., Bernardes, R., and Moraes, L. 2002. Brazil's modelling hopes. The public health perspective. *Water 21*, October, 40–41.

Stephens, C., Pathnaik, R., and Lewin, S. 1994. *This Is My Beautiful Home: Risk Perceptions Towards Flooding and Environment in Low Income Communities*, London School of Hygiene and Tropical Medicine.

Tayler, K. 1996. Low-cost sewerage systems in South Asia, Chapter 4, in *Low-Cost Sewerage* (ed. D.D. Mara), John Wiley and Sons.

Tobiasa, R. O'Keefe, M., Künzlea, R., Gebauera, H., Gründl, H., Morgenrotha, E.,Pronka, W. and Larsen, T.A. 2017. Early testing of new sanitation technology for urban slums: The case of the Blue Diversion Toilet. *Science of the Total Environment, 576*, 264–272.

Vairavamoorthy, K. 2023. Water's global agenda: where do we go from here? *The Source – the Magazine of the International Water Association*, www.thesourcemagazine.org

Wondimu, A. and Alfakih, E. 1998. Urban drainage in Addis Ababa (Ethiopia): Existing situation and improvement ideas. *Fourth International Conference on Developments in Urban Drainage Modelling—UDM '98*, London, 823–830.

Chapter 24

Towards sustainable and resilient urban water management

24.1 INTRODUCTION

This chapter looks to the future and asks whether the urban drainage systems we now have are sustainable and resilient, and how they might need to change. Sustainable development is defined and interpreted in the context of sustainable water management in the next section (24.2). Aspects of what makes an urban drainage system sustainable are presented (Section 24.3) together with promising approaches that are "steps in the right direction" (Section 24.4). Methods of assessing sustainability are explained in Section 24.5. Further to sustainability, this chapter also discusses the planning, design, and adaptation of urban drainage systems in the context of resilience. Section 24.6 provides the key definitions and discusses the relationship between resilience and sustainability. Building resilience is covered in Section 24.7 and assessment methods are discussed in Section 24.8. The chapter concludes with a discussion on urban futures and possible urban water systems that will be needed to support them (24.9).

24.2 SUSTAINABILITY: DEFINITIONS AND OBJECTIVES

24.2.1 Sustainable development

Sustainable development has been on the world agenda since the 1987 World Commission on Environment and Development. The outcome (the *Brundtland Report* [World Commission on Environment and Development, 1987]) offered a viable alternative to the view that the pursuit of economic growth is incompatible with a responsible policy towards the environment. Such an attitude was unacceptable in developing countries, where increasing national wealth was a primary aim, and where any global environmental policy that threatened growth was seen as the rich countries "pulling up the ladder after them." In the context in which the phrase "sustainable development" was coined, it was the inclusion of the word *development* that was particularly significant: it affirmed the right of a country to seek to develop – but in a way that does not compromise opportunities for the future.

Gro Harlem Brundtland defined sustainable development as the development "which meets the needs and aspirations of the present generation without compromising the ability of future generations to meet their own needs." Further reflection has determined that the three main pillars of sustainable development are economic growth, environmental protection, and social equality. Specifically,

- It must be economically viable to provide a continuous flow of goods and services derived from the earth's natural resources

DOI: 10.1201/9781003408635-24

- It must not damage or destroy the basic life support system of our planet: the air, water, and soil, and the biological systems
- It requires developed social systems, at international, national, local, and family levels, to ensure the equitable distribution of the benefits of the goods and services produced.

24.2.2 Sustainability

Even though the key components of sustainability – economy, environment, and society – are widely agreed upon in principle, what they mean in practice is open to interpretation. So, the answer to the question, "What is sustainable?" depends on who is asking the question and why. So although sustainable development, when considered as an *outcome*, is difficult to fully define, when considered as a process of learning and communication, a journey that is constantly reviewed and adapted in the light of new information and knowledge, *more* sustainable development becomes a realistic target (see Box 24.1). Furthermore, a multi-disciplinary, multi-actor approach is particularly beneficial in arriving at agreed-upon solutions, especially in long-term collaborations. This is because the development of a common language, which will allow an informed debate about sustainability to take place, is a long and difficult process (Sharp, 2006). A lack of dialogue can lead to a continuation of "tried and tested" practices and an over-reliance on monetary evaluation to determine appropriate solutions (Brown et al., 2009).

24.2.3 Sustainable urban water management

According to the American Society of Civil Engineers/UN Educational, Scientific, and Cultural Organization (ASCE/UNESCO, 1998), sustainable water systems are "those systems designed and managed to fully contribute to the objectives of society, now and in the future, while maintaining their ecological, environmental and hydrological integrity." Butler and Maksimović (1999) pointed out how the notion of sustainable development has contributed to a rethinking of traditional approaches to urban water systems. They emphasised the importance of public participation, individual responsibility, and water systems that are environmentally friendly, socially acceptable, and financially viable. Generic priority areas were identified.

- *Integration*: acknowledging and exploiting the reality of the urban water system as a unified whole
- *Interaction*: identifying synergistic effects and guarding against antagonism
- *Interfacing*: prioritising the need for engagement with the public and developing a greater urgency concerning the part urban water systems play in the protection of the environment
- *Instrumentation*: acknowledging the key role of sensor technology and the need for greater active control of the system
- *Intelligence*: wider use of decision support tools, data, and information management
- *Interpretation*: wisdom to bring about ordered change and the willingness to think laterally
- *Implementation*: the need to make progress on the ground, including "learning by doing."

More succinctly, Butler et al. (2014) defined sustainability as "the degree to which the [water] system maintains levels of service in the long-term whilst maximising social, economic and environmental goals."

BOX 24.1   A journey into the unknown

The year 1927 saw the first flight across the Atlantic Ocean from New York to Paris. The most striking aspect of Charles Lindbergh's pioneering plane – *The Spirit of St. Louis* – was the fact that it had no front window (Figure 24.1). Lindbergh could not see what lay ahead; he could only see to the sides. He did have a compass to give him a heading, but there were few other navigational aids. This meant he only knew where he had been (the past) and where he was currently (the present), but he did not know or see exactly where he was going (the future).

We can draw a parallel with the journey towards sustainable development – it is a journey into the unknown. Arguably, we have a compass of sustainability principles (society, economy, environment), but we are not certain of our final destination or when we might reach it. History records that Lindbergh was successful in navigating purely by the past and present and extrapolating intelligently into the future. What we can learn is twofold: that the journey of discovery along the route to sustainable development is as important as the final destination, and that the most important step to take is the first one, accepting that some wrong turns may be made along the way (Butler et al., 2010).

Figure 24.1 The Spirit of St Louis.

24.2.4 Sustainability in urban drainage

Butler and Parkinson (1997) were among the first to consider how the general principles of sustainable development and sustainability could be applied specifically to urban drainage. They proposed the following objectives:

- Maintain an effective public health barrier
- Avoid local or distant flooding

- Avoid local or distant degradation/pollution of the environment (water, soil, air)
- Minimise the utilisation of natural resources (water, nutrients, energy/carbon, materials)
- Be reliable in the long term and adaptable to future (as yet unknown) requirements
- Be affordable to the community it serves
- Be socially and culturally acceptable.

It is interesting to reflect on to what extent current urban drainage systems fulfil these objectives (see Problem 2).

24.2.5 Sustainability and climate change

We have already discussed climate change in Section 4.6, including its causes, magnitude, and implications for urban drainage systems. However, are sustainability and climate change the same and if not, how are they related? Sustainability is, by definition, a broader concept which encompasses climate change but is in essence a wider balancing act, which aims to provide enough resources for people to live full, healthy lives, without compromising humanity's future. Climate change influences key natural and human living conditions and thereby also social and economic development. However, society's priorities on sustainable development also influence the GHG emissions that are causing climate change and its vulnerability to its impacts.

24.3 BUILDING SUSTAINABILITY IN URBAN DRAINAGE

24.3.1 Strategies

The search for economically viable solutions with low water, energy/carbon, and maintenance requirements, which are both flexible to change and hygienically acceptable under changing conditions, is a great challenge. Butler and Parkinson (1997) also suggested three fundamental strategies that should be pursued in pursuit of sustainable urban drainage:

- Reduce the reliance on water as a transport medium for waste
- Avoid mixing industrial wastes with domestic wastewater
- Avoid mixing storm runoff with wastewater.

They argued that if priority were given to these strategies, many benefits would be immediately realised, even if they are only introduced singularly or incrementally. Wholesale replacement of existing systems, however, is unlikely to be the more sustainable approach (but see Section 24.9 – Where to next?). The strategies and their potential advantages and disadvantages are summarised in Table 24.1 and discussed in more detail in the following sections.

24.3.2 Water transport

Centralised urban drainage systems arguably utilise water inefficiently. Large quantities of water are abstracted and treated using expensive treatment technology to drinking standards yet are subsequently used to flush faeces and urine from the WC. Not only is this wasteful of a precious resource, but it also promotes unnecessary mixing and dilution of wastes and contamination of previously unpolluted water. End-of-pipe treatment technology is then needed to extract the solid and dissolved pollutants from the liquid component of the waste flow to avoid receiving water pollution.

Table 24.1 Strategies towards sustainable urban drainage

Component of wastewater	Problems	Proposed strategies	Potential advantages	Potential disadvantages
Carriage water	• Unnecessary water consumption • Dilution of wastes • Requires expensive end-of-pipe treatment	• Introduce water conservation/efficiency techniques • Reuse water • Seek alternative means of waste conveyance	• Conserves water resources • Improves efficiency of treatment processes	• Increases possibility of sedimentation in sewers • Health hazards associated with water reuse
Industrial waste	• Disrupts conventional biological treatment • Increases cost of wastewater treatment • Causes accumulation of toxic chemicals in the environment • Renders organic wastes unsuitable for agricultural reuse	• Remove from domestic waste streams • Pre-treat, reduce the concentration of problematic chemicals • Promote alternative industrial processes using biodegradable substances	• Improves treatability of wastewaters • Improves quality of effluents and sludges • Reduces environmental damage • Saves costs associated with the reuse of recovered chemicals	• Costs associated with implementing new practice • Lack of monitoring facilities • May promote illicit waste disposal
Stormwater	• Requires large and expensive sewerage systems • Transient flows disrupt treatment processes • Discharge from overflows causes environmental damage and health hazards • Causes floods	• Utilise overland drainage patterns • Store and use stormwater as a water resource • Provide infiltration ponds, percolation basins, and permeable pavements • Promote ecologically sensitive engineering, for example, constructed wetlands	• Reduces overflow spills • Improves efficiency of treatment • Recharges groundwater • Reduces demand for potable water • Reduces hydraulic capacity requirements of conduits	• Decentralised facilities are harder to monitor • Increases space requirements • Risk of groundwater contamination

Household water consumption can be reduced substantially by a number of means (Butler and Memon, 2006), and comprehensive performance evidence is available (Lawson et al., 2018; Omambala et al., 2011). Box 24.2 gives a good example of an innovative, ultra-low-flush toilet with the potential to significantly reduce domestic water (and indirectly energy) consumption. A key challenge is to ensure that existing systems will operate effectively with much lower dry weather flows (see Box 24.3). Alternative no-flow technologies for the conveyance of sanitary waste, such as dry sanitation systems (Section 23.4.4) or some non-gravity systems (Section 15.7), should be investigated as feasible technical options for the future.

BOX 24.2 Prototype ultra-low-flush toilet

The *Propelair* toilet is a patented ultra-low-flush toilet (ULFT). As can be seen (Figure 24.2), in appearance it is similar to a standard toilet. However, in function and performance, it is very different. It works by locally pressurising air, which is then expelled through the toilet under pressure during the flushing cycle instead of water. Just 1.5 L of water per flushing cycle is used to cleanse the bowl and refill the water seal. Cleansing of the pan is excellent, exceeding the requirements of BS EN 997: 2018. The ULFT can only be flushed when its lid is fully closed and it can be connected to a standard gravity drainage system.

The prototype was thoroughly evaluated to determine the following key aspects: downstream performance, user acceptance, and water and energy savings.

Figure 24.2 Prototype ultra-low-flush toilet. (Courtesy of Garry Moore.)

24.3.2.1 Downstream performance

The ULFT's flushing performance was tested in a full-scale test rig and compared with that of a 6/4 L dual flush toilet. Results showed that the toilet works best with a 50 mm diameter downstream drain (which can be flexible and does not need to have a constant gradient) and that its limiting solid transport distance (LSTD) performance (see Section 9.5.2) exceeds that of a 6 L toilet (Littlewood et al., 2007). A later 6-month trial of 120 Propelair WCs on a university campus noted 95 maintenance issues were reported, equating to 1 in 2700 flushes (0.04%). However, the frequency of incidents decreased after an initial commissioning period and no evidence was found of an increased risk of blockage in the local sewer network (Melville-Shreeve et al., 2021).

24.3.2.2 User acceptance

Figure 24.2 shows one of two toilets installed in an office building, which was monitored over a period of 8 months to gauge functionality, water and energy savings, and user acceptance. During the trial, no incidents of blockage were reported, and there was no interference with the water traps of adjacent toilets. Some 58% of respondents to a questionnaire thought the ULFT was easy to use, and 93% described the flushing performance as being good. However, some users viewed the need to close the toilet as unhygienic and were concerned about not knowing when the lid was closed properly (Millán et al., 2007).

24.3.2.3 Water and energy consumption

When compared with the adjacent conventional toilets (9 L), water saving per flush was 84%. Although the ULFT uses a small amount of energy for each flush, taking into account the energy displaced by not using water, 80% energy savings were also made (Millán et al., 2007).

24.3.3 Mixing of industrial and domestic wastes

The mixing of industrial effluents with domestic wastewater can cause problems for conventional wastewater treatment resulting in the creation of water pollution problems. The reuse potential of nutrients and minerals contained in sludge is diminished by small concentrations of industrial contaminants. In many cases, this results in the disposal of sludge to landfill sites or by incineration, and the waste of a valuable resource. Synthetic organic chemicals and heavy metals can also accumulate in the environment causing problems to ecosystems as they move through the food chain.

Removing the industrial waste component from municipal wastewater is, therefore, important to more sustainable drainage strategies. Dealing with the waste as close to the location of production as possible reduces the problems associated with treating a highly complex mixture of substances and increases the reuse potential of the more valuable components of waste (Hvitved-Jacobsen et al., 1995). Where the complete isolation of an industrial waste stream is impractical, pre-treatment using best available technology to reduce concentrations of the undesirable wastes (be they refractory chemicals or excessively high concentrations of biodegradable waste) is required prior to discharge into the sewer.

Similar considerations and solutions can be extended to substances of concern in domestic wastewater including pharmaceuticals, PFAS, microplastics, and FOG (see Sections 2.3 and 3.5).

BOX 24.3   Droylsden, UK

Droylsden is a largely residential suburb of Manchester. It has a population of about 20,000 and an area of 420 ha. The majority of the catchment is urbanised, although there are substantial undeveloped areas adjacent to the River Medlock and its tributaries. The existing sewerage system is combined, with just a few newer developments having separate sewers. The system drains to a single outfall crossing the river before discharging into a large interceptor sewer.

A detailed theoretical exercise (Parkinson et al., 2005) was carried out to assess the implications on the existing sewers of introducing water-conserving devices throughout the catchment. To do this, a sewer simulation model was developed and verified using *in situ* measured data. Figure 24.3 shows the relative frequency of velocities during dry weather in all the modelled sewers for the base case of 9 L flush toilets. As illustrated, in this system, very few sewers reach the velocities required for self-cleansing (a minimum of 0.6 m/s).

According to model predictions, the application of lower-flush toilets only has a small effect. However, the effect is interesting in that it increases the proportion of pipes that have lower velocities, and therefore decreases those with higher velocities. Thus, fewer sewers will reach the self-cleansing threshold.

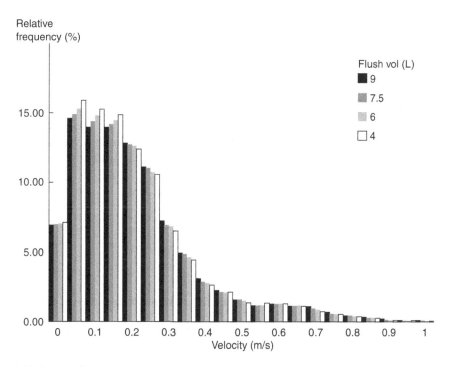

Figure 24.3 Relative frequency of sewer dry weather flow velocities with and without water-conserving toilets.

24.3.4 Mixing of stormwater and wastewater

By disconnecting stormwater from overloaded combined systems, a number of benefits are achieved: flooding frequency and magnitude are reduced, energy costs and carbon emissions are decreased, and CSO discharges are limited. This has the effect of reducing the extent of environmental and public health impact, and the problems caused at treatment works by peak and transient flows are reduced; the potential for use of rainwater for reuse or recreation is also increased.

Isolating storm runoff from wastewater is not a new idea. As discussed in Chapter 1, separate systems were regarded in the 1950s as being the drainage systems of the future and, indeed, are still the current practice in most developed countries. Generally, these systems are more costly to construct, generate higher emissions during manufacture, and yet have not delivered the expected environmental improvements (see further discussion on the pros and cons of combined and separate sewers in Chapter 1, and sewer separation in Section 13.8.2.2). However, it seems that the overall strategy of separating the two components is not at fault, but rather the method through which the strategy has been implemented.

Current best practice advocates separation of storm runoff from other urban wastewaters but, instead of routing stormwater directly into a separate piped system, recommends utilising the natural drainage patterns of the catchment, and/or using SuDS techniques as described in Chapter 11. Many of these methods utilise on-site, small-scale soil infiltration. Where the soil and the quality of the runoff permits, direct infiltration into the ground is preferable in order to recharge groundwater reserves. Where infiltration is not possible or sufficient, the development of natural drainage patterns (see also Chapter 11) offers a range of opportunities for conservation, recreation, and amenity, as well as providing basic flood and pollution control.

24.3.5 Integration

Dixon et al. (2014) argue that in order to meet the needs of a sustainable future, current water infrastructure systems and technologies must be comprehensively reconfigured and/or retrofitted not as individual entities, but by integrating the urban water cycle as a whole and promoting multiple benefits. This should

- Reduce water demand
- Restore urban ecosystems
- Increase resource use efficiency
- Develop new and sustainable resources
- Change water cultures and practices
- Use end-use water qualities
- Increase the flexibility, resilience, and adaptability of water, sanitation, and drainage systems

Not only should this retrofit be technologically based, but it also needs to be in terms of societal values and practices associated with water, its quality, and our intended uses. Figure 24.4 presents a range of retrofit interventions and how they are associated with the main water challenges (drought, flooding, and water pollution), with the emphasis being on multifunctional interventions. For example, rainwater harvesting systems can now be successfully designed to be dual-purpose saving water and mitigating flooding (Melville-Shreeve et al., 2016), so-called *rainwater management*. Substantial and carefully planned instalments of green infrastructure can reduce the impacts of diffuse pollution, surface runoff, and CSOs on receiving waterbodies (Rozos et al., 2013).

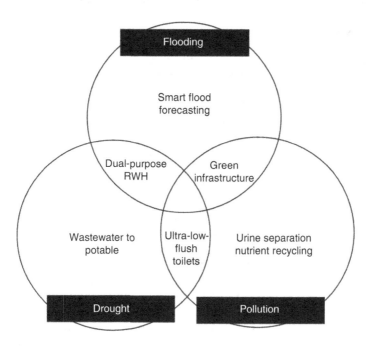

Figure 24.4 Example retrofit interventions to water challenges.

24.3.6 Decentralisation – does size matter?

Over the last few decades, there has been a great deal of interest and activity in researching and indeed promoting small-scale, decentralised approaches as more appropriate solutions than traditional, large-scale, centralised systems (Hesarkazzazi et al., 2022; Otterpohl et al., 1997; Tjandraatmadja et al., 2005). To date, it is not possible to be definitive about whether size does indeed matter, or of the circumstance in which it might. However, the prevailing if not fully substantiated view is that small-scale systems are more sustainable than large-scale ones (Garrido-Baserba et al., 2022, Makropoulos and Butler, 2010). One possible approach to using decentralised systems is discussed in the next section.

24.3.7 Hybrid grey–green systems

The benefits and drawbacks of "grey" centralised sewer systems and "green" decentralised infrastructure (SuDS) have been summarised in Table 17.2 in Chapter 1. Conventional centralised systems (typically combined systems in well-established cities) are struggling to cope with the plethora of challenges they face such as surface flooding, watercourse pollution and public health concerns driven by population growth, urbanisation, climate change and rising user expectations. A promising way forward is to couple the two approaches – grey and green – into one hybrid system, hopefully to build the "best of both worlds." Promising results from modelling studies have been reported by Casal-Campos et al. (2015), Leng et al. (2021) and Wang (2022) broadly indicating (relative to grey-only systems) increased sustainability and the potential for increased robustness. This approach is not a universal panacea, however. Work by Rodriguez et al. (2024) has shown that while incorporating green retrofits into an existing grey infrastructure is effective in increasing system resilience (in terms of CSO discharges), it falls short of fully offsetting the impacts of climate change.

24.4 PROMISING TECHNOLOGIES AND APPROACHES

24.4.1 Grey water recycling

The reuse of grey water (all the wastewater produced in a house except by the WC) potentially reduces the need to use potable water for non-potable applications, with the water effectively being used twice before discharge to the sewer. The major reuse potential is for WC flushing and garden watering, relieving demand on public water supplies and wastewater collection and treatment facilities (Nolde, 2014). The main elements of a grey water reuse system are a collection and distribution pipe network, sufficient storage volume to balance inflows and outflows, and appropriate treatment to render the recycled water fit for the purpose (Gross et al., 2015).

If WC flushing and garden watering demand can be fully satisfied using recycled water, comparison with Table 3.1 indicates that a 36% reduction in demand will be realised. In theory, this demand could easily be met by reusing water first used for personal washing (26%) and clothes washing (12%). System modelling shows that 90% WC water-saving efficiency can be achieved using a storage capacity of approximately 200 L (Dixon et al., 1999a).

Widespread adoption of such systems will inevitably depend on their cost. Retrofitting systems to individual houses is unlikely to be financially attractive, although systems built into new houses may be. Larger-scale applications (e.g., hotels and office blocks) may prove to be more cost-effective (Memon et al., 2005; Ricardo, 2020) or when used synergistically between buildings (Zadeh et al., 2013). There are still some nagging concerns about health issues (Dixon et al., 1999b) and whether systems are carbon-positive or negative (Memon et al., 2007).

In terms of the effect of grey water systems on the sewer systems they connect to, Penn et al. (2013) showed that flow rate, velocity and proportional depth decrease as grey water system penetration system increases. However, sewer blockage rates were not expected to increase significantly.

24.4.2 Nutrient recycling

It is an important fact that, in principle, each person produces in their excreta enough nutrients to grow 250 kg of cereal per year, which is about enough for one person to live on (Drangert, 1998). Of course, some of these nutrients are currently used on agricultural land as treated sludge (biosolids) derived from wastewater treatment plants (WTPs). However, significant quantities of nutrients are discharged through the WTP to the receiving water and so are effectively lost to agriculture. Also, industrial discharges add heavy metals and other contaminants to wastewater, rendering it less usable (Section 24.3.3).

One way to make better (more sustainable) use of wastewater nutrients is to isolate and handle urine separately and use it as a fertiliser. One important reason for this strategy is the fact that urine is the source of around 30% of phosphorus and around 70% of nitrogen in wastewater (Figure 24.5). Additional benefits include the fact that urine is relatively low in heavy metal concentration and also relatively easy to handle, and much less water is needed to flush it. The concept of managing and exploiting excreta and wastewater in this way has come to be known as *ecological sanitation* (or *ecosan*) (Winblad and Simpson-Hébert, 2004).

To isolate and reuse urine requires special no-mix toilets, which can range from the simple to the sophisticated, with choice depending on the socio-economic culture. Box 24.4 describes a case study application of urine-separation technology at a small scale. Lager scale implementation has not been without its difficulties as reported in the Erdos Ecotown

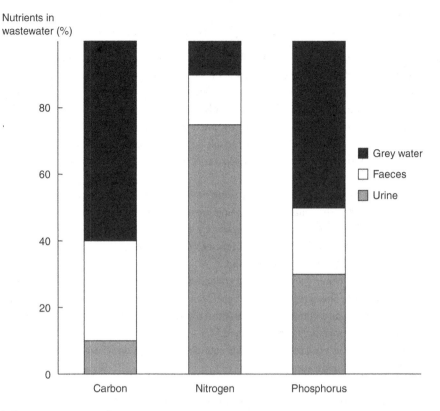

Figure 24.5 Percentage source of nutrients in wastewater.

project in Inner Mongolia, China (McConville et al., 2010). Under more controlled conditions, Jagtapa and Boyera (2020) report on experimental research to demonstrate urine collection in a multi-story building in Arizona, USA, and evaluate the quantity of N, P, K, and water that can be recovered. This was found to be over 90% and the synthesis of value-added products compared well with non-potable water and fertilisers on the market.

Larsen and Guyer (2013) propose three different scenarios for the implementation of urine separation:

- Established industrialised cities
- Rapidly growing cities
- Informal settings

In industrialised settings, for example, they argue it is possible to start immediately. This will decrease the load on existing WTPs, reduce CSO problems and can be continuously rolled out as elements of the centralised system decay or underperform.

24.4.3 Heat recovery

A large amount of thermal energy is effectively wasted as warm wastewater is transported from houses and businesses in sewers to the WTP. As yet, this is largely untapped although there are some examples in the UK (e.g., www.scottishwater.co.uk/about-us/energy-and-

BOX 24.4   Björsbyn, Sweden

Björsbyn is an "ecological" village built in 1994 in northern Sweden. It was designated as such to emphasise that the inhabitants' way of living is deliberately different from traditional urban dwellers, moving towards a more sustainable lifestyle. Interestingly, however, ecological concern was a major reason for moving into the village for only two of the families. (The main attraction was an appealing rural area within a reasonable distance from the city centre.) The village has urine-separating toilets, compost bins for biodegradable organic wastes at each house, and a small outhouse for the collection of paper and glass separated from the solid wastes (Hanaeus et al., 1997).

- Excreta are separated at the source using no-mix toilets (Figure 24.6). The toilets used were equipped with a small collection unit for urine with its own flush-water system (0.1 L per flush). The urine is collected separately and led through a sewer system to a collection tank with 8 months' storage capacity. When the tanks are full, or urine is needed, they are emptied onto nearby farmland.
- The faeces and other grey water are collected and treated by septic tanks followed by infiltration beds. The sludge from the septic tanks is treated locally by a combined sludge drying, freezing, and composting unit. After treatment, the sludge is used as a fertiliser and soil conditioner on agricultural land.
- Stormwater flows via ditches to the surrounding countryside where it is discharged and infiltrates to groundwater.

When studying this system, Hanaeus et al. (1997) concluded that the successful operation of a urine-separation system is particularly dependent on complete separation (to avoid dilution or contamination of the urine). This can be promoted by well-designed toilets, careful design of the collection system, and education of users.

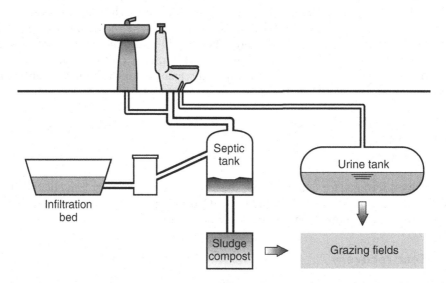

Figure 24.6 The Björsbyn system. (Reproduced from Hanaeus, J. et al. 1997. *Water Science and Technology*, 35[9], 153–160, with permission of publishers Pergamon Press and copyright holders IAWQ.)

sustainability/renewable-energy-technologies/heat-from-waste-water)   and   internationally (Box 24.5) where it has been attempted successfully. Heat recovery from wastewater can be carried out at a variety of scales and locations as depicted in Figure 24.7. These are

A) Individual household appliances (e.g., showers)
B) Individual house or houses
C) In-sewer
D) At the inlet or within the WTP inlet
E) At the WTP outlet prior to the outfall.

The heat available for recovery from sewers (C) is potentially significant due to the high volumes of wastewater collected on a 24/7 basis and the relatively high wastewater temperatures found throughout. In a case study reported by Abdel-Aal et al. (2018), for example, 7–18% of the heat demand for a 79,500 population equivalent Belgian city could be met by sewer heat recovery (although assuming an unrealistic 100% efficient heat exchange system).

Sewer heat recovery systems typically use exchangers set in the invert of the pipes to extract the heat from the wastewater. Heat pumps then boost temperatures to a range useful

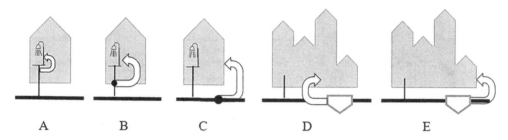

Figure 24.7 Possible locations for heat recovery from wastewater. (Adapted from Bulteau et al. 2019. *Energy from sewers*. Report No. 19/SW/03/1, UKWIR.)

BOX 24.5   Wastewater heat recovery in Vancouver, Canada

The concept for the False Creek Neighbourhood Energy Utility (NEU) was approved by the City of Vancouver in 2006 and began operation during the 2010 Winter Olympics to supply accommodation for the athletes. Approximately 70% of its heat energy needs were satisfied by heat recovered from wastewater. The heat that would otherwise be wasted was extracted from sewer flows using heat exchangers and transferred to a hot water distribution system (Baber, 2010). There is no contact between the wastewater itself and the external hot water system (Figure 24.8).

"It's very similar to geothermal energy," Chris Baber, project manager for the NEU, said of the sewer-heat system. Much as geothermal systems use heat exchangers to extract heat from the soil, the sewer-heat system uses exchangers to extract the otherwise waste heat from the city's sewers. The heat can then be used to warm up buildings and provide hot water.

Initially, the system provided hot water to buildings with a total floor area of nearly 400,000 m². Since 2010, the City Council has expanded the service area extensively and as of 2022, the NEU Energy Centre supplies 3.2 MW of space heating and domestic hot water to over 700,000 m² of residential, commercial, and institutional space. About 2.3 million m² is anticipated to be connected on completion of the project.

https://reshapestrategies.com/case_studies/waste-heat-recovery-at-the-false-creek-neu

https://www.bbc.com/future/article/20240103-sewage-a-low-cost-low-carbon-way-to-warm-homes

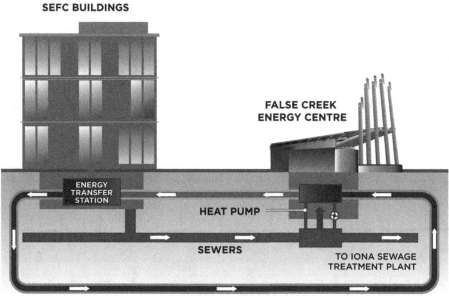

Figure 24.8 Sewer-heat recovery system in Vancouver. (Courtesy of the City of Vancouver.)

for residential space heating and domestic hot water. Compared with other traditional solutions, the energy is

- Renewable
- Locally available
- Associated with significantly reduced GHGs.
- Forms part of the circular economy (see Section 24.9.3)

Disadvantages include

- Relatively unfamiliar technology
- Finding suitable locations
- Concern that recovery systems may have a detrimental impact on the operation of the sewer network and WTPs
- Legal and regulatory issues.

Suitable heat users include large institutional buildings, relevant industrial processes, and district heating schemes.

24.5 ASSESSING SUSTAINABILITY

How can we assess whether one urban drainage option is more sustainable than another? The most common approach is to use *indicators*. Sustainability can be indicated by the performance of the system or by its properties/attributes and this distinction is discussed later in the chapter in the context of resilience assessment (Sections 24.6–24.8). An example of a performance-based approach is that developed by a project named SWARD, Sustainable Water Industry Asset Resource Decisions (Ashley et al., 2004). This project investigated the way sustainability could be included in asset investment decisions. The decision support processes recommended are set out in seven stages:

1. Define objectives
2. Generate options
3. Select criteria
4. Collect data
5. Analyse options
6. Select and implement the preferred option
7. Monitor outcome

The options, considered in this way, can be quite general (e.g., alternative strategies for managing domestic sanitary waste) or quite specific (alternative locations for a WTP facility). Candidate criteria are proposed under four headings: environmental, social, economic, and technical. Recommended "primary criteria" are listed in Table 24.2. Each primary criterion leads to a number of secondary criteria, each with a recommended performance indicator

Table 24.2 SWARD primary criteria

Type	Primary criterion
Environmental	Resource utilisation
	Service provision
	Environmental impact
Social	Impact on risks to human health
	Acceptability to stakeholders
	Participation and responsibility
	Public awareness and understanding
	Social inclusion
Economic	Life-cycle costs
	Willingness to pay
	Affordability
	Financial risk exposure
Technical	Performance of the system
	Reliability
	Durability
	Flexibility and adaptability

Table 24.3 SWARD: Secondary criteria and indicators under the primary criterion resource utilisation

Secondary criteria	Indicator
Water resource use	
– Withdrawal	Annual freshwater withdrawal ÷ annual available volume (%)
– River water quality	Percentage of rivers of good or fair quality
– Nutrients in water	Percentage of river length with greater than guideline nutrient concentrations
Land use	Land area used (km²)
Energy use	
– Energy for water supply	Energy use (kW.h/m³)
– Energy for wastewater treatment	Energy use (kW.h/m³)
Chemical use	
– WTP or on-site (herbicides)	Litres/year
Material use	
– Aggregates, plastics, metals	Total material requirement (tonnes/year)

to be used in the assessment. The proposed secondary criteria and indicators for one of the primary environmental criteria, *resource utilisation*, are given in Table 24.3.

Analysis and comparison of the options using the selected criteria and their indicators require some form of multi-criteria analysis approach. Case studies using this general approach are described by Ashley et al. (2003).

A formal multi-criteria assessment method was subsequently developed in the WaND project – the Project Assessment Tool, based on SWARD indicators (Butler et al., 2010). This was used with a wide variety of stakeholders to promote discourse, to illustrate innovations, and ultimately to provide a means to operationalise sustainability.

Additional indicators could be used in the analysis. These include, for example,

- Operational carbon or GHG emissions (closely related to energy use), discussed in Section 18.10.5
- Embodied carbon (discussed in Section 19.2.3)
- Biodiversity net gain (discussed in Section 11.3.11)
- Social value (e.g., www.socialvalueportal.com)

Urban drainage research is currently focussing on combining multi-criteria approaches to sustainability assessment with that of resilience (e.g., Bakhshipour et al., 2021; Casal-Campos et al., 2018). Resilience assessment is discussed in Section 24.8.

At the city scale, the KWR Watercycle Research Institute has developed an attribute-based standardised indicator framework to support strategic planning of integrated water management called the City Blueprint (Van Leeuwen et al., 2012, 2016). The methodology was further updated and applied in the framework of the EU TRUST project (Van Leeuwen and Marques, 2013). The main objective of the blueprint is to measure and assess a city's sustainability in order to envision, develop, prioritise, and implement interventions and measures towards a *water-sensitive city* (see watersensitivecities.org.au). City Blueprint consists of 26 indicators classified into seven categories: (1) water quality, (2) solid waste treatment, (3) basic water services, (4) wastewater treatment, (5) infrastructure, (6) climate robustness, and (7) governance (Koop and van Leeuwen, 2015). A city's sustainability assessment is provided in the form of a spider graph as depicted in Figure 24.9, while the overall performance is

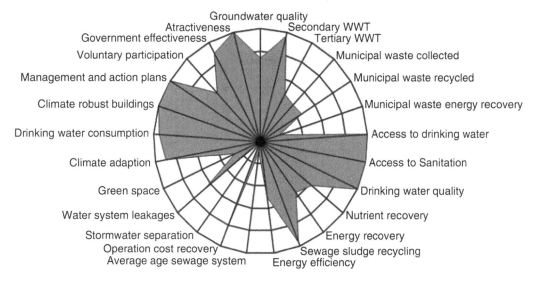

Figure 24.9 Example City Blueprint spider graph.

derived from the geometric mean of all indicators, namely, the Blue City Index. The City Blueprint also provides a set of indicators to assess the trends and pressures on a city (e.g., urbanisation rate, education rate, urban drainage flooding, river peak discharges) to facilitate the determination of limitations and opportunities for urban sustainable water management. The framework has been widely applied across the globe and is particularly useful in terms of benchmarking best practices and comparison between cities.

24.6 RESILIENCE: DEFINITIONS AND OBJECTIVES

24.6.1 Threats

Water services, and especially urban drainage systems, face significant and rising internal and external challenges and pressures (Makropoulos et al., 2016). For instance, flooding can be caused not only by *external* threats such as extreme rainfall events and increasing urbanisation (Mugume and Butler, 2016), but also by *internal* system threats such as equipment malfunction, sewer collapse, and blockages (Mugume et al., 2015). These threats may result in system or component failures whose impacts may lead to adverse consequences for the users of the system (including the environment). This sequence of events and relationships: threat-system-impact-consequences is depicted diagrammatically in the "Safe & SuRe" intervention framework (Butler et al., 2016) shown in Figure 24.10.

24.6.2 Interpretations and applications

To tackle these threats and the challenges they bring, we need to build more *resilient* water systems to enhance their ability to both maintain their level of service and minimise the consequences during unexpected or exceptional loading conditions (Butler et al., 2016; Mugume et al., 2015). Although the concept of *resilience* has only relatively recently been applied to infrastructure systems, there is extensive literature on its definitions and interpretations, which are varied across different scientific fields and types of systems (Mugume et al., 2015;

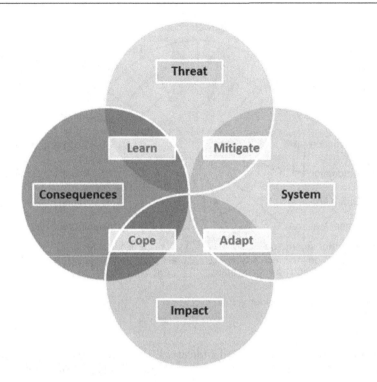

Figure 24.10 The "Safe & SuRe" intervention framework

Pizzol, 2015). The common ground between them points towards the characterisation of resilience as a given performance objective delivered by the properties of the system as a whole, rather than individual elements or units. According to Morecroft et al. (2012), all definitions revolve around two key system performance objectives:

- The amount of disturbance that a system can withstand without changing self-organised processes and structures (Holling, 1973)
- The return time to a stable state following a perturbation (Brede and de Vries, 2009)

Furthermore, Pizzol (2015) highlights three common ideas across different interpretations and applications of resilience:

- Resilience depends on both system elements *and* how they are structured. Specific designs lead to increased resilience (e.g., by increasing the number of connections between elements or their strength – ideas also loosely connected with redundancy or overdesign)
- There is a trade-off between resilience and efficiency, with some natural systems favouring resilience while most human systems favour efficiency
- Resilience is associated with the sustainability of a system, intended as the capacity of systems to maintain their functions in a context of continuous change. From this perspective, resilience is a key attribute or prerequisite of a sustainable system. In many ways, differences between these two terms could be understood as varying in the boundaries of the system in question and the number and scale of interactions of relevance (see Section 24.6.3)

Focussing on engineered water systems, such as urban drainage, work on "Safe and SuRe" Water Management defined resilience as "the degree to which the system minimises level of service failure magnitude and duration over its design life when subject to exceptional conditions" (Butler et al., 2016). These exceptional conditions can be considered the threats or pressures that lead to system failures, such as extreme rainfall events, sewer collapse, or blockage. According to this definition, the goal of resilience is the maintenance of acceptable functionality levels (by withstanding service failure) and the rapid recovery from failure once it occurs (Butler et al., 2016; Lansey, 2012; Park et al., 2013). According to Butler et al. (2014), resilience can be further classified into three broad categories:

1. *General* (attribute-based) *resilience* that refers to the state of the system that enables it to limit failure duration and magnitude to any threat (i.e., all hazards)
2. *Specified* (performance-based) *resilience* that refers to the agreed performance of the system in limiting failure magnitude and duration to a given (known) threat
3. *Technology-based resilience* that refers to the equipment that can improve the preparedness of the user for extreme events (e.g., SuDS).

24.6.3 Resilience and sustainability

As discussed in the previous sections, the key idea behind sustainability is the maintenance or preferably enhancement of broad socially derived goals (social, economic, environmental), not only for now but also for future generations. In the framework of the Safe & SuRe work, sustainability is defined as "the degree to which the system maintains levels of service in the long-term whilst maximising social, economic and environmental goals" (Butler et al., 2014). So, sustainability targets long-term horizon planning and hence requires the design of resilient systems that will be able to cope with any possible future threat or pressure (Scholz et al., 2012). Although resilience lies at the core of sustainability in an operational context (Pickett et al., 2014), the details of the exact relationship are not fully resolved; hence, currently, it is not clear how to maximise sustainability by building resilience (Butler et al., 2014). The conceptual relationship of the three notions is presented schematically in Figure 24.11, where a safe/reliable system should be built upon by resilience and "topped off" with sustainability.

Figure 24.11 The "Safe & SuRe" pyramid.

24.7 BUILDING RESILIENCE IN URBAN DRAINAGE

24.7.1 Strategies

Referring again to the Safe & SuRe framework in Figure 24.10, as well as providing a consistent framework and diagrammatic representation of the relationship between threats and their social, economic, and environmental consequences, it also clarifies the role of the various intervention points between threat and consequence leading to the identification of four broad resilience strategies:

- Mitigation
- Adaptation
- Coping
- Learning

Essentially, threats result in system failures (if mitigation measures are insufficient); system failures result in impacts (if adaptation measures are insufficient); impacts result in consequences (if coping measures are insufficient); and learning aims to embed new knowledge, thereby reducing consequences resulting from a threat. Example approaches of how these strategies might be implemented include the following:

- Scale (e.g., centralised, decentralised, hybrid).
- Low tech (e.g., water butts, water reuse).
- High tech (e.g., smart sensors, real-time control).
- Phased construction (e.g., option portfolios, incremental build)
- Emergency planning (e.g., warning systems, flood resilience measures).
- Policy change (e.g., planning controls, building regulations).
- Behavioural change (e.g., coping strategies, education).

24.7.2 Attributes and interventions

As mentioned in Section 12.6 and defined above, it is helpful to distinguish between resilience attributes and resilience performance. Here we explore what are the broad attributes or properties of a system that could build resilience and to what extent they can actually deliver the system performance necessary. Table 24.4 provides a (non-exhaustive) summary of a number of relevant system characteristics, traits or attributes that have been mentioned in the scientific literature as conferring greater resilience. Each attribute has its own strengths and weaknesses and will be more or less effective depending on the particular threat encountered. Some may actually reduce system resilience and some attributes overlap with others. The question arises, and has yet to be fully determined, to what extent do the example approaches mentioned in the previous section deliver these attributes (see Problem 9) and how effective will they be?

In order to explore the relationship between properties and performance, Sweetapple et al. (2018) used 250 combined sewer network virtual case studies (VCSs) – See Section 21.8 to explore the effects of two "attribute-based" interventions on performance-based measures of resilience. In this case, two specific interventions were used: (1) increased storage capacity [adaptation] and (2) reduced imperviousness [mitigation], both based on consistent design standards. Two example VCSs are shown in Figure 24.12 (a and c) with their appropriate storage interventions (b and d). Two "failure modes" were considered: pipe failure and change in rainfall intensity allowing specified resilience to be calculated across all the virtual networks. Key findings were

- Implementation of attribute-based interventions provided improved resilience in the majority of the 250 cases, however, improvement was sometimes only small and positive results were not guaranteed
- In systems with initially low resilience to rainfall intensity change, increasing distributed storage and reducing imperviousness were both effective interventions and could be expected to provide a significant overall improvement under a range of rainfall magnitudes.

Table 24.4 System resilience attributes (adapted from Nikolopoulos, 2023)

Attribute	Definition	Intervention/Comment
Agility	Capacity of the system to rapidly recover normal operational function	For example, strong communication systems and well-rehearsed emergency plans/procedures.
Buffering capacity	Capacity, in excess of normal operational needs, to call on in an emergency.	For example, sewer stormwater storage tanks.
Diversity	The extent to which multiple distinct functions can be simultaneously used in the system.	For example, different types of SuDS components. Related to flexibility.
Flexibility	Ability to readily reconfigure the system to meet short-term needs.	For example, active system control.
Interconnectivity	Ability to receive support from other systems in case of failures.	Could also be detrimental if cascading failure is caused.
Preparedness	Systems where future (extreme) conditions have been evaluated before they occur and plans are put in place.	For example, resilience assessment. Related to agility.
Redundancy	Availability of multiple substitutable components with similar or overlapping functions	Standardised or mutually compatible components, for example, pumps, pipes.
Robustness	Ability to withstand a given level of stress without loss of function.	For example, adequate design standards. Related to reliability.
Safe failure	Capacity to absorb shocks or stresses sufficiently to avoid catastrophic failure.	

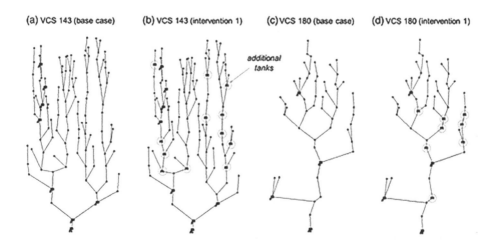

Figure 24.12 Example Virtual Case Studies: additional tanks provided under intervention 1 are circled. (Reprinted from Sweetapple et al., (2018), with permission of the copyright holders, IWA Publishing).

24.8 ASSESSING RESILIENCE

24.8.1 Resilience in strategic planning

Water service providers, tasked to manage urban water and the infrastructure that delivers it to customers (or protects customers from it, as is the case in urban drainage), come face to face with the concept of resilience during the process of strategic planning. To decide between alternative system configurations and large-scale interventions that will change the function of the system in the longer term, providers need to understand the future behaviour and performance of their systems in the longer term, under accidents/incidents and/or extreme events, and under changing external conditions (e.g., societal, economic, or climatic trends) that involve significant uncertainties. In this setting, resilience has come to dominate the discussion around "future-proofing" water systems.

An operational methodological framework to assess the resilience of the complete urban water cycle was developed for the Dutch Water Sector (Makropoulos et al., 2016). Here, resilience is defined as "the degree to which a water system continues to perform under progressively increasing disturbance." The degree of performance in this definition is quantified through *reliability*, which is defined in this work as "the ability of the system to consistently deliver its objectives, considered over a timespan." This extension of the typical definition of the reliability term (Mays, 1989) enables the assessment and operationalisation of the effect of different "failure modes" and pressures on the water system. Progressively increasing disturbance comes from progressively more "stressful" scenarios, not unlike the approach discussed in Section 12.6.2 (Mugume et al., 2015). The result is a form of *stress-testing* of water systems, whereby the system is exposed to progressively increasing disturbance and the change in its behaviour (i.e., performance) is assessed and traced in a *Resilience Profile* graph.

Each point of a typical *Resilience Profile Graph* (Figure 24.13) gives the reliability of a given objective or group of objectives (*y*-axis) under the conditions of a particular stress scenario (*x*-axis). The *x*-axis of the resilience profile graph is constructed as a series of progressively more extreme disturbances in the form of scenarios and is therefore, by definition, an ordinal scale.

It is useful to observe that this graph also provides an assessment of the system's *robustness*, which is here defined as "the extent to which a system can keep on performing within design specifications under increasing stress." In other words, robustness is defined as the

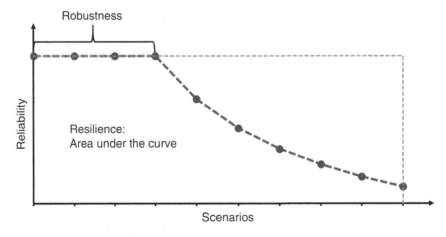

Figure 24.13 Resilience profile graph.

level of pressure that the system can take without failing, while resilience refers to the ability of the system to cope (well or otherwise) with failure. A robust system is also resilient, but the opposite does not necessarily hold: a system can be resilient without being robust.

To scale resilience and robustness to a maximum of 1, the area under the curve is divided by the area of a "completely robust" system, and robustness is divided by the number of points in the resilience profile diagram (i.e., the number of stress scenarios).

Resilience assessment for strategic planning, resulting in the above *Resilience Profile Graph*, comprises the following steps (Makropoulos et al., 2016):

1. Identify and evaluate the main properties of the water system under study to create a baseline scenario. Typical aspects include geography, geomorphology, climate, socio-cultural-economic background, water sources, urban typology-topology, infrastructure, and demand types.
2. Identify and define alternative design philosophies and a set of interventions, including technical and non-technical measures, that are available for the system under study. These alternative configurations are also termed *levels of ambition* (see Makropoulos and Butler, 2010).
3. Set up one or several models (e.g., using the Urban Water Optioneering Tool [UWOT] – see Chapter 21) to simulate the urban water system.
4. Identify the external pressures/threats, which are expected to influence the system, and compose a set of scenarios of increased disturbance.
5. Run each model to the *same scenario*, and evaluate the performance of the system.
6. For each set of interventions, the *resilience profile graph* is composed.
7. To explore total system resilience and receive answers to different questions, one may
   a. Test different interventions to explore which improves the system resilience more.
   b. Test the same interventions under different scenarios to explore system resilience under a specific setting.

An example of using this approach is given in Box 24.6.

BOX 24.6   Resilient Planning of WaterCity

The Resilience Assessment for the strategic planning, discussed previously, was demonstrated using the semi-hypothetical "WaterCity," a typical, but anonymised Dutch City shown in Figure 24.14 (Makropoulos et al., 2018). The scope of this analysis was to provide evidence-based support to investment decisions for future water system configurations. The following three alternative sets of interventions (levels of ambition) were assessed:

1. The current state, *business as usual* (BAU), model of the system (following standard practices of the Dutch Water Sector)
2. *Next step* (NS) interventions that could be implemented tomorrow (or be included in the next 5-year plan and are to a large extent dependent on internal system variables). These interventions include local treatment of grey water at the household level, drinking water production from seawater, drinking water transport bundles, and aquifer storage recovery (ASR) stormwater collection technology for horticulture uses
3. *Further ahead* (FA) interventions that need more time and possibly more investment (which would typically need 10 or more years to be implemented)

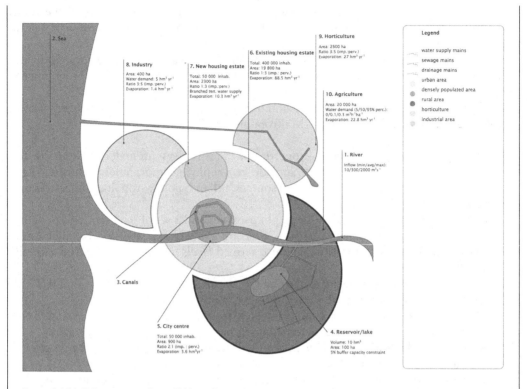

Figure 24.14 Schematic outline of WaterCity.

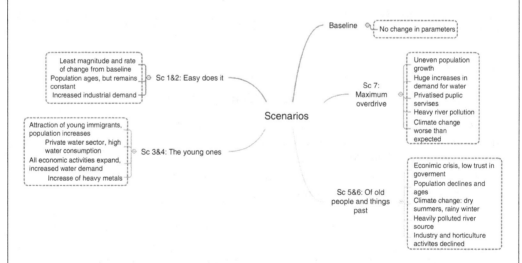

Figure 24.15 Schematic overview of scenarios and parameters.

Each of the above system configurations was tested against a series of increased disturbance scenarios. Each scenario is composed of a set of pressures (model parameters) with varying degrees of magnitude and rates of change. The schematic overview of both scenarios and parameters is presented in Figure 24.15.

The simulation of the urban water cycle for the three system configurations, under alternative scenarios, was performed using UWOT. As a performance measure, the system's reliability towards *potable water coverage* was used, although other reliability metrics may also be used (including metrics directly relevant to urban drainage, such as CSO spills or flooding extent). The *Resilience Profile Graph* along with the trade-off between energy costs and resilience for the three system configurations are depicted in Figure 24.16.

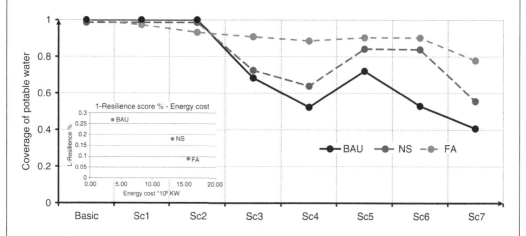

Figure 24.16 Resilience plot graph for BAU, NS, and FA configurations and the trade-off between resilience and energy costs.

24.8.2 Developing approaches

Since the original work of Butler et al. (2014, 2016), research into the definition, quantification, modelling and implications of resilience in urban drainage systems has been developing rapidly. The most promising contemporary resilience research areas include

- Incorporation within the wider integrated wastewater system (e.g., Casal-Campos et al., 2018), including consideration of WTP performance and representation of receiving water quality (see Section 21.4)
- Incorporation within a multi-objective optimisation framework (e.g., Bakhshipour et al., 2021, Leng, et al., 2021) to allow the discovery of optimum (or close to optimum) solutions based, for example, minimising life-cycle costs and maximising "sustainability" (Section 21.3)
- Combination with sustainability (e.g., Bakhshipour et al., 2021; Casal-Campos et al., 2018) to ensure resilient solutions do not compromise sustainability requirements (Section 21.5)
- Consideration of cost-effectiveness (e.g., Xu et al., 2021) to minimise total (operational and capital) costs (Section 19.2.1.1)
- Developing a resilience modelling toolbox (Kamali, et al., 2022)
- Proposing more comprehensive definitions of resilience (e.g., Binesh et al., 2022; Dong et al., 2017) such as specific inclusion of robustness

24.9 WHERE TO NEXT?

24.9.1 Urban futures

Although Niels Bohr was surely correct when he said, "Prediction is very difficult, especially about the future," there is definite merit in assessing what *might* happen in the future to understand what options are feasible and what strategies are required to achieve them. It is sometimes argued that it is impossible to know *anything* about the future so it can be ignored, only to be responded to when it actually happens. Of course, only a little can be known about the future, but that little is important because *some* knowledge of the future is vital in making good decisions. In addition, learning what is knowable about the future enables action in the present that will contribute to achieving a desirable goal. Three main approaches to thinking about the future are commonly adopted: trend, precursor, and scenario analysis, and the latter is discussed here.

Scenario analysis consists of constructing explanatory "futures" such as those used to stress test the urban water system as the *x*-axis of the Resilience Profile described in the previous section. These futures describe broad social, economic, and political changes that *may* rather than *will* occur in the future. To that extent, they are coherent, internally consistent, and plausible, typically including a narrative element and some quantitative indicators (Berkhout et al., 2002). Of the many scenarios produced, typically no attempt is made to predict which, if any, is most likely to occur.

Figure 24.17 provides an example of this approach based on two key drivers of future change: social values (*x*-axis) and systems of governance (*y*-axis). Social values range from individualistic to more community-orientated ideals, including the political and economic implications that arise from them. Governance ranges from autonomy where power remains at the local level, through regional and national levels, to interdependence where globalisation is the norm. Four futures can then be developed that are conveniently placed in the four different quadrants of the axes: world markets, global sustainability, national enterprise, and local stewardship. The "position" of current development is also shown.

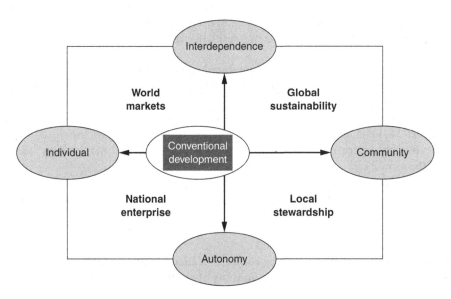

Figure 24.17 Socio-economic futures. (Adapted from OST. 1998. *Environmental Futures*, Department of Trade and Industry/Office of Science and Technology.)

A similar approach, but using different futures, was taken by the International Water Association/International Association on Hydraulic Engineering and Research (IWA/IAHR) Joint Committee on Urban Drainage (www.jcud.org). Their four futures are summarised in Table 24.5. In the first case, the green future, political ecology prevails and the environment is put first. The second future is technology-based and describes a world where engineers are in charge, and presumes that "technological fixes" will prevail. In the third future, the key issue is finance, and widespread privatisation is assumed. The final future is business as usual, which presumes that financial, technical, and political driving forces act together, as today, but not in a particularly coordinated way. Table 24.5 also includes notes on what type of urban drainage system might be prevalent in each future.

While the use of futures does give insights into future possible states and can help guide us to a more sustainable future, it does have its weaknesses:

- Technical solutions do not naturally or necessarily map onto any particular socio-economic future.
- Technical solutions need not be *exclusive* to any particular future but may be used in combination.
- The high sunk asset value, inertia, and conservatism of the water industry mean that any future that is not closely related to the present is unlikely.

Table 24.5 Proposed futures and their urban drainage implications

Future	Characteristics	Urban drainage components
Green	Ecologists are in charge. They have imposed a *balanced integrated approach* attempting to combine both "soft" and "hard" technologies in achieving sustainability of urban systems, including drainage. This scenario may be hard to imagine because, in most developed societies, only the engineers are licensed with the responsibility to build infrastructure. Soft measures include source controls (both legislated and voluntary).	On-site: • Minimisation • SuDS • Reuse and recycling • Existing sewerage has very low flows and is being progressively abandoned or used for other purposes • Engages users (who may subcontract responsibility) Problems occur due to • On-site failures/misuse/abuse • Transition from existing to new systems is costly and slow
Technocratic	Publicly employed engineers are in charge of the water cycle and decide which technical approaches should be used and the ways the systems are managed. This scenario represents the classical approach before water privatisation became the international norm. Application of largely proven technology, coupled with minimum risk, with duplication, and with large safety factors, engineers make sure that the systems do not fail. Long-term planning (Grand Master Plan).	End-of-pipe systems maintained • Centralisation with large treatment plants • Large fully sewered networks • Continues to utilise valuable existing infrastructure assets • Innovation is limited although incremental changes in technology will solve new problems Problems due to • Lack of sustainability and high maintenance costs • Inappropriate for large-scale new developments in the developing world. • Costs rising • Discourages user engagement and responsibility

(Continued)

Table 24.5 (Continued)

Future	Characteristics	Urban drainage components
Privatisation	Investors and economists are in charge, and the mission is cost efficiency and shareholder dividends. This is achieved by control of labour costs and limited investments in both new and existing infrastructure. The main idea of the privatisation scenario is not to find the most effective solution to a technical problem, but to make, or at least minimise loss of, money. In the worst situation, the water service provider might define the level of service.	As above, but motives are profits. • Likely to continue to utilise the serviceability of existing assets as long as possible (operating licence) and minimise investment in new systems • A lot of subcontracting is out of service provisions Problems due to • Service monopolies • Lack of incentive to reduce demand (customer) • Reluctance to innovate due to risk responsibility framework • Poor performance of subcontractors • Mergers/acquisitions/lack of loyalty to staff • Remoteness of end users (customers) • Unlikely that the true costs of service provision can be passed to users
Business as usual	Continuation of what has been done for the past 30 years. Nobody is in charge. This scenario is poorly defined because the current levels of drainage services and approaches vary widely between countries and between jurisdictions. Practically everywhere, it is recognised as (or called) not sustainable, yet it is the most common. Generally because of the large inertia of existing infrastructure in the developed countries, radical change is not feasible.	A confusing mixture of all of the above. • Recognises the need to continue to utilise the valuable existing infrastructure base, but the introduction of innovation is possible • A perceived low-risk approach (always done this way). Unlikely to have many catastrophic failures Problems: • Possibly all of those from above • Unlikely to lead to sustainable systems in time • Passes much of the financial burden to future generations • Will not attract the best people into a career in this area

Source: Schilling, W. 2003. Urban drainage – quo vadis? *Proceedings of XXX IAHR Congress*, Thessaloniki, Greece, 1–17.

That said, more detailed work under the WaND project has been attempted to logically and consistently map the state of urban water systems to six socio-political futures (Makropoulos et al., 2008).

24.9.2 Transition states

Another way of thinking about the future is to consider so-called "transition states." Brown et al. (2009) proposed a water system transitions framework based on a representation of different temporal, ideological, and technological states that cities transition through when moving between different management paradigms, towards more sustainable futures. As shown in Figure 24.18, the framework includes six city states, namely, the Water Supply City, the Sewered City, the Drained City, the Waterways City, the Water Cycle City, and the Water Sensitive City. Each of the six city states is characterised by a distinct shift in the following pillars of institutional practice:

- Cognitive: dominant knowledge, thinking, and skills
- Normative: values and leadership
- Regulative: administration, rules, and systems

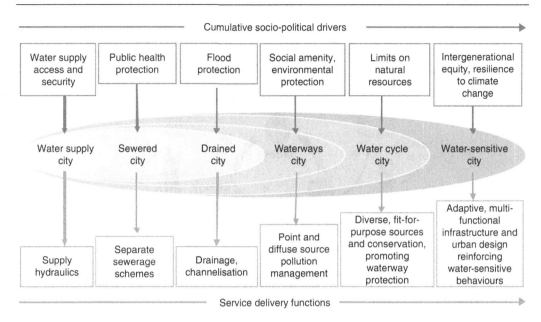

Figure 24.18 Urban water management transition states. (Reproduced from Brown, R. et al. 2009. *Water Science and Technology*, 59[5], 847–855, with permission from the copyright owners, IWA Publishing.)

The "cumulative socio-political drivers" reflect shifts over time in the agreements between communities, businesses, and governments on how water should be managed (also termed a *hydro-social contract*; Lundqvist et al., 2001). The "service delivery functions" represent the key practical concerns and priorities being addressed in those paradigms. This model is useful when applied as a diagnostic tool to cities both to judge the current state and to provide a potential pathway to a more sustainable and resilient future.

24.9.3 Pathways

The final approach to thinking about "Where to next?" is through using the concept of pathways. While it is difficult to be precise about the characteristics of future urban water systems, and more specifically urban drainage, several trends have become clear through research and practice over the last few decades and these could be regarded as pathways leading towards more sustainable systems (Fu and Butler, 2021), namely,

- Decentralisation: discussed in Section 24.3.6
- Greening: discussed in Chapter 11 and Section 24.3.7
- Digitalisation: discussed in Chapter 22
- Circular economy: The design of most urban water systems has followed a linear "take, make, use, and dispose" model. The circular economy model, as applied to water systems, should provide a continuous positive development cycle that preserves and enhances natural capital, optimises resource yields, and maximises resource value at each component of a system by managing finite stocks and renewable flows, closing the resources loop. Figure 24.19 broadly illustrates the interlinkages between resources (the water–energy–materials nexus) and the urban water system, including the role of urban drainage.

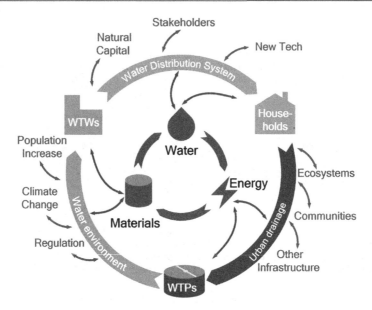

Figure 24.19 Urban water systems and the circular economy

It should be noted that these pathways are not necessarily mutually exclusive but can be interrelated. For example, digitalisation can cut across all other pathways as digital solutions can help us monitor and control green infrastructure and improve resource recovery efficiency, through optimal management of the water–energy–materials nexus. It can also accelerate decentralisation by enabling more autonomous operations of decentralised water systems and remote control, decreasing the need for onsite physical presence, which would be prohibitive at small scales.

PROBLEMS

1 Define *sustainable development*, and interpret its implications in terms of society, economics, and the environment. Illustrate your answer with examples from urban drainage.
2 To what extent do current systems meet the objectives for sustainable urban drainage set out in 24.2.4?
3 Explain what you understand by the term sustainable urban drainage.
4 In what ways is centralised urban drainage unsustainable?
5 Describe three techniques or technologies that claim to be more sustainable, and cite evidence to support the claims.
6 What sustainability indicators would you use to evaluate urban drainage systems and why?
7 Explain what you understand by the term *resilience* and how it can be increased in urban drainage systems.
8 Explain the relationship between resilience, sustainability, and reliability. Discuss possible differences in working definitions and what each one highlights.
9 Choose three key resilience attributes and explain how they can be included in urban drainage systems.
10 Explain the main steps to create a resilience profile graph for strategic planning.

11 Discuss what you understand by urban futures and explain how they can be used. Should an assessment be made of the future that is most likely to occur in the real world?

12 Describe the six urban water management transition states, the socio-political drivers, and their delivery functionalities. Pick a city you know, and explain which state it is currently in.

13 What do you think will be the dominant development pathways for urban water systems over the next decade?

KEY SOURCES

Butler, D., Memon, F.A., Makropoulos, C., Southall, A., and Clarke, L. 2010. *Guidance on Water Cycle Management for New Developments*, CIRIA C690.

Butler, D., Ward, S., Sweetapple, C., Astaraie-Imani, M., Diao, K., Farmani, R., and Fu, G. 2016. Reliable, resilient and sustainable water management: The Safe & SuRe approach. *Global Challenges, 1*(1), 63–77.

Larsen, T.A., Hoffmann, S., Lüthi, C., Truffer, B. and Maurer, M. 2016. Emerging solutions to the water challenges of an urbanizing world. *Science, 352*, 6288.

Larsen, T.A., Uderts, K.M. and Lienert, J. (eds.) 2013. *Source Separation and Decentralization for Wastewater Management*, IWA Publishing.

Makropoulos C., and Butler, D. 2010. Distributed water infrastructure for sustainable communities. *Water Resources Management, 24*(11), 2795–2816.

Maksimović, Č. and Tejada-Guibert, J.A. eds. 2001. *Frontiers in Urban Water Management. Deadlock or Hope?*, IWA Publishing.

REFERENCES

Abdel-Aal, M., Schellart, A., Kroll, S., Mohamed, M. and Tait, S. 2018. Modelling the potential for multi-location in-sewer heat recovery at a city scale under different seasonal scenarios. *Water Research, 145*, 618–630.

American Society of Civil Engineers/UN Educational, Scientific, and Cultural Organization (ASCE/UNESCO). 1998. *Sustainability Criteria for Water Resource Systems*, Cambridge University Press.

Ashley, R.M., Blackwood, D., Butler, D., Davies, J.W., Jowitt, P., and Smith, H. 2003. Sustainable decision making for the UK water industry. *Proceedings of the Institution of Civil Engineers, Engineering Sustainability, 156(ES1)*, 41–49.

Ashley, R., Blackwood, D., Butler, D., and Jowitt, P. 2004. *Sustainable Water Services: A Procedural Guide*, IWA Publishing.

Baber C. 2010. Tapping into waste heat. *Water Environment and Technology, 22(12)*, 40–45.

Bakhshipour, A.E., Dittmer, U., Haghighi, A. and Nowak, W. 2021. Toward sustainable urban drainage infrastructure planning: A combined multiobjective optimization and multicriteria decision-making platform. *Journal of Water Resources Planning and Management, 147*(8), 04021049.

Berkhout, F., Hertin, J., and Jordan, A. 2002. Socio-economic futures in climate change impact assessment: Using scenarios as "learning machines". *Global Environmental Change, 12*, 83–95.

Binesh, N., Sarang, A., Niksokhan, M.H., Rauch, W. and Aronica, G.T. 2022. Quantifying the UDS Hydraulic and Social Resilience to Flooding: An Index-Based Approach vs. a Parameter-Based MCDM Method. *Water, 14*, 2007.

Brede, M. and B. J. M. de Vries. 2009. Networks that optimize a tradeoff between efficiency and dynamical resilience. *Physics Letters A, 373*(43), 3910–3914.

Brown, R., Keath, N., and Wong, T. 2009. Urban water management in cities: Historical, current and future regimes. *Water Science and Technology, 59*(5), 847–855.

BS EN 997: 2018. *WC Pans and WC Suites with Integral Trap.*

Bulteau, G., Hoffman, J., Fairhurst, R. and Orman, N. 2019. *Energy from sewers.* Report No. 19/SW/03/1, UKWIR. https://ukwir.org/energy-from-sewers-0

Butler, D. and Maksimović, Č. 1999. Urban water management—Challenges for the third millennium. *Progress in Environmental Science, 1*(3), 213–235.

Butler, D. and Memon, F.A. eds. 2006. *Water Demand Management,* IWA Publishing.

Butler, D. and Parkinson, J. 1997. Towards sustainable urban drainage. *Water Science and Technology, 35*(9), 53–63.

Butler, D., Farmani, R., Fu, G., Ward, S., Diao, K., and Astaraie-Imani, M. 2014. A new approach to urban water management: Safe and SuRe. *Procedia Engineering, 89,* 347–354.

Casal-Campos, A., Fu, G., Butler, D. and Moore, A. 2015. An integrated environmental assessment of green and gray infrastructure strategies for robust decision making. *Environmental Science and Technology, 49*(14), 8307–8314.

Casal-Campos, A., Sadr, S. M. K., Fu, G. and Butler, D. 2018. Reliable, resilient and sustainable urban drainage systems: An analysis of robustness under deep uncertainty. *Environmental Science and Technology, 52*(16), 9008–9021.

Dixon, A., Butler, D., and Fewkes, A.F. 1999a. Water saving potential of domestic water reuse systems using greywater and rainwater in combination. *Water Science and Technology, 39*(5), 25–32.

Dixon, A., Butler, D., and Fewkes, A.F. 1999b. Guidelines for greywater reuse—Health issues. *CIWEM Water and Environmental Management Journal, 13(5),* 322–326.

Dixon, T., Eames, M., Hunt, M., and Lannon, S. (eds.). 2014. *Urban Retrofitting for Sustainability: Mapping the Transition to 2050,* Routledge.

Dong, X. Guo, H., Zeng, S. 2017. Enhancing future resilience in urban drainage system: Green versus grey infrastructure. *Water Research, 124,* 280–289.

Drangert, J.-O. 1998. Fighting the urine blindness to provide more sanitation options. *Water SA, 24*(2), 157–164.

Fu, G. and Butler, D. 2021. Pathways towards sustainable and resilient urban water systems, in *Water-Wise Cities and Sustainable Water Systems: Concepts, Technologies, and Applications* (eds X. C. Wang and G. Fu), IWA Publishing, 3–24.

Garrido-Baserba, M., Barnosell, I., Molinos-Senante, M., Sedlak, D.L. Rabaey, K., Schraa, O., Verdaguer, M., Rosso, D. and Poch, M. 2022. The third route: A techno-economic evaluation of extreme water and wastewater decentralization. *Water Research, 222,* 118408.

Gross, A., Maimon, A., Alfiya, Y., and Friedler, E. 2015. *Greywater Reuse,* CRC Press.

Hanaeus, J., Hellstorm, D., and Johansson, E. 1997. A study of a urine separation system in an ecological village in northern Sweden. *Water Science and Technology, 35*(9), 153–160.

Hesarkazzazi, S., Bakhshipour, A.E., Hajibabaei, M., Dittmer, U., Haghighi, A. and Sitzenfrei, R. 2022. Battle of centralized and decentralized urban stormwater networks: From redundancy perspective. *Water Research, 222,* 118910.

Holling, C.S. 1973. Resilience and stability of ecological systems. *Annual Review of Ecology and Systematics, 4,* 1–23.

Hvitved-Jacobsen, T., Nielsen, P.H., Larsen, T., and Jensen, N.A. (eds.). 1995. *The sewer as a physical, chemical and biological reactor.* Water Science and Technology series, Book 31, International Water Association, London.

Jagtapa, N.S. and Boyera, T.H. 2020. Urine collection in a multi-story building and opportunities for onsite recovery of nutrients and non-potable water. *Journal of Environmental Chemical Engineering, 8,* 1039564.

Kamali, B., Ziaei, A.N., Beheshti, A. and Farmani, R. 2022. An open-source toolbox for investigating functional resilience in sewer networks based on global resilience analysis. *Reliability Engineering and System Safety, 218,* 108201.

Koop, S.H.A. and van Leeuwen, C.J. 2015. Assessment of the sustainability of water resources management: A critical review of the city blueprint approach. *Journal of Water Resources Management, 29,* 5649–5670.

Lansey, K. 2012. Sustainable, robust, resilient, water distribution systems. *14th Water Distribution Systems Analysis Conference, WDSA 2012,* Adelaide, Australia, 1–18.

Larsen, T.A. and Guyer, W. 2013. Implementation of source separation and decentralization in cities, in *Source Separation and Decentralization for Wastewater Management* (eds T. Larsen, K.M. Udert and J. Lienert), IWA Publishing, 135–150.

Lawson, R., Marshallsay, D., DiFiore, D., Rogerson, S., Meeus, S., Sanders, J. and Horton, B. 2018. *The long term potential for deep reductions in household water demand, Artesia Consulting*, Report AR1206.

Leng, L., Jiaa, H., Chen, A.S., Zhu, D.Z., Xu, T. and Yu, S. 2021. Multi-objective optimization for green-grey infrastructures in response to external uncertainties. *Science of the Total Environment*, 775, 145831.

Littlewood, K., Memon, F.A., and Butler, D. 2007. Downstream implications of ultra-low flush WCs. *Water Practice and Technology*, 2(2), wpt2007037.

Lundqvist, J., Turton, A., and Narain, S. 2001. Social, institutional and regulatory issues, in *Frontiers in Urban Water Management: Deadlock or Hope* (eds C. Maksimovic and J. A. Tejada-Guilbert), IWA Publishing, 344–398.

Makropoulos, C., Memon, F.A., Shirley-Smith, C., and Butler D. 2008. Futures: An exploration of scenarios for sustainable urban water management. *Water Policy, 10(4)*, 345–373.

Makropoulos, C., Nikolopoulos, D., Palmen, L., Kools, S., Segrave, A., Vries, D., Koop, S. 2018. A resilience assessment method for Urban Water Systems. *Urban Water Journal, 15(4)*, 1–13.

Makropoulos, C., Palmen, L., Kools, S., Segrave, A., Vries, D., Koop, S., van Alphen, H.J., Vonk, E., and van Thienen, P. 2016. *Developing water wise cities: A methodological proposition*. KWR Watercycle Research Institute, BTO Report 400695/044.

Mays, L.W. (ed.). 1989. *Reliability Analysis of Water Distribution Systems*. ASCE.

McConville, A., Rosemarin, A., Li, Z. and Flores, A. 2010. Assessing the sustainability of innovations in Urban Ecological Sanitation: Erdo Eco-town Project, in *Water Infrastructure for Sustainable Communities: Chian and the World* (X. Hao, V. Novotny and V. Nelson), IWA Publishing, 311–322.

Melville-Shreeve, P., Cotterill, S., Newman, A. and Butler, D. 2021. Campus study of the impact of ultra-low flush toilets on sewerage networks and water usage. *Water, 13*, 419.

Melville-Shreeve, P., Ward, S., and Butler, D. 2016. Dual-purpose rainwater harvesting system design, in *Sustainable Surface Water Management. A Handbook for SUDS* (eds S.M. Charlesworth and C.A. Booth), Wiley Blackwell, 255–268.

Memon, F.A., Butler, D., Han, W., Liu, S., Makropoulos, C., Avery, L., and Pidou, M. 2005. Economic assessment tool for greywater recycling systems. *Institution of Civil Engineers Engineering Sustainability, 158(ES3)*, 155–161.

Memon, F.A., Zheng, Z., Butler, D., Shirley-Smith, C., Liu, S., Makropoulos, C., and Avery, L. 2007. Life cycle impact assessment of greywater treatment technologies for new developments. *Journal of Environmental Monitoring and Assessment, 129*, 27–35.

Millán, A.M., Memon, F.A., Butler, D., and Littlewood, K. 2007. User perceptions and basic performance of an innovative WC, in *Water Management Challenges in Global Change* (eds B. Ulanicki, K. Vairavamoorthy, D. Butler, P.L.M. Bounds and F.A. Memon), Taylor and Francis, 629–633.

Morecroft, M.D., Crick, H.Q.P., Duffield, S.J., and Macgregor, N.A. 2012. Resilience to climate change: Translating principles into practice. *Journal of Applied Ecology, 49(3)*, 547–551.

Mugume, S.N. and Butler, D. 2016. Evaluation of functional resilience in urban drainage and flood management systems using a global analysis approach. *Urban Water Journal, 14(7)*, 727–736.

Mugume, S.N., Gomez, D.E., Fu, G., Farmani, R., and Butler, D. 2015. A global analysis approach for investigating structural resilience in urban drainage systems. *Water Research, 81*, 15–26.

Nikolopoulos, D. 2023. *Resilience assessment of cyber-physical water systems*, Ph. D. Thesis, National Technical University of Athens.

Nolde, E. 2014. Greywater recycling in buildings, in *Water Efficiency in Buildings: Theory and Practice* (ed K. Adeyeye), Chapter 10, 169–189.

Omambala, I., Russell, N., Tompkins, J., Bremner, S., Millar, R., Rathouse, K., and Cooper, M. 2011. *Evidence Base for Large-Scale Water Efficiency. Phase II Final Report.*

Otterpohl, R., Grottker, M., and Lange J. 1997. Sustainable water and waste management in urban areas. *Water Science and Technology, 35(9)*, 121–133.

OST. 1998. *Environmental Futures*, Department of Trade and Industry/Office of Science and Technology.

Park, J., Seager, T.P., Rao, P.S.C., Convertino, M., and Linkov, I., 2013. Integrating risk and resilience approaches to catastrophe management in engineering systems. *Risk Analysis*, 33(3), 356–367.

Parkinson, J., Schütze, M., and Butler, D. 2005. Modelling the impacts of domestic water conservation on the sustainability of the urban wastewater system. *Chartered Institution of Water and Environmental Management, Water and Environment Journal*, 19(1), 49–56.

Penn, R., Schütze, M. and Friedler, E. 2013. Modelling the effects of on-site greywater reuse and low flush toilets on municipal sewer systems. *Journal of Environmental Management*, 114, 72–83.

Pickett, S.T.A., McGrath, B., Cadenasso, M.L., and Felson, A.J. 2014. Ecological resilience and resilient cities. *Building Research and Information*, 42, 143–157.

Pizzol, M. 2015. Life cycle assessment and the resilience of product systems. *Journal of Industrial Ecology*, 19(2), 296–306.

Ricardo. 2020. *Independent Review of the Costs and Benefits of Rainwater Harvesting and Grey Water Recycling Options in the UK*. Final Report for Waterwise, WEStrategy002.

Rodriguez, M., Fu, G., Butler, D., Yuan, Z. and Cook, L. 2024. The effect of green infrastructure on resilience performance of combined sewer systems under climate change. *Journal of Environmental Management*, 353, 120229.

Rozos, E., Makropoulos, C., and Maksimović, Č. 2013. Rethinking urban areas: An example of an integrated blue-green approach. *Water Science and Technology: Water Supply*, 13(6), 1534–1542.

Schilling, W. 2003. Urban drainage – Quo vadis?. Proceedings of XXX IAHR Congress, Thessaloniki, Greece, 1–17.

Scholz, R.W., Blumer, Y.B., and Brand, F.S. 2012. Risk, vulnerability, robustness, and resilience from a decision-theoretic perspective. *Journal of Risk Research*, 15(3), 313–330.

Sharp, L. 2006. Water demand management in England and Wales: Constructions of the domestic water user. *Journal of Environmental Planning and Management*, 49(6), 869–889.

Sweetapple, C., Fu, G., Farmani, R., Meng, F., Ward, S. and Butler, D. 2018. Attribute-based intervention development for increasing resilience of urban drainage systems. *Water Science and Technology*, 77(6), 1757–1764.

Tjandraatmadja, G., Burn, S., McLaughlin, M., and Biswas, T. 2005. Rethinking urban water systems—Revisiting concepts in urban wastewater collection and treatment to ensure infrastructure sustainability. *Water Science and Technology*, 5(2), 145–154.

Van Leeuwen, K. and Marques, R.C. 2013. *Current state of sustainability of urban water cycle services. Transition to the urban water services of tomorrow (TRUST)*. Report D11.1. https://riunet.upv.es/bitstream/handle/10251/35724/Current_State_of_Sustainability_of_Urban_Water_Cycle_Services.pdf?sequence=1

Van Leeuwen, C.J., Frijns, J., Van Wezel, A., and Van De Ven, F.H.M. 2012. City blueprints: 24 indicators to assess the sustainability of the urban water cycle. *Journal of Water Resources Management*, 26, 2177–2197.

Van Leeuwen, C.J., Koop, S.H.A., and Sjerps, R.M.A. 2016. City Blueprints: Baseline assessments of water management and climate change in 45 cities. *Environment, Development and Sustainability*, 18(4), 1113–1128.

Wang, M., Zhang, Y., Bakhshipour, A.E., Liu, M., Rao, Q. and Lu, Z. 2022. Designing coupled LID–GREI urban drainage systems: Resilience assessment and decision-making framework. *Science of the Total Environment*, 834, 155267.

Winblad, U. and Simpson-Hébert, M. eds. 2004. *Ecological Sanitation, revised and enlarged edition*, Stockholm Environment Institute.

World Commission on Environment and Development. 1987. *Our Common Future*, Oxford University Press.

Xu, Z., Dong, X., Zhao, Y. and Du, P. 2021. Enhancing resilience of urban stormwater systems: cost-effectiveness analysis of structural characteristics. *Urban Water Journal*, 18(10), 850–859.

Zadeh, S.M., Hunt, D.V.L., Lombardi, D.R., and Rogers, C.D.F. 2013. Shared urban greywater recycling systems: Water resource savings and economic investment. *Sustainability*, 5(7), 2887–2912.

Index

Pages in *italics* refer to figures and those in **bold** refer to tables.

Printed in the United States
by Baker & Taylor Publisher Services